S0-CKK-376

VOLUME FIVE HUNDRED AND SEVENTY EIGHT

METHODS IN ENZYMOLOGY

Computational Approaches for Studying Enzyme Mechanism Part B

METHODS IN ENZYMOLOGY

VOLUME FIVE HUNDRED AND SEVENTY EIGHT

METHODS IN ENZYMOLOGY

Computational Approaches for Studying Enzyme Mechanism Part B

Edited by

GREGORY A. VOTH

Department of Chemistry
The University of Chicago
Chicago, Illinois, United States

ELSEVIER

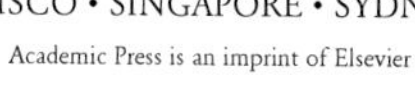

AMSTERDAM • BOSTON • HEIDELBERG • LONDON
NEW YORK • OXFORD • PARIS • SAN DIEGO
SAN FRANCISCO • SINGAPORE • SYDNEY • TOKYO
Academic Press is an imprint of Elsevier

Academic Press is an imprint of Elsevier
50 Hampshire Street, 5th Floor, Cambridge, MA 02139, United States
525 B Street, Suite 1800, San Diego, CA 92101–4495, United States
The Boulevard, Langford Lane, Kidlington, Oxford OX5 1GB, United Kingdom
125 London Wall, London, EC2Y 5AS, United Kingdom

First edition 2016

ISBN: 978-0-12-811107-9
ISSN: 0076-6879

Publisher: Zoe Kruze
Acquisition Editor: Zoe Kruze
Editorial Project Manager: Helene Kabes
Production Project Manager: Magesh Kumar Mahalingam
Cover Designer: Greg Harris

Typeset by SPi Global, India

CONTENTS

CONTRIBUTORS

P.K. Agarwal
Computational Biology Institute, Oak Ridge National Laboratory, Oak Ridge; University of Tennessee, Knoxville, TN, United States

N.A. Baker
Pacific Northwest National Laboratory, Richland, DC; Brown University, Providence, RI, United States

J.L. Baylon
Center for Biophysics and Quantitative Biology; University of Illinois at Urbana-Champaign; Beckman Institute for Advanced Science and Technology, University of Illinois at Urbana-Champaign, Urbana, IL, United States

D. Bellinger
Julius-Maximilians-Universität Würzburg, Institut für Physikalische und Theoretische Chemie, Würzburg, Germany

R.B. Best
Laboratory of Chemical Physics, National Institute of Diabetes and Digestive and Kidney Diseases, National Institutes of Health, Bethesda, MD, United States

J. Blumberger
University College London, London, United Kingdom

S. Bowerman
Center for Molecular Study of Condensed Soft Matter, Illinois Institute of Technology, Chicago, IL, United States

G.R. Bowman
Washington University School of Medicine; Center for Biological Systems Engineering, Washington University School of Medicine, St. Louis, MO, United States

X. Che
College of Chemistry and Molecular Engineering, Beijing National Laboratory for Molecular Sciences; Biodynamic Optical Imaging Center (BIOPIC), Peking University, Beijing, PR China

C. Chennubhotla
University of Pittsburgh, Pittsburgh, PA, United States

P.P.-H. Cheung
The Hong Kong University of Science and Technology, Kowloon, Hong Kong

J.-W. Chu
Institute of Bioinformatics and Systems Biology; Department of Biological Science and Technology; Institute of Molecular Medicine and Bioengineering, National Chiao Tung University, Hsinchu, Taiwan, ROC

T.H. Click
Institute of Bioinformatics and Systems Biology, National Chiao Tung University, Hsinchu, Taiwan, ROC

D. De Sancho
CIC NanoGUNE, Donostia-San Sebastián; Ikerbasque, Basque Foundation for Science, Bilbao, Spain

N. Doucet
INRS—Institut Armand-Frappier, Université du Québec, Laval, QC, Canada

M.W. Dzierlenga
University of Arizona, Tucson, AZ, United States

B. Engels
Julius-Maximilians-Universität Würzburg, Institut für Physikalische und Theoretische Chemie, Würzburg, Germany

Y.Q. Gao
College of Chemistry and Molecular Engineering, Beijing National Laboratory for Molecular Sciences; Biodynamic Optical Imaging Center (BIOPIC), Peking University, Beijing, PR China

B. Ginovska
Pacific Northwest National Laboratory, Richland, WA, United States

M.R. Gunner
City College of New York in the City University of New York, New York, United States

X. He
School of Chemistry and Molecular Engineering, East China Normal University; NYU-ECNU Center for Computational Chemistry at NYU Shanghai, Shanghai, China

X. Huang
The Hong Kong University of Science and Technology; State Key Laboratory of Molecular Neuroscience, Center for System Biology and Human Health, School of Science and Institute for Advance Study, The Hong Kong University of Science and Technology, Kowloon, Hong Kong

P. Imhof
Institute of Theoretical Physics, Free University Berlin, Berlin, Germany

H. Jiang
The Hong Kong University of Science and Technology, Kowloon, Hong Kong

T. Jiang
Center for Biophysics and Quantitative Biology; University of Illinois at Urbana-Champaign; Beckman Institute for Advanced Science and Technology, University of Illinois at Urbana-Champaign, Urbana, IL, United States

P. Mahinthichaichan
University of Illinois at Urbana-Champaign; Beckman Institute for Advanced Science and Technology, University of Illinois at Urbana-Champaign, Urbana, IL, United States

M.A. Martí
FCEN, UBA, Buenos Aires, Argentina

C.G. Mayne
University of Illinois at Urbana-Champaign; Beckman Institute for Advanced Science and Technology, University of Illinois at Urbana-Champaign, Urbana, IL, United States

M.P. Muller
Center for Biophysics and Quantitative Biology; University of Illinois at Urbana-Champaign; Beckman Institute for Advanced Science and Technology; College of Medicine, University of Illinois at Urbana-Champaign, Urbana, IL, United States

C. Narayanan
INRS—Institut Armand-Frappier, Université du Québec, Laval, QC, Canada

F. Pardo-Avila
The Hong Kong University of Science and Technology, Kowloon, Hong Kong

J.M. Parks
Oak Ridge National Laboratory, Oak Ridge; University of Tennessee, Knoxville, TN, United States

N. Raj
Institute of Bioinformatics and Systems Biology, National Chiao Tung University, Hsinchu, Taiwan, ROC

A. Ramanathan
Oak Ridge National Laboratory, Oak Ridge, TN, United States

C.L. Ramírez
FCEN, UBA, Buenos Aires, Argentina

S. Raugei
Pacific Northwest National Laboratory, Richland, WA, United States

A.E. Roitberg
University of Florida, Gainesville, FL, United States

S. Sacquin-Mora
Laboratoire de Biochimie Théorique, CNRS UPR9080, Institut de Biologie Physico-Chimique, Paris, France

S.D. Schwartz
University of Arizona, Tucson, AZ, United States

W.J. Shaw
Pacific Northwest National Laboratory, Richland, WA, United States

M. Shekhar
Center for Biophysics and Quantitative Biology; University of Illinois at Urbana-Champaign; Beckman Institute for Advanced Science and Technology, University of Illinois at Urbana-Champaign, Urbana, IL, United States

F.K. Sheong
The Hong Kong University of Science and Technology, Kowloon, Hong Kong

E. Shinn
Center for Biophysics and Quantitative Biology; University of Illinois at Urbana-Champaign; Beckman Institute for Advanced Science and Technology, University of Illinois at Urbana-Champaign, Urbana, IL, United States

J.C. Smith
Oak Ridge National Laboratory, Oak Ridge; University of Tennessee, Knoxville, TN, United States

E. Tajkhorshid
Center for Biophysics and Quantitative Biology; University of Illinois at Urbana-Champaign; Beckman Institute for Advanced Science and Technology; College of Medicine, University of Illinois at Urbana-Champaign, Urbana, IL, United States

S. Thangapandian
University of Illinois at Urbana-Champaign; Beckman Institute for Advanced Science and Technology, University of Illinois at Urbana-Champaign, Urbana, IL, United States

N. Trebesch
Center for Biophysics and Quantitative Biology; University of Illinois at Urbana-Champaign; Beckman Institute for Advanced Science and Technology, University of Illinois at Urbana-Champaign, Urbana, IL, United States

M.J. Varga
University of Arizona, Tucson, AZ, United States

J.V. Vermaas
Center for Biophysics and Quantitative Biology; University of Illinois at Urbana-Champaign; Beckman Institute for Advanced Science and Technology, University of Illinois at Urbana-Champaign, Urbana, IL, United States

P.-H. Wang
RIKEN Theoretical Molecular Science Laboratory, Wako-shi, Saitama, Japan

X. Wang
Center for Optics & Optoelectronics Research, College of Science, Zhejiang University of Technology, Hangzhou, Zhejiang; School of Chemistry and Molecular Engineering, East China Normal University, Shanghai, China

Y. Wang
Center for Biophysics and Quantitative Biology; University of Illinois at Urbana-Champaign; Beckman Institute for Advanced Science and Technology, University of Illinois at Urbana-Champaign, Urbana, IL, United States

D. Weber
Julius-Maximilians-Universität Würzburg, Institut für Physikalische und Theoretische Chemie, Würzburg, Germany

P.-C. Wen
University of Illinois at Urbana-Champaign; Beckman Institute for Advanced Science and Technology, University of Illinois at Urbana-Champaign, Urbana, IL, United States

J. Wereszczynski
Center for Molecular Study of Condensed Soft Matter, Illinois Institute of Technology, Chicago, IL, United States

L. Yang
College of Chemistry and Molecular Engineering, Beijing National Laboratory for Molecular Sciences; Biodynamic Optical Imaging Center (BIOPIC), Peking University, Beijing, PR China

J. Zhang
College of Chemistry and Molecular Engineering, Beijing National Laboratory for Molecular Sciences; Biodynamic Optical Imaging Center (BIOPIC), Peking University, Beijing, PR China

J.Z.H. Zhang
School of Chemistry and Molecular Engineering, East China Normal University; NYU-ECNU Center for Computational Chemistry at NYU Shanghai, Shanghai, China; New York University, New York, NY, United States

L. Zhang
The Hong Kong University of Science and Technology, Kowloon, Hong Kong

Z. Zhao
Center for Biophysics and Quantitative Biology; University of Illinois at Urbana-Champaign; Beckman Institute for Advanced Science and Technology, University of Illinois at Urbana-Champaign, Urbana, IL, United States

M.I. Zimmerman
Washington University School of Medicine, St. Louis, MO, United States

PREFACE

The computational study of enzyme structure and function has reached exciting and unprecedented levels. This state of affairs is due to a combination of powerful new computational methods, a critical evolution of ideas and insights, the ever-increasing power of computers, and important new experimental results for validation. In these two volumes of *Methods in Enzymology*, many of the leading computational researchers present their latest work, representing a range of state-of-the-art topics in computational enzymology. Generally speaking, the two volumes are divided into two general areas, the first being mostly devoted to the calculation of the free energy barriers and reaction pathways for enzymes—often using powerful quantum mechanics/molecular mechanics (QM/MM) methods—while the second volume contains a broader range of topics, including the role of enzyme dynamics and allostery, electrostatics, ligand binding, and several specific case studies.

The topic of computational enzymology—and the field of computational biophysics and biochemistry in general—has grown enormously since its inception in the early 1970s, with some of these original work being recognized by the 2013 Nobel Prize in Chemistry. It is tempting to conclude that the field is now "mature" and all that remains is for researchers in it to carry out increasingly detailed and accurate set of applications of the powerful computational methods presented herein. However, nothing could be further from the truth. Real enzymes exist and function in highly complex, multiscale biological environments. Often several enzymes function cooperatively, and they can be influenced by, or respond to, their local environment, whether be it a lipid membrane or the crowded cellular interior. In the venerable theories of activated dynamics for condensed-phase chemical kinetics, such as transition state theory, it is tacitly assumed that the chemical reaction dictates the slowest dynamical timescale of the system, thus corresponding to the highest free energy barrier. This basic assumption also allows one to apply these simpler theories to calculate quantities such as the free energy profile, ie, free energy barrier, for a reaction along a chemical pathway in an enzyme (the so-called "potential of mean force"). Yet, in complex biological systems, there are a wide range of timescales associated with *numerous* processes, some of which may be intrinsically coupled to the reactive process of the enzyme. In that light a key question then arises for the future. Can we better understand enzyme kinetics in the larger biological context of the living cell through computation? This challenge awaits us.

There is also the important fact that real biological systems are *not in a state of equilibrium*. Indeed, they can be rather far from it. Much of the standard condensed-phase kinetic theory developed in the last century—and applied in present day computational enzymology—relies on the key notion of the famous fluctuation–dissipation theorem, ie, that the behavior of systems perturbed out of equilibrium can be understood from studies of ones that are actually *in equilibrium*. This so-called "linear response" assumption leads us to powerful mathematical formulas for observables such as kinetic rate constants, as well as the algorithms one uses to compute them, which are based on *equilibrium* molecular dynamics simulation. However, much work remains to be done to develop theories and computational algorithms for enzymes functioning in a nonequilibrium biological context, albeit some important work in that direction, motivated by experiments, has already been initiated.

The great degree of progress to date on the topic of computational enzymology in reflected in these two volumes of *Methods in Enzymology*. Moreover, this field of research continues to evolve at an increasingly rapid pace. The scope of the enzyme systems presently under study, and the elaboration of their complex behaviors, is remarkable. As an example I can point to some of the research completed by talented young theorists as they passed through my own research group (see McCullagh, M., Saunders, M. G., & Voth, G. A. (2014). Unraveling the mystery of ATP hydrolysis in actin filaments. *Journal of the American Chemical Society, 136*, 13053–13058). In this work, QM/MM, molecular dynamics, advanced free energy sampling, and insights from coarse-grained modeling were all combined to explain the origins of the $>10^4$ acceleration of ATP hydrolysis in actin filaments (F-actin) over the free monomeric form (G-actin). This ATP hydrolysis by F-actin, which has been a mystery for years, is critical to the functioning of the actin-based eukaryotic cellular cytoskeleton and now computation has solved it.

Nevertheless, in light of the great remaining challenges described in the earlier paragraphs, it is abundantly clear that much remains to be done to further advance computational enzymology, in some cases even at a qualitative level of basic understanding. It will certainly be both important and fascinating to survey future volume(s) of *Methods in Enzymology* devoted to this topic—perhaps 10 or even 20 years from now—and to celebrate what I am sure will the outcomes from an exciting and continual evolution of this important field of research.

G.A. Voth
The University of Chicago

CHAPTER ONE

Continuum Electrostatics Approaches to Calculating pK_as and E_ms in Proteins

M.R. Gunner*[,1], N.A. Baker[†,‡]
*City College of New York in the City University of New York, New York, United States
[†]Pacific Northwest National Laboratory, Richland, DC, United States
[‡]Brown University, Providence, RI, United States
[1]Corresponding author: e-mail address: mgunner@ccny.cuny.edu

Contents

Abstract

Proteins change their charge state through protonation and redox reactions as well as through binding charged ligands. The free energy of these reactions is dominated by solvation and electrostatic energies and modulated by protein conformational relaxation in response to the ionization state changes. Although computational methods for calculating these interactions can provide very powerful tools for predicting protein charge states, they include several critical approximations of which users should be aware. This chapter discusses the strengths, weaknesses, and approximations of popular computational methods for predicting charge states and understanding the underlying electrostatic interactions. The goal of this chapter is to inform users about applications and potential caveats of these methods as well as outline directions for future theoretical and computational research.

Methods in Enzymology, Volume 578
ISSN 0076-6879
http://dx.doi.org/10.1016/bs.mie.2016.05.052

1. INTRODUCTION

Methods that use continuum electrostatics have been developed to calculate the energies of protein charge states as they change through processes such as residue protonation, redox chemistry, or ion binding. While only a subset of amino acids are titratable, they play key roles in protein function (Bartlett, Porter, Borkakoti, & Thornton, 2002). The model pK_a of an isolated amino acid in aqueous solution (Richarz & Wüthrich, 1975) can be used to calculate the probability that the isolated residue is charged at a given pH. Aspartate (Asp), glutamate (Glu), arginine (Arg), and lysine (Lys) comprise approximately 25% of average proteins and their p$K_{a,\text{sol}}$ values favor their ionization at physiological pHs (Kim, Mao, & Gunner, 2005). The termini of amino acid chains also have model pK_a values that cause them to frequently be ionized. Isolated histidine (His) has a pK_a value near 7, which makes it easy to titrate at physiological pH values. It not surprising that His is highly enriched in active sites (Holliday, Almonacid, Mitchell, & Thornton, 2007). Cysteine (Cys) (Go & Jones, 2013) and tyrosine (Tyr) (Styring, Sjoholm, & Mamedov, 2012) are acids with higher model pK_a values and are therefore less frequently ionized; however, these residues can play important functional roles as proton donors and as redox active sites. The residues in the active sites of proteins are often made of clusters of residues with linked protonation equilibria, leading to "nonideal" titration curves (Ondrechen, Clifton, & Ringe, 2001). A remarkable number of the mutations that lead to cancer involve protonatable residues (Webb, Chimenti, Jacobson, & Barber, 2011). Although nucleic acids (Wong & Pollack, 2010) and phospholipid membranes (Argudo, Bethel, Marcoline, & Grabe, 2016) also have titratable groups and strong electrostatic interactions, this chapter will focus on proteins.

The ionization states of small molecules are important to their function as substrates, cofactors, and control factors. Many protein ligands are charged, with ionization states that can change during the binding process or enzymatic reactions (Dissanayake, Swails, Harris, Roitberg, & York, 2015; Lee, Miller, & Brooks, 2016; Schindler et al., 2000). Cofactors such as NAD or FAD have charged groups such as phosphates that do not participate in reactions but must be bound to the protein for enzyme catalysis. Metal ions are often used by proteins to enhance stability such as in Zn fingers and as participants in redox reactions (Williams, 1997). Biological processes often occur at salt concentrations of 150 mM (or higher) (Bowers & Wiegel, 2011) such that all biomolecules are surrounded by a bath of small ions. The resulting ion cloud interacts with the protein in several ways, including salt-specific protein binding, electrostatic screening, and changing the thermodynamic activity of the protein in solution (Grochowski & Trylska, 2008; Record, Anderson, & Lohman, 1978).

Charged groups have very favorable interactions with water that strongly influence their behavior in aqueous solutions (Ren et al., 2012; Warshel & Russell, 1984). They are often found on the protein exterior surfaces maximizing their interaction with water while ensuring protein solubility and influencing interactions with other biomolecules. Supercharged proteins with total charges in excess of ± 30 e are now used in protein design to prevent aggregation (Lawrence, Phillips, & Liu, 2007). However, an important minority of charges are found within proteins, where they play functional roles.

Protein interiors are not simple hydrophobic environments and thus can tolerate internal charges through favorable electrostatic interactions (Kim et al., 2005; Spassov, Ladenstein, & Karshikoff, 1997). Interaction with other buried charges can form stabilizing ion pairs. Additionally, polar and polarizable groups are also present in protein interiors: the amide backbone dipole moment is larger than water's (Gunner, Saleh, Cross, ud-Doula, & Wise, 2000), many amino acid side chains that are polar or polarizable, and many crystal structures show water molecules and ions in protein interiors (Makarov, Pettitt, & Feig, 2002; Nayal & Di Cera, 1994, 1996). One of the goals of the electrostatic calculation methods described in this chapter is to quantitatively understand the nature of these electrostatic interactions.

This review will describe current methods for computing the charge states of residues and ligands as a function of pH, E_h, or solution salt concentrations. The key reactions are thus:

$$pK_a \quad AH \rightarrow A^- + H^+ \quad \text{or} \quad BH^+ \rightarrow B + H^+ \tag{1a}$$

$$E_m \quad O + n\ e^- + m\ H+ \rightarrow R^{(-n+m)} \tag{1b}$$

$$K_a \quad \text{Protein} + L^{+/-} \rightarrow \text{Protein} : L^{+/-}, \tag{1c}$$

where the species include acids (A), bases (B), ligands (L), oxidized species (O), and reduced species (R). Redox reactions are characterized by the number of electrons (n) and hydrogens (m) transferred. Ligand binding may also be accompanied by changes in protonation of the protein or ligand (Lee et al., 2016). Computational methods for modeling these reactions attempt to predict how the energetics of proton (pK_a), electron (E_m), or ligand (K_d) change as a function of environment (eg, protein interior vs solution). A wide range of experimental data are available for testing the predicted pK_a (Gosink, Hogan, Pulsipher, & Baker, 2014; Stanton & Houk, 2008), E_m (Reedy & Gibney, 2004), and K_d (Gilson et al., 2016) values. The goal of matching specific numerical values allows for rigorous testing of these calculation methodologies.

The pK_a value is the solution pH where the activities of A^- and AH (or B and BH^+) are equal. However, proton/electron/ion-binding affinities depend on the pH-dependent ionization states of other protein residues, thus making the proton affinity and thus the in situ pK_a pH dependent. Therefore, a single pK_a is often insufficient for characterizing the behavior of a titratable residue. Titration curves, describing the charge state as a function of pH can provide valuable information about the energetics influencing protein charge regulation (Webb, Tynan-Connolly, et al., 2011). In addition, as reactions often occur far from the pK_a of the reactant, it is often important to determine the proton affinity at physiological pH (Goyal, Lu, Yang, Gunner, & Cui, 2013). Insight into protein electrostatics is ideally obtained through the (favorable) comparison of experimental and calculated titration curves together with the analysis of the microscopic information (eg, electrostatic potentials, hydrogen-bonding networks, etc.) obtained from computational methods (Nielsen, Gunner, & Garcia-Moreno, 2011).

The calculation of the free energy of a group of charges in a protein or other macromolecule by continuum methods has been reviewed extensively (Alexov et al., 2011; Bashford, 2004; Garcia-Moreno & Fitch, 2004; Gunner & Alexov, 2000; Warshel, Sharma, Kato, & Parson, 2006). Thus, rather than providing a detailed methods review, we will highlight the information and choices that guide continuum electrostatics calculation and discuss emerging strategies for improving these methods.

2. BIOMOLECULAR STRUCTURE AND FLEXIBILITY

Structures are a required starting point for contemporary continuum electrostatic calculations (Berman et al., 2000). Charge state calculations are sensitive to structural details so care should be taken to assess the structural quality using tools now found associated with all structures in the PDB database. However, rigid, single structures are inadequate for accurate charge state calculations due to the importance of changes in flexibility and conformation that occur upon introduction of new charges into a protein.

One of the key choices in charge state modeling involves the degrees of freedom (DOFs) included in the model. In the simplest case of rigid molecules, the only DOFs are the protonation or redox states or the ligand-binding state. Because the protein will move in response to changes in the protonation, redox, or binding states, sampling DOFs for multiple structural "conformers" available to protein or ligand allows more a more "physical" analysis of the process. Thus, continuum electrostatics simulations balance implicit DOFs which, as described later, are approximated by the dielectric constant of the protein ($\varepsilon_{\text{solute}}$) and explicit DOFs.

A protein microstate is a defined choice for each element that has any DOF. Each microstate has an associated energy that is used to generate the thermodynamic averages for titration curves that yield pK_as, E_ms, binding probabilities, etc. The energy G_α of a microstate α can be written as a sum of contributions, which is implicitly summed over each group with DOF:

$$G_\alpha = m_\alpha \mu_\alpha + U_\alpha^{\text{MM}} + U_\alpha^{\text{elec}} + \Delta G_\alpha^{\text{p}} + \Delta G_\alpha^{\text{np}} \tag{2}$$

where $m_\alpha \mu_\alpha$ is the free energy of m_α bound species with chemical potential μ_α, U_α^{MM} is the nonelectrostatic molecular mechanical energy, U_α^{elec} is the electrostatic energy, the U_α terms depend explicitly on the state of the other residues in the microstate, $\Delta G_\alpha^{\text{p}}$ is the polar solvation energy, and $\Delta G_\alpha^{\text{np}}$ is the nonpolar solvation energy. The chemical potential μ_α varies for the quantity of interest; for protonation, it can be written as:

$$\mu_\alpha = \pm k_B T \log(10)(pH - pK_{a,\alpha}) \tag{3}$$

where $k_B T \approx 2.5$ kJ/mol (0.43 pK_a units) is the thermal energy at room temperature and $\log(10) \approx 2.3$. Most energies represent energy differences between the microstate in the protein interior and a reference state in solution. The quantity $pK_{a,\alpha}$ is the model value for the residue in solution: the positive form of the expression is used for acidic sites, and the negative form is used for basic sites. Given this reference value, the other terms in the equation represent an effective shift of the model $pK_{a,\alpha}$ to account for the influence of the protein environment.

The probability of a microstate β is given by the ensemble average

$$p_\beta = \frac{\sum_{\alpha \neq \beta} e^{-G_\alpha / k_B T}}{\sum_\alpha e^{-G_\alpha / k_B T}}. \tag{4}$$

If n groups each sample 2 protonation states, then there are 2^n microstates. If residue i has m_i protonation or steric conformers there are $\prod_i m_i$

microstates where the product runs over all residues with DOF. The high dimensionality of this sum makes it impractical to evaluate for most protein systems. Instead of direct evaluation, p_β is often calculated through limited conformational sampling; eg, via Monte Carlo (MC) simulations (Polydorides & Simonson, 2013; Song, Mao, & Gunner, 2009). Conformational DOFs can range from sampling side-chain rotameric states to relatively inexpensive optimization of steric clashes (Song, 2011) and simple enumeration of different tautomeric forms for the hydrogen position on protonated side chains.

Allowing only dipolar groups to reorient and sample multiple rotameric and tautomer states has significant advantages (Nielsen & Vriend, 2001). Modifying these positions, remodels the hydrogen bond network in response to charge changes, which can provide a significant energetic stabilization of titration events. Note that this form of limited sampling can require ad hoc entropy corrections to compensate for larger numbers of neutral state tautomers or conformers (Song et al., 2009).

One significant approximation in conformational sampling involves the treatment of the intramolecular interactions, which use force fields with only self and pairwise energetics to greatly improve the efficiency when evaluating microstate energies: the U_α terms include only pairwise additive energetics between two groups and are independent of the state of any third group. Evaluation of all pairwise interactions yields an energy matrix of dimension m^2 for m conformers. This pairwise decomposition is possible for less accurate nonpolarizable force fields and algorithms that sample proton positions and side-chain rotameric states. However, more recent polarizable force fields and larger collective protein motions such as backbone displacements generally cannot be represented in this pairwise form. However, while most methods that utilize Monte Carlo sampling make this approximation, it should be recognized that it misses motions that are likely to be important (Richman, Majumdar, & Garcia-Moreno, 2014).

Monte Carlo methods can be used to incorporate side-chain conformer sampling on a rigid protein backbone (Polydorides & Simonson, 2013; Rabenstein, Ullmann, & Knapp, 1998; Song et al., 2009). Such sampling attempts to explicitly evaluate the ensemble average described earlier and thus incorporates a significant amount of side-chain response to charge state changes. However, adding conformational DOFs requires new approximations. In particular, the shape of the protein can change when sampling different side chain or backbone conformations. Calculation costs can increase dramatically if the shape of each microstate is calculated explicitly.

To maintain the cost for calculating the interaction energies of $O(m^2)$ often relies on all conformers being present when the continuum electrostatic pairwise interactions between conformers are determined. This can exaggerate the low dielectric space of the protein. Some early approaches scaled the electrostatic interactions by an empirical screening function (Georgescu, Alexov, & Gunner, 2002). Newer methods correct for dielectric boundary errors due to excess conformers by using information obtained from a small number of calculations with an exact boundary (Song et al., 2009).

Molecular dynamics (MD) calculations can also be used to provide conformations in different protonation states. Given a particular titratable site, two sets of simulations are performed: one with the charged state of the site and a second with the neutral state. For sufficiently small energetic differences between the two protonation states (ie, when linear response theory is valid), these ensembles will substantially overlap and the titration probability can be calculated via simple ensemble averages (Sham, Chu, & Warshel, 1997). MD-based linear response approaches have two key limitations. The first is the computational expense of running $O(2^n)$ MD simulations to sample the n distinct neutral and ionized charge states. Rational choices can help pick consequential protonation states to sample (Meyer & Knapp, 2015; Witham et al., 2011). The second limitation is the underlying linear response assumption requiring the energy difference between neutral and ionized states is small—which is often not true for the important titration events in protein systems (Di Russo, Marti, & Roitberg, 2014).

MD simulations can also be performed in open constant-pH ensembles. Unlike the fixed charge state simulations described earlier, constant-pH methods allow charged sites to exchange protons with the surrounding solution based on the pH of the bulk media, the model pK_a of the site, and the energetics of the conformational ensemble. One class of methods performs MD for 10s of fs followed by a continuum electrostatics pK_a calculation as described earlier to modify the protonation states in the MD trajectory (Baptista, Martel, & Peterson, 1997; Lee, Miller, Damjanovic, & Brooks, 2015; Swails, York, & Roitberg, 2014). Alternatively, protonation states can be changed continuously via λ dynamics (Goh, Hulbert, Zhou, & Brooks, 2014; Khandogin & Brooks, 2005). The primary hurdle to adoption of such continuous-pH MD methods is the difficulty of reaching convergence of the simulations. The use of pH replica exchange has led to significant improvements, but these methods are still not routine for the study of large proteins.

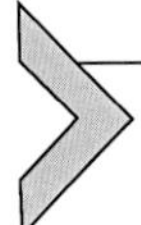

3. SOLVENT MODELS OR: HOW I LEARNED TO STOP WORRYING AND LOVE THE DIELECTRIC COEFFICIENT

Many methods use continuum models of solvation behavior to incorporate the effects of solvent on charging energetics because of the computation effort associated with explicit descriptions of water molecules. The simplest continuum model for electrostatics represents the solvent as a dielectric material, usually with a dielectric coefficient of approximately 80 to represent water. The Poisson equation (Baker, 2004; Nicholls & Honig, 1991) describes polar solvation (electrostatic) energies within this dielectric approximation. This equation can be combined with a nonpolar solvation term that describes the nonelectrostatic contributions from the solvent when conformers with significantly different surface exposure are sampled (Song et al., 2009). For calculations of ligand affinity and for ligand partition coefficients, continuum models generally include a shape-related contribution, to describe the work associated with inserting the uncharged solute into water, and a Lennard-Jones-like term to describe the weak solute–solvent dispersive interactions (Lee et al., 2016). Although small, such weak dispersive forces play an important role in protein solvation and in titration state calculations (Levy, Zhang, Gallicchio, & Felts, 2003; Song et al., 2009; Wagoner & Baker, 2004). Popular models for the cavity energy generally contain a term that scales as the area of the molecule times the surface tension of the solution and often also include a term that multiples solution (hard sphere) pressure with the volume of the solute (Wagoner & Baker, 2006).

Use of the Poisson equation—or other related continuum models—assumes that all polarization in the system (molecule and solvent) is linear, local, and time independent. Linear response implies that, no matter how large the electric field, the system will polarize in a proportional manner. However, given the finite density and polarizability of water and molecular solutes, this assumption is clearly violated at high charge densities and field strengths, such as found near nucleic acids (Lipfert, Doniach, Das, & Herschlag, 2014). Local response implies that system polarization always occurs in the same location as an applied field. However, given the nonzero size and hydrogen-bonding structure of water and most biomolecular species, this assumption is nearly always violated in biologically relevant systems (Mobley, Barber, Fennell, & Dill, 2008; Xie, Jiang, Brune, & Scott, 2012). Finally, the static response assumes no time dependence for molecular polarization. However, even in bulk solvent, this time independence is violated, with the optical dielectric constant of water of 2, which increases to the static value of 80 on the picosecond–nanosecond timescale (Fernandez, Mulev, Goodwin, & Sengers, 1995; Zasetsky, 2011).

Given that nearly all of the assumptions of continuum electrostatics are violated in biologically relevant systems, the reader is probably wondering "why bother?" The answer lies in the power of heuristics. Although there are many arguments about its accuracy at microscopic levels (for example, Kukic et al., 2013; Schutz & Warshel, 2001; Simonson, 2013), the continuum model of water with a dielectric coefficient of 78–80 has proven remarkably useful for a wide range of applications. Likewise, while the ab initio derivation of solute dielectric constants is likely an exercise in futility, several heuristics have been useful in extending the applicability of continuum electrostatics to real biomolecular systems. These heuristics are described later; however, it is *essential* that the users of continuum electrostatics methods are aware that they are using imperfect surrogates for complicated molecular phenomena. In particular, continuum electrostatics calculations should always be benchmarked for accuracy against real experimental data.

Simple—but imperfect—heuristics can be used to guide the selection of a molecular dielectric coefficient value (ε_{solute}). These are presented in Fig. 1 and comprise three basic regimes:

- An ε_{solute} of 2 represents the electronic polarization that will be found in any condensed matter system (Landau, Lifshitš, & Pitaevskiĭ, 1984). This interpretation has an important implication for continuum electrostatics calculations: $\varepsilon_{solute} \geq 2$ should be used for all calculations with nonpolarizable force fields (Leontyev & Stuchebrukhov, 2009).

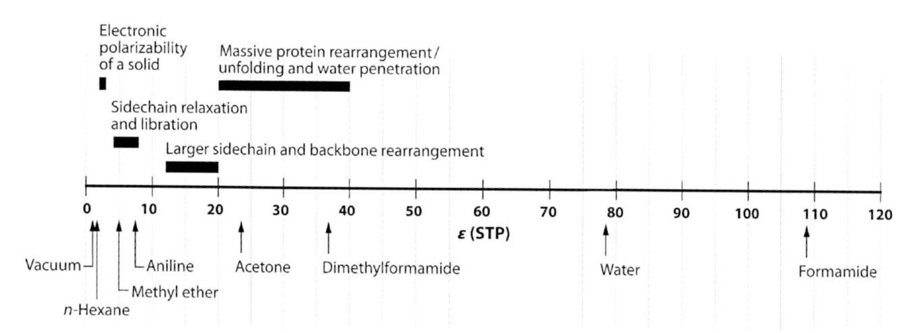

Fig. 1 A summary of dielectric constants of model compounds (*bottom*) and their application to protein continuum electrostatics modeling (*top*).

- An ε_{solute} of 4 has been ascribed to dried proteins and can be interpreted to include a very constrained polarization response of the protein dipoles (Gilson & Honig, 1986). This interpretation has an important implication for continuum electrostatics: an $\varepsilon_{solute} \geq 4$ should be used for all calculations that do not allow backbone rearrangement through MD or Monte Carlo configuration sampling.
- Larger values of ε_{solute} allow more of the dipolar rearrangement of the backbone and side chains to be treated in an averaged manner with a single, compact parameter. Values of $4 < \varepsilon_{solute} < 12$ have been successfully used to predict protein–protein binding and are often attributed to limited side-chain rearrangement. Values of ε_{solute} above 12 are associated with larger scale backbone rearrangement and water penetration. Early continuum electrostatics attempts to model pK_a values in proteins showed that $\varepsilon_{solute} = 20$ gave the best predictive power for calculations using a single protein dielectric constant and a single conformation (Antosiewicz, McCammon, & Gilson, 1994).

Not all continuum electrostatics treatments use a constant dielectric coefficient for the solute interior; some models use larger dielectric values for regions of the protein with greater responses to charge changes. For example, Alexov has varied the dielectric constant based on the atomic packing density (Li, Li, Zhang, & Alexov, 2013). While such variation does not explicitly take into account the chemical nature of the side chains, it does provide a mechanism for modeling internal DOFs through the dielectric coefficient. Because these calculations avoid explicit conformational sampling, they offer the possibility of improved dielectric descriptions with the efficiency of standard continuum electrostatic methods.

Most continuum electrostatics software packages will identify interior cavities large enough to accommodate a water molecule—and many will assign these interior cavities a bulk dielectric value of $\varepsilon_{solute} = 80$. However, the high-dielectric treatment of internal cavities comes with a few important caveats. First, it is difficult to provide a physical justification for a single water molecule having the dielectric behavior of the bulk solvent. Second, this procedure is sensitive to small conformational changes that may cause regions to switch between ε_{solute} and $\varepsilon_{solvent}$. To address this issue, Knapp has explored the effects of modeling the cavities with higher detail using a finer grid, which can accept smaller or less spherical wet regions, which improves the fit to benchmark pK_as (Meyer, Kieseritzky, & Knapp, 2011). Other methods make use of Gaussian dielectric boundaries in the calculation of the Poisson–Boltzmann equation, which also raises the effective internal dielectric constant (Li, Li, & Alexov, 2014; Word & Nicholls, 2011).

As an alternative to high-dielectric models of internal cavities, continuum electrostatics software such as MCCE can include explicit water molecules within the protein (Song, Mao, & Gunner, 2003). The included waters require explicit sampling. They must be optimized for each charge state and the number of waters may change with the charge state. Waters are often found in clusters so this optimization must be performed for multiple water molecules simultaneously. As a result, the inclusion of explicit water molecules can substantially increase the computational expense of the charge state calculation. The pK_as obtained with implicit or explicit waters in the cavities have been found to agree surprisingly well in limited testing.

The various modifications of the methods described earlier all improve the fit to known data essentially by increasing the effective interior dielectric constant. The electrostatic energy of a charge depends on the atomic charge distribution, the radius, and the interior and exterior dielectric constant. Thus, the effective interior dielectric constant can be raised by increasing $\varepsilon_{\text{solute}}$ directly, or by smoothing the dielectric surface, or by enhancing cavities in the interior. The effects of changing these parameters have been explored separately. Without a better sense of exactly how the various parameters interact the search through parameter space remains Balkanized with different laboratories exploring their favorite parameters. However, it should be noted that all of these changes do lead to significant improvement in the correspondence between experimental and calculated values.

4. MODELING ION–SOLUTE INTERACTIONS

Ions are arguably more difficult to model than solvent. The simplest—and most widely used—model of ion behavior is based on Debye–Hückel descriptions of aqueous ions as a diffuse "cloud" that nonspecifically screens electrostatic behavior in solution. The only major determinants of ion behavior in Debye–Hückel-like models are the ion concentration and charge valencies. However, this treatment has extreme limitations in describing realistic protein–ion interactions that often include specific ion binding to protein sites as well as strong dependence on ion species, even for ions with the same charge. To address these issues, some researchers have begun to use models that combine implicit solvent descriptions with explicit simulation of the ions (often via Monte Carlo sampling) (Chen, Marucho,

Baker, & Pappu, 2009; Sharp, Friedman, Misra, Hecht, & Honig, 1995; Song & Gunner, 2009). Nevertheless, many charge state calculations still use Poisson–Boltzmann (PB) methods, which combine a Poisson treatment of the solvent with the Boltzmann Debye–Hückel-like ion description.

5. FORCE FIELD AND PARAMETER CHOICES

The microstate energy calculations described earlier require several different types of parameters to describe molecular mechanics interactions, solvent characteristics, as well as atomic size and charge. The molecular mechanics energies, atomic charges, and solute–solvent Lennard-Jones interactions are often specified by standard molecular simulation force fields such as AMOEBA (Schnieders, Baker, Ren, & Ponder, 2007; Shi et al., 2013), AMBER (Pearlman et al., 1995), or CHARMM (Brooks et al., 2009). These force fields can also be used to specify the solute–solvent boundary through atomic radii; however, custom parameter sets such as PARSE (Tannor et al., 1994) or ZAP (Word & Nicholls, 2011) are generally preferred because they have been optimized to reproduce solvation energies. In addition to atomic radii, the user must specify the algorithm used to determine the shape of the solute–solvent interface. A variety of choices are available for these shape algorithms ranging from simple unions of spheres (Lee & Richards, 1971) or Gaussians (Grant, Pickup, Sykes, Kitchen, & Nicholls, 2007) to heuristic molecular-accessible surfaces (Connolly, 1983) to thermodynamically defined self-consistent solute–solvent interface definitions (Chen, Baker, & Wei, 2010, 2011; Cheng, Dzubiella, McCammon, & Li, 2007). Additionally, the user must choose a function to define the ion-accessible regions around the protein; however, this interface is commonly chosen as an ion-accessible union of spheres with radii equal to the atomic radii plus a nominal ionic radius of 0.2 nm. It is important to note that the optimal choices for radii, charges, and surface definitions are strongly correlated; ie, the radii are often optimized for a specific surface definition (Dong & Zhou, 2002). These many choices of parameters are then presented to a program that will solve the Poisson–Boltzmann equation to provide the solvation energy of individual conformers (within the environment of the protein) and the pairwise interactions between all pairs of conformers. For example, the programs DelPhi (Li, Li, Zhang, & Alexov, 2012) or APBS (Baker, Sept, Joseph, Holst, & McCammon, 2001) have been employed within programs such as MCCE (Song et al., 2009), and Karlsberg (Meyer et al., 2011), DelPhi pK_a (Wang, Li, & Alexov, 2015), and

PDB2PKA (Dolinsky et al., 2007; Olsson, Sondergaard, Rostkowski, & Jensen, 2011) to calculate the equilibrium protonation, redox, and ligand-binding states as a function of the appropriate chemical potential.

Titration state prediction methods must be benchmarked against datasets of in situ pK_as, E_ms, or K_ds to determine their accuracy. There are approximately ≈350 wild-type residues with known pK_as that are used extensively for such benchmarking (Song et al., 2009). These include a large number surface residues where the protein does not significantly influence the proton affinity. A subset of 100 residues has been selected to yield better range of pK_as for training and testing (Stanton & Houk, 2008). The "null model" for charge state prediction assigns the model amino acid pK_a value to all residues in the protein, regardless of their location or interactions. When the null model is used with the 100-residue subset, the RMSD between predicted and experimental pK_a values is ≈1 pH unit. This sets a challenging metric for evaluating the performance of more sophisticated titration prediction methods. For example, the RMSD using modern Monte Carlo methods with continuum electrostatics force field, an $\varepsilon_{\text{solute}}$ between 4 and 8 and addition of conformational sampling or modification of the dielectric boundary and distribution can be between 0.9 and 1.1 (Polydorides & Simonson, 2013; Song et al., 2009; Wang et al., 2015). However, informatics-based methods such as PROPKA3 can do much better while sacrifice the underlying physical interactions for knowledge-based potentials (Olsson, 2011).

The Garcia-Moreno lab has placed >100 mutated residues into the core of Staphylococcal nuclease (Isom, Castaneda, Cannon, & Garcia-Moreno, 2011; Isom, Castaneda, Cannon, Velu, & Garcia-Moreno, 2010; Richman, Majumdar, & Garcia-Moreno, 2015). These residues formed the basis of the only blind challenge; ie, where pK_as were calculated without the experimental value being known (Nielsen et al., 2011). A metaanalysis (Gosink et al., 2014) of the blind predictions found that the RMSD for the null model is ≈3.5, indicating that the pK_as for these residues were very shifted from the model values due to their burial in the protein. Empirical methods such as PROPKA3 (Olsson, 2011) did significantly better than the null model, methods with added conformational sampling did slightly better than the null model, while methods without added sampling did worse. Papers submitted after the pK_as were revealed were able to obtain RMSDs <2 for this challenging dataset, as different modifications were explored once the errors were known. Particular improvement was found for methods that increased the response of the protein; eg, by using more explicit sampling via continuous-pH MD (Wallace et al., 2011), by adding ensembles of structures obtained with MD (Witham et al., 2011) or Rosetta (Song, 2011), through increased side-chain conformation sampling (Gunner, Zhu, & Klein, 2011), by increasing the effective ε_{solute} to implicitly model more internal water (Meyer et al., 2011), or by using a smoother dielectric boundary (Word & Nicholls, 2011). The errors for calculations with rigid backbones were smaller when crystal structures of the mutants were used rather than when the mutation was made in silico. Ensemble models which aggregated all of the predictions using Bayesian model averaging gave the best overall results (Gosink et al., 2014).

6. CONCLUSIONS

The goal of this chapter was to present an overview of computational methods for predicting charge states of proteins with an emphasis on the issues that arise when applying continuum electrostatic methods to these problems. Given the many choices that must be made when applying these computational methods, one of the most important issues for this field is the availability of well-curated experimental data sets for testing computational predictions. The pK_a cooperative is a collaborative activity focused on assembling such data sets, performing blind predictions, and discussing the results as well as how to improve computational predictions (http://pkacoop.org/). All of the methods described earlier can be tuned to provide reasonable agreement with experimental data in a *post*diction setting. However, only a few methods perform with acceptable accuracy (~1 pK_a unit error) in blind challenge predictions. Among these, constant-pH MD methods generated the best predictions—at significantly increased computational expense and the risk of poor convergence of the MD simulations. Thus, computational methods continue to evolve to make the calculation of the energy of charges in protein faster and more accurate while providing increased physical insight into the forces at work. The current methods, despite their limitations, provide guidance as to the proton affinities of sites in proteins as well as the atomic interactions that affect a specific charge in a specific site and thus can be invaluable in getting more understanding of protein structure/function relationships.

ACKNOWLEDGMENTS

M.R.G. gratefully acknowledges the support of Grant MCB1519640 from NSF, as well as National Institute on Minority Health and Health Disparities Grant 8G12MD7603-28 for infrastructure. N.A.B. gratefully acknowledges support from NIH Grants R01GM069702 and R01GM099450.

REFERENCES

Alexov, E., Mehler, E. L., Baker, N., Baptista, A. M., Huang, Y., Milletti, F., ... Word, J. M. (2011). Progress in the prediction of pK_a values in proteins. *Proteins*, *79*, 3260–3275.

Antosiewicz, J., McCammon, J. A., & Gilson, M. K. (1994). Prediction of pH-dependent properties in proteins. *Journal of Molecular Biology*, *238*, 415–436.

Argudo, D., Bethel, N. P., Marcoline, F. V., & Grabe, M. (2016). Continuum descriptions of membranes and their interaction with proteins: Towards chemically accurate models. *Biochimica et Biophysica Acta*, *1858*, 1619–1634. http://dx.doi.org/10.1016/j.bbamem.2016.02.003.

Baker, N. A. (2004). Poisson-Boltzmann methods for biomolecular electrostatics. *Methods in Enzymology*, *383*, 94–118.

Baker, N. A., Sept, D., Joseph, S., Holst, M. J., & McCammon, J. A. (2001). Electrostatics of nanosystems: Application to microtubules and the ribosome. *Proceedings of the National Academy of Sciences of the United States of America*, *98*, 10037–10041.

Baptista, A. M., Martel, P. J., & Peterson, S. B. (1997). Simulation of protein conformational freedom as a function of pH: Constant-pH molecular dynamics using implicit titration. *Proteins: Structure, Function, and Genetics*, *27*, 523–544.

Bartlett, G. J., Porter, C. T., Borkakoti, N., & Thornton, J. M. (2002). Analysis of catalytic residues in enzyme active sites. *Journal of Molecular Biology*, *324*, 105–211.

Bashford, D. (2004). Macroscopic electrostatic models for protonation states in proteins. *Frontiers in Bioscience*, *9*, 1082–1099.

Berman, H. M., Westbrook, J., Feng, Z., Gilliland, G., Bhat, T. N., Weissig, H., ... Bourne, P. E. (2000). The protein data bank. *Nucleic Acids Research*, *28*, 235–242.

Bowers, K. J., & Wiegel, J. (2011). Temperature and pH optima of extremely halophilic archaea: A mini-review. *Extremophiles*, *15*, 119–128.

Brooks, B. R., Brooks, C. L., 3rd., Mackerell, A. D., Jr., Nilsson, L., Petrella, R. J., Roux, B., ... Karplus, M. (2009). CHARMM: The biomolecular simulation program. *Journal of Computational Chemistry*, *30*, 1545–1614.

Chen, Z., Baker, N. A., & Wei, G. W. (2010). Differential geometry based solvation model I: Eulerian formulation. *Journal of Computational Physics*, *229*, 8231–8258.

Chen, Z., Baker, N. A., & Wei, G. W. (2011). Differential geometry based solvation model II: Lagrangian formulation. *Journal of Mathematical Biology*, *63*, 1139–1200.

Chen, A. A., Marucho, M., Baker, N. A., & Pappu, R. V. (2009). Simulations of RNA interactions with monovalent ions. *Methods in Enzymology*, *469*, 411–432.

Cheng, L. T., Dzubiella, J., McCammon, J. A., & Li, B. (2007). Application of the level-set method to the implicit solvation of nonpolar molecules. *The Journal of Chemical Physics*, *127*, 084503.

Connolly, M. L. (1983). Solvent-accessible surfaces of proteins and nucleic acids. *Science*, *221*, 709–713.

Di Russo, N. V., Marti, M. A., & Roitberg, A. E. (2014). Underlying thermodynamics of pH-dependent allostery. *The Journal of Physical Chemistry. B*, *118*, 12818–12826.

Dissanayake, T., Swails, J. M., Harris, M. E., Roitberg, A. E., & York, D. M. (2015). Interpretation of pH-activity profiles for acid-base catalysis from molecular simulations. *Biochemistry*, *54*, 1307–1313.

Dolinsky, T. J., Czodrowski, P., Li, H., Nielsen, J. E., Jensen, J. H., Klebe, G., & Baker, N. A. (2007). PDB2PQR: Expanding and upgrading automated preparation of biomolecular structures for molecular simulations. *Nucleic Acids Research*, *35*, W522–W525.

Dong, F., & Zhou, H. X. (2002). Electrostatic contributions to T4 lysozyme stability: Solvent-exposed charges versus semi-buried salt bridges. *Biophysical Journal*, *83*, 1341–1347.

Fernandez, D., Mulev, Y., Goodwin, A., & Sengers, J. (1995). A database for the static dielectric constant of water and steam. *Journal of Physical and Chemical Reference Data*, *24*, 33.

Garcia-Moreno, E. B., & Fitch, C. A. (2004). Structural interpretation of pH and salt-dependent processes in proteins with computational methods. *Methods in Enzymology*, *380*, 20–51.

Georgescu, R. E., Alexov, E. G., & Gunner, M. R. (2002). Combining conformational flexibility and continuum electrostatics for calculating pK_as in proteins. *Biophysical Journal*, *83*, 1731–1748.

Gilson, M. K., & Honig, B. H. (1986). The dielectric constant of a folded protein. *Biopolymers*, *25*, 2097–2119.

Gilson, M. K., Liu, T., Baitaluk, M., Nicola, G., Hwang, L., & Chong, J. (2016). BindingDB in 2015: A public database for medicinal chemistry, computational chemistry and systems pharmacology. *Nucleic Acids Research*, *44*, D1045–D1053.

Go, Y. M., & Jones, D. P. (2013). The redox proteome. *The Journal of Biological Chemistry*, *288*, 26512–26520.

Goh, G. B., Hulbert, B. S., Zhou, H., & Brooks, C. L., 3rd. (2014). Constant pH molecular dynamics of proteins in explicit solvent with proton tautomerism. *Proteins*, *82*, 1319–1331.

Gosink, L. J., Hogan, E. A., Pulsipher, T. C., & Baker, N. A. (2014). Bayesian model aggregation for ensemble-based estimates of protein pKa values. *Proteins*, *82*, 354–363.

Goyal, P., Lu, J., Yang, S., Gunner, M. R., & Cui, Q. (2013). Changing hydration level in an internal cavity modulates the proton affinity of a key glutamate in cytochrome c oxidase. *Proceedings of the National Academy of Sciences of the United States of America*, *110*, 18886–18891.

Grant, J. A., Pickup, B. T., Sykes, M. J., Kitchen, C. A., & Nicholls, A. (2007). The Gaussian Generalized Born model: Application to small molecules. *Physical Chemistry Chemical Physics*, *9*, 4913–4922.

Grochowski, P., & Trylska, J. (2008). Continuum molecular electrostatics, salt effects, and counterion binding—A review of the Poisson-Boltzmann theory and its modifications. *Biopolymers*, *89*, 93–113.

Gunner, M. R., & Alexov, E. (2000). A pragmatic approach to structure based calculation of coupled proton and electron transfer in proteins. *Biochimica et Biophysica Acta*, *1458*, 63–87.

Gunner, M. R., Saleh, M. A., Cross, E., ud-Doula, A., & Wise, M. (2000). Backbone dipoles generate positive potentials in all proteins: Origins and implications of the effect. *Biophysical Journal*, *78*, 1126–1144.

Gunner, M. R., Zhu, X., & Klein, M. C. (2011). MCCE analysis of the pK_as of introduced buried acids and bases in staphylococcal nuclease. *Proteins*, *79*, 3306–3319.

Holliday, G. L., Almonacid, D. E., Mitchell, J. B., & Thornton, J. M. (2007). The chemistry of protein catalysis. *Journal of Molecular Biology*, *372*, 1261–1277.

Isom, D. G., Castaneda, C. A., Cannon, B. R., & Garcia-Moreno, B. (2011). Large shifts in pKa values of lysine residues buried inside a protein. *Proceedings of the National Academy of Sciences of the United States of America*, *108*, 5260–5265.

Isom, D. G., Castaneda, C. A., Cannon, B. R., Velu, P. D., & Garcia-Moreno, E. B. (2010). Charges in the hydrophobic interior of proteins. *Proceedings of the National Academy of Sciences of the United States of America*, *107*, 16096–16100.

Khandogin, J., & Brooks, C. L., 3rd. (2005). Constant pH molecular dynamics with proton tautomerism. *Biophysical Journal*, *89*, 141–157.

Kim, J., Mao, J., & Gunner, M. R. (2005). Are acidic and basic groups in buried proteins predicted to be ionized? *Journal of Molecular Biology*, *348*, 1283–1298.

Kukic, P., Farrell, D., McIntosh, L. P., Garcia-Moreno, E. B., Jensen, K. S., Toleikis, Z., … Nielsen, J. E. (2013). Protein dielectric constants determined from NMR chemical shift perturbations. *Journal of the American Chemical Society*, *135*, 16968–16976.

Landau, L. D., Lifshitš, E. M., & Pitaevskiĭ, L. P. (1984). *Electrodynamics of continuous media* (2nd ed.). Oxford Oxfordshire; New York: Pergamon.

Lawrence, M. S., Phillips, K. J., & Liu, D. R. (2007). Supercharging proteins can impart unusual resilience. *Journal of the American Chemical Society, 129*, 10110–10112.

Lee, J., Miller, B. T., & Brooks, B. R. (2016). Computational scheme for pH-dependent binding free energy calculation with explicit solvent. *Protein Science, 25*, 231–243.

Lee, J., Miller, B. T., Damjanovic, A., & Brooks, B. R. (2015). Enhancing constant-pH simulation in explicit solvent with a two-dimensional replica exchange method. *Journal of Chemical Theory and Computation, 11*, 2560–2574.

Lee, B., & Richards, F. M. (1971). The interpretations of protein structures: Estimation of static accessibility. *Journal of Molecular Biology, 55*, 379–400.

Leontyev, I. V., & Stuchebrukhov, A. A. (2009). Electronic continuum model for molecular dynamics simulations. *The Journal of Chemical Physics, 130*, 085102.

Levy, R. M., Zhang, L. Y., Gallicchio, E., & Felts, A. K. (2003). On the nonpolar hydration free energy of proteins: Surface area and continuum solvent models for the solute-solvent interaction energy. *Journal of the American Chemical Society, 125*, 9523–9530.

Li, L., Li, C., & Alexov, E. (2014). On the modeling of polar component of solvation energy using smooth Gaussian-based dielectric function. *Journal of Theoretical and Computational Chemistry, 13*(03), 1440002.

Li, C., Li, L., Zhang, J., & Alexov, E. (2012). Highly efficient and exact method for parallelization of grid-based algorithms and its implementation in DelPhi. *Journal of Computational Chemistry, 33*, 1960–1966.

Li, L., Li, C., Zhang, Z., & Alexov, E. (2013). On the dielectric "constant" of proteins: Smooth dielectric function for macromolecular modeling and Its implementation in DelPhi. *Journal of Chemical Theory and Computation, 9*, 2126–2136.

Lipfert, J., Doniach, S., Das, R., & Herschlag, D. (2014). Understanding nucleic acid-ion interactions. *Annual Review of Biochemistry, 83*, 813–841.

Makarov, V., Pettitt, B. M., & Feig, M. (2002). Solvation and hydration of proteins and nucleic acids: A theoretical view of simulation and experiment. *Accounts of Chemical Research, 35*, 376–384.

Meyer, T., Kieseritzky, G., & Knapp, E. W. (2011). Electrostatic pK_a computations in proteins: Role of internal cavities. *Proteins, 79*, 3320–3332.

Meyer, T., & Knapp, E. W. (2015). pK_a values in proteins determined by electrostatics applied to molecular dynamics trajectories. *Journal of Chemical Theory and Computation, 11*, 2827–2840.

Mobley, D. L., Barber, A. E., 2nd., Fennell, C. J., & Dill, K. A. (2008). Charge asymmetries in hydration of polar solutes. *The Journal of Physical Chemistry. B, 112*, 2405–2414.

Nayal, M., & Di Cera, E. (1994). Predicting Ca^{2+}-binding sites in proteins. *Proceedings of the National Academy of Sciences of the United States of America, 91*, 817–821.

Nayal, M., & Di Cera, E. (1996). Valence screening of water in protein crystals reveals potential Na^{+} binding sites. *Journal of Molecular Biology, 256*, 228–234.

Nicholls, A., & Honig, B. (1991). A rapid finite difference algorithm utilizing successive over-relaxation to solve the Poisson-Boltzmann equation. *Journal of Combinatorial Chemistry, 12*, 435–445.

Nielsen, J. E., Gunner, M. R., & Garcia-Moreno, B. E. (2011). The pK_a cooperative: A collaborative effort to advance structure-based calculations of pK_a values and electrostatic effects in proteins. *Proteins, 79*, 3249–3259.

Nielsen, J. E., & Vriend, G. (2001). Optimizing the hydrogen-bond network in Poisson-Boltzmann equation-based pK_a calculations. *Proteins: Structure, Function, and Genetics, 43*, 403–412.

Olsson, M. H. (2011). Protein electrostatics and pK_a blind predictions; contribution from empirical predictions of internal ionizable residues. *Proteins*, *79*, 3333–3345.

Olsson, M. H., Sondergaard, C. R., Rostkowski, M., & Jensen, J. H. (2011). PROPKA3: Consistent treatment of internal and surface residues in empirical pKa predictions. *Journal of Chemical Theory and Computation*, *7*, 525–537.

Ondrechen, M., Clifton, J., & Ringe, D. (2001). THEMATICS: A simple computational predictor of enzyme function from structure. *Proceedings of the National Academy of Sciences of the United States of America*, *98*, 12473–12478.

Pearlman, D. A., Case, D. A., Caldwell, J. W., Ross, W. S., Cheatham, T. E., III, DeBolt, S., … Kollman, P. (1995). AMBER a package of computer programs for applying molecular mechanics, normal mode analysis, molecular dynamics and free energy calculations to simulate the structural and energetic properties of molecules. *Computer Physics Communications*, *91*, 1–41.

Polydorides, S., & Simonson, T. (2013). Monte Carlo simulations of proteins at constant pH with generalized Born solvent, flexible sidechains, and an effective dielectric boundary. *Journal of Computational Chemistry*, *34*, 2742–2756.

Rabenstein, B., Ullmann, G. M., & Knapp, E.-W. (1998). Calculation of protonation patterns in proteins with structural relaxation and molecular ensembles-application to the photosynthetic reaction center. *European Biophysics Journal*, *27*, 626–637.

Record, M. T., Anderson, C. F., & Lohman, T. M. (1978). Thermodynamic analysis of ion effects on binding and conformational equilibria of proteins and nucleic-acids—Roles of ion association or release, screening, and ion effects on water activity. *Quarterly Reviews of Biophysics*, *11*, 103–178.

Reedy, C. J., & Gibney, B. R. (2004). Heme protein assemblies. *Chemical Reviews*, *104*, 617–649.

Ren, P., Chun, J., Thomas, D. G., Schnieders, M. J., Marucho, M., Zhang, J., & Baker, N. A. (2012). Biomolecular electrostatics and solvation: A computational perspective. *Quarterly Reviews of Biophysics*, *45*, 427–491.

Richarz, R., & Wüthrich, K. (1975). Carbon-13 NMR chemical shifts of the common amino acid residues measured in aqueous solutions of the linear tetrapeptides H-Gly-Gly-X-L-Ala-OH. *Biopolymers*, *17*, 2133–2141.

Richman, D. E., Majumdar, A., & Garcia-Moreno, E. B. (2014). pH dependence of conformational fluctuations of the protein backbone. *Proteins*, *82*, 3132–3143.

Richman, D. E., Majumdar, A., & Garcia-Moreno, E. B. (2015). Conformational reorganization coupled to the ionization of internal lys residues in proteins. *Biochemistry*, *54*, 5888–5897.

Schindler, T., Bornmann, W., Pellicena, P., Miller, W. T., Clarkson, B., & Kuriyan, J. (2000). Structural mechanism for STI-571 inhibition of abelson tyrosine kinase. *Science*, *289*, 1938–1942.

Schnieders, M. J., Baker, N. A., Ren, P., & Ponder, J. W. (2007). Polarizable atomic multipole solutes in a Poisson-Boltzmann continuum. *The Journal of Chemical Physics*, *126*, 124114.

Schutz, C. N., & Warshel, A. (2001). What are the "dielectric constants" of proteins and how to validate electrostatic models? *Proteins: Structure, Function, and Genetics*, *44*, 400–417.

Sham, Y. Y., Chu, Z. T., & Warshel, A. (1997). Consistent calculations of pK_a's of ionizable residues in proteins: Semi-microscopic and microscopic approaches. *The Journal of Physical Chemistry*, *101*, 4458–4472.

Sharp, K. A., Friedman, R. A., Misra, V., Hecht, J., & Honig, B. (1995). Salt effects on polyelectrolyte-ligand binding: Comparison of Poisson-Boltzmann, and limiting law/counterion binding models. *Biopolymers*, *36*, 245–262.

Shi, Y., Xia, Z., Zhang, J., Best, R., Wu, C., Ponder, J. W., & Ren, P. (2013). The polarizable atomic multipole-based AMOEBA force feld for proteins. *Journal of Chemical Theory and Computation*, *9*, 4046–4063.

Simonson, T. (2013). What is the dielectric constant of a protein when its backbone is fixed? *Journal of Chemical Theory and Computation*, *9*, 4603–4608.
Song, Y. (2011). Exploring conformational changes coupled to ionization states using a hybrid Rosetta-MCCE protocol. *Proteins*, *79*(12), 3356–3363. http://dx.doi.org/10.1002/prot.23146.
Song, Y., & Gunner, M. R. (2009). Using multi-conformation continuum electrostatics to compare chloride binding motifs in a-amylase, human serum albumin, and Omp32. *Journal of Molecular Biology*, *387*, 840–856.
Song, Y., Mao, J., & Gunner, M. R. (2003). Calculation of proton transfers in bacteriorhodopsin bR and M intermediates. *Biochemistry*, *42*, 9875–9888.
Song, Y. F., Mao, J. J., & Gunner, M. R. (2009). MCCE2: Improving protein pK_a calculations with extensive side chain rotamer sampling. *Journal of Computational Chemistry*, *30*, 2231–2247.
Spassov, V. Z., Ladenstein, R., & Karshikoff, A. D. (1997). Optimization of the electrostatic interactions between ionized groups and peptide dipoles in proteins. *Protein Science*, *6*, 1190–1196.
Stanton, C. L., & Houk, K. N. (2008). Benchmarking pK_a prediction methods for residues in proteins. *Journal of Chemical Theory and Computation*, *4*, 951–966.
Styring, S., Sjoholm, J., & Mamedov, F. (2012). Two tyrosines that changed the world: Interfacing the oxidizing power of photochemistry to water splitting in photosystem II. *Biochimica et Biophysica Acta*, *1817*, 76–87.
Swails, J. M., York, D. M., & Roitberg, A. E. (2014). Constant pH replica exchange molecular dynamics in explicit solvent using discrete protonation states: Implementation, testing, and validation. *Journal of Chemical Theory and Computation*, *10*, 1341–1352.
Tannor, D. J., Marten, B., Murphy, R., Friesner, R. A., Sitkoff, D., Nicholls, A., … Goddard, W. A., III. (1994). Accurate first principles calculation of molecular charge distributions and solvation energies from *ab initio* quantum mechanics and continuum dielectric theory. *Journal of the American Chemical Society*, *116*, 11875–11882.
Wagoner, J., & Baker, N. A. (2004). Solvation forces on biomolecular structures: A comparison of explicit solvent and Poisson-Boltzmann models. *Journal of Computational Chemistry*, *25*, 1623–1629.
Wagoner, J. A., & Baker, N. A. (2006). Assessing implicit models for nonpolar mean solvation forces: The importance of dispersion and volume terms. *Proceedings of the National Academy of Sciences of the United States of America*, *103*, 8331–8336.
Wallace, J. A., Wang, Y., Shi, C., Pastoor, K. J., Nguyen, B.-L., Xia, K., & Shen, J. K. (2011). Toward accurate prediction of pK_a values for internal protein residues: The importance of conformational relaxation and desolvation energy. *Proteins*, *79*(12), 3364–3373. http://dx.doi.org/10.1002/prot.23080.
Wang, L., Li, L., & Alexov, E. (2015). pK_a predictions for proteins, RNAs, and DNAs with the Gaussian dielectric function using DelPhi pK_a. *Proteins*, *83*, 2186–2197.
Warshel, A., & Russell, S. T. (1984). Calculations of electrostatic interactions in biological systems and in solutions. *Quarterly Reviews of Biophysics*, *17*, 283–422.
Warshel, A., Sharma, P. K., Kato, M., & Parson, W. W. (2006). Modeling electrostatic effects in proteins. *Biochimica et Biophysica Acta*, *1764*, 1646–1676.
Webb, B. A., Chimenti, M., Jacobson, M. P., & Barber, D. L. (2011). Dysregulated pH: A perfect storm for cancer progression. *Nature Reviews. Cancer*, *11*, 671–677.
Webb, H., Tynan-Connolly, B. M., Lee, G. M., Farrell, D., O'Meara, F., Sondergaard, C. R., … Nielsen, J. E. (2011). Remeasuring HEWL pK_a values by NMR spectroscopy: Methods, analysis, accuracy, and implications for theoretical pK_a calculations. *Proteins*, *79*, 685–702.
Williams, R. J. (1997). The natural selection of the chemical elements. *Cellular and Molecular Life Sciences*, *53*, 816–829.

Witham, S., Talley, K., Wang, L., Zhang, Z., Sarkar, S., Gao, D., … Alexov, E. (2011). Developing hybrid approaches to predict pK_a values of ionizable groups. *Proteins*, *79*, 3389–3399.

Wong, G. C., & Pollack, L. (2010). Electrostatics of strongly charged biological polymers: Ion-mediated interactions and self-organization in nucleic acids and proteins. *Annual Review of Physical Chemistry*, *61*, 171–189.

Word, J. M., & Nicholls, A. (2011). Application of the Gaussian dielectric boundary in Zap to the prediction of protein pK_a values. *Proteins*, *79*, 3400–3409.

Xie, D. X., Jiang, Y., Brune, P., & Scott, L. R. (2012). A fast solver for a nonlocal dielectric continuum model. *SIAM Journal on Scientific Computing*, *34*, B107–B126.

Zasetsky, A. Y. (2011). Dielectric relaxation in liquid water: Two fractions or two dynamics? *Physical Review Letters*, *107*, 117601.

CHAPTER TWO

Path Sampling Methods for Enzymatic Quantum Particle Transfer Reactions

M.W. Dzierlenga[1], M.J. Varga[1], S.D. Schwartz[2]
University of Arizona, Tucson, AZ, United States
[2]Corresponding author: e-mail address: sschwartz@email.arizona.edu

Contents

Abstract

The mechanisms of enzymatic reactions are studied via a host of computational techniques. While previous methods have been used successfully, many fail to incorporate the full dynamical properties of enzymatic systems. This can lead to misleading results in cases where enzyme motion plays a significant role in the reaction coordinate, which is especially relevant in particle transfer reactions where nuclear tunneling may occur. In this chapter, we outline previous methods, as well as discuss newly developed dynamical methods to interrogate mechanisms of enzymatic particle transfer reactions. These new methods allow for the calculation of free energy barriers and kinetic isotope effects (KIEs) with the incorporation of quantum effects through centroid molecular dynamics (CMD) and the full complement of enzyme dynamics through transition path sampling (TPS). Recent work, summarized in this chapter, applied the method for calculation of free energy barriers to reaction in lactate dehydrogenase (LDH) and yeast alcohol dehydrogenase (YADH). We found that tunneling plays an insignificant role in YADH but

[1] These authors contributed equally to this work.

Methods in Enzymology, Volume 578
ISSN 0076-6879
http://dx.doi.org/10.1016/bs.mie.2016.05.028

plays a more significant role in LDH, though not dominant over classical transfer. Additionally, we summarize the application of a TPS algorithm for the calculation of reaction rates in tandem with CMD to calculate the primary H/D KIE of YADH from first principles. We found that the computationally obtained KIE is within the margin of error of experimentally determined KIEs and corresponds to the KIE of particle transfer in the enzyme. These methods provide new ways to investigate enzyme mechanism with the inclusion of protein and quantum dynamics.

1. INTRODUCTION

The determination of reaction rates is a crucial cornerstone in the study of enzymatic reaction mechanism. In particular, noting how the rate constant changes with respect to isotopic substitution, called a kinetic isotope effect (KIE), can provide a level of mechanistic detail that is otherwise inaccessible. Reaction rates and KIEs have classically been found using a host of experimental techniques which have some limitations, sometimes leading to ambiguous or confusing results. Experimental determination of KIEs currently relies on three overarching classes of methods: noncompetitive, competitive, and equilibrium perturbation. Noncompetitive methods, also known as direct comparison methods, provide the method for the determination of the KIEs on V itself, in addition to V/K, where V is the maximum rate achieved by the enzyme, and K is the substrate concentration at which the rate is half of V. While this is an advantage over other methods, noncompetitive methods are limited in sensitivity compared to other methods, and thus are mostly useful only for large isotope effects, such as primary H/D effects. Competitive and equilibrium perturbation methods alleviate the sensitivity issues of noncompetitive methods, allowing the determination of effects on the scale of 1–3%, but can only determine the KIE on V/K and require a great deal of care in their experimental setup and execution (Cook & Cleland, 2007).

One important use of KIEs is to provide information about the contribution of nuclear tunneling in enzymatic reactions. The main experimental marker of tunneling in particle transfer reactions is an elevated Swain–Schaad exponent (Cha, Murray, & Klinman, 1989). The Swain–Schaad equation relates the KIE of different isotopes, often H/D to H/T, and has an exponential relationship between the two KIEs (Swain, Stivers, Reuwer, & Schaad, 1958). Using only the masses, one can determine a semiclassical lower limit for the exponent, which disregards tunneling effects

(Swain et al., 1958). Elevation above this semiclassical limit has been used to argue for the presence of nuclear tunneling in transfer reactions (Cha et al., 1989; Saunders, 1985). However, the source of this elevation may also be explained by other factors, such as differences in the enzyme's tunneling ready state structure, ie, the ensemble of active site structures at the point at which the particle tunnels, between the isotopically substituted enzymes (Roston & Kohen, 2013).

These experimental ambiguities necessitate methods which can provide complimentary detail while circumventing their limitations. This is the role of computational methods in the study of enzymatic reaction mechanisms. These methods provide the atomistic detail to clarify ambiguous results and to simplify the study of complex systems. For example, the atomistic detail provided by molecular dynamics (MD) simulations allows one to examine the effects on and by individual atoms as opposed to the bulk average effects usually obtained through experimental techniques. Additionally, alteration in the structure, such as through amino acid point mutations or isotopic substitutions, can be performed with relative ease in most of the popular molecular mechanics software packages. Computational methods have the ability to apply these advantages to large enzymatic systems while being able to predict experimental results.

One main computational tool is the calculation of free energy differences between the different states through which the reaction progresses. These free energy differences can be then used to distinguish between different paths the reaction has the ability to pass through, intuit the effects of quantum dynamics, or, through the use of the transition state theory (TST), calculate the rate of reaction. In the following section, we outline some of the most commonly used computational techniques to obtain changes in free energy, reaction rates, and KIEs, along with some of their limitations. Following the introduction of these well-known methods, we detail dynamical methods which can avoid some of these limitations. These dynamical methods are especially suited for the calculation of KIEs and other comparisons between similar enzymatic systems.

1.1 Established Methods for Computational Calculation of Enzymatic Free Energy Barriers and Rates

One method for calculating the free energy barrier of any transition is to find the potential of mean force (PMF) (Kirkwood, 1935, 1936; Onsager, 1933). The PMF, $F(\xi) = -\beta \ln Q(\xi)$, where, β is $(kT)^{-1}$ throughout this chapter,

ξ is the reaction coordinate, and $Q(\xi)$ is the canonical partition function as a function of the reaction coordinate, provides the free energy change for a trajectory obtained by Monte Carlo or MD simulations. For a reaction with low free energy barriers, those on the order of kT, a standard MD simulation will be able to sample the region of interest in a reasonable timeframe and the free energy can be obtained. Of course, most transitions of interest, including most chemical reactions, have barriers much higher than this, which will not be properly sampled by MD simulations without special methods to study rare events.

One such method for the study of rare events is umbrella sampling (Kästner, 2011; Torrie & Valleau, 1977). This method uses a biasing potential to force the system to sample all the regions of interest, even the relatively high energy regions between meta-stable states. In umbrella sampling, the transition is parameterized by a prechosen reaction coordinate and the average force is calculated, via separate simulations, at a variety of reaction coordinate values using the biasing potential. These separate simulations can then be combined and the change in free energy relative to a reference structure obtained by subtracting the biasing potential from the obtained free energy of the windows. Alternate methods to combine windows and obtain a free energy include umbrella integration (Kästner & Thiel, 2005) and the weighted histogram analysis method (Kumar, Rosenberg, Bouzida, Swendsen, & Kollman, 1992).

Umbrella sampling has a long history of successful use but uses a few approximations that can be inappropriate in certain systems. For example, the reaction coordinate must be known, or more usually assumed, in order to bias the system along it. For very simple processes, eg, the movement of a solute through the surface between two immiscible liquids, this may be easy; the distance of the molecule from the surface may be an appropriate reaction coordinate. However, for complex transitions, including all but the simplest chemical reactions, the reaction coordinate is almost impossible to predict a priori. Using chemical intuition, one can create an order parameter that denotes how the system moves along the transition of interest, but all the parts of the reaction coordinate that are not included in the biasing potential are averaged over in the calculation. This reduces the accuracy of the result, and, if the motions not included in the estimated reaction coordinate are important, possibly even an incorrect or misleading result. This issue can be especially prevalent in enzymatic reactions, where many studies have shown that protein dynamics plays a significant enough role in the reaction that it would be difficult to guess without using a method to interrogate the protein directly (Antoniou, Ge, Schramm, & Schwartz, 2012; Basner & Schwartz, 2005; Crehuet & Field, 2007; Kipp, Silva, & Schramm, 2011; Quaytman & Schwartz, 2007; Silva, Murkin, & Schramm, 2011).

Even if the reaction coordinate is chosen correctly, the umbrella sampling methodology includes the implicit proposal that the motion along the reaction coordinate and the motion along all the other coordinates in the system are completely separable. This is the equilibrium solvation limit from Grote–Hynes theory (Hynes, 1985). In solution phase reactions, this may be an acceptable assumption, but in enzymatic reactions, there is the possibility of protein motions that are correlated to the reaction coordinate and also on the same timescale as barrier crossing. In cases where this type of motion may be important, eg, in enzymatic systems where there may be a promoting vibration, it is not appropriate to average over these motions and doing so affects the accuracy of the results.

In addition to umbrella sampling, there are many other methods to obtain the PMF. These include thermodynamic integration (Straatsma & Berendsen, 1988), also known as blue moon sampling (Ciccotti, Kapral, & Vanden-Eijnden, 2005), and steered MD (Park & Schulten, 2004). Free energy perturbation (FEP) methods are also used, which differ from umbrella sampling and steered MD in that the system is immediately biased to the product state (Zhang, Liu, & Yang, 2000). FEP methods rely on Jarzynski's equality, $\Delta F = -\beta^{-1} \ln \exp(-\beta W)$, where ΔF is the change in free energy, W is the total work performed on the system, and the $\langle \ldots \rangle$ denote an ensemble average, to determine the free energy (Jarzynski, 1997).

Free energy methods are often used within the context of the TST to obtain the rates of reaction. TST is a framework which allows for the computation of the rate of transition between two areas of phase space and was first suggested by Horiuti (1938). The rate of reaction from TST is

$$\nu_S^{\mathrm{TST}} = \sqrt{\frac{2}{\pi\beta}} Q^{-1} \int_S e^{-\beta V(x)} \mathrm{d}\sigma(x), \tag{1}$$

where ν_S^{TST} is the rate with dividing surface S, Q is the partition function of reactants, $V(x)$ is the potential energy, and $\mathrm{d}\sigma(x)$ is the surface element on S (Vanden-Eijnden & Tal, 2005). As the TST equation relies on the knowledge of the free energy, it inherits any problems that arise in the free energy calculation. Another issue is that calculation of a KIE involves the combination of two separate rate calculations resulting in a large error in the KIE if one of the rate calculations has a large error. This issue has previously been considered by Major and Gao (2007), who circumvented the issue by utilizing a mass perturbation approach in an umbrella sampling methodology which determines the KIE in a single calculation.

2. TRANSITION PATH SAMPLING: A NEW PARADIGM FOR THE STUDY OF ENZYMATIC MECHANISM

While the previously outlined techniques are limited in that they map complex dynamical problems onto static approximations, the transition path sampling (TPS) method was developed to study the dynamics of rare events, such as enzymatic reactions, without these approximations (Bolhuis, Dellago, & Chandler, 1998; Dellago, Bolhuis, & Chandler, 1998, 1999; Dellago, Bolhuis, Csajka, & Chandler, 1998). TPS is a dynamical, Monte Carlo technique which samples reactive trajectory space. As TPS is a dynamical method, it allows for the study of systems without a priori knowledge of a reaction coordinate, making it particularly useful for high-dimensional systems such as enzymes.

2.1 A Statistical Method for Studying Enzymatic Reactions

Assume the existence of a two state system, with states R and P, reactants and products in a chemical reaction, or any two meta-stable states. If the typical equilibrium fluctuations of the system do not kick the system out of either state, then the states are stable and long-lived and transitions between the two are rare. This transition can be described by an order parameter, which need not be the exact reaction coordinate, but merely some metric by which the transition can be parameterized. An initial reactive trajectory is obtained, typically by implementing harmonic constraints between donor and acceptor, in transfer reactions, but other methods can be used. From this initial reactive trajectory, a random timeslice is chosen, with momenta p and coordinates q. The momenta at this timeslice are then perturbed randomly, in accordance with a Boltzmann distribution, to create a new set of momenta applied to the original coordinates, which are then propagated forward and backward in time to create a new trajectory. This new trajectory is accepted or rejected via the Metropolis criterion (Bolhuis, Chandler, Dellago, & Geissler, 2002). In a canonical ensemble, the acceptance probability for a new trajectory n, generated from an old trajectory o, is $P_{\mathrm{acc}}^{o\rightarrow n} = \min\left(1, \exp\left\{-\beta\left[\mathcal{H}\left(x_0^n\right) - \mathcal{H}\left(x_0^o\right)\right]\right\} h_{\mathrm{R}}\left(x_0^n\right) H_{\mathrm{P}}\left(x_0^n\right)\right)$, where h_{R} and H_{P} are Heaviside functions for commitment to the reactant and visitation to the product well, respectively. In a microcanonical ensemble, where the system energy is fixed, this acceptance probability simplifies to $P_{acc}^{o \to n} = h_R(x_0^n) H_P(x_0^n))$, which is unity if the trajectory starts in the reactant well and ends in the product well and zero otherwise (Bolhuis et al., 1998). Most uses of TPS in enzymatic systems utilize microcanonical TPS (Dzierlenga, Antoniou, & Schwartz, 2015; Masterson & Schwartz, 2013, 2014). If the new trajectory is rejected, the process is repeated from the last accepted trajectory, but if the new trajectory is rejected, the process continues from the new trajectory. By iteratively adding trajectories to this ensemble, with each new trajectory created with a slice from the previous reactive trajectory, a reactive trajectory ensemble which spans all energetically available paths from reactants to products is created.

TPS only provides an algorithm with which to sample trajectory space and needs a potential energy method to simulate the system in question. In theory, any method can be used, but within the context of particle transfer reactions, quantum mechanics/molecular mechanics (QM/MM) (Warshel & Levitt, 1976) is the most oft used method. A quantum mechanical portion is important in the study of these particle transfer reactions because classical MD methods do not allow for structural changes in chemical bonds, a critical component for any chemical reaction. In particular, the techniques described herein utilize the MD package Chemistry at Harvard Molecular Dynamics (CHARMM) (Brooks et al., 1983, 2009) with a semiempirical quantum region, a PM3 method (Stewart, 1989a, 1989b) modified to better replicate biological zinc (Brothers, Suarez, Deerfield, & Merz, 2004), linked together using the generalized hybrid orbital (GHO) method (Gao, Amara, Alhambra, & Field, 1998). The GHO method provides a well-defined potential energy surface for hybrid QM/MM systems by placing hybrid orbitals on atoms on the QM/MM boundary region, providing better modeling at the QM/MM barrier.

Beyond this flexibility in the choice of force field, TPS can also be used with a variety of propagation methods. One recent development has been the use of normal mode centroid molecular dynamics (CMD) (Cao & Voth, 1994a, 1994b, 1994c, 1994d, 1994e) to include approximate quantum effects in the propagation of the transferring particle in TPS simulations (Antoniou & Schwartz, 2009; Dzierlenga et al., 2015). Full descriptions of the inclusion of this method into TPS can be found in Antoniou and Schwartz (2009), and a short description of this method follows. Each quantum particle is described by a ring of beads connected by harmonic potentials. The motion of quantum particle i is described by the motion of the centroid of the beads, which is propagated according to

$$m_i\ddot{\boldsymbol{R}}_i = \langle \boldsymbol{F}_i(\boldsymbol{R}_i, \ldots, \boldsymbol{R}_N)\rangle_c, \tag{2}$$

where m_i is the mass of the particle, $\boldsymbol{R}_i$ is the centroid position, $\boldsymbol{F}_i$ is the force on the centroid, and the angle brackets denote a path integral average. This path integral average is weighted by a factor of $\exp(-\beta V_{\text{eff}})$ where V_{eff} is the effective bead potential,

$$V_{\text{eff}}\left(r_i^{\beta}\right) = \sum_{i=1}\sum_{\beta=i}\left[\frac{1}{2}k_i\left(r_i^{\beta} - r_i^{\beta+1}\right)^2 + \frac{1}{B}V\left(r_i^{\beta}\right)\right], \tag{3}$$

A normal mode transformation is applied to separate faster motions of the bead from the relatively slower motion of the centroid (Eq. 4). The effective potential then becomes,

$$V_{\text{eff}}\left(\boldsymbol{R}_i; q_\alpha^i\right) = \sum_{i=1}^{N}\sum_{\beta=i}^{B-1}\left[\frac{1}{2}m_\alpha^i\left(\omega_\alpha^i q_\alpha^i\right)^2 + \frac{1}{B}V\left(\boldsymbol{R}_i; q_\alpha^i\right)\right], \tag{4}$$

In addition to separating the motions into two timescales, this transformation allows for the path integral average force to be replaced by the instantaneous force of the propagation of the centroid. Within TPS, this requires a multistep propagation; during the energy calculation at each timestep, the CMD code calculates the force of the quantum particle over several shorter bead propagation timesteps, and this then replaces the CHARMM calculated force of the quantum particle. It should be noted that though CMD allows for the approximate inclusion of quantum effects on the transferring particle, it does so at the price of increased computation time.

2.2 TPS Ensemble Analysis

Examination of the TPS ensemble directly, without the use of additional analysis methods, can provide several types of useful information. One common piece of information gathered from these raw ensembles are distances, often donor–acceptor or transferring particle-donor or -acceptor distances (Dzierlenga et al., 2015). These distances are useful to determine timescales of the reaction, such as the timescale of the transfer event itself, as well as to shed light on timing of the motion of the active site and other nearby residues.

In addition to straightforward examinations of the TPS ensemble, there are also methods which analyze ensembles of trajectories to provide extended detail on the reaction mechanism. One of these methods is the calculation of the committor, which allows for a strict determination of the progress of the reaction more rigorously than a geometrically defined order parameter. This methodology was first developed to study ion-pair recombination (Onsager, 1938) and then repurposed to examine TPS ensembles (Bolhuis et al., 2002). To determine the commitment probability at a point in the trajectory, a series of new trajectories are initiated with random momenta from that point in configuration space. After some period of time, each new trajectory is checked for whether it is in the reactant or product region of configuration space. The fraction of trajectories that have ended in the product region is called the commitment probability, and the point at which the commitment probability is equal to one-half is defined as the transition state. At the transition state, there is an equal probability of proceeding to either the reactants or the products. The set of transition states for a TPS ensemble lies along the transition state surface for the reaction, which is called the separatrix (Bolhuis et al., 2002).

Once a set of several transition states is obtained for a TPS ensemble, committor distribution analysis can be used to determine which motions in the system play significant roles in the reaction coordinate. For this analysis technique, constrained dynamics is begun from a transition state. If the constrained residues are a good approximation of the true reaction coordinate, then the constrained dynamics will travel along the separatrix and will be restricted from alternative motion. Then, the commitment probability is sampled along the constrained dynamics trajectory. After obtaining a distribution of committors via constrained dynamics from a number of transition states with a given reaction coordinate, the distribution of committors is analyzed. If the reaction coordinate constrained during dynamics was close to the true reaction coordinate, the structure during dynamics stayed close to the transition state surface, and the distribution of committors will be sharply peaked about one-half commitment probability. If the presumed reaction coordinate does not accurately account for the true reaction coordinate, then the dynamic trajectories will quickly fall off of the transition state surface and the graph of the committor distribution will not peak at 0.5.

Another technique to analyze the trajectories in a TPS ensemble is essential dynamics (ED), also called principal component analysis (Amadei, Linssen, & Berendsen, 1993). ED was developed to separate the protein degrees of freedom that are constrained from the protein motions that are able to undergo coherent conformational shifts during dynamics. This is done by first diagonalizing the covariance matrix. Then the resulting eigenvectors can be grouped by their eigenvalues into a small set that contains the coherent conformational changes the protein undergoes, and a set containing the protein degrees of freedom that change little during dynamics. This is a useful tool for the analysis of protein dynamics because it substantially decreases the dimensionality of obtained data, sifting out the degrees of freedom that are essentially constrained during the trajectory.

3. NEW METHODS FOR CALCULATION OF VALUES RELEVANT TO ENZYME MECHANISM

3.1 Free Energy Probes of Nuclear Tunneling

Nuclear tunneling in enzymatic reactions, especially in reactions involving hydrogen transfer where the particle is small enough to often have a significant tunneling contribution, is of interest to those who are involved in determining the mechanisms of enzymes. This is partly because the involvement of nuclear quantum dynamics has been historically neglected from the study of enzyme mechanism (Cha et al., 1989).

To examine the effects of quantum dynamics in enzymatic reactions computationally, we devised a method to examine microscopic free energy changes with respect to the transferring particle that happen along the course of the reaction. By integrating the force on the transferring particle, we obtain the work on the transferring particle from the other atoms in the system. The average of this change in reversible work is the free energy for this step in the reaction in the reference frame of the particle. It is important to note that this free energy change is not the free energy for the entire reaction or for the chemical step of the reaction, but only the free energy change for the transfer of the particle between atoms in the reactive conformation. The probability of a particle to tunnel through a barrier is related to both the height and the width of the barrier. It is not the free energy barrier of the full reaction that is relevant to particle tunneling, but the barrier that exists as the particle transfers, which can be found through the calculation of the work done on the particle.

Unfortunately, a simple integration over the particle path is not sufficient for the barrier to be fully exposed. Even during the particle transfer, the particle does not move directly between the donor and acceptor but also moves against the sides of the potential pocket. To extract the reactive motion from the extraneous motions, a number of steps are taken. The first is to project the motion of the transferring particle onto the donor–acceptor axis. Since the particle is transferring between these two particles, almost all the reactive motion of the transferring particle will be along this axis. The second is to transform the coordinates of the system such that the hydride work is not affected by the slight movement or rotation of the active site within the protein. This is done by transforming the coordinates so that the donor carbon is at the origin.

The calculation of free energy barriers in this way is useful because it allows for a comparison of the barrier to particle transfer in TPS ensembles. However, when using the method, one must resist the temptation to draw larger conclusions from the method than are appropriate. For instance, putting a free energy obtained from this method into the TST formalism and using it to calculate a rate of reaction would be incorrect, because the free energy calculated using this method is explicitly a free energy difference for the transfer of the particle of interest, not a free energy that corresponds to the full energy barrier for the reaction. However, if one is interested in the barrier to particle transfer to investigate tunneling or to compare two ensembles where the difference is expected to be almost exclusively in the particle transfer, for example, a KIE, then the method is appropriate.

3.1.1 Application of Work Calculation to Hydride Transfer

The calculation of free energy from the work during particle transfer was applied to yeast alcohol dehydrogenase (YADH) and human heart lactate dehydrogenase (LDH) to test for hydride tunneling in these systems. YADH has a long history of use as a model enzyme in research (Ganzhorn, Green, Hershey, Gould, & Plapp, 1987; Hayes & Velick, 1954; Klinman, 1972) and was one of the first enzymes to be used in probes of nuclear tunneling in enzymatic reactions (Klinman, 1976). The seminal study which began to highlight the importance of nuclear tunneling in enzymes used mixed labeling in a competitive experiment to obtain the Swain–Schaad exponents in YADH (Cha et al., 1989). When this exponent was found to be above the semiclassical value, they hypothesized that hydride tunneling played a significant role in the reaction. Nuclear quantum dynamics have subsequently been suspected to play a role in several other enzymatic reactions, including liver alcohol dehydrogenase, soybean lipoxygenase, copper amine oxidase, and others (Klinman, 2009; Layfield & Hammes-Schiffer, 2014; Machleder, Pineda, & Schwartz, 2010; Truhlar, 2010).

Swain–Schaad exponents are a relatively indirect method of addressing questions of quantum dynamics, and competitive labeling experiments can be muddled due to issues of kinetic complexity, as mentioned previously. To further investigate the issue, we used the free energy barrier calculation from the work of particle transfer to examine tunneling in this enzyme from a different perspective. To determine the contribution of tunneling, we created TPS ensembles with different propagation schemes: an ensemble using a traditional semiempirical QM/MM method that cannot capture nuclear tunneling effects, called the classical ensemble, and an ensemble applying the CMD method to the transferring particle to include these effects, called the CMD1 ensemble. By comparing the free energy barriers between the two ensembles, we were able to deduce the role of tunneling effects in the enzyme. In addition to these two ensembles, we also obtained an ensemble with CMD applied to three particles, the transferring hydride and the other hydrogen atoms bonded to the donor and acceptor carbon, called the CMD3 ensemble. This was done in order to see if there is quantum coupling between the hydrogen in the secondary position and the transferring particle. We also generated an ensemble with CMD applied to only the transferring particle which was substituted for a deuteride, called the CMD-D ensemble, to estimate the primary KIE value in the enzyme.

As can be seen in Table 1, the average value for the classical ensemble is very low; low enough that tunneling is not necessary for reaction. This is further corroborated by the difference between the classical and CMD1 ensemble, which is a difference that is more in line with the inclusion of zero-point energy in the CMD1 ensemble than to tunneling effects. Further evidence for the lack of tunneling is the nonsignificant difference in the distance between the minimum donor and acceptor distance. If tunneling played a major role in the reaction, then one would expect that in the ensembles with quantum dynamics, tunneling would allow for an increase

Table 1 Average Properties of TPS Ensembles with Different Propagation Schemes for the Reaction Catalyzed by YADH

Method	Free Energy Barrier (kcal/mol)	D–A Distance[a] (Å)
Classical	0.97	2.78 ± 0.06
CMD[b]	0.28	2.72 ± 0.09
CMD3[c]	0.32	2.69 ± 0.27
CMD-D[d]	1.05	2.70 ± 0.04

[a]Average minimum in the D–A distance near the reaction.
[b]CMD applied to the transferring hydride.
[c]CMD applied to the secondary hydrogens as well as the transferring hydride.
[d]CMD applied to the transferring deuteride which replaces the hydride in the reaction.

in the donor–acceptor distance by allowing for larger barriers to be surmounted. However, we found no significant difference between the barriers in the CMD1 and CMD3 ensembles. This indicates that quantum coupling between the three hydrogen is not playing a significant role, at least in the energy required for hydride transfer.

In order to compare with an experimental measure, we used the experimental primary H/D KIE to find an energy barrier difference between the two isotopes, under the approximation that the isotopic substitution affects only the free energy barrier, which we then compared to the energy difference of the CMD1 ensemble and CMD-D ensemble. We found that the experimental measured primary deuterium KIE, 3.4 (Klinman, 1976), corresponded to an energy difference of 0.73 kcal/mol, while the difference between our ensembles is 0.77 kcal/mol, a surprisingly close comparison.

In LDH, two different TPS ensembles were generated, one without quantum dynamical propagation, called the classical ensemble, and one with CMD applied to the transferring proton and hydride, called the CMD ensemble. In LDH, unlike in YADH, the proton and hydride transfer at approximately the same time and must both be included in the reaction. The results of the free energy barrier calculation in LDH, shown in Table 2, are somewhat higher than in YADH. The classical barrier was found to be within the range of energies where tunneling can play a role, though it is relatively low. Further evidence that quantum tunneling plays a role in LDH is found in the average donor–acceptor distance, which is significantly lower in the classical ensemble than in the CMD ensemble. This is what one would expect in a situation where tunneling plays a role in the reaction, in that quantum effects allow the particle transfer to occur over a longer distance. It is important to note that the free energy difference has a large range between trajectories within an ensemble, due to the sensitivity to small differences in the trajectories. The distribution of transfer barriers is shown in Fig. 1.

Table 2 Average Properties of TPS Ensembles with Different Propagation Schemes for the Reaction Catalyzed by LDH

Method	Free Energy Barrier (kcal/mol)	D–A Distance[a] (Å)
Classical	7.77	2.77 ± 0.02
CMD[b]	3.61	2.85 ± 0.03

[a]Average minimum in the D–A distance near the reaction.
[b]CMD applied to the transferring hydride and transferring proton.

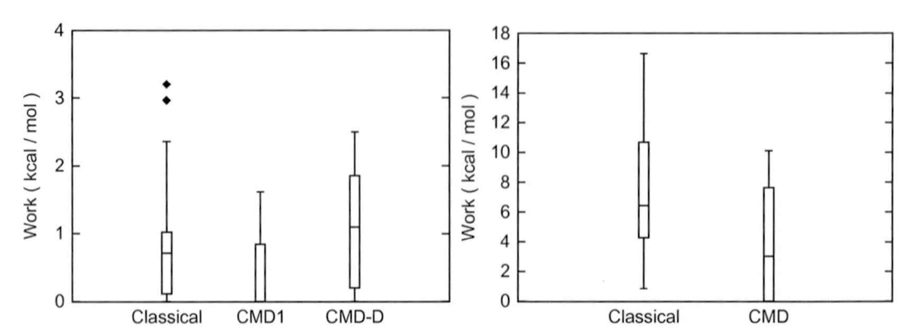

Fig. 1 *Box* and *whiskers plots* made from the barriers to reaction in YADH (*left*) and LDH (*right*). The *center line* marks the median; the *upper* and *lower* ends *of the box* mark the first and third quartiles; the *whiskers* mark the last datum inside the 1.5 times the interquartile range; outside this range, data are marked by *solid squares*.

3.2 KIEs of Quantum Particle Transfer from TPS

In recent work, we developed a method to calculate KIEs from first principle path sampling calculations (Varga & Schwartz, 2016). As in the earlier described study, TPS and CMD were utilized to circumvent limitations of other techniques dealing with the transfer of quantum particles. These methods, combined with an algorithm developed by the Chandler group (Dellago, Bolhuis, &, Chandler, 1998; Dellago et al., 1999; Dellago, Bolhuis, Csajka, et al., 1998), provide a powerful technique for the calculation of KIEs of enzymatic quantum particle transfer reactions.

3.2.1 Rate Calculation Algorithm

Formulation of the rate calculation algorithm starts from a foundation of Bennett–Chandler theory relating the rate of transition between two metastable states to a correlation function (Bennett, 1977; Chandler, 1978). The time derivative of this correlation function, $C(t)$, is the reaction rate constant in the region in which the time derivative of $C(t)$ plateaus, ie, in the steady state. $C(t)$ can be calculated directly with ensemble averages of Heaviside functions for the reactant and product wells, h_r and h_p (Eq. 5)

$$C(t) = \frac{\langle h_{\mathrm{R}}(0) h_{\mathrm{P}}(t) \rangle}{\langle h_{\mathrm{R}}(0) \rangle}. \tag{5}$$

Factorization of $C(t)$ splits the equation into two terms (Eq. 6), where $\langle h_{\mathrm{P}}(t) \rangle_{\mathrm{RP}}$ and $\langle h_{\mathrm{P}}(t') \rangle_{\mathrm{RP}}$ are the ensemble average values of $h_{\mathrm{P}}(t)$ and $h_{\mathrm{P}}(t')$ in the ensemble $F_{\mathrm{RP}}(x_0, T) \equiv \rho(x_0) h_{\mathrm{R}}(x_0) H_{\mathrm{P}}(x_0, T)$, with $H_{\mathrm{P}}(x_0, T)$

denoting trajectories which visit the product well in the interval $[0:T]$. $C(t')$ is a correlation function at a specific time t' in the interval $[0:t)$.

$$k(t) = \frac{d}{dt}C(t) = \frac{\langle \dot{h_P}(t) \rangle_{RP}}{\langle h_P(t') \rangle_{RP}} \times C(t') \quad (6)$$

Thus, the computationally intensive process of calculating $C(t)$ directly at every time t has been simplified to two smaller, more manageable calculations: the calculation of the first term, $h_p(t)$, via one TPS simulation to a plateau region, and the calculation of $C(t')$ via the method described later.

The calculation of $C(t')$ requires a more intensive calculation than the first term. For trajectories starting in the reactant region, R, at time $t=0$, the distribution of the order parameter, λ, at time t is,

$$P(\lambda, t) = \frac{\int dx_0 \rho(x_0) h_R(x_0) \delta[\lambda - \lambda(x_t)]}{\int dx_0 \rho(x_0) h_R(x_0)}. \quad (7)$$

In a rare event system, this distribution is small in the product region, and direct calculation is not feasible. To avoid this difficult, direct calculation, we first define a series of overlapping windows $W[i]$, where $x \in W[i] \rightarrow \lambda_{min}[i] \leq \lambda(x) \leq \lambda_{max}[i]$ and the union of all windows $W[i]$ yields the entirety of phase space. The distribution of the order parameter in the windows $W[i]$ is then calculated separately,

$$P(\lambda, t; i) = \frac{\int dx_0 \rho(x_0) h_R(x_0) h_{W[i]}(x_t) \delta[\lambda - \lambda(x_t)]}{\int dx_0 \rho(x_0) h_R(x_0) h_{W[i]}(x_t)}. \quad (8)$$

By matching these separate distribution functions, using the overlap windows of each window, where available, we can obtain the full distribution function, $P(\lambda, t)$. The correlation function above, $C(t')$, is then calculated by integration of the product region of this full histogram (Eq. 9).

$$C(t') = \int_{\lambda_{min}}^{\lambda_{max}} d\lambda \, P(\lambda, t) \quad (9)$$

By multiplying these two terms, $\langle \dot{h_P}(t) \rangle_{RP} / \langle h_P(t') \rangle_{RP}$ and $C(t')$, the reaction rate constant can then be determined. This method has previously been used to determine the rate of reaction in small systems, such as the dissociation of sodium chloride at air–water (Wick, 2009) and organic-aqueous (Wick & Dang, 2010) interfaces, and to determine the contribution of mechanical stress in the base-catalyzed hydrolysis of tetraglycine (Xia, Bronowska, Cheng, & Gräter, 2011).

3.2.2 Application of Modified Algorithm to YADH

In recent work, the rate algorithm with CMD was applied to the calculation of the primary H/D KIE in YADH (Varga & Schwartz, 2016). This is the first application of the Chandler rate algorithm to an enzyme, as well as the first application in tandem with CMD. These simulations required the generation of more than 20,000 trajectories of which approximately 1800 were accepted. As mentioned previously, calculation of the first term proceeds via a normal TPS simulation. Reactive trajectories populated a cumulative histogram of $\langle h_p(t) \rangle$ (Fig. 2).

From this, one can see the linear region formed around 800 timesteps, or 400 fs. This linear region, when fit to a linear regression, yielded a slope corresponding to the value of the first term, $9.901 \times 10^{-4}\ \text{fs}^{-1}$ and $1.033 \times 10^{-3}\ \text{fs}^{-1}$ for the hydride and deuteride systems, respectively. Second term computations started from a trajectory from the first term calculation as a starting trajectory. Each system's trajectory was split into multiple windows based on order parameter, 7 for the hydride system and 9 for the deuteride system (Tables 3 and 4).

As can be seen in Table 3, windows 4 and 5 were difficult to populate with trajectories, with acceptance ratios of less than 5%. Similarly, windows 5, 6, 7, and 8 were difficult to populate in the deuteride system (Table 4).

This is likely because these windows lay on or around the free energy barrier, making it unlikely for the system to linger in those windows. Fig. 3 shows the completed and normalized histograms, with Hermite interpolated fits, and the product regions of the histograms with their fits. The second term value of the full rate equation is then calculated through integration of the product region fits. Combined with the first term values, we obtained a calculated primary H/D KIE of 5.22 for the conversion of benzyl alcohol to benzaldehyde in YADH.

As discussed earlier, the primary H/D KIE for the conversion of benzyl alcohol to benzaldehyde in YADH is 3.4, determined through noncompetitive KIE studies (Klinman, 1976). The calculated result is within the margin of error of the experimentally determined result. Due to the nature of these experimental studies, the KIE is determined from the Michaelis complex to the first irreversible step (Simon & Palm, 1966),

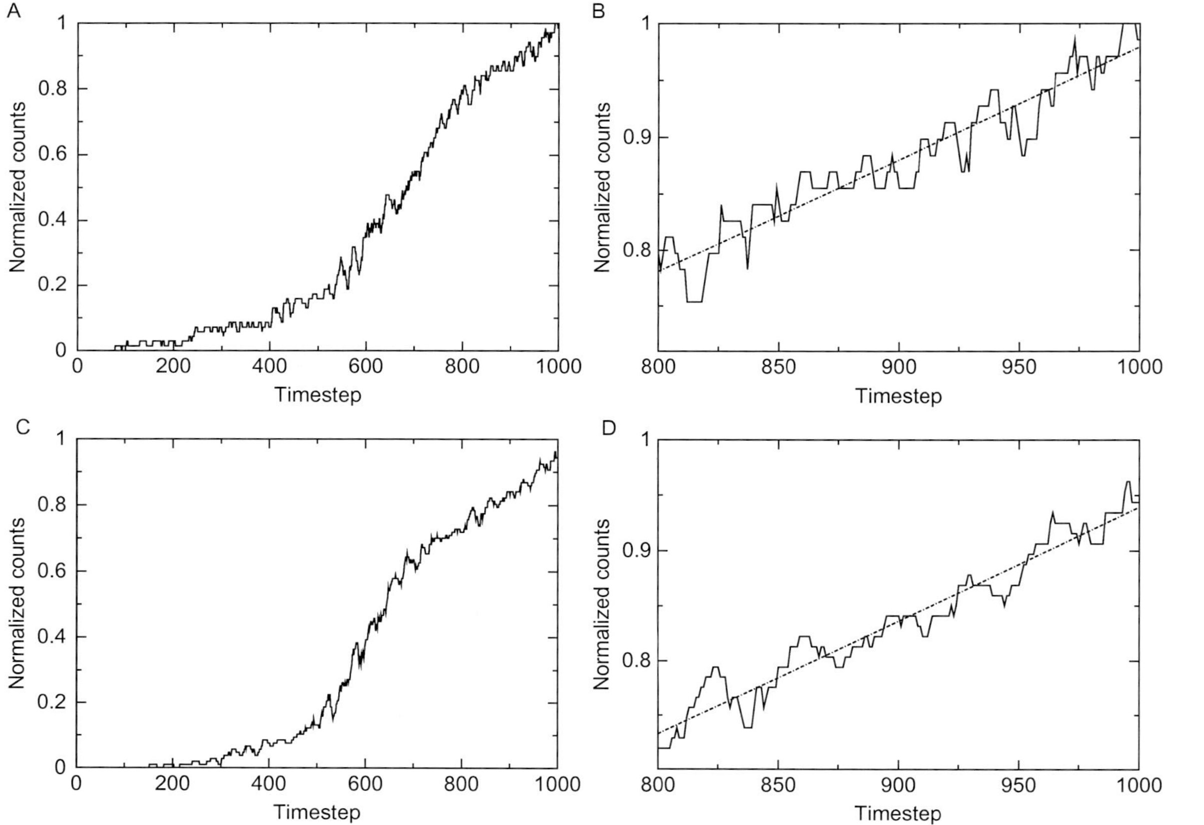

Fig. 2 Cumulative histograms for the calculation of the first term for hydride (A) and deuteride (C) systems, normalized to unity integral. Fits of the linear region, from 800 to 1000 timesteps (400–500 fs) are shown in *dashed lines* for hydride (B) and deuteride (D) systems. The slope of these linear fits are the values for the first term of the rate calculation.

Table 3 Windows and Their Member Trajectories for the Calculation of $C(t')$ in the Hydride system

Window	Range of Order Parameter (λ)	Accepted Trajectories	Total Trajectories	Acceptance Ratio (%)
1	0.200–0.360	216	493	43.8
2	0.350–0.410	179	487	36.8
3	0.400–0.460	167	1087	15.4
4	0.450–0.510	68	2066	3.10
5	0.500–0.560	56	2072	2.70
6	0.550–0.610	111	1303	8.51
7	0.600–0.900	125	811	15.4

Note the low acceptance ratio near the barrier, in windows 4, 5, and 6.

Table 4 Windows and Their Member Trajectories for the Calculation of $C(t')$ in the Deuteride System

Window	Range of Order Parameter (λ)	Accepted Trajectories	Total Trajectories	Acceptance Ratio (%)
1	0.170–0.260	201	517	38.9
2	0.250–0.310	150	506	29.6
3	0.300–0.350	113	481	23.5
4	0.350–0.400	138	744	18.5
5	0.400–0.450	40	885	4.52
6	0.450–0.510	46	2513	1.83
7	0.500–0.560	59	2451	2.35
8	0.550–0.610	94	2630	3.57
9	0.600–0.900	101	1458	6.9

Note the low acceptance ratio near the barrier, in windows 5, 6, 7, and 8.

the transition state in the case of YADH (Klinman, 1976). In an enzymatic reaction, a conformational search occurs to reach a reactive conformation from the Michaelis complex. However, the timescales of TPS trajectories require the simulation to start at or near the reactive conformation, excluding this conformational search. This extra step included in the experimental studies has the ability to mask a larger KIE on the transfer event alone, in the case where the conformational search is not affected by isotopic substitution (Schramm, 2011).

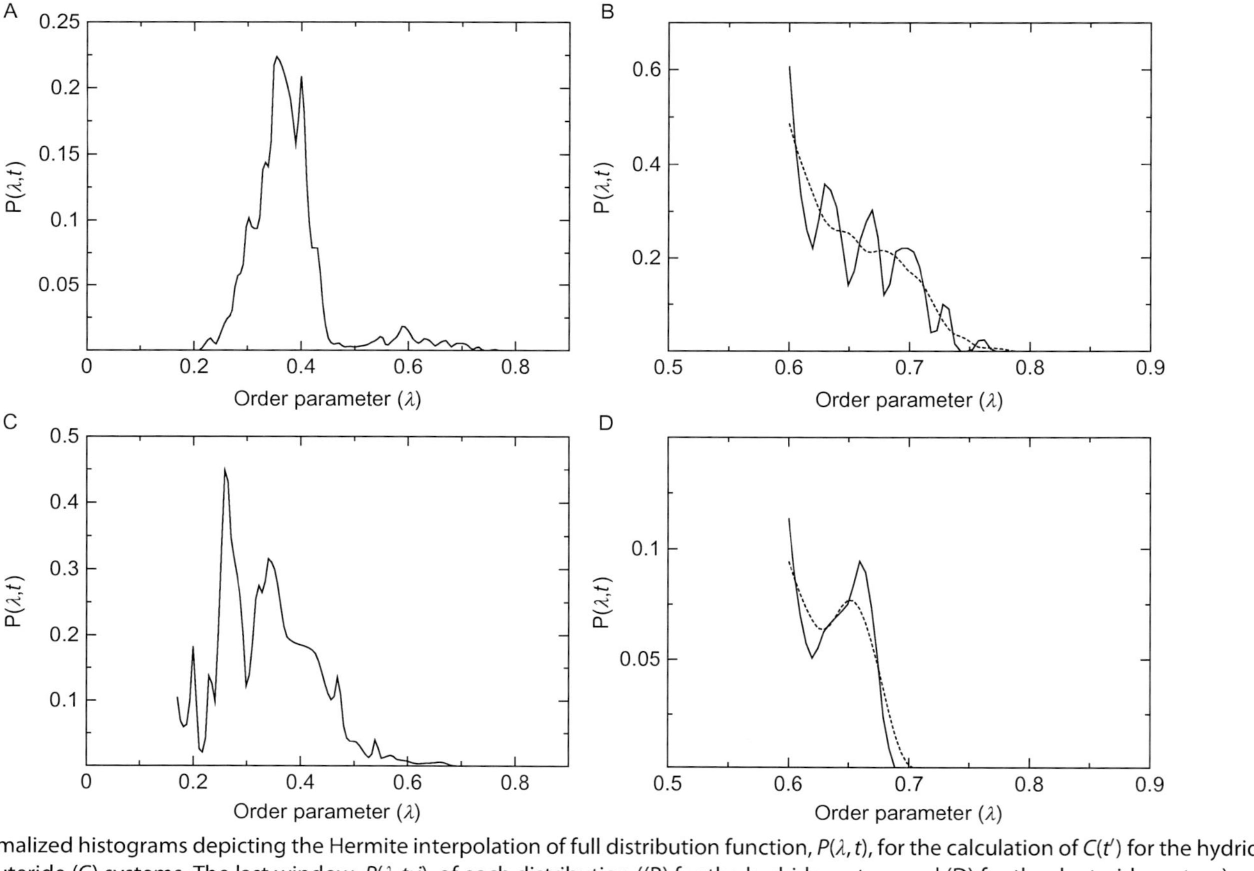

Fig. 3 Normalized histograms depicting the Hermite interpolation of full distribution function, $P(\lambda, t)$, for the calculation of $C(t')$ for the hydride (A) and deuteride (C) systems. The last window, $P(\lambda, t; i)$, of each distribution ((B) for the hydride system and (D) for the deuteride system) was denoted as the product region and was smoothed with a Gaussian function (shown in the *dashed line*) which was then integrated to determine $C(t')$.

4. CONCLUSION

We have summarized newly developed methods which combine a first principles sampling method, TPS, with quantum propagation of transferring particles, CMD. These methods were successfully applied to the calculation of free energy barriers in YADH and LDH and the primary H/D KIE in LDH. The inclusion of full protein and quantum dynamics into the calculation of free energy barriers and KIEs is a crucial stepping stone toward a more refined computational view of enzymatic mechanism.

REFERENCES

Amadei, A., Linssen, A. B. M., & Berendsen, H. J. C. (1993). Essential dynamics of proteins. *Proteins*, *17*, 412–425.

Antoniou, D., Ge, X., Schramm, V. L., & Schwartz, S. D. (2012). Mass modulation of protein dynamics associated with barrier crossing in purine nucleoside phosphorylase. *Journal of Physical Chemistry Letters*, *3*, 3538–3555.

Antoniou, D., & Schwartz, S. D. (2009). Approximate inclusion of quantum effects in transition path sampling. *The Journal of Chemical Physics*, *131*(22), 224111.

Basner, J. E., & Schwartz, S. D. (2005). How enzyme dynamics helps catalyze a reaction in atomic detail: A transition path sampling study. *Journal of the American Chemical Society*, *127*(40), 13822–13831.

Bennett, C. H. (1977). Molecular dynamics and transition state theory: the simulation of infrequent events. In R. E. Christofferson (Ed.), *ACS symposium series*: *Vol. 46. Algorithms for chemical computations* (pp. 63–97). Washington, DC: American Chemical Society.

Bolhuis, P. G., Chandler, D., Dellago, C., & Geissler, P. L. (2002). Transition path sampling: Throwing ropes over rough mountain passes, in the dark. *Annual Review of Physical Chemistry*, *53*(1), 291–318.

Bolhuis, P. G., Dellago, C., & Chandler, D. (1998). Sampling ensembles of deterministic transition pathways. *Faraday Discussions*, *110*, 421–436.

Brooks, B. R., Brooks, C. L., Mackerell, A. D., Nilsson, L., Petrella, R. J., Roux, B., … Karplus, M. (2009). CHARMM: The biomolecular simulation program. *Journal of Computational Chemistry*, *30*(10), 1545–1614.

Brooks, B. R., Bruccoleri, R. E., Olafson, B. D., States, D. J., Swaminathan, S., & Karplus, M. (1983). CHARMM: A program for macromolecular energy, minimization, and dynamics calculations. *Journal of Computational Chemistry*, *4*(2), 187–217.

Brothers, E. N., Suarez, D., Deerfield, D. W., & Merz, K. M. (2004). PM3-compatible zinc parameters optimized for metalloenzyme active sites. *Journal of Computational Chemistry*, *25*(14), 1677–1692.

Cao, J., & Voth, G. A. (1994a). The formulation of quantum statistical mechanics based on the Feynman path centroid density. I. Equilibrium properties. *The Journal of Chemical Physics*, *100*(7), 5093–5104.

Cao, J., & Voth, G. A. (1994b). The formulation of quantum statistical mechanics based on the Feynman path centroid density. II. Dynamical properties. *The Journal of Chemical Physics*, *100*(7), 5106–5117.

Cao, J., & Voth, G. A. (1994c). The formulation of quantum statistical mechanics based on the Feynman path centroid density. III. Phase space formalism and analysis of centroid molecular dynamics. *The Journal of Chemical Physics*, *101*(7), 6157–6167.

Cao, J., & Voth, G. A. (1994d). The formulation of quantum statistical mechanics based on the Feynman path centroid density. IV. Algorithms for centroid molecular dynamics. *The Journal of Chemical Physics*, *101*(7), 6168–6183.
Cao, J., & Voth, G. A. (1994e). The formulation of quantum statistical mechanics based on the Feynman path centroid density. V. Quantum instantaneous normal mode theory of liquids. *The Journal of Chemical Physics*, *101*(7), 6184–6192.
Cha, Y., Murray, C. J., & Klinman, J. P. (1989). Hydrogen tunneling in enzyme reactions. *Science*, *243*(4896), 1325–1330.
Chandler, D. (1978). Statistical mechanics of isomerization dynamics in liquids and the transition state approximation. *The Journal of Chemical Physics*, *68*, 2959.
Ciccotti, G., Kapral, R., & Vanden-Eijnden, E. (2005). Blue moon sampling, vectorial reaction coordinates, and unbiased constrained dynamics. *ChemPhysChem*, *6*(9), 1809–1814.
Cook, P. F., & Cleland, W. W. (2007). *Enzyme kinetics and mechanism*. New York City: Taylor & Francis.
Crehuet, R., & Field, M. J. (2007). A transition path sampling study of the reaction catalyzed by the enzyme chorismate mutase. *The Journal of Physical Chemistry. B*, *111*(20), 5708–5718.
Dellago, C., Bolhuis, P. G., & Chandler, D. (1998a). Efficient transition path sampling: Application to Lennard-Jones cluster rearrangements. *The Journal of Chemical Physics*, *108*(22), 9236–9245.
Dellago, C., Bolhuis, P. G., & Chandler, D. (1999). On the calculation of reaction rate constants in the transition path ensemble. *The Journal of Chemical Physics*, *110*(14), 6617–6625.
Dellago, C., Bolhuis, P. G., Csajka, F. S., & Chandler, D. (1998b). Transition path sampling and the calculation of rate constants. *The Journal of Chemical Physics*, *108*(5), 1964–1977.
Dzierlenga, M. W., Antoniou, D., & Schwartz, S. D. (2015). Another look at the mechanisms of hydride transfer enzymes with quantum and classical transition path sampling. *The Journal of Physical Chemistry Letters*, *6*(7), 1177–1181.
Ganzhorn, A. J., Green, D. W., Hershey, A. D., Gould, R. M., & Plapp, B. V. (1987). Kinetic characterization of yeast alcohol dehydrogenases. Amino acid residue 294 and substrate specificity. *The Journal of Biological Chemistry*, *262*(8), 3754–3761.
Gao, J., Amara, P., Alhambra, C., & Field, M. J. (1998). A generalized hybrid orbital (GHO) method for the treatment of boundary atoms in combined QM/MM calculations. *The Journal of Physical Chemistry. A*, *102*(24), 4714–4721.
Hayes, J. E., & Velick, S. F. (1954). Yeast alcohol dehydrogenase: Molecular weight, coenzyme binding, and reaction equilibria. *The Journal of Biological Chemistry*, *207*(2), 225–244.
Horiuti, J. (1938). On the statistical mechanical treatment of the absolute rate of chemical reaction. *Bulletin of the Chemical Society of Japan*, *13*(1), 210–216.
Hynes, J. T. (1985). Chemical reaction dynamics in solution. *Annual Review of Physical Chemistry*, *36*, 573–597.
Jarzynski, C. (1997). Nonequilibrium equality for free energy differences. *Physical Review Letters*, *78*(14), 2690–2693.
Kästner, J. (2011). Umbrella sampling. *Wiley Interdisciplinary Reviews: Computational Molecular Science*, *1*(6), 932–942.
Kästner, J., & Thiel, W. (2005). Bridging the gap between thermodynamic integration and umbrella sampling provides a novel analysis method: "Umbrella integration" *The Journal of Chemical Physics*, *123*(14), 144104.
Kipp, D. R., Silva, R. G., & Schramm, V. L. (2011). Mass-dependent bond vibrational dynamics influence catalysis by HIV-1 protease. *Journal of the American Chemical Society*, *133*(48), 19358–19361.

Kirkwood, J. G. (1935). Statistical mechanics of fluid mixtures. *The Journal of Chemical Physics, 3*(1935), 300–313.
Kirkwood, J. G. (1936). Statistical mechanics of liquid solutions. *Chemical Reviews, 19*(3), 275–307.
Klinman, J. P. (1972). The mechanism of enzyme-catalyzed reduced nicotinamide adenine dinucleotide-dependent reductions. Substituent and isotope effects in the yeast alcohol dehydrogenase reaction. *The Journal of Biological Chemistry, 247*(24), 7977–7987.
Klinman, J. P. (1976). Isotope effects and structure-reactivity correlations in the yeast alcohol dehydrogenase reaction. A study of the enzyme-catalyzed oxidation of aromatic alcohols. *Biochemistry, 15*(9), 2018–2026.
Klinman, J. P. (2009). An integrated model for enzyme catalysis emerges from studies of hydrogen tunneling. *Chemical Physics Letters, 471*(4-6), 179–193.
Kumar, S., Rosenberg, J. M., Bouzida, D., Swendsen, R. H., & Kollman, P. A. (1992). The weighted histogram analysis method for free-energy calculations on biomolecules. I. The method. *Journal of Computational Chemistry, 13*(8), 1011–1021.
Layfield, J. P., & Hammes-Schiffer, S. (2014). Hydrogen tunneling in enzymes and biomimetic models. *Chemical Reviews, 114*(7), 3466–3494.
Machleder, S. Q., Pineda, J. R. E. T., & Schwartz, S. D. (2010). On the origin of the chemical barrier and tunneling in enzymes. *Journal of Physical Organic Chemistry, 23*(7), 690–695.
Major, D. T., & Gao, J. (2007). An integrated path integral and free-energy perturbation—Umbrella sampling method for computing kinetic isotope effects of chemical reactions in solution and in enzymes. *Journal of Chemical Theory and Computation, 3*, 949–960.
Masterson, J. E., & Schwartz, S. D. (2013). Changes in protein architecture and subpicosecond protein dynamics impact the reaction catalyzed by lactate dehydrogenase. *The Journal of Physical Chemistry. A, 117*(32), 7107–7113.
Masterson, J. E., & Schwartz, S. D. (2014). The enzymatic reaction catalyzed by lactate dehydrogenase exhibits one dominant reaction path. *Chemical Physics, 442*(17), 132–136.
Onsager, L. (1933). Theories of concentrated electrolytes. *Chemical Reviews, 13*(1), 73–89.
Onsager, L. (1938). Initial recombination of ions. *Physical Review, 54*(8), 554–557.
Park, S., & Schulten, K. (2004). Calculating potentials of mean force from steered molecular dynamics simulations. *The Journal of Chemical Physics, 120*(13), 5946.
Quaytman, S. L., & Schwartz, S. D. (2007). Reaction coordinate of an enzymatic reaction revealed by transition path sampling. *Proceedings of the National Academy of Sciences of the United States of America, 104*(30), 12253–12258.
Roston, D., & Kohen, A. (2013). A critical test of the "tunneling and coupled motion" concept in enzymatic alcohol oxidation. *Journal of the American Chemical Society, 135*(37), 13624–13627.
Saunders, W. H., Jr. (1985). Calculations of isotope effects in elimination reactions. New experimental criteria for tunneling in slow proton transfers. *Journal of the American Chemical Society, 107*(1), 164–169.
Schramm, V. L. (2011). Enzymatic transition states, transition-state analogs, dynamics, thermodynamics, and lifetimes. *Annual Review of Biochemistry, 80*, 703–732.
Silva, R. G., Murkin, A. S., & Schramm, V. L. (2011). Femtosecond dynamics coupled to chemical barrier crossing in a Born-Oppenheimer enzyme. *Proceedings of the National Academy of Sciences of the United States of America, 108*(46), 18661–18665.
Simon, H., & Palm, D. (1966). Isotope effects in organic chemistry and biochemistry. *Angewandte Chemie, International Edition, 5*(11), 920–933.
Stewart, J. J. P. (1989a). Optimization of parameters for semiempirical methods I. Method. *Journal of Computational Chemistry, 10*(2), 209–220.
Stewart, J. J. P. (1989b). Optimization of parameters for semiempirical methods II. Applications. *Journal of Computational Chemistry, 10*(2), 221–264.

Straatsma, T. P., & Berendsen, H. J. C. (1988). Free energy of ionic hydration: Analysis of a thermodynamic integration technique to evaluate free energy differences by molecular dynamics simulations. *The Journal of Chemical Physics*, *89*(9), 5876.

Swain, C. G., Stivers, E. C., Reuwer, J. F., Jr., & Schaad, L. J. (1958). Use of hydrogen isotope effects to identify the attacking nucleophile in the enolization of ketones catalyzed by acetic acid. *Journal of the American Chemical Society*, *80*(21), 5885–5893.

Torrie, G. M., & Valleau, J. P. (1977). Nonphysical sampling distributions in Monte Carlo free-energy estimation: Umbrella sampling. *Journal of Computational Physics*, *23*(2), 187–199.

Truhlar, D. G. (2010). Tunneling in enzymatic and nonenzymatic hydrogen transfer reactions. *Journal of Physical Organic Chemistry*, *23*(7), 660–676.

Vanden-Eijnden, E., & Tal, F. A. (2005). Transition state theory: Variational formulation, dynamical corrections, and error estimates. *The Journal of Chemical Physics*, *123*(18), 184103.

Varga, M. J., & Schwartz, S. D. (2016). Enzymatic kinetic isotope effects from first-principles path sampling calculations. *Journal of Chemical Theory and Computation*, *12*(4), 2047–2054.

Warshel, A., & Levitt, M. (1976). Theoretical studies of enzymic reactions: Dielectric, electrostatic and steric stabilization of the carbonium ion in the reaction of lysozyme. *Journal of Molecular Biology*, *103*(2), 227–249.

Wick, C. D. (2009). NaCl dissociation dynamics at the air–water interface. *The Journal of Physical Chemistry. C*, *113*(6), 2497–2502.

Wick, C. D., & Dang, L. X. (2010). Computational investigation of the influence of organic-aqueous interfaces on NaCl dissociation dynamics. *The Journal of Chemical Physics*, *132*(4), 044702.

Xia, F., Bronowska, A. K., Cheng, S., & Gräter, F. (2011). Base-catalyzed peptide hydrolysis is insensitive to mechanical stress. *The Journal of Physical Chemistry. B*, *115*(33), 10126–10132.

Zhang, Y., Liu, H., & Yang, W. (2000). Free energy calculation on enzyme reactions with an efficient iterative procedure to determine minimum energy paths on a combined *ab initio* QM/MM potential energy surface. *The Journal of Chemical Physics*, *112*(8), 3483–3492.

CHAPTER THREE

Accurate Calculation of Electric Fields Inside Enzymes

X. Wang*,†, X. He†,‡,1, J.Z.H. Zhang†,‡,§,1

*Center for Optics & Optoelectronics Research, College of Science, Zhejiang University of Technology, Hangzhou, Zhejiang, China
†School of Chemistry and Molecular Engineering, East China Normal University, Shanghai, China
‡NYU-ECNU Center for Computational Chemistry at NYU Shanghai, Shanghai, China
§New York University, New York, NY, United States
[1]Corresponding authors: e-mail address: xiaohe@phy.ecnu.edu.cn; zhzhang@phy.ecnu.edu.cn

Contents

Abstract

The specific electric field generated by a protease at its active site is considered as an important source of the catalytic power. Accurate calculation of electric field at the active site of an enzyme has both fundamental and practical importance. Measuring site-specific changes of electric field at internal sites of proteins due to, eg, mutation, has been realized by using molecular probes with C═O or C≡N groups in the context of vibrational Stark effect. However, theoretical prediction of change in electric field inside a protein based on a conventional force field, such as AMBER or OPLS, is often inadequate. For such calculation, quantum chemical approach or quantum-based polarizable or polarized force field is highly preferable. Compared with the result from conventional force field, significant improvement is found in predicting experimentally measured mutation-induced electric field change using quantum-based methods, indicating that quantum effect such as polarization plays an important role in accurate description of electric field inside proteins. In comparison, the best theoretical prediction comes from fully quantum mechanical calculation in which both polarization and inter-residue charge transfer effects are included for accurate prediction of electrostatics in proteins.

Methods in Enzymology, Volume 578
ISSN 0076-6879
http://dx.doi.org/10.1016/bs.mie.2016.05.043

1. INTRODUCTION

Most life processes involve enzymatic reactions (Bugg, 2001). Understanding how enzymes work is one of the most important fields of research. Although the origin of the enormous catalytic power of enzymes is still under active debate, it is generally understood that enzymes reduce the energy barrier for the specific chemical reaction by producing a more favorable electrostatic environment for the reaction than that in the solution, a proposal widely investigated by various theoretical and experimental studies (Fried, Bagchi, & Boxer, 2014; Warshel et al., 2006; Xiang, Duan, & Zhang, 2011). Fried et al. pointed out that there is a clear correlation between the magnitude of the electric field at the active site and the catalytic rate enhancement in ketosteroid isomerase, and they raised the possibility of predicting enzyme's reaction rate by exploring the magnitude of the electric field at the active site (Fried et al., 2014). To experimentally measure the electrostatic field inside proteins, a vibrational Stark effect (VSE) spectroscopy method has been developed (Fried, Bagchi, & Boxer, 2013; Fried et al., 2014; Park, Andrews, Hu, & Boxer, 1999; Saggu, Levinson, & Boxer, 2011, 2012; Steffen, Lao, & Boxer, 1994; Suydam, Snow, Pande, & Boxer, 2006; Webb & Boxer, 2008; Xu, Cohen, & Boxer, 2011) in which a molecular probe group such as C=O or C≡N is introduced to the specific site in proteins.

The vibrational frequency shift ($\Delta\tilde{\upsilon}_{\mathrm{PG}}$) of the probe group (PG) and the change of electric field acting on it ($\Delta\vec{F}$) usually show an approximate linear relationship as follows (Fried et al., 2013, 2014; Steffen et al., 1994; Webb & Boxer, 2008):

$$hc\Delta\tilde{\upsilon}_{\mathrm{PG}} = -\Delta\vec{\mu}_{\mathrm{PM}} \cdot \Delta\vec{F} \tag{1}$$

where h is Planck constant, c is the speed of light, $\Delta\tilde{\upsilon}_{\mathrm{PG}}$ is the observed frequency shift of the probe group, $\Delta\vec{F}$ is the change of electric field acting on the probe group, and $\Delta\vec{\mu}_{\mathrm{PM}}$ is the Stark tuning rate of a PG-containing probing molecule, representing the change of the dipole moment between the ground state and the excited state of the vibrational transition of the probe. $\Delta\vec{\mu}_{\mathrm{PM}}$ is different for different PG-containing probing molecules and can be obtained by linear fitting from vibrational Stark spectrum of the probe group based from Eq. (1).

It is worth noting that Eq. (1) may not be strictly correct in all cases. A previous study (Choi, Oh, Lee, Lee, & Cho, 2008) based on ab initio calculations by Choi et al. demonstrated that when there are hydrogen bond interactions between the nitrile and its surroundings (water molecules), the linear relationship between the VSE shift and the change of the electric field is no longer valid due to the change of hydrogen-bonding network around the nitrile. They also mentioned that it is rooted in electron interactions (beyond the electrostatics) between the water molecules and the nitrogen atom's lone pair orbital and nitrile's π orbitals. Experimental works (Saggu et al., 2011, 2012) along with ab initio calculations by Saggu et al. confirmed that only for weak hydrogen bonds, the linear relationship of Eq. (1) is valid for probe groups O–H, N–H, and S–H. In spite of this, the C=O probe shows good linear relationship even in the standard hydrogen-bonding environment (with water molecules) and can be used to measure electrostatic fields in both hydrogen-bonding and nonhydrogen-bonding environment (Choi & Cho, 2011; Fried et al., 2013). Furthermore, high-level ab initio frequency calculations on polar diatomic molecules by Sowlati-Hashjin and Matta demonstrated that quadratic term of the vibrational energy expression in an external homogenous electric field makes nonnegligible contribution to the frequency shift brought from large electric fields (dozens of MV/cm), which could also invalidate the simple linear relationship between the frequency shift and the electric field (Sowlati-Hashjin & Matta, 2013).

A conventional way of predicting the electrostatics of proteins is computational simulation based on empirical force fields. Nevertheless, previous studies have demonstrated that empirical force fields are generally inadequate in accurate theoretical calculation of electrostatic properties of proteins and the linear vibrational Stark shift (Suydam et al., 2006; Wang, He, & Zhang, 2013; Webb & Boxer, 2008). Calculations based on ab initio methods could more accurately capture the electrostatic properties of molecules, as verified by theoretical studies of linear vibrational Stark shift using probes such as O–H, N–H, and S–H in different solvents (Saggu et al., 2011, 2012). However, a key issue in ab initio calculation of large molecules is the scaling problem. At the Hartree–Fock (HF) and density functional theory (DFT) levels, the conventional high-power scaling is $O(N^3)$ (N denotes the size of the system). The scaling of post-HF methods (such as second-order Møller–Plesset perturbation theory (MP2), coupled-cluster (CC) method that includes single and double excitations (CCSD)) is even higher ($\mathrm{MP2} \sim O(N^5)$ and $\mathrm{CCSD} \sim O(N^6)$). Performing standard quantum chemistry calculation for proteins that contain thousands of atoms plus the solvent environment is computationally impractical. Various linear-scaling approaches (Goedecker, 1999; He & Merz, 2010; Strain, Scuseria, & Frisch, 1996; Yang, 1991) have been developed over the past decades, among which the fragmentation approach has become one of the most important methods in extending quantum mechanical methods to large systems (Chung et al., 2015; Collins & Bettens, 2015; Gordon, Fedorov, Pruitt, & Slipchenko, 2012; Raghavachari & Saha, 2015). There are several advantages in fragmentation approach for electronic structure calculation of large systems, such as easy implementation of parallelization without extensively modifying the existing QM programs and straightforward application at all levels of ab initio electronic structure theories.

In this chapter, fragment-based QM computational methods based on MFCC (molecular fractionation with conjugate caps method) ansatz and its extended version EE-GMFCC (electrostatically embedded generalized molecular fractionation with conjugate caps method) are described. By comparing the computed electric fields inside proteins using conventional empirical force fields, polarized force field, and the EE-GMFCC method, respectively, with experimentally measured vibrational Stark shift, the accuracies of theoretical predictions of the electrostatics inside proteins are discussed.

2. THEORETICAL METHODS

2.1 Fragment Approach to Quantum Calculation of Protein–Ligand Interaction

The MFCC method was initially developed for calculating protein–ligand (protease–substrate) interaction energies using ab initio methods (Zhang & Zhang, 2003). The main idea of the MFCC approach is to divide a protein into a series of amino-acid fragments by cutting at the peptide bond (see Fig. 1) or other bonds. For a center residue (CR), a pair of conjugate caps (concaps) at the cutting location is introduced to saturate the dangling bonds and also mimic the local chemical environment of the cutoff fragment. Using the fragmentation scheme, the interaction energy of the protein–ligand can be computed by separate calculations of individual fragments interacting with the ligand. The pair of concaps is fused to form proper molecular species such that the doubly counted interaction energy between caps of the fragments and the ligand can be subtracted.

In the MFCC scheme, there are $N-2$ capped fragments and $N-3$ concaps for a protein system containing N residues and a ligand. The interaction energy for such a protein–ligand system ($E_{\mathrm{P-L}}$) is given by the following expression:

$$E_{\mathrm{P-L}} = \sum_{i=1}^{N-2} E_{\mathrm{F}_i-\mathrm{L}} - \sum_{i=1}^{N-3} E_{\mathrm{CC}_i-\mathrm{L}} - \sum_{i=1}^{N-2} E_{\mathrm{F}_i} + \sum_{i=1}^{N-3} E_{\mathrm{CC}_i} - E_{\mathrm{L}} \quad (2)$$

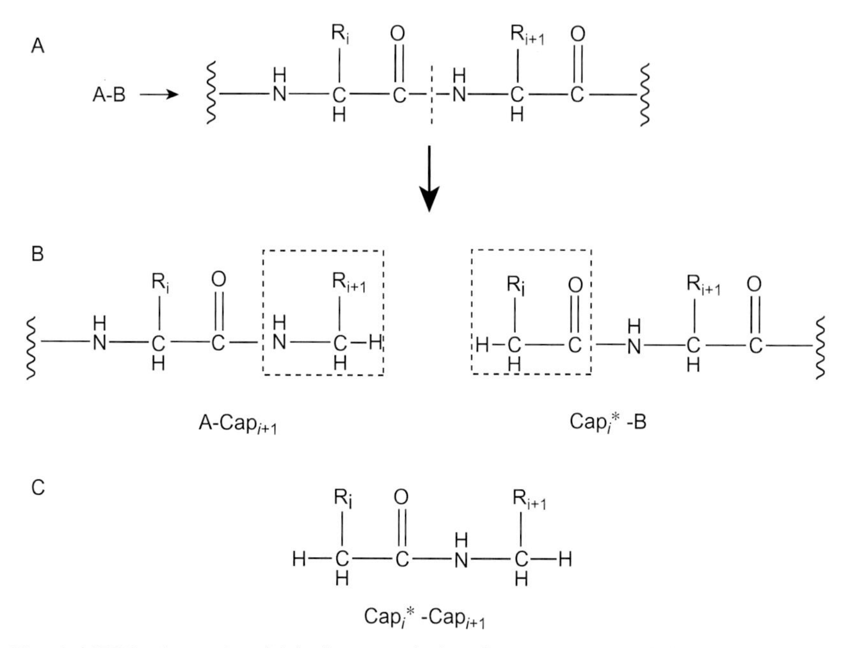

Fig. 1 MFCC scheme in which the peptide bond is cut in (A) and the fragments are capped with Cap_{i+1} and its conjugate Cap_i^* in (B), where i represents the index of the ith amino acid in the given protein. The atomic structure of concap is shown in (C). The concap is defined as the $\text{Cap}_i^* - \text{Cap}_{i+1}$ fused molecular species.

where $E_{\text{F}_i-\text{L}}$ represents the total energy of the ith capped fragment with ligand and $E_{\text{CC}_i-\text{L}}$ represents the total energy of the ith concap and ligand. E_{F_i} and E_{CC_i} is the self-energy of the ith capped fragment and ith concap, respectively. The MFCC method can accurately reproduce the ab initio interaction energy of protein-ligand with a low computational cost, which has been validated in previous works (He, Zhu, Wang, Liu, & Zhang, 2014; Zhang & Zhang, 2003). Here, the comparison between the full system M062X/6-311G** and the results based on MFCC for calculation of the interaction energy between Efavirenz and a polypeptide which contains 19 residues (a model system extracted from HIV-1 reverse transcriptase (RT)) is shown in Fig. 2.

2.2 Fragment Quantum Calculation of Protein in Solvent

For protein in solvent, the MFCC calculation of protein can be coupled with a solvation model such as Polarizable Continuum Model (PCM) or Poisson–Boltzmann (PB) model. In the MFCC-PB protocol, the PB solver Delphi (Rocchia et al., 2002) is used to calculate induced charges (*gray points* in Fig. 3) on the solute–solvent interface. The protein is then divided

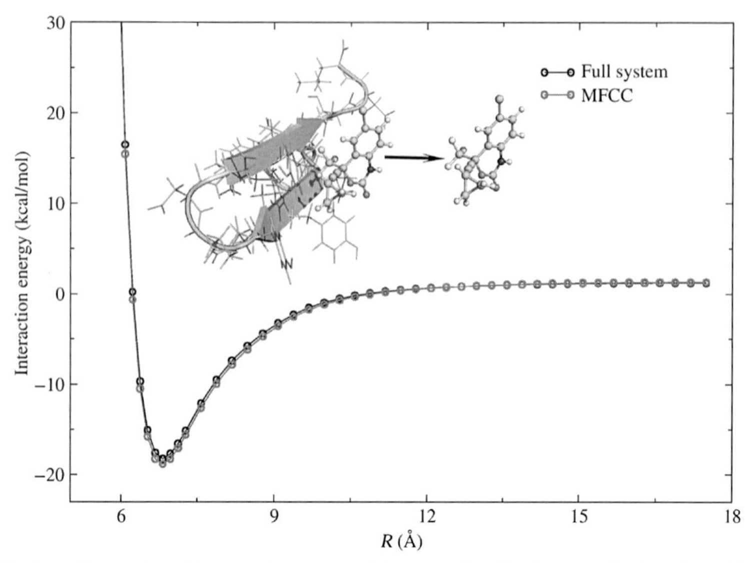

Fig. 2 One-dimensional interaction potential curves for Efavirenz and a fraction of HIV-1 reverse transcriptase (Asn175 to Leu193 of chain A) extracted from RT-Efavirenz complex (PDB code: 1FKO). The interaction energy is calculated at the M062X/6-311G** level along the direction from the geometric center of Efavirenz to that of the polypeptide. *Reprinted with permission from He, X., Zhu, T., Wang, X., Liu, J., & Zhang, J. Z. H. (2014). Fragment quantum mechanical calculation of proteins and its applications.* Accounts of Chemical Research, *47(9), 2748–2757. doi: 10.1021/ar500077t. Copyright 2014 American Chemical Society.*

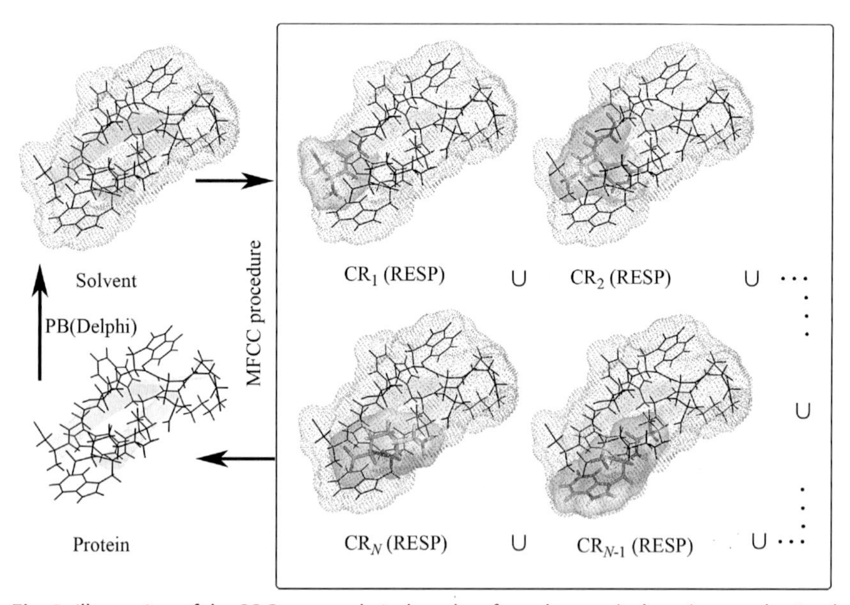

Fig. 3 Illustration of the PPC approach. Induced surface charges (solvent) were obtained by solving Poison–Boltzmann equation using Delphi. Other part of the protein was treated as background charges in QM calculation of center residue (CR; *colored surfaces*). Each fragment calculation was performed in parallel. (See the color plate.)

into capped amino-acid fragments. In QM calculation of the ith fragment (*colored sticks* in Fig. 3) using DFT at the B3LYP/6-31G* level, the induced charges (solvent) and the rest of the protein fragments (*gray lines* in Fig. 3) are represented by the background charges. The electrostatic potentials are saved, and a standard two-stage restrained electrostatic potential (RESP) fitting procedure is used to fit effective point charges of the protein. In the first stage, all atoms are allowed to vary, and then in the second stage, only charges of all degenerate hydrogen atoms are refitted and constrained to have equal charge. The charge transfer is localized and only allowed within each individual amino acid, and thus each residue is kept charge neutral. The newly fitted atomic charges are passed to the next calculation of charge fitting. This process (see Fig. 3) is iterated until the corrected reaction field energy calculated with Delphi is converged and its variations are smaller than a certain criterion. More details can be found in Ji, Mei, and Zhang (2008).

The fragment quantum calculation of protein in solvent also enables us to derive the polarized protein-specific charge (PPC) for molecular dynamics simulation with polarization (Ji & Mei, 2014; Ji et al., 2008). PPCs are fitted to the electrostatic potentials by fragment quantum mechanical calculations using an iterative approach. In the PPC fitting procedure for a given protein structure, the hydrogen atoms are added using LEaP module (Case et al., 2005) of the Amber program and then a series of minimizations using the Amber ff99SB force field (Cornell et al., 1995) were carried out to remove bad contacts. The minimized structure is then used to calculate PPCs by the fragment method. PPC shows many advantages in MD simulations of protein properties due to the inclusion of electrostatic polarization (Duan, Mei, Zhang, & Zhang, 2009; Ji et al., 2008; Ji & Zhang, 2009, 2012; Li, Zhang, & Mei, 2014; Lu, Mei, Zhang, & Zhang, 2010; Mei, Li, Zeng, & Zhang, 2012; Tong, Ji, Mei, & Zhang, 2009; Tong, Mei, Li, Ji, & Zhang, 2010). PPC could also provide better prediction of mutation-induced vibrational Stark shift (from the change of electric fields) than conventional force fields as demonstrated in Section 2.5.1.

2.3 EE-GMFCC Method for Calculation of Enzyme Energy

The EE-GMFCC method is an extended version of the original MFCC approach (Wang, Liu, Zhang, & He, 2013). In the framework of MFCC (see Fig. 4A), the total energy of a given protein that contains N residues

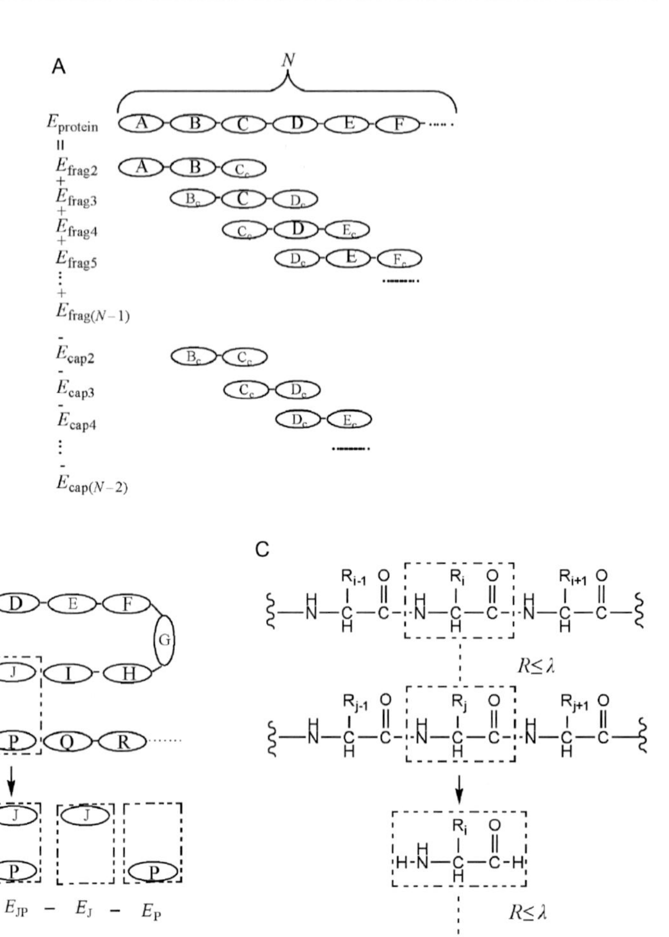

Fig. 4 (A) $N-2$ fragments and $N-3$ caps for a given single-chain protein that contains N residues. (B) Representation of nonneighboring residues J and P whose closest distance is less than a predefined threshold λ and two-body representation for residues J and P. (C) The generalized concap (Gconcap) scheme and the atomic structure of the Gconcap.

can be approximately expressed as the sum of the capped fragments minus that of the conjugate caps (concap),

$$E_{\text{MFCC}} = \sum_{i=2}^{N-1} E(\text{Cap}_{i-1}^{*} F_i \text{Cap}_{i+1}) - \sum_{i=2}^{N-2} E(\text{Cap}_i^{*} \text{Cap}_{i+1}) \quad (3)$$

where i is the index for the ith residue, $E(\text{Cap}_{i-1}^{*} F_i \text{Cap}_{i+1})$ is the self-energy of the ith residue F_i with a pair of caps. $E(\text{Cap}_i^{*} \text{Cap}_{i+1})$ is the self-energy of the *i*th concap. The double counting of energy of caps is deducted in the form of self-energy of concaps. Eq. (3) does not include interactions between nonneighboring residues.

In the EE-GMFCC method, two key modifications are introduced for the total energy calculation of the proteins. (1) The ab initio calculation for each capped fragment is embedded in the electrostatic field of the point charges representing the remaining amino acids in the protein, which accounts for the electronic polarization effect of the protein environment. (2) To better capture the QM interactions for nonneighboring fragments that are spatially in close contact (see Fig. 4B), generalized concaps (Gconcaps) are introduced (see Fig. 4C). The total energy of a protein (with N residues) using the EE-GMFCC method is expressed as

$$
\begin{aligned}
E_{\text{EE-GMFCC}} = & \sum_{i=2}^{N-1} \widetilde{E}\left(\text{Cap}_{i-1}^{*} F_i \text{Cap}_{i+1}\right) - \sum_{i=2}^{N-2} \widetilde{E}\left(\text{Cap}_{i}^{*} \text{Cap}_{i+1}\right) \\
& + \sum_{\substack{i,j>i+2, \\ \left|R_i - R_j\right| \le \lambda}} \left(\widetilde{E}_{ij} - \widetilde{E}_i - \widetilde{E}_j\right)_{\text{QM}} \\
& - \left\{ \sum_{k,l} \sum_{m,n} \frac{q_{m(k)} q_{n(l)}}{R_{m(k)n(l)}} - \sum_{\substack{i,j>i+2, \\ \left|R_i - R_j\right| \le \lambda}} \sum_{m',n'} \frac{q_{m'(i)} q_{n'(j)}}{R_{m'(i)n'(j)}} \right\}
\end{aligned}
\tag{4}
$$

where I and j represent the residue pairs and $\tilde{E}$ (different from E of Eq. 3) denotes the sum of the self-energy of the fragment and the interaction energy between the fragment and background charges of the remaining system. The $\widetilde{E}_{ij} - \widetilde{E}_i - \widetilde{E}_j$ represents the two-body QM interaction energy between residues i and j whose closest distance is less than a predefined threshold λ. The interaction energy doubly counted in the first three terms of Eq. (4) is deducted by the pairwise charge–charge interactions approximately in the last term, where $q_{n(l)}$ represents the point charge of the nth atom in the fragment l, and $k = (\text{H})\text{Cap}_{i-1}^{*} - \text{NH}$ (where (H) denotes that the added hydrogen atom on the left is excluded) and $l = \text{CO} - A_{i+2}A_{i+3}\cdots A_N$ ($i = 2, 3, \ldots, N-2$, A_i denotes the ith residue).

The total energies of several proteins with up to 1142 atoms were calculated using EE-GMFCC at the HF/6-31G* level, and the overall mean unsigned error of EE-GMFCC is only a few kcal/mol with reference to the conventional full system calculations. The deviations of EE-GMFCC energies for small proteins at the DFT and MP2 levels are also within a few kcal/mol away from those obtained by the full system calculations (Wang, Liu, et al., 2013). The computational scale for EE-GMFCC is $O(N)$ at the HF and MP2 levels, in contrast to $O(N^3)$ and $O(N^5)$ for traditional HF and MP2 calculation on the entire system (see Fig. 5; Wang, Liu, et al., 2013).

2.4 Quantum Calculation of Electrostatics Inside Enzyme

The EE-GMFCC method has been applied in calculation of electrostatic solvation energy of proteins (Jia et al., 2013), geometry optimization and vibrational spectrum simulation of proteins (Liu, Zhang, & He, 2016), ab initio molecular dynamics (AIMD) simulations for proteins (Liu, Zhu, Wang, He, & Zhang, 2015), and protein–ligand binding affinities (Liu, Wang, Zhang, & He, 2015). Here, we focus on its application in the calculation of electrostatic properties of proteins. The electrostatic potential of the system at any point $\vec{r}$ can be obtained by using the following equation,

$$V(\vec{r}) = \sum_A \frac{Z_A}{\left|\vec{R}_A - \vec{r}\right|} - \int \frac{\rho(\vec{r}\,')d\vec{r}\,'}{\vec{r}\,' - \vec{r}} \tag{5}$$

where $\rho(\vec{r}\,')$ is the electron density distribution, and Z_A is the charge of the nucleus A, located at $\vec{R}_A$. In the framework of the EE-GMFCC method, the electrostatic potential of a protein at any point $\vec{r}$ can be derived from linear combination of the electrostatic potentials of each fragment as follows (Wang, Zhang, & He, 2015):

$$\begin{aligned} V(\vec{r}) = & \sum_{i=2}^{N-1} V(\vec{r})\left(\mathrm{Cap}_{i-1}^{*} A_i \mathrm{Cap}_{i+1}\right) - \sum_{i=2}^{N-2} V(\vec{r})\left(\mathrm{Cap}_{i}^{*}\mathrm{Cap}_{i+1}\right) \\ & + \sum_{\substack{i,j>i+2, \\ |R_i - R_j \le \lambda|}} \left[V(\vec{r})_{ij} - V(\vec{r})_i - V(\vec{r})_j\right]_{\mathrm{QM}} \end{aligned} \tag{6}$$

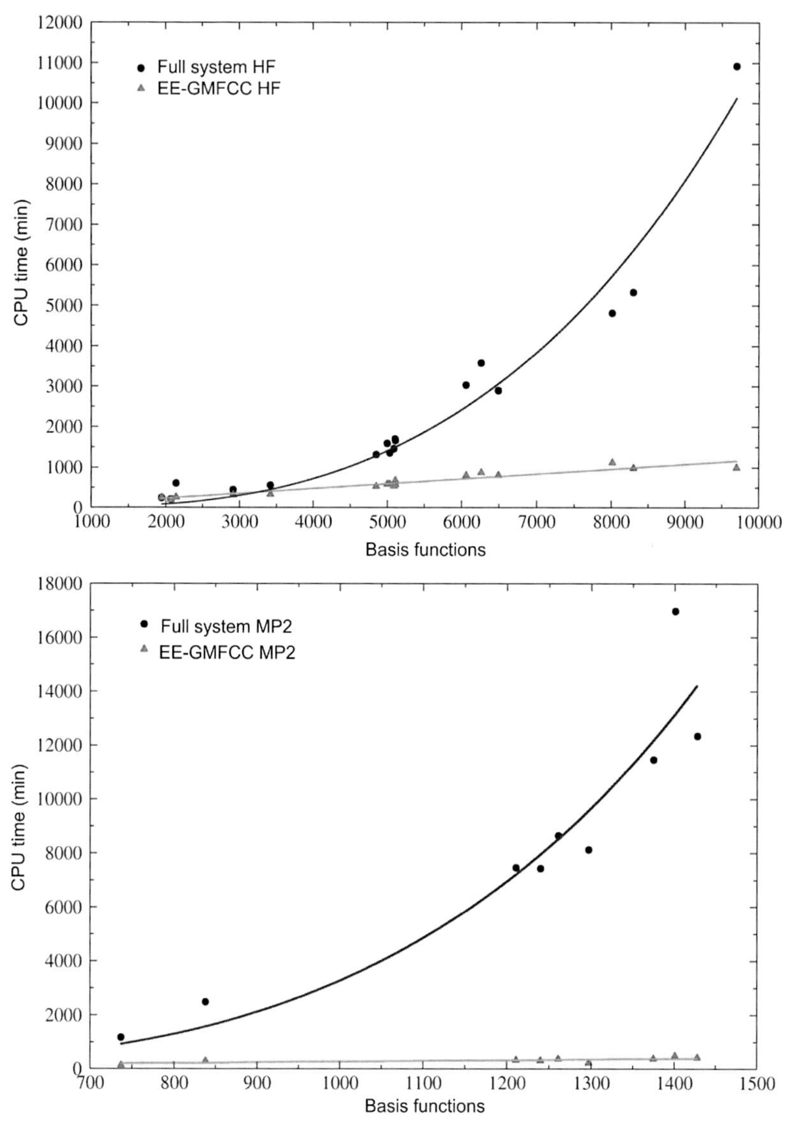

Fig. 5 CPU time for the conventional full system and EE-GMFCC calculations as a function of the number of basis functions at the HF/6-31G* (*top*) and MP2/6-31G* (*bottom*) levels. *Reprinted with permission from Wang, X., Liu, J., Zhang, J. Z. H., & He, X. (2013). Electrostatically embedded generalized molecular fractionation with conjugate caps method for full quantum mechanical calculation of protein energy.* The Journal of Physical Chemistry A, 117*(32), 7149–7161. doi: 10.1021/jp400779t. Copyright 2013 American Chemical Society.*

Similarly, one can obtain the electric field based on the EE-GMFCC method,

$$\vec{F}(\vec{r}) = \sum_{i=2}^{N-1} \vec{F}(\vec{r})(\mathrm{Cap}_{i-1}^{*} A_i \mathrm{Cap}_{i+1}) - \sum_{i=2}^{N-2} \vec{F}(\vec{r})\left(\mathrm{Cap}_{i}^{*} \mathrm{Cap}_{i+1}\right) + \sum_{\substack{i,j>i+2,\\ |R_i - R_j| \leq \lambda}} \left[\vec{F}(\vec{r})_{ij} - \vec{F}(\vec{r})_{i} - \vec{F}(\vec{r})_{j}\right]_{\mathrm{QM}} \tag{7}$$

The subscripts in Eqs. (6) and (7) are the same as those in Eq. (4). The quantum effects such as electronic polarization and charge transfer are included in the calculation of the electrostatic potential and electric field for proteins.

Alternatively, the atomic charges of classical force fields could be used for direct calculations of electrostatic potential at any point using the following expression,

$$\mathrm{V}(\vec{r}) = \frac{1}{4\pi\varepsilon_{\mathrm{eff}}} \sum_{i} \frac{q_i}{|\vec{r} - \vec{r_i}|} \tag{8}$$

where i runs over all atoms of protein, $\vec{r_i}$ and q_i are the position vector and the charge of atom i, respectively. $\varepsilon_{\mathrm{eff}}$ is the effective dielectric constant. The electric field could be obtained straightforwardly using the following equation:

$$F(\vec{r}) = \frac{1}{4\pi\varepsilon_{\mathrm{eff}}} \sum_{i} \frac{q_i}{|\vec{r} - \vec{r_i}|^3} (\vec{r} - \vec{r_i}) \tag{9}$$

However, the electronic polarization and charge transfer effects are missing in the traditional force field. To investigate these effects, the electrostatic potentials and fields at the points around two globular proteins (PDB ID: 1BHI (591 atoms) and PDB: 2KCF (571 atoms)) were calculated with the Amber ff99SB, PPC, and EE-GMFCC approaches. Correlations of the molecular electrostatic potential (MEP) (in a.u.) between full system HF/6-31G* calculation and Amber99SB force field (*black*), PPC (*red* (*gray* in the print version), fitting at the HF/6-31G* level), EE-GMFCC (*blue*; *dark gray* in the print version), respectively, for 1BHI and 2KCF are shown in Fig. 6A and C. One can see from the figure that the calculated electrostatic potential based on the Amber99SB force field has the largest root-mean-square deviation (RMSD) of MEP from full system calculation. The RMSD

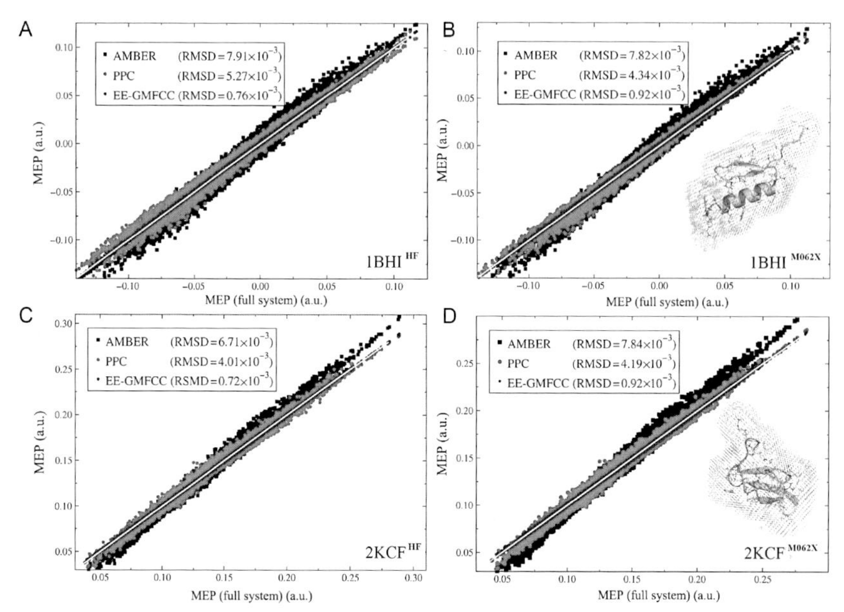

Fig. 6 Correlations of the molecular electrostatic potential (MEP) (in a.u.) between full system QM calculations and Amber99SB force field (*black*), PPC (*red*; *gray* in the print version), EE-GMFCC (*blue*; *dark gray* in the print version), respectively, for 1BHI (*top*) and 2KCF (*bottom*) at the HF/6-31G* (A, C) and M062X/6-31G* (B, D) level. The grid points show where electrostatic potentials and electric fields were calculated. *Reprinted from Wang, X., Zhang, J. Z., & He, X. (2015). Quantum mechanical calculation of electric fields and vibrational Stark shifts at active site of human aldose reductase.* The Journal of Chemical Physics, 143*(18), 184111, with the permission of AIP Publishing.*

of MEP based on PPC is reduced as compared to the Amber99SB result for both proteins. MEPs based on EE-GMFCC are in excellent agreement with the results from full system calculations. As shown in Fig. 6B and D, MEPs based on EE-GMFCC/M062X/6-31G* also give good agreement with the results from full system M062X/6-31G* calculations. There is an order of magnitude improvement in RMSD for EE-GMFCC method as compared to both Amber99SB and PPC results. As for electric field calculation (see Fig. 7), an order of magnitude improvement in RMSD by EE-GMFCC is also observed as compared to those point charge models. These results show that the restraint used in the AMBER and PPC charge fitting protocol, which ensures that each residue is in charge neutral state, would lead to inaccuracies in describing the electrostatic properties for proteins. This underscores the importance of the electronic polarization and charge transfer effect in capturing the electrostatics in proteins.

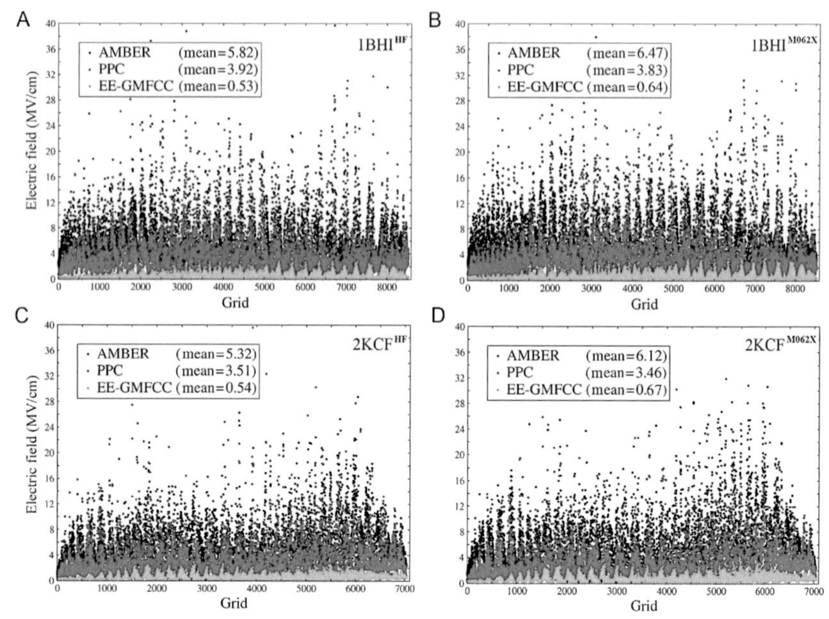

Fig. 7 The *green* (*light gray* in the print version) points indicate the deviation between the electric fields calculated by EE-GMFCC and full system QM results at the HF/6-31G* (A, C) and M062X/6-31G* (B, D) level for system 1BHI (*upper panel*) or 2KCF (*lower panel*). The AMBER results (*black dots*) and PPC results (*red* (*dark gray* in the print version) *dots*) are also plotted in all panels. The mean value donates the mean unsigned error over all the grid points. *Reprinted from Wang, X., Zhang, J. Z., & He, X. (2015). Quantum mechanical calculation of electric fields and vibrational Stark shifts at active site of human aldose reductase.* The Journal of Chemical Physics, 143*(18), 184111, with the permission of AIP Publishing.*

2.5 Electric Field at the Active Site of Enzyme

2.5.1 Effect of Electronic Polarization on Electrostatics in Proteins

A previous experimental study by Suydam et al. (2006) utilized the nitrile-containing inhibitor IDD743 (see Fig. 8A) to probe the change of electric field near the active site of hALR2 brought from point mutations. $\Delta\mu_{\mathrm{PM}}$ for the IDD743 is 0.041D by calibration of the external electric field. By defining the nitrile axis pointing form carbon toward nitrogen, one can obtain $\Delta\mu_{\mathrm{PM}}/hc = -0.69\mathrm{cm}^{-1}/(\mathrm{MV/cm})$. Therefore, the observed vibrational frequency shift of the nitrile group between the mutated hALR2 and the wild-type (WT) protein ($\widetilde{\nu}_{\mathrm{mutant}} - \widetilde{\nu}_{\mathrm{WT}}$) and the change in the electric field along the nitrile axis between the mutant and WT ($F_{\|,\mathrm{mutant}} - F_{\|,\mathrm{WT}}$) has the following relationship,

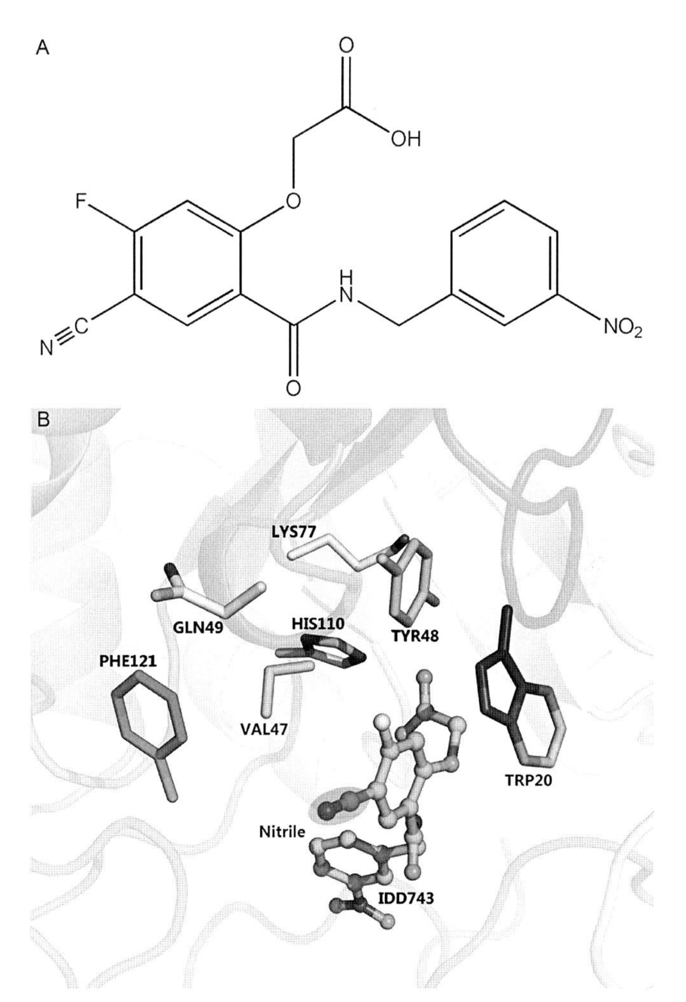

Fig. 8 (A) Chemical structures IDD743. (B) Structural model of the hALR2/NADP$^+$/IDD743 complex. IDD743 and hALR2 are shown in ball-and-stick and cartoon models, respectively. The *pink ellipse* highlights the position of the nitrile probe. The side chains of eight mutated residues are shown as sticks. All the mutated residues are within 11 Å from the midpoint of the nitrile of IDD743, where the electric field was calculated. *Adapted with permission from Wang, X. W., He, X., & Zhang, J. Z. H. (2013). Predicting mutation-induced Stark shifts in the active site of a protein with a polarized force field.* Journal of Physical Chemistry A, 117*(29), 6015–6023. doi:10.1021/Jp312063h. Copyright 2013 American Chemical Society.* (See the color plate.)

$$\Delta\widetilde{\upsilon}_{\mathrm{C\equiv N}} = \left[0.69\mathrm{cm}^{-1}/(\mathrm{MV/cm})\right] \cdot \Delta F_{\parallel} \tag{10}$$

Suydam et al. concluded that MD simulation and ensemble averaging play important roles in accurate prediction of the change in electric field for proteins (Suydam et al., 2006). However, standard force fields do not include electronic polarization arising from the protein environment and inadequate in producing accurate electrostatic potential of proteins (Gascón, Leung, Batista, & Batista, 2006; Menikarachchi & Gascón, 2008; Sandberg, Rudnitskaya, & Gascón, 2012).

MD simulations employing different charge models (namely, PPC and Amber ff99SB) were carried out to calculate the changes of electric fields near the active site of hALR2 resulting from point mutations. In experiment by Suydam et al. (2006), eight hALR2 mutations (Y48F, H110A, W20Y, V47N, F121E, V47D, q49R, and K77M) were selected, and the electric field at the midpoint of the nitrile bond was calculated using the Delphi program and projected onto the direction pointing form the carbon to the nitrogen along the nitrile bond. The change of calculated electric field ($F_{\|,\mathrm{mutant}} - F_{\|,\mathrm{WT}}$) was obtained and compared with experimental results. The PPC and Amber ff99SB for wild-type and eight mutants were used in calculations of electric field based on their respective MD trajectories. The eight mutations of hALR2 were created using the mutagenesis tool in the PyMOL program (DeLano, 2012). Force field parameters from the generalized Amber force field (Wang, Wolf, Caldwell, Kollman, & Case, 2004) and HF/6-31G* RESP charges were used for $NADP^+$ and IDD743 in MD simulations based on Amber ff99SB, while RESP charges for $NADP^+$ and IDD743 were refitted at the B3LYP/6-31G* level in the simulations based on PPC. All ab initio calculations were carried out using the Gaussian09 program (Frisch et al., 2010).

Fig. 9 shows the cross sections of electrostatic potential in the plane of phenyl ring of IDD743 calculated using both PPC and Amber ff99SB based on the same energy-minimized wild-type hALR2/IDD743 complex. Fig. 9A and B shows the calculated electrostatic potential using PPC and Amber ff99SB, respectively, where the same dielectric constant of 1 ($\varepsilon = 1$) was adopted. The interior dielectric constant of 1 in proteins is appropriate in the calculation of electrostatic potentials based on PPC (Schutz & Warshel, 2001), while for the mean-field-like Amber99SB force field, the interior dielectric constant of protein is an uncertain parameter. Usually, $\varepsilon = 4$ is expected to mimic the average protein environment and was adopted by Duan et al. in charge fitting for each type of residue in the Amber ff03 force field (Duan et al., 2003). To examine the ε-dependence of the calculated results based on Amber ff99SB, the calculated electrostatic potentials with $\varepsilon = 2$ and 4 were plotted in Fig. 9C and D.

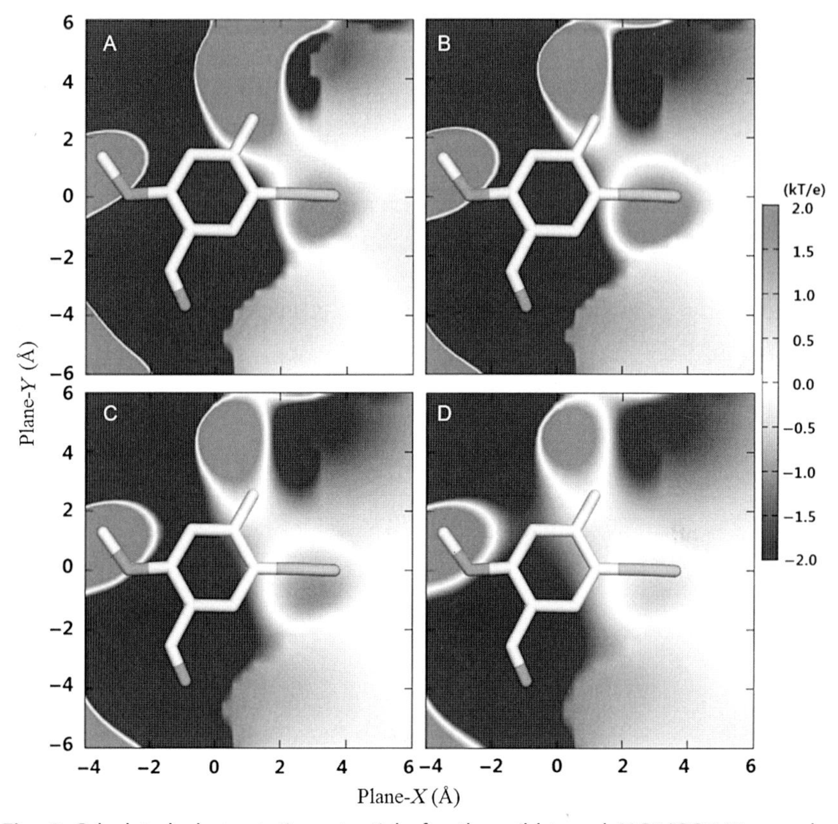

Fig. 9 Calculated electrostatic potentials for the wild-type hALR2/IDD743 complex using the PPC and Amber ff99SB models. Electrostatic potentials are plotted on the plane of the phenyl ring. (A) Calculated electrostatic potential based on the PPC model, with the interior dielectric constant of 1. (B–D) Calculated electrostatic potentials based on the ff99SB charge model, with the interior dielectric constant of (B) 1, (C) 2, and (D) 4. (1 kT/eÅ = 2.57 MV/cm at 298 K, where *k* is the Boltzmann constant, *T* is the temperature, and *e* is the charge of an electron.) *Reprinted with permission from Wang, X. W., He, X., & Zhang, J. Z. H. (2013). Predicting mutation-induced Stark shifts in the active site of a protein with a polarized force field.* Journal of Physical Chemistry A, 117*(29), 6015–6023. doi:10.1021/Jp312063h. Copyright 2013 American Chemical Society.*

There are four red regions (showing positive electrostatic potentials) in all panels. As shown in Fig. 9A and B, the *red* (*gray* in the print version) area around the fluorine atom (marked in aquamarine) clearly increases under PPC as compared to that under Amber ff99SB. The primary contribution to the positive electrostatic potential arises from nearby positively charged residue LYS77, and the primary contribution to the negative electrostatic potential originates from TRP20, followed by TRY48. The deviation of atomic charges for these three residues (namely, LYS77, TRP20, TRY48) between PPC and Amber ff99SB ranges from −0.129 to 0.126 *e*. Therefore, electronic polarization contributes to the electrostatic potential inside proteins.

Two mutants K77M and V47D, which give significant vibrational frequency shifts relative to wild type (−2.2 and −1.6 cm^{-1}, respectively), were selected to investigate the effect on the calculated electric field resulting from electronic polarization. The electric fields at the midpoint of nitrile bond were calculated based on PPC and Amber ff99SB with the dielectric constant of 1. The calculated electric fields as a function of simulation time based on PPC and Amber ff99SB for mutants K77M, V47D, and wild type are shown in Fig. 10. One can see from the figure that the change of

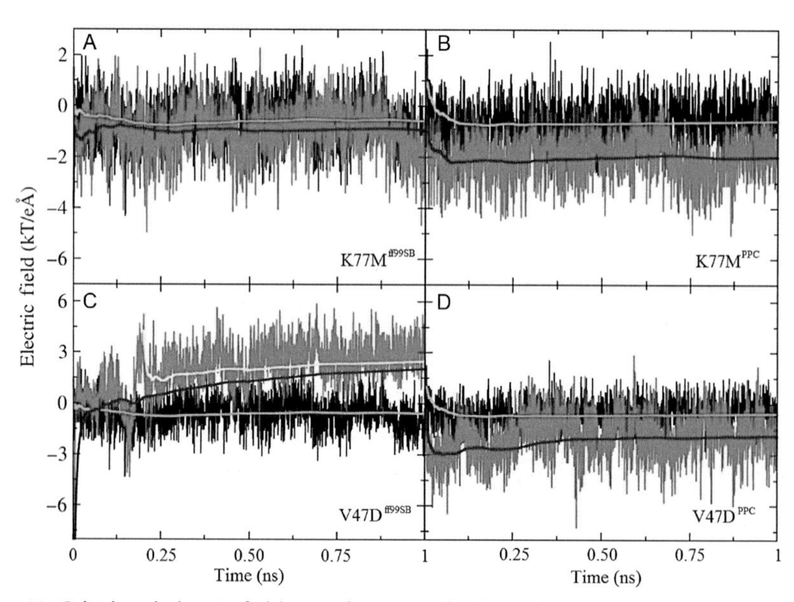

Fig. 10 Calculated electric fields as a function of MD simulation time for the wild type (*black line*) and two mutants (K77M and V47D, *red* (*gray* in the print version) *line*) of hALR2. The *left* (A, C) and *right* (B, D) *columns* represent the calculated results based on the Amber ff99SB and PPC models, respectively. The time-averaged electric fields for the wild type and mutants are shown in *green* (*light gray* in the print version) and *blue* (*dark gray* in the print version) *lines*, respectively. In addition, a *yellow* (*white* in the print version) *line* that shows the time-averaged electric field for the last 800 ps is plotted for the mutant V47D based on the Amber force field. The interior dielectric constant was set to 1 for the electric field calculations. *Reprinted with permission from Wang, X. W., He, X., & Zhang, J. Z. H. (2013). Predicting mutation-induced Stark shifts in the active site of a protein with a polarized force field.* Journal of Physical Chemistry A, 117*(29), 6015–6023. doi:10.1021/Jp312063h. Copyright 2013 American Chemical Society.*

time-averaging electric field for K77M (with reference to wild type) calculated by PPC is larger than that from Amber ff99SB calculation. The converted vibrational Stark shift based on the calculated electric field (ensemble average over 1600 snapshots from the final 800 ps simulation) based on PPC is $-2.5\ cm^{-1}$, which exhibits better agreement with the experimental value of $-2.2\ cm^{-1}$. In contrast, the frequency shift of $-0.7\ cm^{-1}$ predicted by Amber ff99SB is clearly underestimated. As for V47D, the predicted change of electric field using Amber ff99SB is $5.4\ cm^{-1}$, which shows an opposite direction to the electric field obtained from measured vibrational Stark shift. On the contrary, the change of electric field calculated by PPC is $-2.1\ cm^{-1}$, which is in accord with experimental value of $-1.6\ cm^{-1}$.

The coordinates of side-chain atoms of mutated residues MET77 of K77M and ASP47 of V47D were projected onto the plane of the phenyl ring of IDD743, and the evolution of their relative positions during MD simulation was shown in Fig. 11. The initial positions of corresponding atoms for 1 ns MD simulations were enclosed in triangles. As shown in Fig. 10C, the calculated electric field using Amber ff99SB from the initial structure of V47D is lower than −7.0 kT/eÅ and then dramatically increases to positive after 150 ps. Correspondingly, Fig. 11C shows that the side-chain atoms of ASP47 moved away from the initial positions and sampled around a new geometric center, which causes the dramatic change in calculated electric field. However, the relative position of ASP47 did not change much from the initial structure in PPC simulation (see Fig. 11D), and hence the change of time-averaged electric field for V47D from the initial structure using PPC was much smaller than that using Amber ff99SB (see Fig. 10D). In contrast, Fig. 11A and B shows that the evolution of relative positions between the side-chain atoms of MET77 and the phenyl ring of IDD743 is close to their initial positions in MD simulations under both PPC and Amber ff99SB. As a result, the calculated electric field does not show dramatic changes (see Fig. 10A and B). Therefore, we conclude that the electronic polarization effect would not only impact the direct calculation of the electric field for a given protein structure (static effect) but also offer a better conformational sampling for ensemble-averaging calculation (dynamic effect). The dynamic effect also plays an important role in accurate prediction of electric fields in proteins.

The predicted change of electric field resulting from eight mutants relative to the wild type with OPLS-UA, Amber ff99SB, and PPC was presented in Fig. 12. As shown in Fig. 12, the results based on PPC are in better

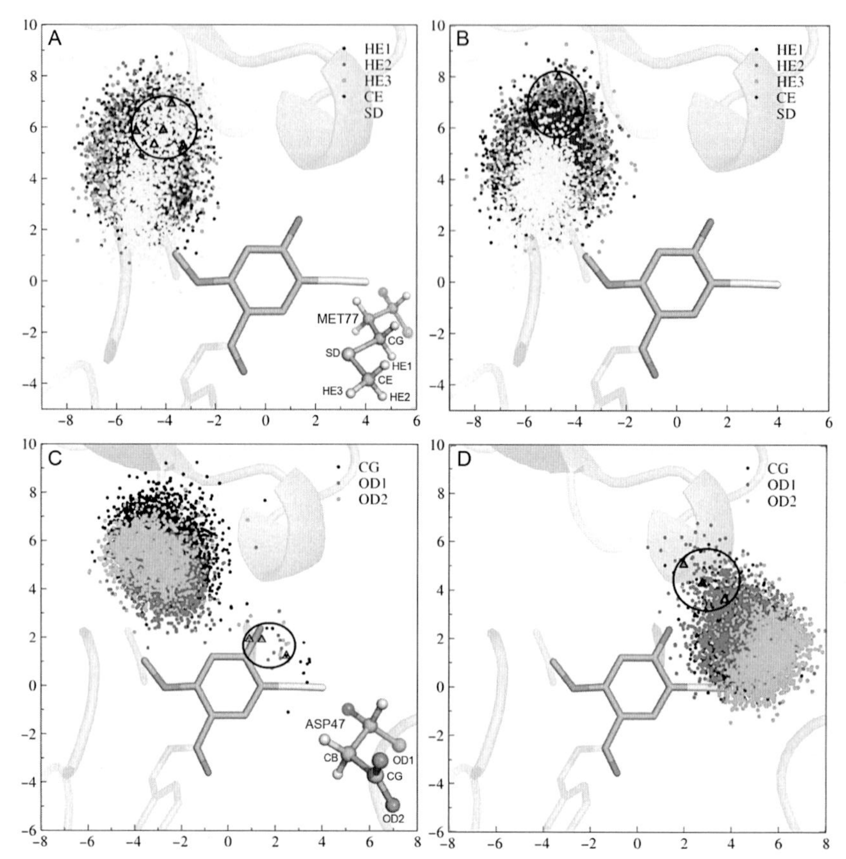

Fig. 11 Distribution of projected coordinates of selected atoms of MET77 (*top*) and ASP47 (*bottom*) onto the plane of the phenyl ring of IDD743 during MD simulations based on the Amber ff99SB (A, C) and PPC (B, D) models. The *triangles* in the *black circles* represent the initial projected positions of the corresponding atoms. *Reprinted with permission from Wang, X. W., He, X., & Zhang, J. Z. H. (2013). Predicting mutation-induced Stark shifts in the active site of a protein with a polarized force field.* Journal of Physical Chemistry A, 117*(29), 6015–6023. doi:10.1021/Jp312063h. Copyright 2013 American Chemical Society.* (See the color plate.)

agreement with the measured vibrational Stark shifts than those calculated by Amber ff99SB, especially for the point mutations (namely, Y48F, V47D, K77M, and V47N), which caused relatively larger electric field changes. Because the electronic polarization effect of the protein and solvent is explicitly incorporated in PPC, we conclude that the electronic polarization effect is crucial to accurate prediction of electric field in proteins.

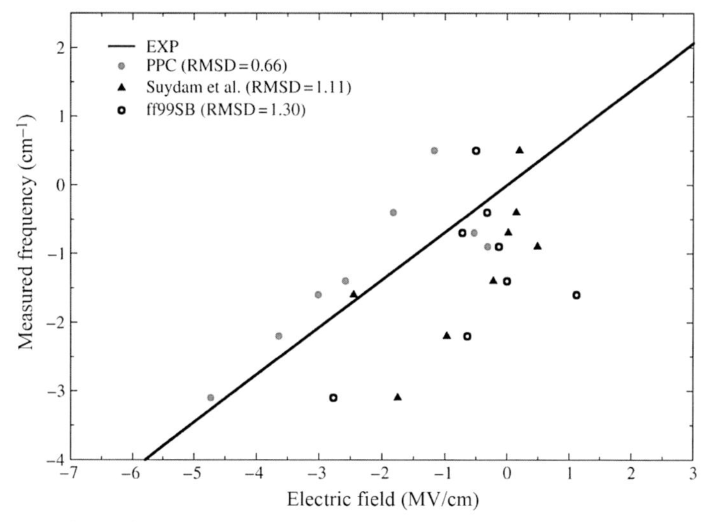

Fig. 12 Correlation between the observed changes in frequency and the calculated changes in electric field brought from eight mutations relative to wild-type hALR2. The *symbols* in order from *top* to *bottom* denote the mutants H110A, Q49R, W20Y, F121E, Y48F, V47D, K77M, and V47N, respectively. The *black line* represents the linear correlation between changes in frequency and changes in electric field observed from experiment (see Eq. 10). *Reprinted with permission from Wang, X. W., He, X., & Zhang, J. Z. H. (2013). Predicting mutation-induced Stark shifts in the active site of a protein with a polarized force field.* Journal of Physical Chemistry A, 117*(29), 6015–6023. doi:10.1021/Jp312063h. Copyright 2013 American Chemical Society.*

2.5.2 Effect of Charge Transfer

For measuring the change of electric field at the active site of hALR2 brought by point mutations, another nitrile-containing inhibitor (denotes the nitrile-containing inhibitor (LIG)) was designed (see Fig. 13A) by Webb and Boxer (2008). By calibration of the external electric field on the nitrile group, they obtained $\Delta\vec{\mu}_{\mathrm{LIG}}/hc = -0.77\ \mathrm{cm}^{-1}/(\mathrm{MV/cm})$ for LIG. LIG is different from IDD743, in which the nitrile group substitutes to a new site. When LIG binds to hALR2, the nitrile would be deeply buried in the binding pocket of the protein which forms an anisotropic environment (see Fig. 13B). Many kinds of hydrophobic or polar residues around the nitrile probe form a complex cavity, where the charge transfer effect among residues in close contact would be significant.

MD simulations and calculations of electric fields were performed for wild-type and four mutated hALR2 structures (namely, T113V,

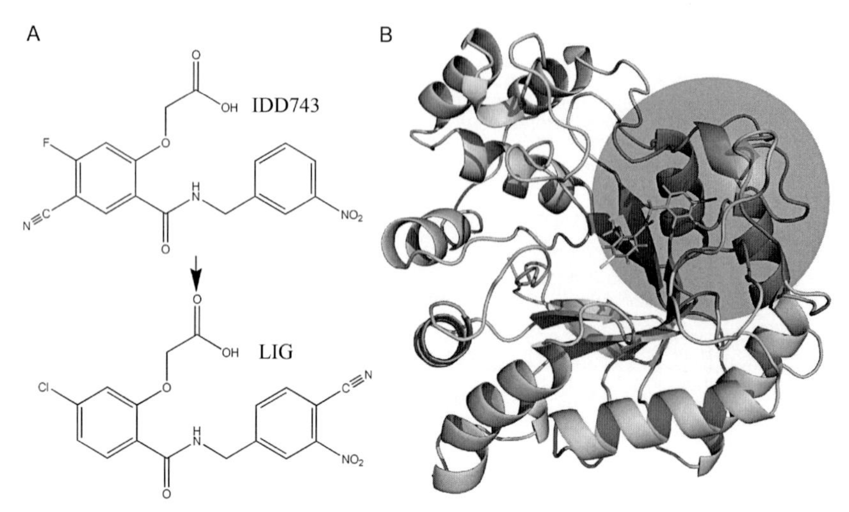

Fig. 13 (A) Chemical structures of IDD743 and the modified ligand (referred as LIG). (B) The complex structure of LIG bound to hALR2. The nitrile is colored in *red* (*gray* in the print version). *Blue* (*gray* in the print version) *sphere* indicates that 10 Å threshold was used in calculation of electric fields at the midpoint of nitrile group for two-body QM calculations (see text for more details).

T113A, T113A_C303N, T113A_C303D$^-$) using PPC, respectively. The electric field at the midpoint of the nitrile was calculated using the following equation:

$$\vec{F}\left(\vec{r}_{\mathrm{C\equiv N}}\right)=\frac{1}{4\pi\varepsilon_{\mathrm{eff}}}\sum_{i}\sum_{j\in i}\frac{q_{ij}}{\left|\vec{r}_{\mathrm{C\equiv N}}-\vec{r}_{ij}\right|^{3}}\left(\vec{r}_{\mathrm{C\equiv N}}-\vec{r}_{ij}\right) \tag{11}$$

where i (represents the residue number) runs over all residues of the protein and the cofactor $NADP^+$, j is the atom number in residue i, q_{ij} and $\vec{r}_{ij}$ are the charge and position vector of atom j in residue i, respectively, $\vec{r}_{\mathrm{C\equiv N}}$ denotes as the position vector at the midpoint of nitrile, and $\varepsilon_{\mathrm{eff}}$ is the effective dielectric constant. The calculated electric field $\vec{F}\left(\vec{r}_{\mathrm{C\equiv N}}\right)$ was projected along the nitrile bond and the positive direction was selected from the carbon to the nitrogen atom. The structure of the binding pocket, where the nitrile is deeply buried, ensures that water molecules and ions could not enter the binding site. Water and ions were not found within 5 Å from the midpoint of nitrile during 3 ns MD simulation for the wild-type hALR2 and its mutants, and thus the influence of solvent effect was not considered for this particular system. $\varepsilon_{\text{eff}} = 2$ was chosen for the calculation of electrostatics of this protein complex (Sandberg et al., 2012; Schutz & Warshel, 2001).

The time-averaged electric field was obtained based on 2000 snapshots generated from the last 2 ns of 3 ns MD simulation. The calculated changes of electric field brought from four mutations using PPC with a dielectric constant of 1 and 2 are both shown in Table 1. The overall RMSD of the calculated electric field using PPC with ε_{eff} of 2 is 9.38 MV/cm with reference to the experimental results, better than 11.95 MV/cm using $\varepsilon_{\text{eff}} = 1$. Nevertheless, the deviation from the experiment is still significant, which indicates that only including electrostatic polarization still does not fill the gap between the experimental measurement and theoretical prediction for electric fields.

The EE-GMFCC method was also utilized in calculation of the change of electric fields caused by mutations of this system. For hALR2 complexed with LIG and cofactor $NADP^+$, the electric field at the

Table 1 Calculated Electric Fields from EE-GMFCC and PPC

Model	EE-GMFCC^S	EE-GMFCC^T	PPC ($\xi_{\text{eff}} = 2.0$)$_{MD}$	PPC ($\xi_{\text{eff}} = 1.0$)	EXP
Wild type	−3.74	−3.50	−4.69	−9.38	—
ΔT113V	0.39	−0.39	6.24	12.48	−4.5
ΔT113A	−7.27	−7.59	3.52	7.04	−8.6
ΔT113A_C303N	−16.57	−17.66	−2.76	−5.52	−10.4
ΔT113A_C303D$^-$	−4.15	−6.28	−1.73	−3.46	−7.3
RMSD	4.29	4.23	9.38	11.95	
Mean unsigned error	3.88	3.35	9.01	10.34	

PPC denotes the ensemble-averaging electric field based on the PPC charges derived from the native structure. EE-GMFCC^S denotes the calculated electric field using the single-snapshot approach. EE-GMFCC^T denotes averaged electric field over three snapshots (see text for more details). The calculated electric field for the wild type (WT) is chosen as the reference and has been subtracted from the calculated electric field for each mutant to give the change of the electric field. Experimental values were obtained from Webb and Boxer (2008). All values are reported in MV/cm.
For comparison, the results for PPC based on the effective dielectric constant of 2.0 and 1.0 are both presented in the table.

midpoint of the nitrile could be calculated using EE-GMFCC as follows (Wang et al., 2015):

$$\begin{aligned}\vec{F}\left(\vec{r}_{\mathrm{C\equiv N}}\right) = & \sum_{i=2}^{N-1} \vec{F}\left(\vec{r}_{\mathrm{C\equiv N}}\right)\left(\mathrm{Cap}_{i-1}^{*} A_i \mathrm{Cap}_{i+1}\right) \\ & - \sum_{i=2}^{N-2} \vec{F}\left(\vec{r}_{\mathrm{C\equiv N}}\right)\left(\mathrm{Cap}_{i}^{*} \mathrm{Cap}_{i+1}\right) \\ & + \sum_{\substack{i,j>i+2, \\ |R_i - R_j| \leq \lambda}} \left[\vec{F}\left(\vec{r}_{\mathrm{C\equiv N}}\right)_{ij} - \vec{F}\left(\vec{r}_{\mathrm{C\equiv N}}\right)_{i} - \vec{F}\left(\vec{r}_{\mathrm{C\equiv N}}\right)_{j}\right]_{\mathrm{QM}} \\ & + \left[\vec{F}_{\mathrm{cofactor}}\left(\vec{r}_{\mathrm{C\equiv N}}\right)\right]_{\mathrm{QM}} \\ & + \sum_{\mathrm{LIG}_i} \sum_{|R_{\mathrm{LIG}_i} - R_j| < \lambda} \left[\vec{F}\left(\vec{r}_{\mathrm{C\equiv N}}\right)_{\mathrm{LIG}_i, j} - \vec{F}\left(\vec{r}_{\mathrm{C\equiv N}}\right)_{\mathrm{LIG}_i} - \vec{F}\left(\vec{r}_{\mathrm{C\equiv N}}\right)_{j}\right]_{\mathrm{QM}}\end{aligned} \tag{12}$$

$\vec{F}\left(\vec{r}_{\mathrm{C\equiv N}}\right)\left(\mathrm{Cap}_{i-1}^{*} A_i \mathrm{Cap}_{i+1}\right)$ and $\vec{F}\left(\vec{r}_{\mathrm{C\equiv N}}\right)\left(\mathrm{Cap}_{i}^{*} \mathrm{Cap}_{i+1}\right)$ denote the calculated electric field at the midpoint of nitrile generated from the *i*th capped fragments and the *i*th conjugate caps, respectively. $\left[\vec{F}\left(\vec{r}_{\mathrm{C\equiv N}}\right)_{ij} - \vec{F}\left(\vec{r}_{\mathrm{C\equiv N}}\right)_{i} - \vec{F}\left(\vec{r}_{\mathrm{C\equiv N}}\right)_{j}\right]_{\mathrm{QM}}$ represents the two-body correction (at the QM level) to the calculated electric field between residues *i* and *j*, whose closest distance is less than a predefined distance threshold λ ($\lambda = 4$ Å was chosen). $\left[\vec{F}_{\mathrm{cofactor}}\left(\vec{r}_{\mathrm{C\equiv N}}\right)\right]_{\mathrm{QM}}$ represents the calculated electric field of the cofactor NADP^+. The contribution to the electric field by the nitrile-containing ligand itself is supposed to be invariant in both the wild type and the mutants. Hence, in Eq. (12), LIG itself does not make direct contribution to the calculated electric field when only the change of the electric field is considered here. The last term in Eq. (12), which is a two-body contribution, is employed to cover the quantum effect between ligands (cofactor NADP^+ and nitrile inhibitor) and their neighboring residues in the complex. The contribution to the change of the electric field at the midpoint of nitrile from distant two-body terms (two nonneighboring residues that are both beyond 10 Å from LIG in the protein) is very small. A cutoff distance of 10 Å (the *blue* (*gray* in the print version) sphere in Fig. 13B) was used to reduce the computational cost for the two-body correction in Eq. (12). The contributions to the calculated electric field from all atoms of the protein and $NADP^+$ were taken into account at the QM level, while the solvent effect was not included.

The computational cost of QM calculation on multiple configurations of a large protein system is still expensive even with the fragmentation approach. The single-snapshot approach was used to calculate the electric field with the EE-GMFCC method. In this approach, the time-averaged electric field was first calculated using the PPC model, and a snapshot that gave the electric field closest to the time-averaged value was selected for the EE-GMFCC calculation and the results based on this single snapshot is denoted as EE-$GMFCC^S$ in Table 1. It is worth noting that conformational sampling at the QM level is not included in this protocol in the calculation of electric field but is implicitly included to some extent at molecular mechanical level by matching the computed average value. To better account for conformational sampling, two more snapshots that give the electric fields which were next closest to the time-averaged values were selected for calculating electric field and the results over the three structures are denoted as EE-$GMFCC^T$ in Table 1. One can see from the table that EE-GMFCC gives much better agreement with experiment than PPC with RMSD of 4.29 MV/cm. MUE of EE-GMFCC is reduced to 3.88 MV/cm from 9.01 MV/cm calculated by PPC. Furthermore, the RMSD and MUE of EE-$GMFCC^T$ (average result over three snapshots) are 4.23 and 3.35 MV/cm, respectively, which are in the best agreement with experimental observation. These results demonstrate that for accurate description of the intra-protein's electrostatics in proteins, QM calculation including both the electronic polarization and inter-residue charge transfer effect is needed.

ACKNOWLEDGMENTS

This work was supported by the National Natural Science Foundation of China (Grant nos. 21433004, 21303057, and 11547164) and Shanghai Putuo District (Grant no. 2014-A-02). X.H. is also supported by the Specialized Research Fund for the Doctoral Program of Higher Education (Grant no. 20130076120019) and the Fundamental Research Funds for the Central Universities. We thank the Supercomputer Center of East China Normal University for providing us computational time.

REFERENCES

Bugg, T. D. (2001). The development of mechanistic enzymology in the 20th century. *Natural Product Reports*, *18*(5), 465–493.

Case, D. A., Cheatham, T. E., III, Darden, T., Gohlke, H., Luo, R., Merz, K. M., Jr., ... Woods, R. J. (2005). The Amber biomolecular simulation programs. *Journal of Computational Chemistry*, *26*(16), 1668–1688. http://dx.doi.org/10.1002/jcc.20290.

Choi, J.-H., & Cho, M. (2011). Vibrational solvatochromism and electrochromism of infrared probe molecules containing C≡O, C≡N, C=O, or C−F vibrational chromophore. *The Journal of Chemical Physics*, *134*(15), 154513.

Choi, J. H., Oh, K. I., Lee, H., Lee, C., & Cho, M. (2008). Nitrile and thiocyanate IR probes: Quantum chemistry calculation studies and multivariate least-square fitting analysis. *The Journal of Chemical Physics*, *128*(13), 134506. http://dx.doi.org/10.1063/1.2844787.

Chung, L. W., Sameera, W. M., Ramozzi, R., Page, A. J., Hatanaka, M., Petrova, G. P., … Morokuma, K. (2015). The ONIOM method and its applications. *Chemical Reviews*, *115*(12), 5678–5796. http://dx.doi.org/10.1021/cr5004419.

Collins, M. A., & Bettens, R. P. (2015). Energy-based molecular fragmentation methods. *Chemical Reviews*, *115*(12), 5607–5642. http://dx.doi.org/10.1021/cr500455b.

Cornell, W. D., Cieplak, P., Bayly, C. I., Gould, I. R., Merz, K. M., Ferguson, D. M., … Kollman, P. A. (1995). A second generation force field for the simulation of proteins, nucleic acids, and organic molecules. *Journal of the American Chemical Society*, *117*(19), 5179–5197.

DeLano, W. L. (2012). *The PyMOL molecular graphics system; version 1.5.0.1*. San Carlos, CA: DeLano Scientific.

Duan, L. L., Mei, Y., Zhang, Q. G., & Zhang, J. Z. (2009). Intra-protein hydrogen bonding is dynamically stabilized by electronic polarization. *The Journal of Chemical Physics*, *130*(11), 115102.

Duan, Y., Wu, C., Chowdhury, S., Lee, M. C., Xiong, G., Zhang, W., … Lee, T. (2003). A point-charge force field for molecular mechanics simulations of proteins based on condensed-phase quantum mechanical calculations. *Journal of Computational Chemistry*, *24*(16), 1999–2012.

Fried, S. D., Bagchi, S., & Boxer, S. G. (2013). Measuring electrostatic fields in both hydrogen-bonding and non-hydrogen-bonding environments using carbonyl vibrational probes. *Journal of the American Chemical Society*, *135*(30), 11181–11192. http://dx.doi.org/10.1021/Ja403917z.

Fried, S. D., Bagchi, S., & Boxer, S. G. (2014). Extreme electric fields power catalysis in the active site of ketosteroid isomerase. *Science*, *346*(6216), 1510–1514.

Frisch, M. J., Trucks, G. W., Schlegel, H. B., Scuseria, G. E., Robb, M. A., Cheeseman, J. R., … Pople, J. A. (2010). *Gaussian 09, Revision B.01*. Wallingford CT: Gaussian, Inc.

Gascón, J. A., Leung, S. S., Batista, E. R., & Batista, V. S. (2006). A self-consistent space-domain decomposition method for QM/MM computations of protein electrostatic potentials. *Journal of Chemical Theory and Computation*, *2*(1), 175–186.

Goedecker, S. (1999). Linear scaling electronic structure methods. *Reviews of Modern Physics*, *71*(4), 1085.

Gordon, M. S., Fedorov, D. G., Pruitt, S. R., & Slipchenko, L. V. (2012). Fragmentation methods: A route to accurate calculations on large systems. *Chemical Reviews*, *112*(1), 632–672.

He, X., & Merz, K. M., Jr. (2010). Divide and conquer Hartree–Fock calculations on proteins. *Journal of Chemical Theory and Computation*, *6*(2), 405–411.

He, X., Zhu, T., Wang, X., Liu, J., & Zhang, J. Z. H. (2014). Fragment quantum mechanical calculation of proteins and its applications. *Accounts of Chemical Research*, *47*(9), 2748–2757. http://dx.doi.org/10.1021/ar500077t.

Ji, C., & Mei, Y. (2014). Some practical approaches to treating electrostatic polarization of proteins. *Accounts of Chemical Research*, *47*(9), 2795–2803.

Ji, C. G., Mei, Y., & Zhang, J. Z. H. (2008). Developing polarized protein-specific charges for protein dynamics: MD free energy calculation of pK(a) shifts for Asp(26)/Asp(20) in thioredoxin. *Biophysical Journal*, *95*(3), 1080–1088. http://dx.doi.org/10.1529/biophysj.108.131110.

Ji, C. G., & Zhang, J. Z. H. (2009). Electronic polarization is important in stabilizing the native structures of proteins. *Journal of Physical Chemistry B*, *113*(49), 16059–16064. http://dx.doi.org/10.1021/Jp907999e.

Ji, C. G., & Zhang, J. Z. (2012). Effect of interprotein polarization on protein–protein binding energy. *Journal of Computational Chemistry*, *33*(16), 1416–1420.

Jia, X., Wang, X., Liu, J., Zhang, J. Z., Mei, Y., & He, X. (2013). An improved fragment-based quantum mechanical method for calculation of electrostatic solvation energy of proteins. *The Journal of Chemical Physics*, *139*(21), 214104.

Li, Y., Zhang, J. Z., & Mei, Y. (2014). Molecular dynamics simulation of protein crystal with polarized protein-specific force field. *The Journal of Physical Chemistry B*, *118*(43), 12326–12335.

Liu, J., Wang, X., Zhang, J. Z., & He, X. (2015). Calculation of protein–ligand binding affinities based on a fragment quantum mechanical method. *RSC Advances*, *5*(129), 107020–107030.

Liu, J., Zhang, J. Z., & He, X. (2016). Fragment quantum chemical approach to geometry optimization and vibrational spectrum calculation of proteins. *Physical Chemistry Chemical Physics*, *18*(3), 1864–1875. http://dx.doi.org/10.1039/c5cp05693d.

Liu, J., Zhu, T., Wang, X., He, X., & Zhang, J. Z. (2015). Quantum fragment based ab initio molecular dynamics for proteins. *Journal of Chemical Theory and Computation*, *11*(12), 5897–5905.

Lu, Y., Mei, Y., Zhang, J. Z., & Zhang, D. (2010). Communications: Electron polarization critically stabilizes the Mg2+ complex in the catalytic core domain of HIV-1 integrase. *The Journal of Chemical Physics*, *132*(13), 131101.

Mei, Y., Li, Y. L., Zeng, J., & Zhang, J. Z. (2012). Electrostatic polarization is critical for the strong binding in streptavidin-biotin system. *Journal of Computational Chemistry*, *33*(15), 1374–1382.

Menikarachchi, L. C., & Gascón, J. A. (2008). Optimization of cutting schemes for the evaluation of molecular electrostatic potentials in proteins via moving-domain QM/MM. *Journal of Molecular Modeling*, *14*(6), 1–9.

Park, E. S., Andrews, S. S., Hu, R. B., & Boxer, S. G. (1999). Vibrational stark spectroscopy in proteins: A probe and calibration for electrostatic fields. *Journal of Physical Chemistry B*, *103*(45), 9813–9817. http://dx.doi.org/10.1021/Jp992329g.

Raghavachari, K., & Saha, A. (2015). Accurate composite and fragment-based quantum chemical models for large molecules. *Chemical Reviews*, *115*(12), 5643–5677. http://dx.doi.org/10.1021/cr500606e.

Rocchia, W., Sridharan, S., Nicholls, A., Alexov, E., Chiabrera, A., & Honig, B. (2002). Rapid grid-based construction of the molecular surface and the use of induced surface charge to calculate reaction field energies: Applications to the molecular systems and geometric objects. *Journal of Computational Chemistry*, *23*(1), 128–137. http://dx.doi.org/10.1002/Jcc.1161.

Saggu, M., Levinson, N. M., & Boxer, S. G. (2011). Direct measurements of electric fields in weak OH-π hydrogen bonds. *Journal of the American Chemical Society*, *133*(43), 17414–17419. http://dx.doi.org/10.1021/Ja2069592.

Saggu, M., Levinson, N. M., & Boxer, S. G. (2012). Experimental quantification of electrostatics in X-H center...π hydrogen bonds. *Journal of the American Chemical Society*, *134*(46), 18986–18997. http://dx.doi.org/10.1021/Ja305575t.

Sandberg, D. J., Rudnitskaya, A. N., & Gascón, J. A. (2012). QM/MM prediction of the stark shift in the active site of a protein. *Journal of Chemical Theory and Computation*, *8*(8), 2817–2823. http://dx.doi.org/10.1021/Ct300409t.

Schutz, C. N., & Warshel, A. (2001). What are the dielectric "constants" of proteins and how to validate electrostatic models? *Proteins: Structure, Function, and Bioinformatics*, *44*(4), 400–417.

Sowlati-Hashjin, S., & Matta, C. (2013). The chemical bond in external electric fields: Energies, geometries, and vibrational Stark shifts of diatomic molecules. *The Journal of Chemical Physics*, *139*(14), 144101. http://dx.doi.org/10.1063/1.4820487.

Steffen, M. A., Lao, K., & Boxer, S. G. (1994). Dielectric asymmetry in the photosynthetic reaction center. *Science*, *264*(5160), 810–816. http://dx.doi.org/10.1126/science.264.5160.810.

Strain, M. C., Scuseria, G. E., & Frisch, M. J. (1996). Achieving linear scaling for the electronic quantum Coulomb problem. *Science*, *271*(5245), 51.

Suydam, I. T., Snow, C. D., Pande, V. S., & Boxer, S. G. (2006). Electric fields at the active site of an enzyme: Direct comparison of experiment with theory. *Science*, *313*(5784), 200–204.

Tong, Y., Ji, C. G., Mei, Y., & Zhang, J. Z. H. (2009). Simulation of NMR data reveals that proteins' local structures are stabilized by electronic polarization. *Journal of the American Chemical Society*, *131*(24), 8636–8641. http://dx.doi.org/10.1021/Ja901650r.

Tong, Y., Mei, Y., Li, Y. L., Ji, C. G., & Zhang, J. Z. (2010). Electrostatic polarization makes a substantial contribution to the free energy of avidin–biotin binding. *Journal of the American Chemical Society*, *132*(14), 5137–5142.

Wang, X. W., He, X., & Zhang, J. Z. H. (2013). Predicting mutation-induced stark shifts in the active site of a protein with a polarized force field. *Journal of Physical Chemistry A*, *117*(29), 6015–6023. http://dx.doi.org/10.1021/Jp312063h.

Wang, X., Liu, J., Zhang, J. Z. H., & He, X. (2013). Electrostatically embedded generalized molecular fractionation with conjugate caps method for full quantum mechanical calculation of protein energy. *The Journal of Physical Chemistry A*, *117*(32), 7149–7161. http://dx.doi.org/10.1021/jp400779t.

Wang, J. M., Wolf, R. M., Caldwell, J. W., Kollman, P. A., & Case, D. A. (2004). Development and testing of a general amber force field. *Journal of Computational Chemistry*, *25*(9), 1157–1174. http://dx.doi.org/10.1002/Jcc.20035.

Wang, X., Zhang, J. Z., & He, X. (2015). Quantum mechanical calculation of electric fields and vibrational Stark shifts at active site of human aldose reductase. *The Journal of Chemical Physics*, *143*(18), 184111.

Warshel, A., Sharma, P. K., Kato, M., Xiang, Y., Liu, H., & Olsson, M. H. (2006). Electrostatic basis for enzyme catalysis. *Chemical Reviews*, *106*(8), 3210–3235. http://dx.doi.org/10.1021/cr0503106.

Webb, L. J., & Boxer, S. G. (2008). Electrostatic fields near the active site of human aldose reductase: 1. New inhibitors and vibrational stark effect measurements. *Biochemistry*, *47*(6), 1588–1598. http://dx.doi.org/10.1021/Bi701708u.

Xiang, Y., Duan, L., & Zhang, J. Z. H. (2011). Protein's electronic polarization contributes significantly to its catalytic function. *The Journal of Chemical Physics*, *134*(20), 205101. http://dx.doi.org/10.1063/1.3592987.

Xu, L., Cohen, A. E., & Boxer, S. G. (2011). Electrostatic fields near the active site of human aldose reductase: 2. New inhibitors and complications caused by hydrogen bonds. *Biochemistry*, *50*(39), 8311–8322. http://dx.doi.org/10.1021/bi200930f.

Yang, W. (1991). Direct calculation of electron density in density-functional theory. *Physical Review Letters*, *66*(11), 1438.

Zhang, D. W., & Zhang, J. Z. H. (2003). Molecular fractionation with conjugate caps for full quantum mechanical calculation of protein-molecule interaction energy. *The Journal of Chemical Physics*, *119*(7), 3599–3605. http://dx.doi.org/10.1063/1.1591727.

CHAPTER FOUR

Molecular Dynamics Studies of Proton Transport in Hydrogenase and Hydrogenase Mimics

B. Ginovska[1], S. Raugei, W.J. Shaw[1]
Pacific Northwest National Laboratory, Richland, WA, United States
[1]Corresponding authors: e-mail address: bojana.ginvoska@pnnl.gov; wendy.shaw@pnnl.gov

Contents

Abstract

There is extensive interest in hydrogenases based on their ability to rapidly and efficiently interconvert H_2 with protons and electrons, and their (typically) superior function relative to molecular mimics. Understanding the function of enzymes is one approach to implementing design features to make better catalysts and is an approach we have implemented in our work. In this review, we will discuss our efforts to develop design principles from enzymes, with specific focus on proton transport. We will also present computational studies of the mimics we have investigated with similar methodologies. We will discuss the mechanisms used by small scaffolds on molecular mimics which in many cases are surprisingly similar to those used by nature, while in other cases, computational analysis allowed us to reveal an unexpected role. Computational methods provide one of the best ways, and in some cases, the only

Methods in Enzymology, Volume 578
ISSN 0076-6879
http://dx.doi.org/10.1016/bs.mie.2016.05.044

way, to gain insight into the mechanistic details of enzymes. In this review, we illustrate the general computational strategy we used to study the proton pathway of [FeFe]-hydrogenase, and the similar strategy to investigate small molecules. We present the main results we obtained and how our computational work stimulated or worked in concert with experimental investigations. We also focus on estimation of errors and convergence of properties in the simulations. These studies demonstrate the powerful results that can be obtained by the close pairing of experimental and theoretical approaches.

1. HYDROGENASES

Hydrogenases are used in nature to store and produce energy (Fontecilla-Camps, Volbeda, Cavazza, & Nicolet, 2007; Vincent, Parkin, & Armstrong, 2007). Depending upon the organism and the environment, some hydrogenases are biased to produce hydrogen and some hydrogenases favor reducing hydrogen, although in most cases they operate reversibly. There are two types of hydrogenases that produce or oxidize hydrogen, the [FeFe]-hydrogenases, so named due to the bimetallic active site containing two iron atoms, and the [NiFe]-hydrogenases, which have a bimetallic active site containing an iron and a nickel (Lubitz, Ogata, Rudiger, & Reijerse, 2014). There have been many studies on molecular mimics of hydrogenase that have focused on the active site and have provided important insight (Caserta, Roy, Atta, Artero, & Fontecave, 2015; Dubois, 2014; Gloaguen & Rauchfuss, 2009; Simmons, Berggren, Bacchi, Fontecave, & Artero, 2014).

While the active site is the location where the chemical reaction occurs, many features of the protein scaffold contribute significantly to the activity of the enzyme (Dementin et al., 2006; Hamdan et al., 2012; Ragsdale, 2006). The focus of our studies has been on understanding the role of the outer coordination sphere of the enzyme and on molecular complexes (Dutta, DuBois, Roberts, & Shaw, 2014; Dutta et al., 2013; Ginovska-Pangovska, Dutta, Reback, Linehan, & Shaw, 2014; Jain, Buchko, et al., 2012; Jain et al., 2011; Jain, Reback, et al., 2012; Lense, Dutta, Roberts, & Shaw, 2014; Lense et al., 2012; Reback et al., 2014, 2013; Shaw, 2012). The two areas of focus we will cover in this review include proton transport and understanding the contribution of the environment around the active site (Darmon et al., 2015; Ginovska-Pangovska, Dutta, et al., 2014; Jain et al., 2011; Jain, Reback, et al., 2012; Reback et al., 2014, 2013). We have approached understanding the role of these two contributions of the protein scaffold by evaluating both the enzyme and molecular mimics with an outer coordination sphere. Because the enzymes are so complex, lessons are sometimes more easily understood in the molecular complexes, resulting in the knowledge from molecular complexes being used to understand the enzymes, as well as the other way around. This dual approach of studying both enzymes and molecular complexes, therefore, becomes a powerful strategy in developing better catalysts.

2. PROTON TRANSPORT

One of the needed features to achieve the interconversion of H_2 with protons and electrons is an efficient proton pathway. Ultimately designing proton pathways into synthetic systems is needed for effective proton transport; consequently, extracting design principles from enzymes with well-functioning proton pathways would provide a basis upon which to begin building synthetic pathways. Proton transport is an essential aspect in many parts of biology (Wraight, 2006). It serves to balance charge across membranes, adjust pH, and, in some cases, deliver protons for reactivity for energy storage or release. Despite their large size relative to electrons, protons can be transported across great lengths; however, this requires a series of proton transfer residues and/or water molecules to be positioned to facilitate that transport. While some proton pathways have been studied extensively (Wraight, 2006) for instance, bacteriorhodopsin (Wickstrong, Dods, Royant, & Neutze, 2015), cyctochrome *c* oxidase (Yoshikawa, Muramoto, & Shinzawa-Itoh, 2011), and bacterial reaction centers (Okamura, Paddock, Graige, & Feher, 2000), clear design principles for proton channels that reveal the number of residues needed, the distance between residues, the number of pathways in a given molecule, or the relative pK_a values of the residues in the pathway are not well established. These basic design principles are of interest to energy transformation and storage devices such as fuel cells and electrocatalysts for multiproton/multielectron reactions (Diat & Gebel, 2008).

Both the [NiFe]-hydrogenases and the [FeFe]-hydrogenases in the hydrogenase family of enzymes have a proton transport channel to facilitate substrate and product delivery, and each has been studied to varying degrees both experimentally and computationally (Cornish, Gartner, Yang, Peters, & Hegg, 2011; Fontecilla-Camps et al., 2007; Galvan, Volbeda, Fontecilla-Camps, & Field, 2008; Lubitz et al., 2014; Nicolet, Piras, Legrand, Hatchikian, & Fontecilla-Camps, 1999; Sumner & Voth,

2012; Teixeira, Soares, & Baptista, 2008; Tran et al., 2011; Vincent et al., 2007). Understanding the contribution of proton transport supported by the *Clostridium pastuerianum* [FeFe]-hydrogenase scaffold has been a focus of our work to enable the development of proton transport pathways for molecular catalysts (Dutta et al., 2014, 2013; Ginovska-Pangovska, Dutta, et al., 2014; Ginovska-Pangovska, Ho, et al., 2014; Lense et al., 2014). Experimental studies of the proton pathway have been conducted for [FeFe]-hydrogenase by Hegg's group (Cornish et al., 2011) and revealed a pathway using four residues, Cys299, Glu279, Ser319, and Glu282, and a single water molecule. The Cys299 was also confirmed by Knorzer et al. (2012). The conclusion that these residues were critical for the proton pathway was based on the nearly complete loss of activity in both directions when any of these residues was mutated. Mutations of residues which contributed to other proposed channels either did not influence catalysis (ie, Ser298) (Cornish et al., 2011) or were found to be important in stabilization of the active site lacking a direct role on the catalytic mechanism (ie, Lys358; Knorzer et al., 2012), providing further evidence supporting the proposed pathway, *Pathway 1* in Fig. 1.

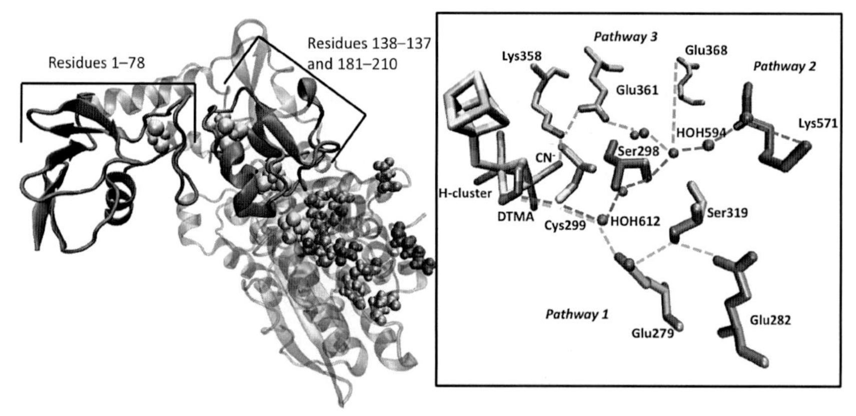

Fig. 1 Proton pathways in *Clostridium pastuerianum* [FeFe]-hydrogenase proposed based on experimental and theoretical studies. *Pathway 1* (*cyan*) has been experimentally demonstrated to almost fully block catalytic activity when any one of the four residues is mutated to a nonhydrogen bonding residue. *Pathway 2* (*blue*) and *Pathway 3* (*orange*), which are largely water-based pathways, are additional proposed pathways based on sequence analysis. *Reprinted from Ginovska-Pangovska, B., Ho, M.-H., Linehan, J. C., Cheng, Y., Dupuis, M., Raugei, S., & Shaw, W. J. (2014). Molecular dynamics study of the proposed proton transport pathways in [FeFe]-hydrogenase.* Biochimica et Biophysica Acta, 1837, *131–138 with permission from Elsevier.* (See the color plate.)

Computational studies of [FeFe]-hydrogenase, mostly based on molecular dynamics simulations, have provided support for the above channel (Ginovska-Pangovska, Ho, et al., 2014; Hong, Cornish, Hegg, & Pachter, 2011; Long, King, & Chang, 2014; Sode & Voth, 2014). The groups of Pachter and Chang both investigated proton movement, and within the context of the assumptions that were made, both found proton transport to the active site to be a highly favorable process (Hong et al., 2011; Long et al., 2014). More recently, Voth and colleagues suggested a different pathway for the second proton, whereby a structural change is the result of reduction of the active site, opening a water channel which has favorable energetics to transfer the second proton (McCullagh & Voth, 2013; Sode & Voth, 2014). [NiFe]-hydrogenases have been extensively studied experimentally. It is suggested that a glutamic residue near the active site plays an important role in proton transport as its mutation results in complete loss in activity (Dementin et al., 2004). The groups of Baptista and Voth have performed a number of computational studies on [NiFe]-hydrogenase and suggested a number of plausible pathways (Sumner & Voth, 2011; Teixeira et al., 2008). However, their experimental validation is yet to be provided.

3. COMPUTATIONAL STRATEGY

In any computational study, certain assumptions have to be made to make the problem tractable, and certain approaches were used to achieve that goal. In particular, in the study of hydrogenases and hydrogenase mimics, several things have been considered to address the challenge of proton movement: hydrogen bonding, pK_a, proton movement, and protons coupling with electrons. Each of these can be studied explicitly or simplified to provide insight into these challenging systems. An example is the approach proposed by Voth's group based on the multistate empirical valence bond (MSEVB) method (Voth, 2006), which allows for bond making and bond breaking to occur, enabling more realistic simulations (Voth, 2006). Each method that has been implemented has advantages and disadvantages, allowing something unique to be learned from each type of approach. Our studies, like others, have been based on MD simulations, exploiting empirical potential energy functions (Ponder & Case, 2003; Price & Brooks, 2002). MD simulations are a truly powerful tool to sample canonical equilibrium configurations, and we will show how they provide unique insight into our understanding of the process of proton delivery to the active site of [FeFe]-hydrogenase and hydrogenase mimics. However,

MD simulations are affected by a variety of errors, and results must be analyzed taking them into account. This is a crucial step for the correct interpretation of the data and often is not sufficiently discussed in the literature. Apart from obvious issues associated with the level of theory used to describe the interatomic interactions, there is an error associated with the finite simulation time. Indeed, to have a meaningful statistical sampling (averaging) of the properties of interest, the simulation time must be far longer than the slowest motions modulating these properties. This time is often not known a priori, and in the case of slow-varying motions practically impossible to achieve with direct MD simulation techniques. For this reason, it is important to perform a careful analysis of the statistical error associated with the computed quantities. Utilizing methodology such as replica exchange molecular dynamics (REMD) can become important to accurately sample the configuration space (Cheng, Cui, Hornak, & Simmerling, 2005; Nymeyer, Gnanakaran, & García, 2004), an approach we used to provide insight into small peptide-based outer coordination spheres (Cheng et al., 2005; Nymeyer et al., 2004; Reback et al., 2014, 2013). With the REMD methodology, a number of parallel simulations are being executed simultaneously at different temperatures (replicas), and at regular time intervals, the structures of the geometries are exchanged, so different initial configurations are introduced for each simulation. This approach rapidly explores the configuration space that is accessible to the system.

Proton movement between residues requires bond breaking and forming, adding new degrees of complexity to the computational study. This process can be studied using either ab initio or semiempirical approaches. Even though semiempirical approaches are computationally less demanding, generally they are also less accurate. When computational resources are available, ab initio approaches are the best choice (Blomberg, Borowski, Himo, Liao, & Siegbahn, 2014; Gao et al., 2006). However, in the case of processes with activation barriers greater than the thermal energy $K_B T$ of the system, direct MD simulations are not applicable as the barrier crossing will be a rare event on the time scale of the simulation (Free Energy Calculations Theory and Applications in Chemistry and Biology, 2007). Overcoming this issue requires use of advanced simulation techniques, such as metadynamics (Laio & Parrinello, 2002), as described later. Furthermore, because of the large system sizes, the computational cost can be reduced even more by employing a hybrid quantum mechanics/molecular mechanics description of the system, where only a subset of atoms are treated at a quantum level, and the rest of the environment is represented by a classical force field.

Finally, detailed quantum calculations (density functional theory, DFT) can be used for sufficiently small systems to explore each step in a given reaction mechanism, allowing for refinement of reaction bottlenecks. Below we provide an example of such calculation for the hydrogenase mimics.

4. PROTON TRANSPORT IN HYDROGENASE

4.1 Classical MD Simulations

To understand proton movement in *C. pastuerianum* [FeFe]-hydrogenase, we started with the foundational hypothesis that a well-structured hydrogen bond network is required for efficient proton movement (Agmon, 1995; Bonin, Costentin, Robert, Saveant, & Tard, 2012; Hammes-Schiffer, 2011; Knight & Voth, 2012; Macarie & Ilia, 2014; Wraight, 2006; Yan & Xie, 2012). Therefore, with this hypothesis, rather than starting with proton movement itself, which requires simplifications in either the size of the system or how that system is described to minimize the computational demand of the calculations, we chose to evaluate hydrogen bonding patterns (Ginovska-Pangovska, Ho, et al., 2014). Given the experimental studies suggesting the pathway for proton movement in [FeFe]-hydrogenase (pathway 1), we evaluated this proton pathway with the active site in the Fe(I)Fe(I) redox state. In an effort to understand if there was a single Grotthuss-like concerted movement of protons or if there were multiple individual proton steps that were uncoupled from one another, we studied six different protonation states. This included altering where the resting proton was located from Glu279, Glu282, and the dithiomethylamine (DTMA) on the active site, how many protons were in the channel, and whether or not the hydrogen bonding pattern was optimized for H_2 production or H_2 oxidation (Fig. 2). We also evaluated each of the mutants that were investigated experimentally. In this work, we will largely focus on the results for the different protonation states to describe the methodology and will discuss the mutants that were studied only when they are essential to understanding the proton transport process.

4.2 Classical MD Methodology Used to Study Hydrogenase

4.2.1 Force Field Parameters

Beginning with an accurate force field for the active site was an essential foundation for these calculations. The force field parameters for the standard amino acid residues in the MD simulations were taken from the CHARMM

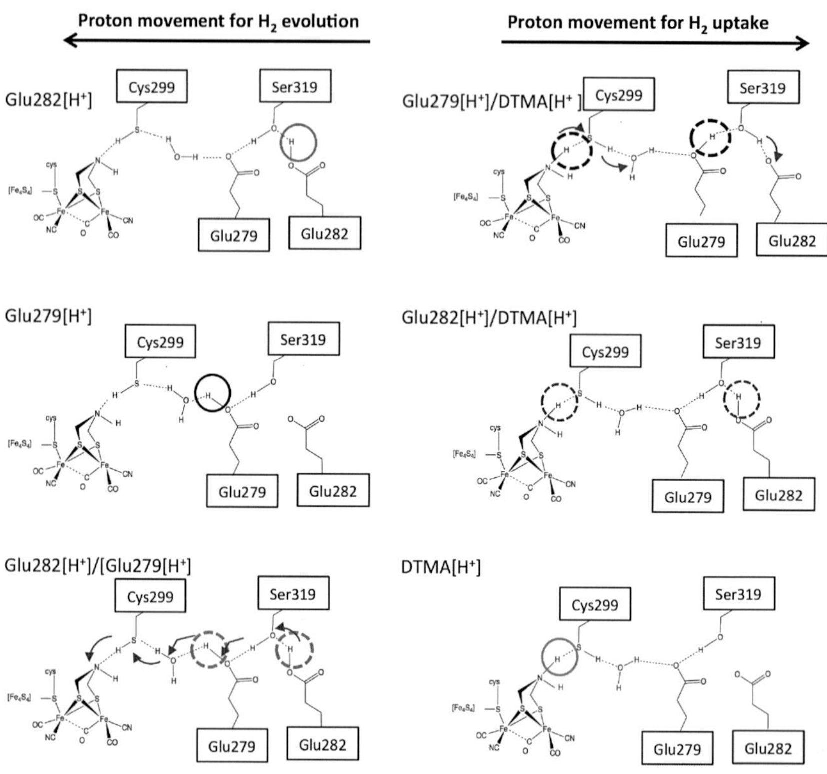

Fig. 2 Six protonation states of the proton pathway in [FeFe]-hydrogenase were studied. H_2 production is shown on the *left*. There are two states with a single proton, either Glu282 (*top*), Glu279 (*middle*), and one state with two protons, on both Glu282 and Glu279 (*bottom*). H_2 oxidation (*right*) also has three protonation states, two with two protons and one with one proton. The states with two protons have the protons located at either Glu279 and DTMA (*top*) or Glu282 and DTMA (*middle*). The state with a single proton has the proton location on the DTMA (*bottom*). The concerted movement of the protons is shown with *blue* (*dark gray* in the print version) *arrows* for protonation states Glu282[H^+]/Glu279[H^+] and Glu279[H^+]/DTMA[H^+]. *Adapted from Ginovska-Pangovska, B., Ho, M.-H., Linehan, J. C., Cheng, Y., Dupuis, M., Raugei, S., & Shaw, W. J. (2014). Molecular dynamics study of the proposed proton transport pathways in [FeFe]-hydrogenase.* Biochimica et Biophysica Acta, 1837, *131–138 with permission from Elsevier.*

(MacKerell et al., 1998) force field, and the parameters for the active site and the FeS cubanes were taken from Chang and Kim (2009). Because our simulations used a different oxidation state of the active site from the Fe(I)Fe(II) oxidation state used by Chang et al., reoptimization of the charges was needed. The active site was modeled in the Fe(I)Fe(I) oxidation state, with a vacant coordination site at the distal Fe atom optimizing the structure for

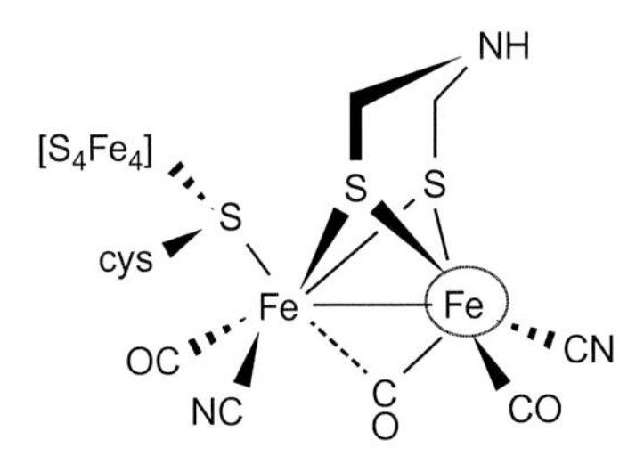

Fig. 3 Active site of the [FeFe]-hydrogenase, referred to as the H-cluster. The distal Fe is *circled.*

H_2 evolution (circled in *red* (*gray* in the print version) in Fig. 3). Reoptimization was accomplished by calculating the charges on a model of the active site including both the 2Fe-subcluster and the 4Fe4S cubane covalently bound to it, as shown in Fig. 3. The charges on the 4Fe4S cubane were constrained to the values determined by Chang (2011) and new charges were calculated only for the 2Fe-subcluster. Because of the antiferromagnetic nature of the clusters, obtaining the correct electronic description to the ground state of the 2Fe-subcluster required explicit inclusion of the 4Fe4S cubane in the QM model. We also found that the commonly used restrained electrostatic potential (RESP) procedure (Bayly, Cieplak, Cornell, & Kollman, 1993) gave nonphysical values for the charges of the Fe atoms for the active site as found for other cofactors or molecular catalysts containing transition metals buried inside. Alternatively, we calculated the charges for the Fe atoms using the CP2K program (VandeVondele et al., 2005) and the scheme proposed by Blöchl (1995) within a DFT scheme (Goedecker, Teter, & Hutter, 1996; Perdew, Burke, & Ernzerhof, 1996, 1997; Valiev et al., 2010; VandeVondele & Hutter, 2007). The remaining charges were calculated using the DFT-RESP approach as implemented in the NWChem program (Schafer, Horn, & Ahlrichs, 1992; Valiev et al., 2010).

In addition to containing a hydrogen bond to the distal Fe atom, the active site structure used for development of the original force field parameters has a nonprotonated DTMA bridge. In order to simulate the protonated DTMA bridge, we adjusted the equilibrium parameters of the force field for the active site (bonds, angles, and dihedrals), based on two gas-phase optimized DFT structures of H-cluster models: one with one proton on the nitrogen atom of the DTMA bridge, and second structure with two protons on the nitrogen (DTMA [H^+]). Protonation states of all of the amino acids were based on a standard pH of 7.0. To evaluate the pK_a of the residues based on the crystal structure, we used an empirical approach implemented in the PROPKA program (Li, Robertson, & Jensen, 2005).

4.3 Evaluation of the Protein Dynamics

The stability of the enzyme during the MD simulations was evaluated by calculating the root-mean square displacement (RMSD) (Fig. 4), and root-mean square fluctuation (RMSF; Fig. 5), of the backbone atoms.

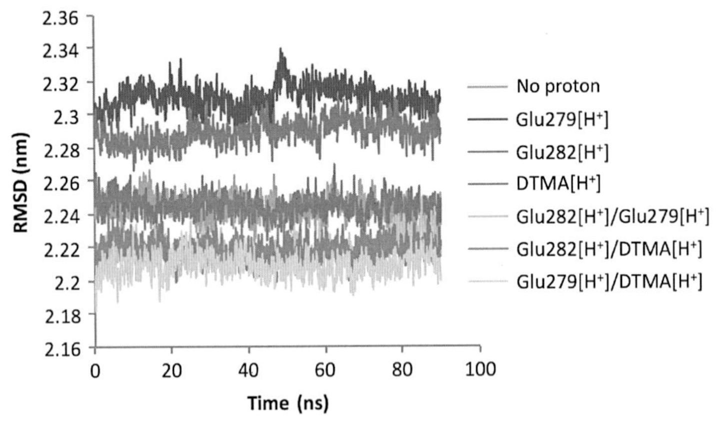

Fig. 4 RMSD of the backbone for the trajectories from the simulations for protonation state Glu279[H$^+$], Glu282[H$^+$], DTMA[H$^+$], Glu279[H$^+$]/Glu282[H$^+$], Glu282[H$^+$]/DTMA[H$^+$], and Glu279[H$^+$]/DTMA[H$^+$] as a function of time. The enzyme maintains its stability and does not undergo any significant conformational changes.

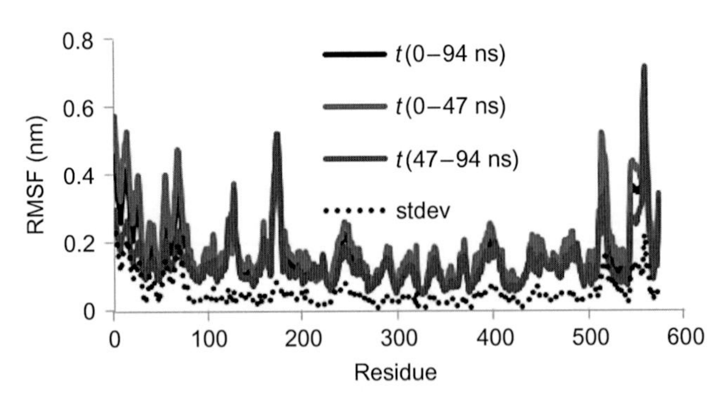

Fig. 5 Analysis of the convergence of the backbone RMSF for a representative trajectory (native enzyme in protonation state Glu282[H$^+$]). *Reprinted from Ginovska-Pangovska, B., Ho, M.-H., Linehan, J. C., Cheng, Y., Dupuis, M., Raugei, S., & Shaw, W. J. (2014). Molecular dynamics study of the proposed proton transport pathways in [FeFe]-hydrogenase.* Biochimica et Biophysica Acta, 1837, *131–138 with permission from Elsevier.*

The RMSD analysis of the enzyme shows values less than 1.5 Å, consistent with insignificant conformational changes for the duration of the simulation (~90 ns). No significant change in the enzyme structure was observed by changing the protonation state or introducing mutations in Pathway 1.

The RMSF analysis is used to determine the flexibility of specific residues in the protein. It is defined as the standard deviation of the displacement of a residue (typically calculated using the C_α position as in our case) in the trajectory from a reference value calculated from either an average structure or the crystallographic or NMR structure. In our case, we used the average structure. The calculated RMSF values are used to calculate the *B*-factors (Debye–Waller factors) (Debye, 1913; Waller, 1923).

4.3.1 Error Analysis

Customary block (time) averaging of the properties of interest (hydrogen bonds, RMSF of the backbone, and the average length of the water wires) indicated that the simulation time was long enough to have reliable estimates (see Fig. 5 for a representative example of this analysis). Within this approach, the average value of the property of interest is calculated from the MD trajectory for different lengths of simulation time. The trajectory of length *t* is split into blocks with different block sizes (*t*/2, *t*/4, *t*/8, etc.), and the property and the standard deviation are calculated for the different block sizes. As the block size increases, the standard deviation should approach an asymptotic value, indicating convergence of the property. Fig. 5 demonstrates the block average analysis for RMSF for blocks of *t*, *t*/2, and *t*/4. Fig. 6 shows the error analysis of the RMSD values in Fig. 4 as a function on block sizes varying from 0.2 to 15 ns.

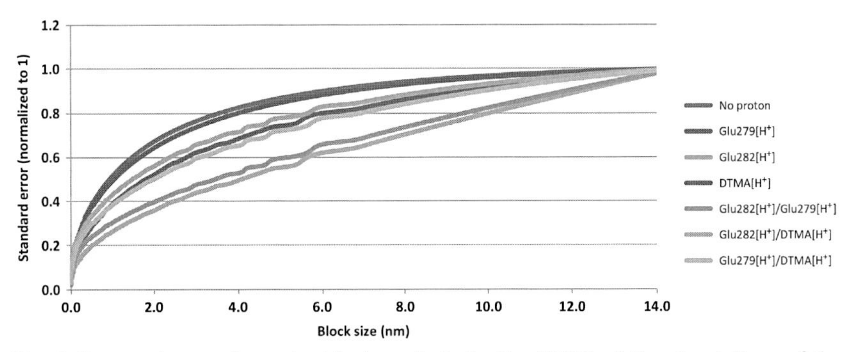

Fig. 6 Error estimates from the block analysis for the RMSD of the simulations of the studied protonation states of [FeFe]-hydrogenase.

4.4 Hydrogen Bond Analysis

4.4.1 Hydrogen Bond Interaction in the Proton Pathway

Analysis of the hydrogen bonding between the residues in the proton pathway can provide insight into the optimal arrangement of the residues and the water for facile transfer of protons. To explore these effects, we analyzed the hydrogen bonding patterns in the proton pathway for the different protonation states of Glu279, Glu282, and the bridging DTMA (Fig. 2). Two groups were considered hydrogen bonding if the following conditions were met: if the distance between the hydrogen bond donor atom X of R_1-XH and the hydrogen bond acceptor atom Y of R_2-Y, where X and Y = O for Glu, Ser, and water, S for Cys, or N for DTMA, was less than 3.5 Å, and the angle X–H…Y was larger than 120 degrees.

The resulting data for the six protonated states discussed here, as well as the unprotonated state, are shown in Fig. 7. In general, there are minimal changes to the hydrogen bonding pattern as a function of protonation state. In all protonation states, there is a significant hydrogen bond between Glu279 and Ser319, in many cases over 100%, reflecting the ability of the serine to hydrogen bond to either oxygen in glutamic acid. This was also observed in the mutations and was interpreted as a stabilizing feature critical to maintaining the proton pathway (Ginovska-Pangovska, Ho, et al., 2014).

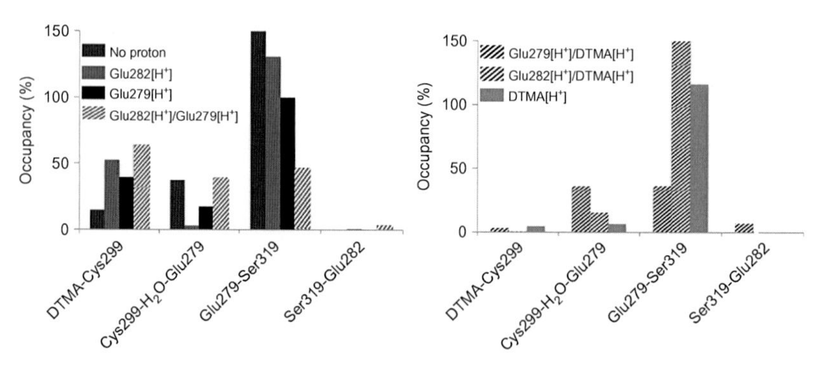

Fig. 7 Occupancy of hydrogen bonds between the residues in *Channel 1*. The *blue* (*dark gray* in the print version) *column* is the hydrogen bond occupancy in the unprotonated channel. The hydrogen bond occupancies of the three H_2 evolution protonation states are shown on the *left* and the three H_2 oxidation states are shown on the *right*. The columns for Cys299-H_2O-Glu279 show the simultaneous occupancy of hydrogen bonding between Glu279-H_2O and H_2O-Cys299. These results suggest a stable hydrogen bonding core between Ser319 and Glu279, with more flexible hydrogen bonding at the ends of the channel.

For protonation states aligned with hydrogen production, there is a high occupancy of hydrogen bonds between the active site (DTMA) and Cys299, and frequently, moderate hydrogen bonding between Cys299, the water molecule, and Glu279. Rarely is hydrogen bonding observed in the residues at the mouth of the pathway, between residues Ser319 and Glu282, likely due to the possibility for these residues to interact with the surface water. Of the four protonation states for H_2 production, the most atypical hydrogen bonding pattern is the doubly protonated species, Glu282[H^+]/Glu279 [H^+], which has ~50% hydrogen bonding from the active site through Ser319, more consistent than for any of the other protonation states. This protonation state would favor a fully concerted proton transfer mechanism and the even distribution of hydrogen bonds may be an indication that this is a preferred pathway.

For protonation states favoring H_2 oxidation, only the Cys299 to water to Glu279 has any significant occupancy other than the highly occupied Glu279 to Ser319. In this case, it is again the protonation state that would result in concerted proton transfer, Glu279[H^+]/DTMA[H^+], that is atypical of the series, where significantly more hydrogen bonding is seen in the Cys299 to water to Glu279 hydrogen bond, while significantly less is observed between Glu279 and Ser319. The implications of this change in hydrogen bond pattern are not fully understood, but the observation that protonation states that favor a concerted mechanism are unique in both directions does suggest a different role or mechanism for those hydrogen bonding patterns.

4.4.2 Hydrogen Bond Interaction with Nearby Residues

In the process of evaluating the various protonation states and mutant enzymes, we discovered that two additional residues, Ser320 and Arg286 near the entrance of the proton pathway (Fig. 8), might also play a role in proton transfer (Ginovska-Pangovska, Ho, et al., 2014). This hypothesis stemmed from the observation that Glu282 was rarely hydrogen bonded into the channel, but instead was hydrogen bonded to either Arg286 or Ser320. This was especially evident when Glu282 is unprotonated, allowing it to form a stable salt-bridge with the positively charged Arg286. This appeared to be preventing protonation of Glu282 and we hypothesized that mutating these residues would enhance catalytic activity. We demonstrated that modifying Arg286 to Leu resulted in a more negative electrostatic potential at the mouth of the proton pathway, which would be expected to create a more favorable entry point for protons. Additional experimental and computational studies are ongoing to evaluate the role of these two residues.

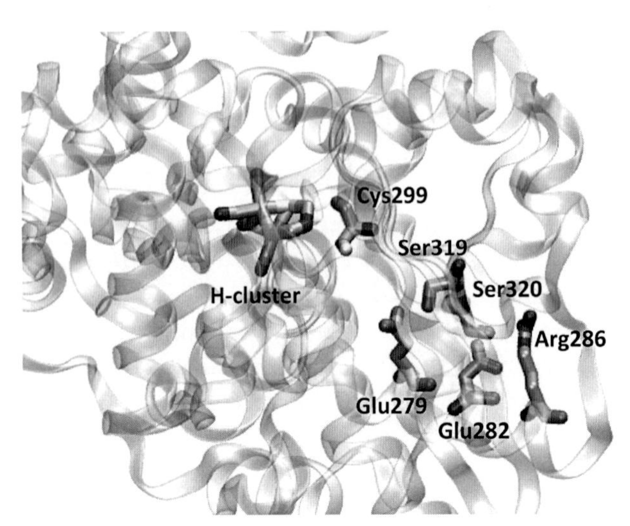

Fig. 8 Proton pathway and active site in [FeFe]-hydrogenase, showing the proposed entrance to the proton pathway as well as the relative orientations of Ser320, Arg286, and Glu282 to each other and to the surface of the protein. (See the color plate.)

4.4.3 Error Analysis for Hydrogen Bonding

Error analysis of the hydrogen bonding was also done using the block averaging approach described earlier. A representative example for the singly protonated state Glu279[H^+] is given in Fig. 9.

4.5 The Role of Dynamics in Enzymatic Activity

To further our understanding of the effect of protonation state on the enzyme, we performed an analysis to correlate the protonation state with the flexibility of the protein scaffold. We found that the dynamics of the protein scaffold were highly sensitive to the protonation state of the proton pathway. In particular, the regions that are proposed to bind the redox partners have a wide range of dynamics as a function of the number of protons in the channel as well as the location (Fig. 10). While this does not provide direct evidence for coupled proton and electron transfer, the mobility observed in the redox partner binding regions as protons are located in different regions of the pathway demonstrates that the regions responsible for electron transfer respond to the presence of protons and may in fact suggest a coupled response.

4.5.1 Covariance Analysis

To more quantitatively evaluate the motion in the electron partner binding regions with protonation state, we performed a covariance analysis to

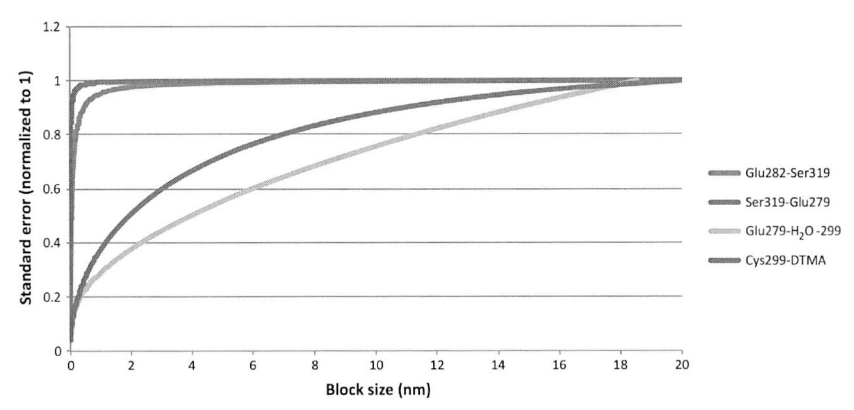

Fig. 9 Representative block error analysis for the hydrogen bond occupancy (Fig. 5) for protonation state Glu279[H^+].

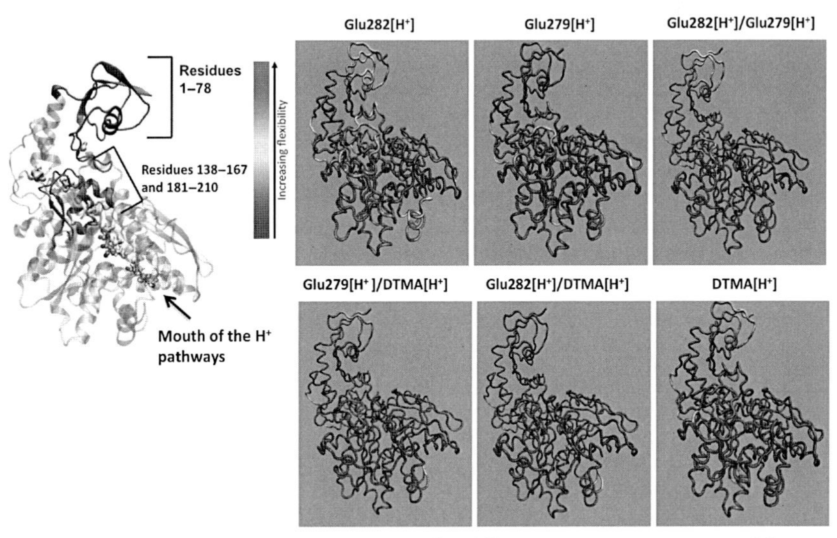

Fig. 10 Flexibility of hydrogenase backbone for different protonation states of the proton pathway. *Left*: Residues for the three domains involved in redox partner binding for [FeFe]-hydrogenase are shown in *red* (*dark gray* in the print version) (residues 1–78, 138–167, and 181–210). *Right*: Effects of the protonation state on the flexibility of the enzyme for the H_2 evolution system (*top panels*) and H_2 oxidation (*bottom panels*). Regions of high flexibility are in *red* (*dark gray* in the print version) and *green* (*light gray* in the print version). Note the varying flexibility in the protein in the redox partner binding region and in the mouth of the channel as a function of protonation state. *Adapted with permission from Ginovska-Pangovska, B., Ho, M.-H., Linehan, J. C., Cheng, Y., Dupuis, M., Raugei, S., & Shaw, W. J. (2014). Molecular dynamics study of the proposed proton transport pathways in [FeFe]-hydrogenase.* Biochimica et Biophysica Acta, 1837, *131–138.*

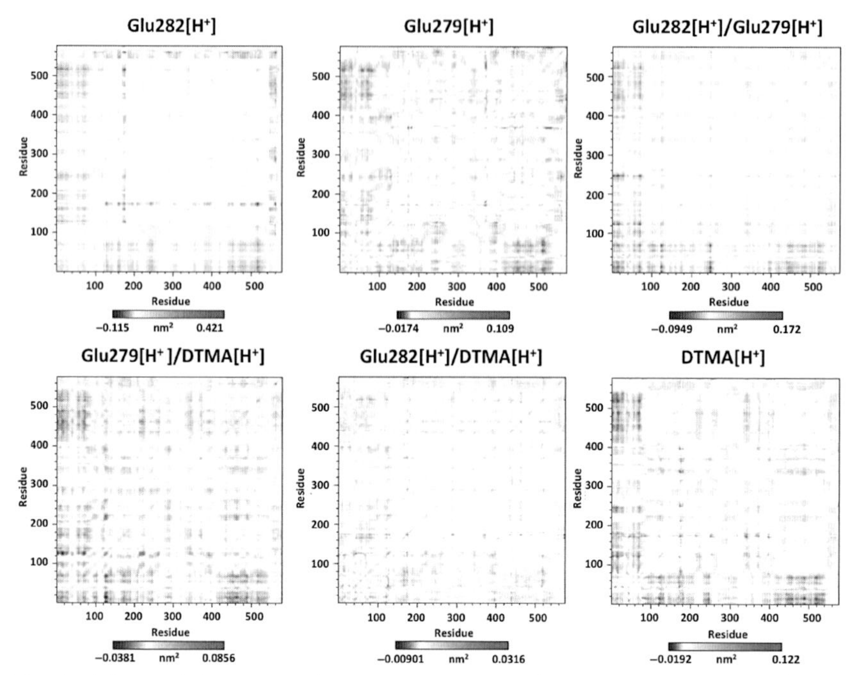

Fig. 11 Effects of the protonation state on the correlated motion between different regions of the enzyme in the H_2 evolution system (*top*) and H_2 oxidation system (*bottom*). *White areas* show no correlation, *red* (*gray* in the print version) *areas* show correlated motion, and *blue* (*dark gray* in the print version) areas show anticorrelated motion. The covariance maps were generated following the motion of the backbone C_α atoms. *Adapted with permission from Ginovska-Pangovska, B., Ho, M.-H., Linehan, J. C., Cheng, Y., Dupuis, M., Raugei, S., & Shaw, W. J. (2014). Molecular dynamics study of the proposed proton transport pathways in [FeFe]-hydrogenase.* Biochimica et Biophysica Acta, 1837, *131–138.*

determine which regions of the enzyme have motions that are correlated and/or anticorrelated. The resulting correlated motions are shown in *red* (*gray* in the print version) on the maps in Fig. 11, and the anticorrelated motions are shown in *blue* (*dark gray* in the print version). The white regions represent the areas of low correlation. The most interesting feature of these maps is the dominance of anticorrelated motion compared to correlated motion for all protonation states. Additionally, throughout all protonation states, we observe a change of intensity in the correlations, but always in the same region, consistent with an interaction with site specificity. Results from the covariance analysis do not predict causality of the dynamics, but it is evidence of a possible allosteric regulation existing between the proton transfer and the electron transfer pathways.

5. PROTON TRANSPORT IN HYDROGENASE MIMICS

Hydrogenases have provided inspiration for the design of molecular electrocatalyts for oxidation and production of H_2 (Caserta et al., 2015; Dubois, 2014; Gloaguen & Rauchfuss, 2009; Rakowski DuBois & DuBois, 2009; Simmons et al., 2014). Indeed, work conducted by our group and others has indicated that controlling proton movement into and out of the active site is one of the most important catalytic steps. Consistent with this observation, hydrogenases are able to move protons efficiently. As we discussed earlier, the precisely controlled nature of proton movement in hydrogenases is dictated by finely tuned free energy differences between the various protonated intermediates connecting the catalytic center and the exterior of the enzyme. Using these ideas as guiding principles, improved electrocatalysts for oxidation and production of H_2 have been proposed in the past few years. In this section, we review some of the computational and theoretical studies that helped us understand how to achieve this result by creating a bioinspired environment around a catalytic metal center that promotes fast proton transfer to/from the active site and at the same time proves concepts inferred from the enzymes. Below, we analyze key issues that limit proton delivery to/from molecular catalysts. Then we illustrate how the addition of key functionalities around the catalytic core improves proton movement.

5.1 Bottlenecks for Proton Movement in Hydrogenase Mimics

Protonation (deprotonation) of molecular catalysts is a complicated process where an acid (base) associates with the catalyst, delivers (removes) the proton, and then dissociates. The precise delivery of protons and the rate of protonation/deprotonation are regulated in a nontrivial way by the combination of three factors: steric effects, hydrogen bonding, and electrostatic interactions between the catalyst and the exogenous proton donor/ acceptor. In particular, in nonaqueous environments, computations indicate that steric hindrance and solvation/desolvation processes are the major detrimental contributions to catalytic rates. Our studies suggest that the presence of well-positioned protic residues around the catalytic center in [FeFe]-hydrogenases largely reduces these penalties as discussed in the following two sections.

As an illustrative example, in Fig. 12 we show the free energy for the deprotonation of a protonated intermediate in the oxidation of H_2 catalyzed

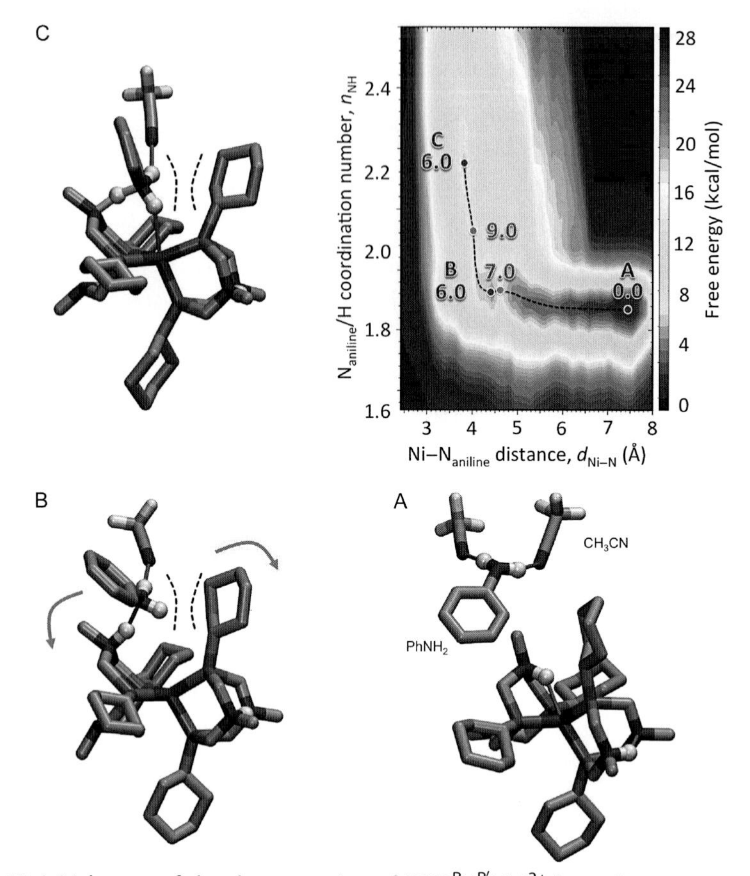

Fig. 12 Initial steps of the deprotonation of $[Ni(P^{R}_{2}N^{R'}_{2}H)_{2}]^{2+}$ by aniline in acetonitrile solutions: base association and proton transfer. *Top, right panel*: Free energy for the binding of aniline as a function of the distance (d_{Ni-N}) between the Ni center and the N atom of aniline, and as a function of the fractional number (n_{NH}) of hydrogen atoms around the N atom of aniline (coordination number) obtained from hybrid QM/MM metadynamics simulations. Catalyst and base were described within the DFT framework, and the solvent and counterions with classical force field. Local minima and transition states are marked with *blue* and *red circles*, respectively. Free energies at minima and transition states are also reported using the same colors. The *dotted lines* indicate the lowest-free energy pathway. Representative configurations of the minima on the free energy surface are shown in (A)–(C). Hydrogen bond interactions are shown as *red lines*; in (C) the *green sticks* indicate that the H atom is shared between the two N atoms. The *red arrows* in (C) indicate the overall movement of the pendant amine and the proximal cyclohexyl group upon binding of the base. In all of the figures, acetonitrile molecules hydrogen bonded to aniline are also shown. For clarity, nonprotic H atoms on the Ni complex and aniline are not shown. *Reprinted with permission from O'Hagan, M., Ho, M. H., Yang, J. Y., Appel, A. M., Rawkowski DuBois, M., Raugei, S., … Bullock, R. M. (2012). Proton delivery and removal in [Ni((P2N2R)-N-R')](2) hydrogen production and oxidation catalysts.* Journal of the American Chemical Society, *134(47), 19409–19424. doi:10.1021/Ja307413x. Copyright 2012 American Chemical Society.* (See the color plate.)

by the nickel complex $[Ni(P_2^RN_2^{R'})_2]^{2+}$, where $P_2^RN_2^{R'}$ = 1,5-diaza-3,7-diphosphacyclooctane ligands, R = cyclohexyl (Cy), and R′ = methyl (Me) (O'Hagan et al., 2012). In this family of catalysts, the pendant amine in the second coordination sphere of the metal center, which facilitates proton movement, mimics the function of the pendant amine of the azadithiolate bridge in the [FeFe]-hydrogenase.

Fig. 12 shows that the binding of aniline is an endergonic process as a consequence of the steric clash between the base and the cyclohexyl substituent on the P atom of the ligand opposite to the protonated pendant amine (symbolically indicated with dashed lines in B and C). Using smaller bases and ligand substituents mitigates this steric penalty, but it also changes the overall energetics of the catalytic process. Controlling these changes to favor catalysis is the subject of intense ongoing research activity. Additional studies have also revealed that the reaction medium plays an important role regulating proton delivery. In particular, water, even in small amounts, favors the intermolecular proton movement acting as a proton shuttle between the catalyst and the exogenous proton donor/acceptor (Ho et al., 2015).

Computations also showed that depending on the nature of the metal center, proton transfer between the metal and the pendant amine can represent an additional catalytic bottleneck. In Ni-based catalysts, this proton transfer is very easy as the pK_a of the Ni–H matches or is lower than that of the protonated pendant amine (O'Hagan et al., 2011; Raugei et al., 2012). In the case of Fe-based catalysts, this intramolecular proton movement represents or contributes to the catalytic bottleneck, as the iron center is far more basic than the pendant amine (Darmon et al., 2015; Darmon et al., 2014). Various computational insights on how to improve both intra- and intermolecular proton movement using appropriate catalyst modifications are discussed later.

5.2 Hydrogenase Mimics for H_2 Oxidation

Functional mimics of [FeFe]-hydrogenase have been very successful in optimizing the performance of the active site and the second coordination sphere. Recently, Darmon et al. have demonstrated that optimizing proton movement by tuning the outer coordination sphere can increase the rate of H_2 oxidation two orders of magnitude, without increasing the overpotential for the $(Cp^{C_5F_4N})Fe(P^{Et_2}N^{(CH_2)_3NMe_2}P^{Et_2})(Cl)$ complex (**1**), compared to the simpler $(Cp^{C_5F_4N})Fe(depp)(X)$ and $(Cp^{C_5F_4N})Fe(P^{Et_2}N^{Me}P^{Et_2})(X)$ catalysts (Fig. 13; Darmon et al., 2015). The outer coordination spheres are elements of the scaffold on the molecular catalyst that can influence catalysis but do not directly interact with the metal (Ginovska-Pangovska, Dutta,

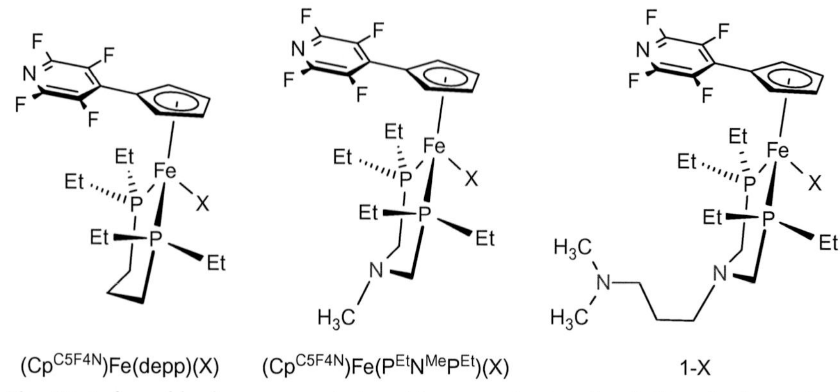

Fig. 13 Fe-based hydrogenase mimics with no pendant amine (*left*), a single pendant amine (*middle*), and two pendant amines (*right*). *Reprinted from reference Darmon, J. M., Kumar, N., Hulley, E. B., Weiss, C. J., Raugei, S., Bullock, R. M., & Helm, M. L. (2015). Increasing the rate of hydrogen oxidation without increasing the overpotential: A bio-inspired iron molecular electrocatalyst with an outer coordination sphere proton relay.* Chemical Science, 6, *2737–2745 with permission from the Royal Society of Chemistry.*

et al., 2014; Shaw, 2012). Studying the effects of proton relays in the mimics allows for accurate DFT calculations due to the reduced size of the system relative to enzymes, resulting in the determination of both reaction energies and barriers to reaction.

5.2.1 DFT calculations

All calculations were carried out using the *Gaussian* 09 suite of programs (Frisch et al., 2009). The geometries for the stable intermediates and transition states were optimized in solution with the B3P86 functional (Perdew, 1986; Perdew et al., 1997), employing the Stuttgart–Dresden relativistic effective core potential for Fe and its associated basis set (SDD) (Andrae, Häußermann, Dolg, Stoll, & Preuß, 1990) and the 6-31G* basis set for all the atoms. The solvent, fluorobenzene, was described implicitly using the SMD continuum model of solvation (Marenich, Cramer, & Truhlar, 2009). Each step in the catalytic mechanism was identified, and the most favorable reaction pathway was established.

The catalytic cycle shows stepwise deprotonation. For both steps, the most favorable pathway results in deprotonation from the ring pendant amine and deprotonation from the ring pendant amine from the NMe_2 in the outer coordination sphere. Computational studies for mimics without the outer coordination sphere amine groups show that the reaction bottleneck is the deprotonation of the Fe(III) hydride (Darmon et al., 2015), an

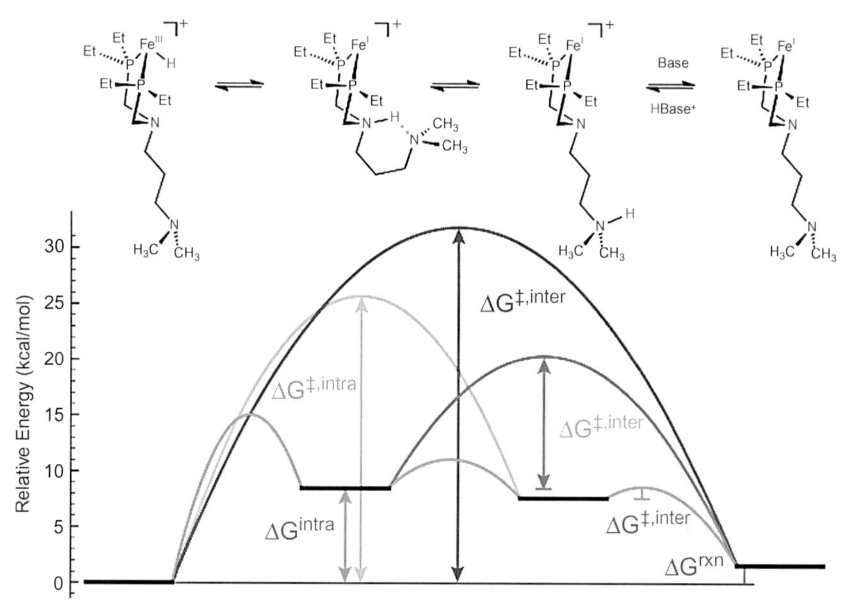

Fig. 14 Free energy diagram for the key intra- and intermolecular deprotonation steps of $[(Cp^{C_5F_4N})Fe(P^{Et_2}N^{(CH_2)_3NMe_2}P^{Et_2})H]^+$ by *N*-methyl-pyrrolidine as obtained from density functional theory calculations. *Red* (*gray* in the print version) *lines*: lowest-free energy deprotonation pathway (upper reaction scheme); *blue* (*dark gray* in the print version) *line*: direct deprotonation of the iron(III) hydride by the exogenous base; *orange* (*light gray* in the print version) *line*: deprotonation of the iron(III) hydride by the outer coordination sphere pendant amine; *green* (*light gray* in the print version) *line*: deprotonation of the second coordination sphere pendant amine by exogenous base. The $Cp^{C_5F_4N}$ substituent has been omitted for clarity. *Reprinted from Darmon, J. M., Kumar, N., Hulley, E. B., Weiss, C. J., Raugei, S., Bullock, R. M., & Helm, M. L. (2015). Increasing the rate of hydrogen oxidation without increasing the overpotential: A bio-inspired iron molecular electrocatalyst with an outer coordination sphere proton relay.* Chemical Science, 6, *2737–2745 with permission from the Royal Society of Chemistry.*

intermediate formed after the first deprotonation. For this process, shown in Fig. 14, addition of an outer coordination sphere pendant amine makes the intramolecular proton transfer from the metal center to the second coordination sphere amine more favorable reducing the basicity of the second coordination sphere amine compared to the complex without the outer coordination sphere, and reducing the barrier to deprotonation (*red* (*gray* in the print version) pathway in Fig. 14). Alternative pathways for the deprotonation of the Fe(III) hydride by either the exogenous base, or directly by the outer coordination sphere NMe_2 (shown in *blue* (*dark gray* in the print version), *orange* (*light gray* in the print version), and *green* (*light gray* in the print version) in Fig. 14), demonstrate the importance of the three-membered proton pathway in achieving the lowest barrier and, consequently, the fastest rates.

5.3 Hydrogenase Mimics for H_2 Production

5.3.1 Outer Coordination Sphere

Another platform we have been using to understand outer coordination sphere effects is the $[Ni(P^R_2N^{R'}_2)_2]^{2+}$ platform, where R is cyclohexyl (Cy) for H_2 oxidation and phenyl (Ph) for H_2 production, and R′ is an amino acid, a dipeptide, or a 10-amino acid peptide (Ginovska-Pangovska, Dutta, et al., 2014). The extensive study of the active site (Dubois, 2014; Rakowski DuBois & DuBois, 2009; Shaw, Helm, & DuBois, 2013; Wilson et al., 2006) allows this molecule to serve as an ideal core upon which to understand effects of an outer coordination sphere. In studies to understand the role of hydrophobic and hydrophilic amino acid functional groups on catalytic activity, we used the $[Ni(P^{Ph}_2N^{Dipeptide}_2)_2]^{2+}$ complexes and used H_2 production activity to evaluate the effect of individual groups (Jain et al., 2011; Jain, Reback, et al., 2012; Reback et al., 2013). The dipeptide consisted of a nonnatural amino acid (propionic acid) to link the complex to a natural amino acid, including alanine ester, serine ester, phenylalanine ester, tyrosine ester, glutamic acid, glutamic acid ester, lysine, lysine ester, aspartic acid, and aspartic acid ester (Fig. 15). We found that basic and acidic groups result in rates as much as five times faster than the parent complexes lacking

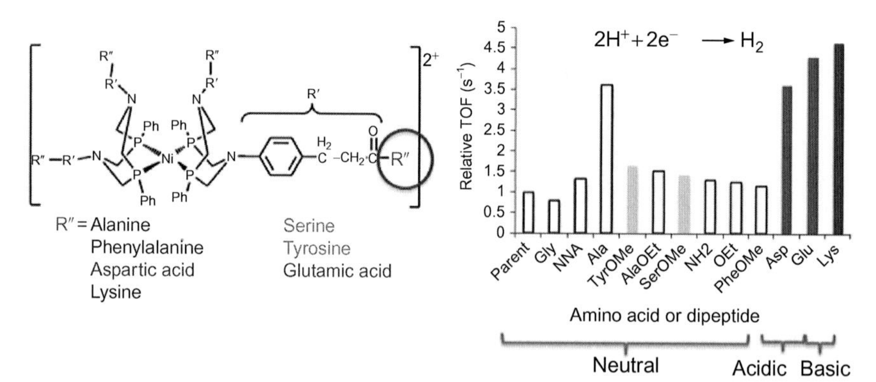

Fig. 15 $[Ni(P^{Ph}_2N^{Dipeptide}_2)_2]^{2+}$ complexes were studied for H_2 production to evaluate the role of functional groups from a simple outer coordination *sphere*. The largest impact was from the addition of charged groups, found from molecular dynamics to be due to concentrating substrate around the active site. Simulations were carried out in acetonitrile/water solutions with water concentration ranging from 0 to 3.0 *M*.

the second amino acid, while polar and nonpolar amino acids only increased the rates by ~10%. MD studies were used to get adequate sampling of these molecular complexes and showed that the influence the charged amino acids had was to concentrate substrate around the active site. Concentrating substrate is a mechanism often proposed for enzymes and demonstrates the possibility of mimicking the protein scaffold even with a very simple outer coordination sphere.

5.3.2 Replica Exchange Molecular Dynamics

The flexibility was studied using REMD. For this system, we used 20 replicas, covering a temperature range from 280 to 930 K. Each simulation was ~95 ns long. The force field parameters for the complex were obtained as reported earlier. The solvent, acetonitrile, was treated implicitly within the generalized Born model (Onufriev, Bashford, & Case, 2004).

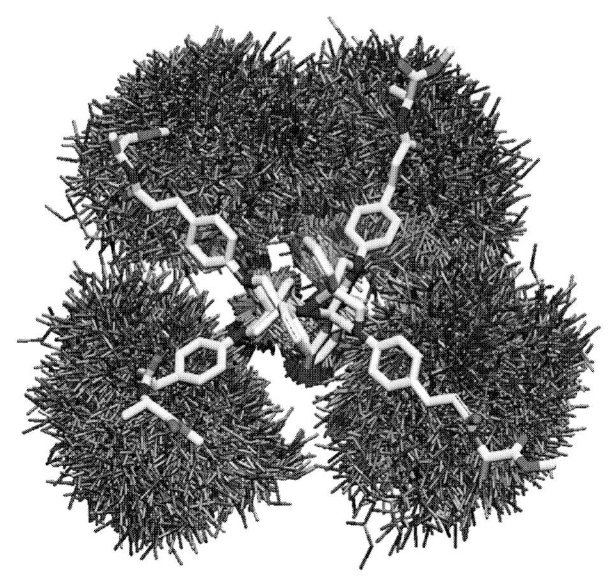

Fig. 16 A static $[Ni(P_2^{Ph}N_2^{Dipeptide})_2]^{2+}$ complex is shown overlaid on top of the full molecular dynamics trajectory. While the dipeptide did result in enhanced catalytic activity, the significant mobility of the dipeptides demonstrates that there is too much mobility in this outer coordination *sphere* to provide the interactions so important in the functions of enzymes. *Adapted from reference Reback, M. L., Ginovska-Pangovska, B., Ho, M.-H., Jain, A., Squier, T. C., Raugie, S., … Shaw, W. J. (2013). The role of a dipeptide outer-coordination sphere on H2-production catalysts: influence on catalytic rates and electron transfer.* Chemistry—A European Journal, 19, *1928–1941 with permission from John Wiley and Sons.* (See the color plate.)

We also used umbrella sampling to evaluate the possibility of the amino acids to deliver protons directly to the pendant amine (Reback et al., 2013). We found that this delivery was structurally possible; however, there was no energy minimum with either the backbone carboxyl/ester or the side chain amine to interact with the pendant amine, suggesting that direct proton delivery was likely a minor contribution to the overall rate, unlike that observed for the Fe-based complex discussed earlier. The lack of a stable interaction was largely due to the extensive flexibility of the dipeptides and the rings in the core complex, resulting in the dipeptides existing across a large region of phase space, shown in Fig. 16. This observation pointed to the need for a more structurally stable outer coordination sphere if the hydrogen bonding interactions observed in enzymes were to be replicated in molecular complexes. We addressed the high flexibility by attaching a 10-mer β-hairpin and investigating the stability with REMD, which revealed significantly limited dynamics with a hinge motion around the three carbon bond (Reback et al., 2014).

6. SUMMARY

Using molecular dynamics studies, we have provided significant insight into the proton transfer mechanism used in *C. pastuerianum* [FeFe]-hydrogenase. In addition to characterizing which pathway was likely the dominant pathway, we were also able to shed light on the nature of the hydrogen bonding as a function of protonation state and the correlation of the protonation state with dynamics in the redox partner binding region. The computational methods used here predicted two residues that might be significant contributors to proton transport, and subsequent experimental and computational studies are ongoing to better understand proton transport in [FeFe]-hydrogenase. Similar methodologies applied to hydrogenase mimics have revealed the importance of pK_a matching and overcoming steric hindrance in enhancing rates, even with a minimal pathway. Further, they show that substrates can be concentrated with a simple scaffold, mimicking the effects of the enzyme scaffold while keeping the overall size of the complex small, a feature that will be advantageous for practical use. This powerful combination of experiment and theory is the most robust way to provide meaningful insight and predictions and is an approach that needs to be emphasized as we continue to investigate proton transport in large molecules and attempt to apply these concepts to synthetic catalysts.

ACKNOWLEDGMENTS

The work reviewed here was funded by the DOE Office of Science Early Career Research Program through the Office of Basic Energy Sciences (W.J.S., B.G.), the DOE Office of Science Office of Basic Energy Sciences (W.J.S., SR), and the Center for Molecular Electrocatalysis, an Energy Frontier Research Center funded by the US DOE, Office of Science, Office of Basic Energy Sciences (S.R.). Computational resources were provided by W. R. Wiley Environmental Molecular Science Laboratory (EMSL), a national scientific user facility sponsored by the Department of Energy's Office of Biological and Environmental Research located at Pacific Northwest National Laboratory, and a portion of the research was performed using PNNL Institutional Computing at Pacific Northwest National Laboratory. Pacific Northwest National Laboratory is operated by Battelle for the US Department of Energy.

REFERENCES

Agmon, N. (1995). The Grotthuss mechanism. *Chemical Physics Letters*, *244*(5–6), 456–462. http://dx.doi.org/10.1016/0009-2614(95)00905-J.

Andrae, D., Häußermann, U., Dolg, M., Stoll, H., & Preuß, H. (1990). Energy-adjusted ab initio pseudopotentials for the second and third row transition elements. *Theoretica Chimica Acta*, 77, 123–141.

Bayly, C. I., Cieplak, P., Cornell, W., & Kollman, P. a. (1993). A well-behaved electrostatic potential based method using charge restraints for deriving atomic charges: The RESP model. *The Journal of Physical Chemistry*, *97*, 10269–10280.

Blöchl, P. E. (1995). Electrostatic decoupling of periodic images of plane-wave-expanded densities and derived atomic point charges. *Journal of Chemical Physics*, *103*(17), 7422–7428.

Blomberg, M. R. A., Borowski, T., Himo, F., Liao, R.-Z., & Siegbahn, P. E. M. (2014). Quantum chemical studies of mechanisms for metalloenzymes. *Chemical Reviews*, *114*, 3601–3658.

Bonin, J., Costentin, C., Robert, M., Saveant, J.-M., & Tard, C. (2012). Hydrogen-bond relays in concerted proton-electron transfers. *Accounts of Chemical Research*, *45*, 372–381.

Caserta, G., Roy, S., Atta, M., Artero, V., & Fontecave, M. (2015). Artificial hydrogenases: Biohybrid and supramolecular systems for catalytic hydrogen production or uptake. *Current Opinion in Chemical Biology*, *25*, 36–47.

Chang, C. H. (2011). Computational chemical analysis of [FeFe] hydrogenase H-cluster analogues to discern catalytically relevant features of the natural diatomic ligand configuration. *Journal of Physical Chemistry A*, *115*(31), 8691–8704. http://dx.doi.org/10.1021/Jp112296d.

Chang, C. H., & Kim, K. (2009). Density functional theory calculation of bonding and charge parameters for molecular dynamics studies on [FeFe] hydrogenases. *Journal of Chemical Theory and Computation*, *5*(4), 1137–1145. http://dx.doi.org/10.1021/Ct800342w.

Cheng, X., Cui, G., Hornak, V., & Simmerling, C. (2005). Modified replica exchange simulation methods for local structure refinement. *The Journal of Physical Chemistry B*, *109*, 8220–8230.

Cornish, A. J., Gartner, K., Yang, H., Peters, J. W., & Hegg, E. L. (2011). Mechanism of proton transfer in [FeFe]-hydrogenase from Clostridium pasteurianum. *Journal of Biological Chemistry*, *286*(44), 38341–38347. http://dx.doi.org/10.1074/Jbc.M111.254664.

Darmon, J. M., Kumar, N., Hulley, E. B., Weiss, C. J., Raugei, S., Bullock, R. M., & Helm, M. L. (2015). Increasing the rate of hydrogen oxidation without increasing the overpotential: A bio-inspired iron molecular electrocatalyst with an outer coordination sphere proton relay. *Chemical Science*, *6*, 2737–2745.

Darmon, J. M., Raugei, S., Liu, T., Hulley, E. B., Weiss, C. J., Bullock, R. M., & Helm, M. L. (2014). Iron complexes for the electrocatalytic oxidation of hydrogen: Tuning primary and secondary coordination spheres. *ACS Catalysis*, *4*, 1246–1260.

Debye, P. (1913). Interferenz von Röntgenstrahlen und Wärmebewegung. *Annalen der Physik (in German)*, *348*, 49–92.

Dementin, S., Belle, V., Bertrand, P., Guigliarelli, B., Adryanczyk-Perrier, G., De Lacey, A. L., … Leger, C. (2006). Changing the ligation of the distal [4Fe4S] cluster in NiFe hydrogenase impairs inter- and intramolecular electron transfers. *Journal of the American Chemical Society*, *128*(15), 5209–5218. http://dx.doi.org/10.1021/Ja060233b.

Dementin, S., Burlat, B., Lacey, A. L. D., Pardo, A., Adryanczyk-Perrier, G., Guigliarelli, B., … Rousset, M. (2004). A glutamate is the essential proton transfer gate during the catalytic cycle of the [NiFe] hydrogenase. *The Journal of Biological Chemistry*, *279*, 10508–10513.

Diat, O., & Gebel, G. (2008). Proton channels. *Nature Materials*, 7, 13–14.

Dubois, D. L. (2014). Development of molecular electrocatalysts for energy storage. *Inorganic Chemistry*, *53*(8), 3935–3960.

Dutta, A., DuBois, D. L., Roberts, J. A. S., & Shaw, W. J. (2014). Amino acid modified Ni catalyst exhibits reversible H2 oxidation/production over a broad pH range at elevated temperatures. *Proceedings of the National Academy of Sciences of the United States of America*, *111*, 16286–16291.

Dutta, A., Lense, S., Hou, J., Engelhard, M., Roberts, J. A. S., & Shaw, W. J. (2013). Minimal proton channel enables H2 oxidation and production with a water soluble nickel-based catalyst. *Journal of the American Chemical Society*, *135*, 18490–18496.

Fontecilla-Camps, J. C., Volbeda, A., Cavazza, C., & Nicolet, Y. (2007). Structure/function relationships of [NiFe]- and [FeFe]-hydrogenases. *Chemical Reviews*, *107*, 4273–4303.

Frisch, M. J., Trucks, G. W., Schlegel, H. B., Scuseria, G. E., Robb, M. A., Cheeseman, J. R., … Fox, D. J. (2009). *Gaussian 09, Revision E.01*. Wallingford, CT: Gaussian, Inc.

Pohorille, C. C. A. (Ed.), (2007). *Free energy calculations: Theory and applications in chemistry and biology*. Berlin Heidelberg: Springer-Verlag.

Galvan, I. F., Volbeda, A., Fontecilla-Camps, J. C., & Field, M. J. (2008). A QM/MM study of proton transport pathways in a [NiFe]-hydrogenase. *Proteins: Structure, Function & Bioinformatics*, *73*, 195–203.

Gao, J., Ma, S., Major, D. T., Nam, K., Pu, J., & Truhlar, D. G. (2006). Mechanisms and free energies of enzymatic reactions. *Chemical Reviews*, *106*, 3188–3209.

Ginovska-Pangovska, B., Dutta, A., Reback, M., Linehan, J. C., & Shaw, W. J. (2014). Beyond the active site: The impact of the outer coordination sphere on electrocatalysts for hydrogen production and oxidation. *Accounts of Chemical Research*, *47*(8), 2621–2630. http://dx.doi.org/10.1021/ar5001742.

Ginovska-Pangovska, B., Ho, M.-H., Linehan, J. C., Cheng, Y., Dupuis, M., Raugei, S., & Shaw, W. J. (2014). Molecular dynamics study of the proposed proton transport pathways in [FeFe]-hydrogenase. *Biochimica et Biophysica Acta*, *1837*, 131–138.

Gloaguen, F., & Rauchfuss, T. B. (2009). Small molecule mimics of hydrogenases: Hydrides and redox. *Chemical Society Reviews*, *38*, 100–108.

Goedecker, S., Teter, M., & Hutter, J. (1996). Separable dual-space Gaussian pseudopotentials. *Physical Review B*, *54*(3), 1703–1710. http://dx.doi.org/10.1103/Physrevb.54.1703.

Hamdan, A., Dementin, S., Liebgott, P. P., Gutierrez-Sanz, O., Richaud, P., De Lacey, A. L., … Leger, C. (2012). Understanding and tuning the catalytic bias of hydrogenase. *Journal of the American Chemical Society*, *134*(23), 8368–8371. http://dx.doi.org/10.1021/Ja305217.

Hammes-Schiffer, S. (2011). Current theoretical challenges in proton-coupled electron transfer: Electron-proton nonadiabaticity, proton relays, and ultrafast dynamics. *Journal of Physical Chemistry Letters*, *2*, 1410–1416.

Ho, M.-H., O'Hagan, M., Dupuis, M., DuBois, D. L., Bullock, R. M., Shaw, W. J., & Raugei, S. (2015). Water-assisted proton delivery and removal in bio-inspired hydrogen production catalysts. *Dalton Transactions*, *44*(24), 10969–10979.

Hong, G., Cornish, A. J., Hegg, E. L., & Pachter, R. (2011). On understanding proton transfer to the biocatalytic [Fe-Fe](H) sub-cluster in [Fe-Fe] H(2)ases: QM/MM MD simulations. *Biochimica et Biophysica Acta-Bioenergetics*, *1807*(5), 510–517. http://dx.doi.org/10.1016/J.Bbabio.2011.01.011.

Jain, A., Buchko, G. W., Reback, M. L., O'Hagan, M., Ginovska-Pangovska, B., Linehan, J. C., & Shaw, W. J. (2012). Active hydrogenation catalyst with a structured, peptide-based outer coordination sphere. *ACS Catalysis*, *2*, 2114–2118.

Jain, A., Lense, S., Linehan, J. C., Raugei, S., Cho, H., DuBois, D. L., & Shaw, W. J. (2011). Incorporating peptides in the outer-coordination sphere of bioinspired electrocatalysts for hydrogen production. *Inorganic Chemistry*, *50*(9), 4073–4085. http://dx.doi.org/10.1021/ic1025872.

Jain, A., Reback, M. L., Lindstrom, M. L., Thogerson, C. E., Helm, M. L., Appel, A. M., & Shaw, W. J. (2012). Investigating the role of the outer-coordination sphere in [Ni(P(Ph)2N(PhR)2)2]2+ hydrogenase mimics. *Inorganic Chemistry*, *51*(12), 6592–6602. http://dx.doi.org/10.1021/ic300149x.

Knight, C., & Voth, G. A. (2012). The curious case of the hydrated proton. *Accounts of Chemical Research*, *45*, 101–109.

Knorzer, P., Silakov, A., Foster, C. E., Armstrong, F. A., Lubitz, W., & Happe, T. (2012). Importance of the protein framework for catalytic activity of [FeFe]-hydrogenases. *Journal of Biological Chemistry*, *287*(2), 1489–1499. http://dx.doi.org/10.1074/Jbc.M111.305797.

Laio, A., & Parrinello, M. (2002). Escaping free-energy minima. *Proceedings of the National Academy of Sciences of the United States of America*, *99*, 12562–12566.

Lense, S., Dutta, A., Roberts, J. A. S., & Shaw, W. J. (2014). A proton channel allows a hydrogen oxidation catalyst to operate at a moderate overpotential with water acting as a base. *Chemical Communications*, *50*(7), 792–795.

Lense, S., Ho, M. H., Chen, S. T., Jain, A., Raugei, S., Linehan, J. C., ... Shaw, W. (2012). Incorporating amino acid esters into catalysts for hydrogen oxidation: Steric and electronic effects and the role of water as a base. *Organometallics*, *31*(19), 6719–6731. http://dx.doi.org/10.1021/Om300409y.

Li, H., Robertson, A. D., & Jensen, J. H. (2005). Very fast empirical prediction and rationalization of protein pKa values. *Proteins*, *61*, 704–721.

Long, H., King, P. W., & Chang, C. H. (2014). Proton transport in Clostridium pasteurianum [FeFe]-hydrogenase I: A computational study. *The Journal of Physical Chemistry B*, *118*, 890–900.

Lubitz, W., Ogata, H., Rudiger, O., & Reijerse, E. (2014). Hydrogenases. *Chemical Reviews*, *114*(8), 4081–4148. http://dx.doi.org/10.1021/Cr4005814.

Macarie, L., & Ilia, G. (2014). Synthesis and polymerization of vinylphosphonic acid. *RSC Polymer Chemistry Series*, *11*, 51–67.

MacKerell, A. D., Bashford, D., Bellott, M., Dunbrack, R. L., Evanseck, J. D., Field, M. J., ... Karplus, M. (1998). All-atom empirical potential for molecular modeling and dynamics studies of proteins. *Journal of Physical Chemistry B*, *102*(18), 3586–3616.

Marenich, A. V., Cramer, C. J., & Truhlar, D. G. (2009). Universal solvation model based on the generalized born approximation with asymmetric descreening. *The Journal of Chemical Theory and Computation*, *5*, 2447–2464.

McCullagh, M., & Voth, G. A. (2013). Unraveling the role of the protein environment for [FeFe]-hydrogenase: A new application of coarse-graining. *Journal of Physical Chemistry B*, *117*, 4062–4071.

Nicolet, Y., Piras, C., Legrand, P., Hatchikian, C. E., & Fontecilla-Camps, J. C. (1999). Desulfovibrio desulfuricans iron hydrogenase: The structure shows unusual coordination to an active site Fe binuclear center. *Structure with Folding & Design*, 7(1), 13–23. http://dx.doi.org/10.1016/S0969-2126(99)80005-7.

Nymeyer, H., Gnanakaran, S., & García, A. E. (2004). Atomic simulations of protein folding using the replica exchange algorithm. *Methods in Enzymology*, *383*, 119–149.

O'Hagan, M., Ho, M. H., Yang, J. Y., Appel, A. M., Rawkowski DuBois, M., Raugei, S., …Bullock, R. M. (2012). Proton delivery and removal in [Ni((P2N2R)-N-R')](2) hydrogen production and oxidation catalysts. *Journal of the American Chemical Society*, *134*(47), 19409–19424. http://dx.doi.org/10.1021/Ja307413x.

O'Hagan, M., Shaw, W. J., Raugei, S., Chen, S., Yang, J. Y., Kilgore, U. J.,… Bullock, R. M. (2011). Moving protons with pendant amines: Proton mobility in a nickel catalyst for oxidation of hydrogen. *Journal of the American Chemical Society*, *133*, 14304–14312.

Okamura, M. Y., Paddock, M. L., Graige, M. S., & Feher, G. (2000). Proton and electron transfer in bacterial reaction centers. *Biochimica et Biophysica Acta (BBA)—Bioenergetics*, *1458*, 148–163.

Onufriev, A., Bashford, D., & Case, D. A. (2004). Exploring protein native states and large-scale conformational changes with a modified generalized born model. *Proteins: Structure, Function & Bioinformatics*, *55*, 383–394.

Perdew, J. P. (1986). Density-functional approximation for the correlation energy of the inhomogeneous electron gas. *Physical Review B: Condensed Matter and Materials Physics*, *33*, 8822–8824.

Perdew, J. P., Burke, K., & Ernzerhof, M. (1996). Generalized gradient approximation made simple. *Physical Review Letters*, 77(18), 3865–3868. http://dx.doi.org/10.1103/Physrevlett.77.3865.

Perdew, J. P., Burke, K., & Ernzerhof, M. (1997). Generalized gradient approximation made simple (vol 77, pg 3865, 1996). *Physical Review Letters*, *78*(7), 1396. http://dx.doi.org/10.1103/Physrevlett.78.1396.

Ponder, J. W., & Case, D. A. (2003). Force fields for protein simulations. *Advances in Protein Chemistry*, *66*, 27–85.

Price, D. J., & Brooks, C. L. B., III. (2002). Modern protein force fields behave comparably in molecular dynamics simulations. *The Journal of Computational Chemistry*, *23*, 1045–1057.

Ragsdale, S. W. (2006). Metals and their scaffolds to promote difficult enzymatic reactions. *Chemical Reviews*, *106*, 3317–3337.

Rakowski DuBois, M., & DuBois, D. L. (2009). The roles of the first and second coordination spheres in the design of molecular catalysts for H_2 production and oxidation. *Chemical Society Reviews*, *38*, 62–72. http://dx.doi.org/10.1039/b801197b.

Raugei, S., Chen, S., Ho, M.-H., Ginovska-Pangovska, B., Rousseau, R. J., Dupuis, M., …Bullock, R. M. (2012). The role of pendant amines in the breaking and forming of molecular hydrogen catalyzed by nickel complexes. *Chemistry A European Journal*, *18*, 6493–6506.

Reback, M. L., Buchko, G. W., Kier, B. L., Ginovska-Pangovska, B., Xiong, Y., Lense, S., …Shaw, W. J. (2014). Enzyme design from the bottom up: an active nickel electrocatalyst with a peptide outer coordination sphere. *Chemistry A European Journal*, *20*, 1510–1514. http://dx.doi.org/10.1002/chem.201303976.

Reback, M. L., Ginovska-Pangovska, B., Ho, M.-H., Jain, A., Squier, T. C., Raugie, S., … Shaw, W. J. (2013). The role of a dipeptide outer-coordination sphere on H2-production catalysts: influence on catalytic rates and electron transfer. *Chemistry A European Journal*, *19*, 1928–1941.

Schafer, A., Horn, H., & Ahlrichs, R. (1992). Fully optimized contracted Gaussian-basis sets for atoms Li to Kr. *Journal of Chemical Physics*, *97*(4), 2571–2577.

Shaw, W. J. (2012). The outer-coordination sphere: Incorporating amino acids and peptides as ligands for homogeneous catalysts to mimic enzyme function. *Catalysis Reviews: Science and Engineering*, *54*, 489–550.

Shaw, W. J., Helm, M. L., & DuBois, D. L. (2013). A modular, energy-based approach to the development of nickel containing molecular electrocatalysts for hydrogen production and oxidation. *Biochimica et Biophysica Acta (BBA)—Bioenergetics*, *1827*, 1123–1139.

Simmons, T. R., Berggren, G., Bacchi, M., Fontecave, M., & Artero, V. (2014). Mimicking hydrogenases: From biomimetics to artificial enzymes. *Coordination Chemistry Reviews*, *270–271*, 127–150.

Sode, O., & Voth, G. (2014). Electron transfer activation of a second water channel for proton transport in [FeFe]-hydrogenase. *Journal of Chemical Physics*, *141*, 1–9.

Sumner, I., & Voth, G. A. (2011). Proton transport pathways in [NiFe]-hydrogenase. *The Journal of Physical Chemistry B*, *116*, 2917–2926.

Sumner, I., & Voth, G. A. (2012). Proton transport pathways in [NiFe]-hydrogenase. *Journal of Physical Chemistry B*, *116*, 2917–2926.

Teixeira, V. H., Soares, C. M., & Baptista, A. M. (2008). Proton pathways in a [NiFe]-hydrogenase: A theoretical study. *Proteins: Structure, Function & Bioinformatics*, *70*, 1010–1022.

Tran, P. D., Goff, A. L., Heidkamp, J., Jousselme, B., Guillet, N., Palacin, S., … Artero, V. (2011). Noncovalent modification of carbon nanotubes with pyrene-functionalized nickel complexes: Carbon monoxide tolerant catalysts for hydrogen evolution and uptake. *Angewandte Chemie*, *50*, 1371–1374.

Valiev, M., Bylaska, E. J., Govind, N., Kowalski, K., Straatsma, T. P., Van Dam, H. J. J., … de Jong, W. (2010). NWChem: A comprehensive and scalable open-source solution for large scale molecular simulations. *Computer Physics Communications*, *181*(9), 1477–1489. http://dx.doi.org/10.1016/J.Cpc.2010.04.018.

VandeVondele, J., & Hutter, J. (2007). Gaussian basis sets for accurate calculations on molecular systems in gas and condensed phases. *Journal of Chemical Physics*, *127*(11), 114105. http://dx.doi.org/10.1063/1.2770708.

VandeVondele, J., Krack, M., Mohamed, F., Parrinello, M., Chassaing, T., & Hutter, J. (2005). QUICKSTEP: Fast and accurate density functional calculations using a mixed Gaussian and plane waves approach. *Computer Physics Communications*, *167*(2), 103–128. http://dx.doi.org/10.1016/J.Cpc.2004.12.014.

Vincent, K. A., Parkin, A., & Armstrong, F. A. (2007). Investigating and exploiting the electrocatalytic properties of hydrogenases. *Chemical Reviews*, *107*, 4366–4413.

Voth, G. A. (2006). Computer simulation of proton solvation and transport in aqueous and biomolecular systems. *Accounts of Chemical Research*, *39*(2), 143–150. http://dx.doi.org/10.1021/Ar0402098.

Waller, I. (1923). Zur Frage der Einwirkung der Wärmebewegung auf die Interferenz von Röntgenstrahlen. *Zeitschrift für Physik A (in German)*, *17*, 398–408 (in German).

Wickstrong, C., Dods, R., Royant, A., & Neutze, R. (2015). Bacteriorhodopsin: Would the real structural intermediates please stand up? *Biochimica et Biophysica Acta*, *1850*, 536–553.

Wilson, A. D., Newell, R. H., McNevin, M. J., Muckerman, J. T., DuBois, M. R., & DuBois, D. L. (2006). Hydrogen oxidation and production using nickel-based molecular catalysts with positioned proton relays. *Journal of the American Chemical Society*, *128*(1), 358–366. http://dx.doi.org/10.1021/Ja056442y.

Wraight, C. A. (2006). Chance and design—Proton transfer in water, channels and bioenergetic proteins. *Biochimica et Biophysica Acta*, *1757*, 886–912.

Yan, L., & Xie, L. (2012). Molecular dynamics simulations of proton transport in proton exchange membranes based on acid-base complexes. *Molecular Interactions*, 335–360.

Yoshikawa, S., Muramoto, K., & Shinzawa-Itoh, K. (2011). Proton-pumping mechanism of cytochrome c oxidase. *Annual Review of Biophysics*, *40*, 205–223.

CHAPTER FIVE

Modeling Mercury in Proteins☆

J.M. Parks*,†,1, J.C. Smith*,†,1
*Oak Ridge National Laboratory, Oak Ridge, TN, United States
†University of Tennessee, Knoxville, TN, United States
[1]Corresponding authors: e-mail address: parksjm@ornl.gov; smithjc@ornl.gov

Contents

Abstract

Mercury (Hg) is a naturally occurring element that is released into the biosphere both by natural processes and anthropogenic activities. Although its reduced, elemental form Hg(0) is relatively nontoxic, other forms such as Hg^{2+} and, in particular, its methylated form, methylmercury, are toxic, with deleterious effects on both ecosystems and humans. Microorganisms play important roles in the transformation of mercury in the environment. Inorganic Hg^{2+} can be methylated by certain bacteria and archaea

☆This manuscript has been authored by UT-Battelle LLC under Contract No. DE-AC05-00OR22725 with the US Department of Energy. The US Government retains and the publisher by accepting the article for publication acknowledges that the US Government retains a nonexclusive paid-up irrevocable worldwide license to publish or reproduce the published form of this manuscript or allow others to do so for United States Government purposes. The Department of Energy will provide public access to these results of federally sponsored research in accordance with the DOE Public Access Plan (http://energy.gov/downloads/doe-public-access-plan).

Methods in Enzymology, Volume 578
ISSN 0076-6879
http://dx.doi.org/10.1016/bs.mie.2016.05.041

to form methylmercury. Conversely, bacteria also demethylate methylmercury and reduce Hg^{2+} to relatively inert Hg(0). Transformations and toxicity occur as a result of mercury interacting with various proteins. Clearly, then, understanding the toxic effects of mercury and its cycling in the environment requires characterization of these interactions. Computational approaches are ideally suited to studies of mercury in proteins because they can provide a detailed molecular picture and circumvent issues associated with toxicity. Here, we describe computational methods for investigating and characterizing how mercury binds to proteins, how inter- and intraprotein transfer of mercury is orchestrated in biological systems, and how chemical reactions in proteins transform the metal. We describe quantum chemical analyses of aqueous Hg(II), which reveal critical factors that determine ligand-binding propensities. We then provide a perspective on how we used chemical reasoning to discover how microorganisms methylate mercury. We also highlight our combined computational and experimental studies of the proteins and enzymes of the *mer* operon, a suite of genes that confer mercury resistance in many bacteria. Lastly, we place work on mercury in proteins in the context of what is needed for a comprehensive multiscale model of environmental mercury cycling.

1. INTRODUCTION

1.1 Mercury Toxicity

It is well known that mercury (Hg) is toxic to living organisms. Hg^{2+} and the monomethylated form of Hg, methylmercury (CH_3HgR, where R is an anionic ligand), are particularly dangerous to living organisms. Their toxicity results from the ability to bind strongly to various biomolecules and disrupt their normal function. Methylmercury is also able to cross the blood–brain barrier and is therefore a neurotoxin. It is generally accepted that Hg has no known beneficial function in any living organism, although that view has been challenged recently (Gregoire & Poulain, 2016).

Proteins play important roles in the environmental processes that transform Hg from one form to another. They are also important contributors to Hg toxicity in humans. Notably, Hg^{2+} has an extremely high affinity for thiol(ate) groups, which are present in the amino acid cysteine in numerous proteins. Hg^{2+} can also outcompete other metals, such as Zn and Fe, in the binding sites of metalloproteins or metalloenzymes. Therefore, understanding how Hg interacts with proteins is of both fundamental and practical importance. Specific Hg–protein interactions play key roles in binding to specific functional groups, reduction/oxidation and methylation/demethylation.

1.2 Inorganic Hg Chemistry

Before beginning a discussion of the interaction of Hg with proteins, it is important to provide a brief summary of the aqueous-phase chemistry of inorganic Hg. The most relevant oxidation states are Hg^0 and Hg^{2+}, so we limit the present discussion to these two forms. Elemental Hg(0) is not particularly toxic, but is readily oxidized to highly reactive and highly toxic Hg^{2+}. Hg^{2+} does not generally exist as an isolated dication in aqueous solution. Rather, it is essentially always bound to two or more anionic or neutral ligands. Among the most strongly interacting ligands are thiolates (RS^-), as found in the amino acid cysteine and the cellular redox mediator glutathione. Other low molecular weight thiolates, such as hydrosulfide (HS^-), also bind Hg^{2+} with high affinity. Hg^{2+} typically binds two or three thiolates to form bis or tris species of the form $[Hg^{II}(SR)_n]^{2-n}$, $n=2$ or 3. Selenides, which are chemically similar to but generally less abundant than sulfides, also interact strongly with Hg^{2+}. In the absence of thiolates, Hg^{2+} binds other functional groups, such as hydroxides, halides, carboxylates, and amines, with affinities that span many orders of magnitude in terms of binding constants. In complex biological systems and natural environments, Hg encounters numerous chemical species. Although thermodynamic quantities (ie, solvation and ligand-binding free energies) have been measured for many Hg-containing molecules, there is still a fairly large degree of uncertainty in many of these values. It is therefore important to quantify Hg interactions in a consistent way so that Hg speciation in complex systems can be understood at a fundamental level.

1.3 Quantum Chemistry

Quantum chemistry is a natural means of determining the driving forces behind the speciation of Hg^{2+}. Density functional theory (DFT) is the standard method of choice for quantum chemistry because it can provide accurate structures and energies provided that a suitable density functional and basis set are used. However, wave function methods, such as MP2, can provide similar or better accuracy than DFT. It is also critical to describe relativistic effects adequately in Hg-containing systems. We have found that small-core, quasi-relativistic effective core potentials (ECPs) perform quite well in this regard.

Because our emphasis is on aqueous speciation and reactivity, it is particularly important to account properly for the effects of solvation. It is common to carry out quantum chemical calculations in combination with a polarizable continuum model (PCM) representation of the solvent. However, continuum approaches are known to be unable to describe strong, short-range, anisotropic interactions with the solvent. A well-known example of the limitations of continuum solvation is the inability to describe explicit hydrogen bonding between the solute and solvent, but other types of direct interactions such as Hg–O coordination can also play important roles. Thus, it is becoming increasingly more common to include a relatively small number ($n=1$–18) explicit water molecules as part of the supersolute to describe short-range solute–solvent interactions and implicit (ie, PCM) solvation to describe long-range, bulk solvation effects. Such an approach is often referred to as a mixed discrete-continuum method.

1.4 Hydration Free Energies

In a recent study, we used dispersion-corrected DFT and MP2 calculations in combination with a mixed discrete-continuum solvation approach to compute hydration free energies for the group 12 divalent metal cations Zn^{2+}, Cd^{2+}, and Hg^{2+} and the anions SH^-, OH^-, Cl^-, and F^- (Riccardi, Guo, Parks, Gu, Liang, et al., 2013; Riccardi, Guo, Parks, Gu, Summers, et al., 2013). Long-range solvation effects were described by the SMD solvent model (Marenich, Cramer, & Truhlar, 2009). As is commonly done, we used thermodynamic cycles to compute the hydration free energies, $\Delta G_{aq}{}^*$. In such an approach, (super)solute geometries are optimized in the gas phase, and then solvation energies for each reactant and product are computed as the difference in energy between the gas-phase and PCM-solvated molecules. Including explicit solvent molecules complicates the matter by introducing additional degrees of freedom, and care must be taken to account properly for standard states. More thorough discussions of the challenges associated with mixed discrete-continuum solvation have been covered elsewhere (Bryantsev, Diallo, & Goddard, 2008; Riccardi, Guo, Parks, Gu, Liang, et al., 2013; Riccardi, Guo, Parks, Gu, Summers, et al., 2013).

We considered various schemes for selecting the optimal number of explicit water molecules to include along with the solute, and we found that the use of a constant number for a given ion type (ie, metal cation or inorganic ion) provided the greatest accuracy. For the metal cations and anions, inclusion of 10 and 8 explicit water molecules, respectively, was found to provide the lowest errors compared to experimentally measured relative hydration free energies. Although hybrid DFT performed quite well in this context, the best agreement with experiment came from MP2, which yielded a standard deviation of 2.3 kcal/mol. In another recent DFT study (Afaneh, Schreckenbach, & Wang, 2014), the use of four or five explicit water molecules to saturate the first solvation shell of the solutes was recommended in the context of mixed discrete-continuum calculations. Similar accuracy to our approach was obtained.

1.5 Hg–Ligand Binding Free Energies

Based on the accuracy obtained for the computed (relative) hydration free energies, we extended the approach to Hg–ligand binding free energies (Riccardi, Guo, Parks, Gu, Summers, et al., 2013). In that work, we considered three hydrochalcogenide ligands (OH^-, SH^-, and SeH^-) and three halide ligands (F^-, Cl^-, and Br^-). Binding of these ligands to Hg^{2+} proceeds as follows:

$$Hg^{2+} + L_1 + L_2 \leftrightharpoons HgL_1L_2.$$

Using the dispersion-corrected (Grimme, Antony, Ehrlich, & Krieg, 2010; Grimme, Ehrlich, & Goerigk, 2011) B3PW91 hybrid density functional (Becke, 1993; Perdew & Wang, 1992) (ie, B3PW91-D3), along with small-core ECPs (Peterson & Puzzarini, 2005) and the SMD solvent model (Marenich et al., 2009), aqueous relative binding free energies ($\Delta\Delta G_{aq}{}^*$) were computed and compared with experimental values. Here, we found that it was sufficient to include only two explicit water molecules for each species to obtain highly accurate results (see later). In this case, we obtained a mean signed error of 0.7 kcal/mol and a standard deviation of 0.8 kcal/mol (Fig. 1). The calculations reproduced the correct trend, $Hg(SH)_2 > Hg(OH)_2 > HgBr_2 \simeq Hg(OH)Cl > HgCl_2$, in the experimentally determined aqueous binding affinities.

With quantum chemistry it is also possible to dissect individual energetic contributions to gain insight into the underlying physical phenomena that govern ligand-binding behavior. Solvation effects can be addressed by comparing processes in the gas and aqueous phases with varying descriptions of the solvent. In this case, that approach led to somewhat surprising results. We examined the behavior of the computed binding free energies in four stages. First, we considered only gas-phase species. We then added one explicit water molecule, then two explicit water molecules, and finally two explicit water molecules plus continuum solvation (Fig. 2).

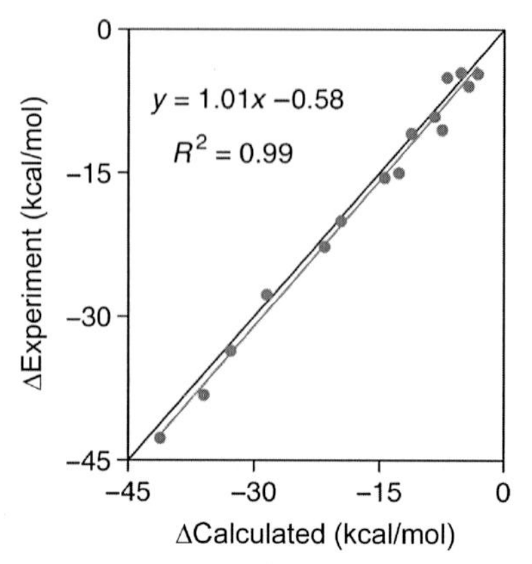

Fig. 1 Experimental and theoretical binding free energy differences for a set of neutral HgR_2 complexes. *Adapted with permission from Riccardi, D., Guo, H.-B., Parks, J. M., Gu, B., Summers, A. O., Miller, S. M., … Smith, J. C. (2013). Why mercury prefers soft ligands.* Journal of Physical Chemistry Letters, 4, *2317–2322. doi:10.1021/jz401075b. Copyright 2013 American Chemical Society.*

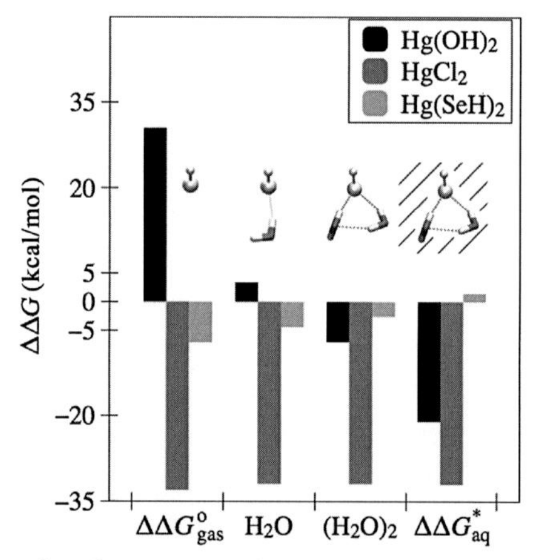

Fig. 2 Differences in binding free energy relative to $Hg(SH)_2$ with varying descriptions of the solvent. *Left* to *right*: gas phase, one and two explicit water molecules, and two explicit water molecules with SMD continuum solvation. The optimized geometries for SH^-, $H_2O{\cdot}SH^-$, and $(H_2O)_2{\cdot}SH^-$ are also shown. The *shaded region* denotes continuum solvation. *Reprinted with permission from Riccardi, D., Guo, H.-B., Parks, J. M., Gu, B., Summers, A. O., Miller, S. M., … Smith, J. C. (2013). Why mercury prefers soft ligands.* Journal of Physical Chemistry Letters, 4, *2317–2322. doi:10.1021/jz401075b. Copyright 2013 American Chemical Society.*

In aqueous solution, formation of $Hg(SH)_2$ is 22.8 kcal/mol more favorable than $Hg(OH)_2$, and our calculations reproduce this free energy difference. On the other hand, in the gas phase the formation of $Hg(SH)_2$ from its constituent ions in the gas phase is ~30 kcal/mol *less* favorable than $Hg(OH)_2$. Inclusion of a single water molecule makes the binding free energies nearly equal, and the formation of $Hg(SH)_2$ becomes more favorable than $Hg(OH)_2$ when only two water molecules are included. When two water molecules and continuum solvation are included, the experimental binding free energy difference between the two complexes is reproduced almost exactly (Fig. 2). Thus, solvation effects are responsible for a 52 kcal/mol change in the relative binding free energies between these two molecules upon transferring them from the gas phase to aqueous solution, and the majority of this effect is imparted by water molecules that interact strongly with the solutes.

2. MICROBIAL INTERACTIONS WITH HG

We now turn to modeling mercury in proteins. Microorganisms are key players in the global cycling of Hg. Various bacteria and archaea are involved in Hg toxification, detoxification, or both. Whereas some microbes methylate Hg, others demethylate it. Some can both methylate and demethylate it. Here, we provide a synopsis of some of the important proteins and enzymes that mediate the transformation of Hg species. We then discuss the computational approaches we have used to provide mechanistic insight into these processes.

2.1 Bacterial Hg Resistance

Many aerobic bacteria are able to thrive in Hg-contaminated environments because they possess the Hg resistance, or *mer*, operon. The prototypical, broad-spectrum *mer* operon encodes several proteins and enzymes. In addition to a transcriptional regulator, MerR, Hg-resistant bacteria produce various membrane-bound transporter proteins (MerC, MerT, and others), a periplasmic metallochaperone (MerP), and two key enzymes (MerB and MerA). Our studies have focused specifically on the two enzymes MerB and MerA, and the transcriptional regulator, MerR.

A common theme in the *mer* proteins is the ubiquity of pairs of Cys residues that bind Hg^{2+} and orchestrate its transfer throughout the cell until it is ultimately reduced to Hg^0 and no longer poses a threat to the cellular machinery.

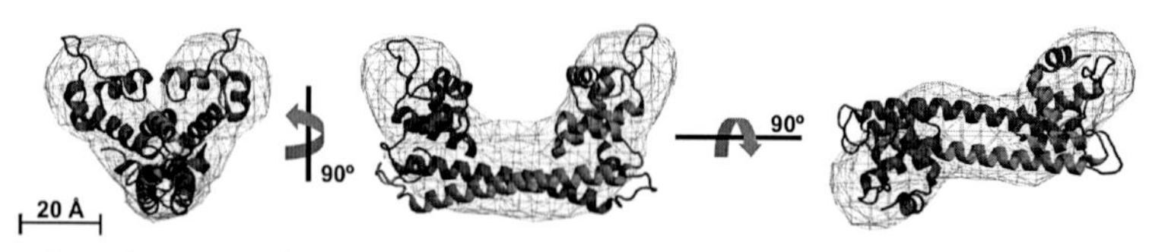

Fig. 3 Best-fitting conformation of Hg(II)-MerR obtained from MD simulations superimposed onto the three-dimensional molecular envelope (*blue mesh*) as determined by SAXS. *Reprinted from Guo, H.-B., Johs, A., Parks, J. M., Olliff, L., Miller, S. M., Summers, A. O., … Smith, J. C. (2010). Structure and conformational dynamics of the metalloregulator MerR upon binding of Hg(II).* Journal of Molecular Biology, 398, *555–568. doi: 10.1016/j.jmb.2010.03.020. Copyright 2010, with permission from Elsevier.* (See the color plate.)

2.2 MerR

The bacterial metalloregulator MerR controls transcription of the *mer* operon in Hg-resistant bacteria. Although related proteins had been characterized by X-ray crystallography, MerR had proven difficult to crystallize. Using small-angle X-ray scattering (SAXS) and molecular dynamics (MD) simulation on a homology of the protein, we sought to determine its structure and conformational dynamics in solution to gain insight into possible mechanisms of transcriptional control (Fig. 3). From the MD simulations, interdomain motions on the ~10 ns time scale were found for the two DNA-binding domains of the homodimeric protein. Thus, Hg^{2+}-bound MerR exhibits conformational dynamics that induce conformational changes in the *mer* operator DNA such that transcription is initiated by RNA polymerase. Recent work performed elsewhere confirmed that the structures obtained with SAXS and MD are in good agreement with X-ray crystal structures of Hg(II)-bound MerR (Chang, Lin, Zou, Huang, & Chan, 2015).

2.3 MerB

The organomercurial lyase, MerB, cleaves Hg–C bonds in organomercurials according to the following reaction:

$$\left[\mathrm{R}-\mathrm{Hg}^{\mathrm{II}}\right]^{+} \rightarrow \mathrm{Hg}^{2+} + \mathrm{R}-\mathrm{H},$$

where R is an alkyl or aryl substituent, with the most prevalent organomercurial species being methylmercury. X-ray crystal structures of Hg-free and Hg^{2+}-bound MerB have been solved (Lafrance-Vanasse, Lefebvre, Di Lello, Sygusch, & Omichinski, 2009), and the active site contains two Cys and one Asp residue (Fig. 4). All three of these residues are required for activity (Parks et al., 2009; Pitts & Summers, 2002).

A common method for simulating enzyme catalysis is the quantum mechanical/molecular mechanical (QM/MM) approach (Senn & Thiel,

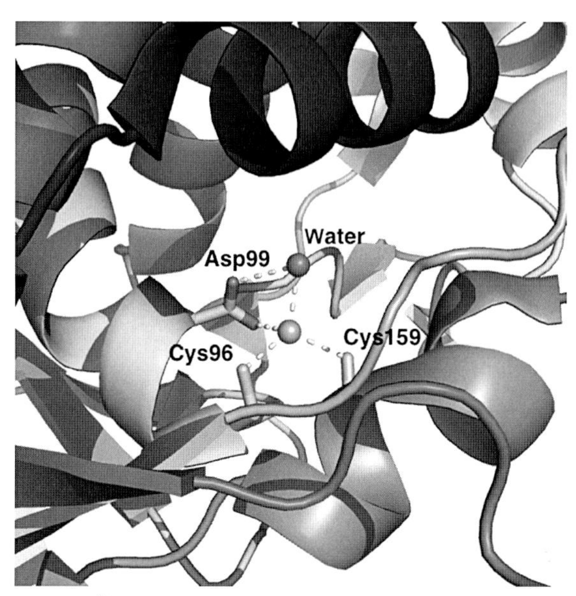

Fig. 4 Active site of Hg^{2+}-bound MerB (PDB entry 3F2F (Lafrance-Vanasse et al., 2009)). *Adapted with permission from Parks, J. M., Guo, H., Momany, C., Liang, L. Y., Miller, S. M., Summers, A. O., & Smith, J. C. (2009). Mechanism of Hg–C protonolysis in the organomercurial lyase MerB.* Journal of the American Chemical Society, 131*(37), 13278–13285. doi:10.1021/Ja9016123. Copyright 2009 American Chemical Society.* (See the color plate.)

2009; Warshel & Levitt, 1976). Alternatively, QM-only cluster models (Siegbahn & Himo, 2011) have been used successfully for numerous enzymes, particularly transition metal-containing metalloenzymes. MerB is not a metalloenzyme in the sense that it uses a metal cofactor (because methylmercury is instead the substrate), and Hg is not a transition metal. However, we recognized that there are certain parallels between MerB and metalloenzymes. In such a case, coordination chemistry tends to be the most important aspect to consider. Thus, we used a quantum chemical cluster approach to study Hg–C cleavage in MerB (Parks et al., 2009).

To construct the active-site model, the Cartesian coordinates of the side chains of Cys96, Asp99, and Cys159 (*Escherichia coli* numbering) along with Hg^{2+} and a single crystallographic water molecule were extracted from the X-ray crystal structure of Hg^{2+}-bound MerB. Missing hydrogens were added and the model was manually modified to include a $[CH_3Hg]^+$ substrate rather than the Hg^{2+} product. We performed hybrid DFT optimizations with the B3PW91 density functional (Becke, 1993; Perdew & Yue, 1986), a suitable ECP for Hg (Andrae, Haussermann, Dolg, Stoll, & Preuss, 1990), and a polarizable continuum representation to describe the

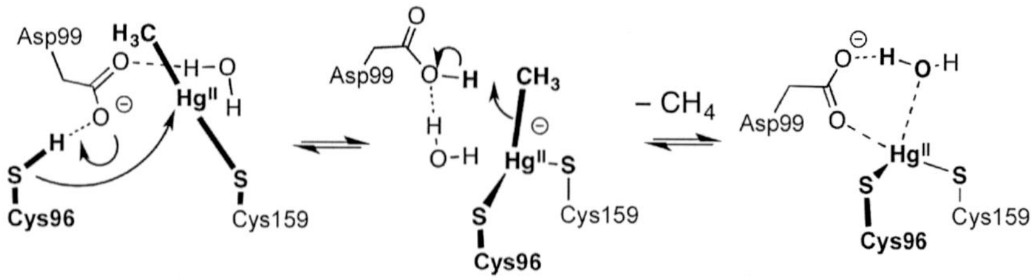

Asp99 H3C O H O Hg^II H H S Cys96 S Cys159 Asp99 O H CH3 H O-H Hg^II-S S Cys159 Cys96 - CH4 Asp99 O H O H Hg^II S S Cys96 Cys159

Fig. 5 Proposed mechanism of MerB as deduced from quantum chemical cluster calculations with polarizable continuum solvation.

protein/solvent environment around the active site. As is common in similar studies, a dielectric constant of 4 was used to describe the mostly hydrophobic active site. Optimized geometries of various reactant, intermediate, transition state, and products were then used to map out the energetic landscape of the reaction and identify energetically favorable reaction mechanisms.

These simple QM cluster models were able to reproduce the experimentally observed trend in enzymatic turnover (ie, k_{cat}) for the three organomercurial substrates considered. Furthermore, the computed activation free energies were all within about 1 kcal/mol of the experimental values derived from k_{cat} using transition state theory. From the calculations, the most energetically favorable reaction mechanism emerged (Fig. 5), which we summarize here. Initially, CH_3Hg^+ is coordinated to only one of the active-site Cys residues and the other Cys remains a neutral thiol. The second Cys residue then coordinates to methylmercury to form a trigonal species, and transfers its proton to Asp99. Coordination by two Cys residues weakens the Hg–C bond substantially, and the Asp is ideally positioned to donate its proton to the leaving group carbon. The result is a protonolytic cleavage of the Hg–C bond, with Hg^{2+} remaining bound in the active site as observed crystallographically. Thus, MerB accomplishes the demethylation of methylmercury by quenching an otherwise highly reactive carbanion through protonation as it begins to form. The calculations showed that coordination of the methylmercury substrate by two Cys thiolates is sufficient to activate the Hg–C bond, redistributing electron density into the methyl leaving group and away from the proton on the catalytic acid, Asp99.

2.4 MerA

After methylmercury has been demethylated by MerB to form Hg^{2+}, it is reduced by the mercuric reductase, MerA, an NADP-dependent flavoenzyme. Although numerous MerA variants exist, our studies have focused on those variants that consist of a homodimeric catalytic core domain with each monomer bearing an N-terminal metallochaperone-like domain, NmerA, connected to the catalytic core domain by a flexible and disordered peptide linker ~30 residues in length. This form of MerA contains multiple pairs of Cys residues that sequentially bind Hg^{2+} and orchestrate its handoff to the next pair.

Initially, Hg^{2+} is bound to NmerA through two highly conserved Cys residues. Because NmerA is linked to the core domain, the two subunits remain in relatively close proximity to the core. In one study (Johs et al., 2011), we used a combination of SAXS and small-angle neutron scattering along with MD simulation to characterize the flexibility of full-length MerA. Using a simple rigid-body MD approach, we generated a large number of linker conformations of the NmerA domains relative to the core. We then computed X-ray scattering profiles and compared them with the experimentally determined profile. The best fits to the experimental data were obtained when a small ensemble of disordered linker conformations was used, suggesting that the NmerA domains and linkers are highly flexible and sample large regions of conformational space.

The dynamic nature of the NmerA metallochaperone domains may facilitate the efficient capture and subsequent transfer of Hg^{2+} to the core domain for subsequent reduction. In a novel approach to protein–protein docking (Johs et al., 2011), we also used SAXS/SANS and MD to reveal how NmerA docks to the core domain to accomplish the interdomain transfer of Hg^{2+}. Scattering profiles were obtained for a variant of the enzyme with a disulfide linkage connecting one Cys on NmerA to another Cys at the C-terminus of the core domain. Computational models of the system in various interdomain orientations were generated and subjected to MD simulation. As for the previous models, scattering profiles were computed for each model and compared with the experimental data. In this case, a single MD conformation was sufficient to match the experimental scattering profile, which revealed the most likely docking configuration between the two domains (Fig. 6).

In subsequent work, neutron spin-echo (NSE) spectroscopy was combined with all-atom, explicitly solvated and coarse-grained MD simulation of MerA (Hong et al., 2014) (Fig. 7). NSE permits large-scale motions of proteins to be probed directly. In both the experiments and the simulations, the NmerA domains were found to interact with the core domain primarily through electrostatic interactions. The observed compact but dynamic

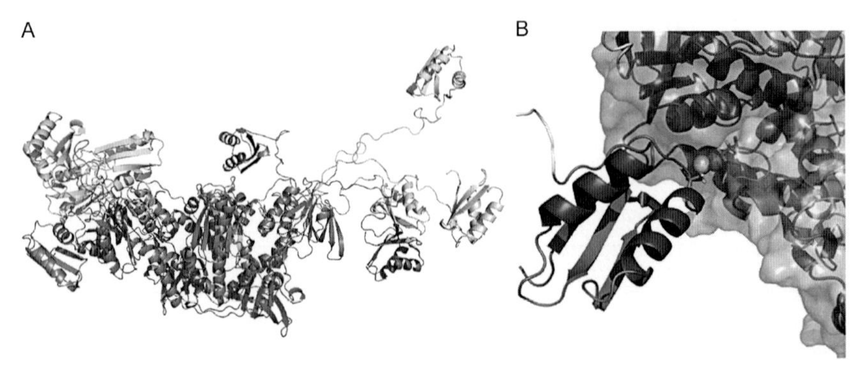

Fig. 6 (A) Representative ensemble of conformations obtained from rigid-body MD simulation. Including multiple conformations improve the agreement with experimental SAXS data compared to using only a single, best-fitting conformation. (B) Docking orientation of NmerA to the catalytic core of MerA as determined from small-angle scattering and MD simulation. *Reprinted from Johs, A., Harwood, I. M., Parks, J. M., Nauss, R. E., Smith, J. C., Liang, L. Y., & Miller, S. M. (2011). Structural characterization of intramolecular Hg^{2+} transfer between flexibly linked domains of mercuric ion reductase.* Journal of Molecular Biology, 413*(3), 639–656. doi:10.1016/J.Jmb.2011.08.042. Copyright 2011, with permission from Elsevier.*

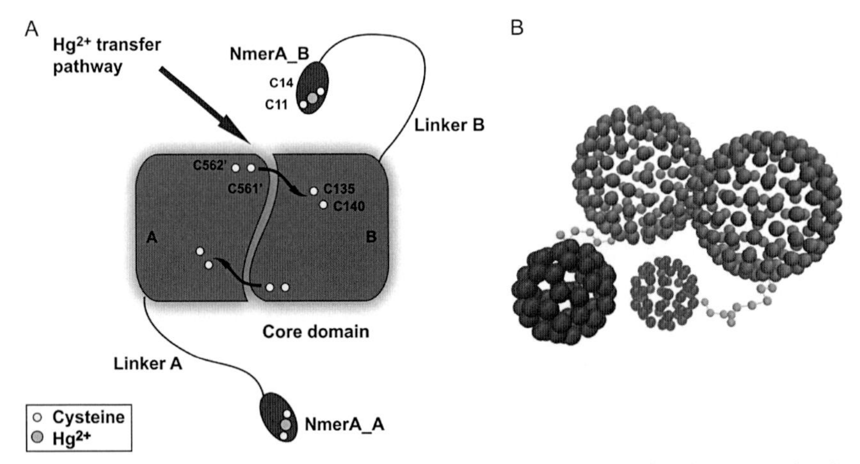

Fig. 7 (A) Schematic illustration of the MerA homodimer and scheme for intramolecular Hg^{2+} transfer. Hg^{2+} is first bound by the NmerA cysteines (C11 and C14), and is then delivered to the C-terminal cysteines of the other monomer (C561′ and C562′). Lastly, Hg^{2+} is transferred to a pair of cysteines in the catalytic site of the core (C135 and C140) where it is then reduced to Hg^0. (B) Simplified coarse-grained model for MerA with a rigid two-sphere core (*red*) and small spheres for the NmerA domains (*blue*), which are connected to the core by flexible linkers (*green*). *Reprinted from Hong, L., Sharp, M. A., Poblete, S., Biehl, R., Zamponi, M., Szekely, N., … Smith, J. C. (2014). Structure and dynamics of a compact state of a multi-domain protein, the mercuric ion reductase.* Biophysical Journal, 107, *2014, 393–400. doi: 10.1016/j.bpj.2014.06.013. Copyright 2014, from Elsevier.* (See the color plate.)

arrangement of the domains may facilitate transfer of Hg^{2+} from NmerA to the core, where it is then reduced to Hg^0.

2.5 Intramolecular Hg^{2+} Transfer

Having established the role of conformational dynamics in MerA, we now discuss the chemistry of Hg^{2+} transfer in MerA. Binding of Hg^{2+} by two Cys thiolates is thermodynamically extremely favorable. Hg^{2+} and $[RHg(II)]^+$ species have extremely high affinities for thiols, with K_{form} values for Hg bis thiolate complexes on the order of $\sim 10^{35}$–10^{40} M^{-2} (Stricks & Kolthoff, 1953). However, these complexes can undergo rapid exchange between thiols (Cheesman, Arnold, & Rabenstein, 1988). That is, one of the thiolates can readily dissociate, provided that a third thiolate first coordinates to Hg. Thus, a transient, trigonal $[Hg^{II}(Cys)_3]^-$ species is expected to be important in Hg^{2+} transfer reactions. Also, acid–base chemistry in which thiols are deprotonated to generate nucleophilic thiolates, or coordinated thiolates are protonated to generate neutral leaving groups, can enhance the rates of Hg^{2+} transfer among pairs of thiols. QM/MM calculations were used to identify a possible mechanism for the intramolecular Hg^{2+} transfer in MerA (Lian et al., 2014), which can be considered a prototype for Hg transfer in other *mer* protein and enzymes. Specifically, an X-ray crystal structure of the catalytic core of MerA with Hg^{2+} bound to the C-terminal Cys pair was used as a starting point for simulating Hg^{2+} transfer from the surface of the protein to the buried, inner pair of Cys residues in the active site (Lian et al., 2014).

From the computed Hg^{2+} transfer pathway (Lian et al., 2014), we note that Hg^{2+} is always paired with two or more thiolates. In both the initial (Hg bound to Cys558′ and Cys 559′) and final (Hg bound to Cys136 and Cys141) states (Fig. 8), the $Hg(Cys)_2$ complexes bear a neutral charge. As Hg^{2+} is transferred from the solvent-exposed protein surface to the buried catalytic site, a proton is transferred in the opposite direction. Thus, a net negative charge is transferred over ~8 Å. The key mechanistic insight from the simulations is that Hg^{2+} transfer is facilitated by coupling the competitive binding of pairs of Cys residues with the proton affinities of the thiolates. These principles are expected to be general to other proteins and enzymes of the *mer* operon and metal ion trafficking proteins in other biological systems.

Once Hg^{2+} is bound to the active-site Cys pair, NADPH transfers hydride to FAD to form $FADH^-$, which reduces the Cys–Hg(II)–Cys complex to produce Hg^0. Because it no longer has any appreciable affinity for Cys thiolates or other functional groups, Hg^0 simply diffuses out of the bacterial cell.

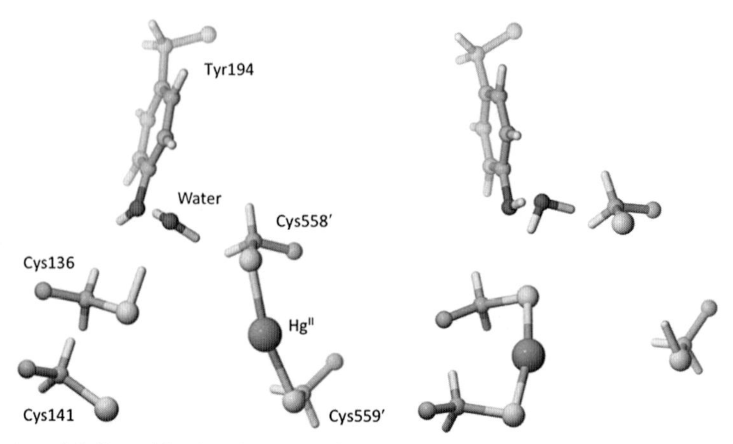

Fig. 8 Initial (*left*) and final (*right*) states from QM/MM simulations of the intramolecular transfer of Hg^{2+} from the C-terminal Cys pair on the surface of the protein to the Cys pair in the active site of MerA. Only the atoms in the QM subsystem are shown. *Adapted with permission from Lian, P., Guo, H. B., Riccardi, D., Dong, A. P., Parks, J. M., Xu, Q., … Guo, H. (2014). X-ray Structure of a Hg^{2+} complex of mercuric reductase (MerA) and quantum mechanical/molecular mechanical study of Hg^{2+} transfer between the C-terminal and buried catalytic site cysteine pairs.* Biochemistry, *53(46), 7211–7222. doi:10.1021/bi500608u. Copyright 2014 American Chemical Society.*

2.6 Hg Methylation

In the late 1960s, it was shown that anaerobic microorganisms could produce methylmercury from inorganic Hg. However, the genetic and biochemical basis for the reaction remained elusive for more than four decades. In the early to mid-1990s, Hg methylation was shown to be an enzyme-catalyzed process (Choi, Chase, & Bartha, 1994a) involving a protein with a corrinoid, ie, vitamin B12-like, cofactor (Choi & Bartha, 1993). The methylation process was also shown to be associated with the reductive acetyl-CoA, or Wood–Ljungdahl (WL) pathway (Choi, Chase, & Bartha, 1994b), which is involved in acetogenesis in anaerobic bacteria (Ragsdale & Pierce, 2008). Numerous bacteria possess the WL pathway, but only a small number were known to be Hg methylators, and there was no obvious phylogenetic pattern to suggest which organisms could and could not perform the reaction. It was unclear which specific protein or proteins were required to methylate Hg.

To identify the genes and corresponding proteins responsible for Hg methylation, we approached the problem by considering the chemistry that would be required if a corrinoid protein were involved. There is only one corrinoid protein in the WL pathway, and this protein, called the corrinoid iron–sulfur protein (CFeSP), accepts and then transfers CH_3^+ in the biosynthesis of acetyl-CoA. In CFeSP, there is no axial cobalt ligand *trans* to the methyl group, and this configuration is known to favor CH_3^+ transfer. The dominant form of Hg in cells is Hg^{2+} bound to two thiolates (Colombo, Ha, Reinfelder, Barkay, & Yee, 2013). It is not possible to form a Hg–C bond between Hg^{2+} and CH_3^+ because there are simply not enough electrons. A more plausible reaction leading to Hg–C bond formation would involve a methyl carbanion, CH_3^-, which can in principle be delivered by a methylated corrinoid. Such a protein must be different from CFeSP to perform the required chemistry (as CFeSP transfers CH_3^+), so we hypothesized that a corrinoid protein that could methylate Hg^{2+} would need to have an electron-donating group in the axial position opposite the methyl group. We hoped that this protein might be similar enough in sequence to the CFeSP that we could search for such a protein in sequence databases. Thus, we performed a BLAST search with the sequence of CFeSP against the translated genome of a known Hg-methylating organism. Although CFeSP is not present in this organism, a portion of a single protein sequence encoded in the entire genome was a match, and this sequence corresponded to the corrinoid-binding domain of CFeSP. We then found that this newly identified protein sequence is encoded in the genomes of all known Hg-methylating bacteria, but absent in all nonmethylators. X-ray crystal structures have been determined for CFeSP (Svetlitchnaia, Svetlitchnyi, Meyer, & Dobbek, 2006), so we generated a homology model of the corrinoid-binding domain. To our surprise, the model revealed a strictly conserved Cys residue positioned ideally for coordination to cobalt on the lower face of the cofactor in what we refer to as a "Cys-on" cobalt binding configuration (Fig. 9A). A Cys thiolate is expected to donate electron density into the cofactor, stabilizing the Co^{3+} oxidation state and facilitating transfer of CH_3^- to a Hg^{2+}-containing substrate molecule.

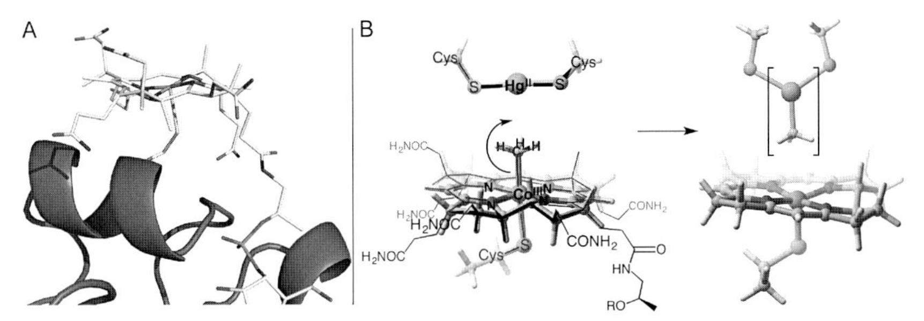

Fig. 9 (A) Predicted "Cys-on" binding of the corrinoid cofactor in a homology model of HgcA. (B) Quantum chemical model used to investigate methyl transfer from HgcA to a $Hg^{II}(Cys)_2$ substrate. The resulting methylmercury product is shown in *square brackets*.

Corrinoids are known to require low-potential electrons to achieve the reduction of Co to the Co^{1+} oxidation state. Intriguingly, a second gene encoding a ferredoxin is always present nearby in the genomes.

Up to this point, the genes and proteins proposed to be involved in Hg methylation had yet not been tested and verified experimentally. It is important to acknowledge the Herculean experimental effort performed by our experimental collaborators that was needed to confirm the predictions. No genetic system for deleting genes in any of the predicted Hg-methylating bacteria, so one had to be developed essentially from scratch. Fortunately, the predictions were ultimately confirmed (Parks et al., 2013), so it was worth the effort. When either gene was deleted, the bacteria no longer produced methylmercury. When the genes were reintroduced into the genome, methylation activity was restored. We now refer to the corrinoid protein as HgcA and the ferredoxin as HgcB, and the genes encoding these proteins have been found in the genomes of numerous bacteria and archaea.

In subsequent work, we used quantum chemical models of HgcA and computed reaction free energies for the transfer of a methyl group to a $Hg^{II}(SR)_2$ substrate (Fig. 9B) (Zhou, Riccardi, Beste, Smith, & Parks, 2014). We compared the favorability of transferring either a methyl radical or methyl carbanion, and found that carbanion transfer was more favorable when (continuum) solvation effects were included. Furthermore, we showed that the axial Cys ligand indeed promotes carbanion transfer compared to a His ligand, the latter of which is common in corrinoid proteins. Site-directed mutagenesis experiments provided further support for Co–Cys coordination in HgcA and its role in methylation. Replacement of Cys with Ala completely abolished Hg methylation activity in vivo (Smith et al., 2015).

2.7 Hg Uptake

Hg exists in numerous forms and in natural environments. Often, these molecules come into contact with microbial cells and interact with functional groups on cell surfaces leading to adsorption, ligand exchange, and redox transformation. Identifying the particular Hg-containing molecules that are bioavailable for cellular uptake into these cells is an active area of research, but $Hg^{II}(Cys)_2$ and related complexes are known to be assimilated readily by aerobic bacteria (Schaefer & Morel, 2009). It has been proposed that Hg is mistaken for other metals by membrane-bound transporters (Schaefer et al., 2011), but no specific transporters have yet been identified.

It is not known how Hg is transported throughout the cell once it has entered, but thiol-containing small molecules or proteins are expected to play important roles.

3. FUTURE PERSPECTIVES

Ten years ago the application of computational chemistry to studies of Hg interacting with proteins was practically nonexistent. Now, however, many of the tools are in place to derive a comprehensive description of Hg cycling not only in proteins but also in the natural environment.

Predictive understanding of environmental Hg cycling is lacking, due in part to the complexity of microbial biochemistry, but also to our inability to model and predict Hg biogeochemistry in terrestrial surface and subsurface systems. However, Hg cycling at the scale of an entire watershed or even large systems ultimately depends on the type of molecular-scale chemistry we have described here. What is required to extend to larger scales is a comprehensive computational catalog of thermodynamic quantities sufficient to describe Hg speciation in any given setting or environment. This data would then be used as input to mesoscale models, in which, typically, the gas, solution, and solid phases are assumed to be well-mixed continua in each representative volume, equilibrium chemical reactions are represented by mass action equations, kinetic reactions are described by ordinary differential equations, and transport is simulated by the advection–diffusion equation. Obtaining the data required for accurate continuum-scale modeling and assembling them in a multiscale simulation framework will be required to obtain predictive understanding of biotic and abiotic Hg transport and transformation in the environment.

In the environment, inorganic Hg is methylated by bacteria and archaea to form particularly toxic methylmercury, and we now know the proteins responsible for this process. We note that the Hg methylation conundrum was solved mainly by chemical reasoning, without the need for expensive computation. However, the details of the reaction and the complete biochemical pathway involved remain to be worked out. Much work is needed to arrive at a complete understanding of microbial uptake, intracellular transport and transformation, and export of Hg.

ACKNOWLEDGMENTS

We thank Anne Summers, Susan Miller, Liyuan Liang, Alex Johs, Demian Riccardi, Jing Zhou, Steve Tomanicek, Hao-Bo Guo, Tamar Barkay, Baohua Gu, Dwayne Elias,

Scott Brooks, Mircea Podar, Steve Brown, Richard Hurt Jr., Xiangping Yin, Romain Bridou, Steve Smith, and Judy Wall for fruitful collaborations over the past several years. This research was supported by the US Department of Energy (DOE), Office of Science, Office of Biological and Environmental Research, through the Mercury Scientific Focus Area at Oak Ridge National Laboratory (ORNL) and the Subsurface Biogeochemical Research (SBR) program at the University of Tennessee Knoxville and ORNL through Grant DE-SC0004895 from the US Department of Energy (DOE). ORNL is managed by UT-Battelle, LLC, for the US Department of Energy under contract DE-AC05-00OR22725.

REFERENCES

Afaneh, A. T., Schreckenbach, G., & Wang, F. Y. (2014). Theoretical study of the formation of mercury (Hg^{2+}) complexes in solution using an explicit solvation shell in implicit solvent calculations. *The Journal of Physical Chemistry B, 118*(38), 11271–11283. http://dx.doi.org/10.1021/jp5045089.

Andrae, D., Haussermann, U., Dolg, M., Stoll, H., & Preuss, H. (1990). Energy-adjusted ab initio pseudopotentials for the 2nd and 3rd row transition elements. *Theoretica Chimica Acta*, 77(2), 123–141. http://dx.doi.org/10.1007/Bf01114537.

Becke, A. D. (1993). Density-functional thermochemistry. 3. The role of exact exchange. *The Journal of Chemical Physics, 98*(7), 5648–5652. http://dx.doi.org/10.1063/1.464913.

Bryantsev, V. S., Diallo, M. S., & Goddard, W. A. (2008). Calculation of solvation free energies of charged solutes using mixed cluster/continuum models. *The Journal of Physical Chemistry B, 112*(32), 9709–9719. http://dx.doi.org/10.1021/jp802665d.

Chang, C. C., Lin, L. Y., Zou, X. W., Huang, C. C., & Chan, N. L. (2015). Structural basis of the mercury(II)-mediated conformational switching of the dual-function transcriptional regulator MerR. *Nucleic Acids Research, 43*(15), 7612–7623. http://dx.doi.org/10.1093/nar/gkv681.

Cheesman, B. V., Arnold, A. P., & Rabenstein, D. L. (1988). Nuclear magnetic-resonance studies of the solution chemistry of metal-complexes. 25. Hg(thiol)3 complexes and Hg(Ii)-thiol ligand-exchange kinetics. *Journal of the American Chemical Society, 110*(19), 6359–6364. http://dx.doi.org/10.1021/Ja00227a014.

Choi, S.-C., & Bartha, R. (1993). Cobalamin-mediated mercury methylation by *Desulfovibrio desulfuricans* LS. *Applied and Environmental Microbiology, 59*, 290–295.

Choi, S.-C., Chase, T., Jr., & Bartha, R. (1994a). Enzymatic catalysis of mercury methylation by *Desulfovibrio desulfuricans* LS. *Applied and Environmental Microbiology, 60*, 1342–1346.

Choi, S.-C., Chase, T., Jr., & Bartha, R. (1994b). Metabolic pathways leading to mercury methylation in *Desulfovibrio desulfuricans* LS. *Applied and Environmental Microbiology, 60*, 4072–4077.

Colombo, M. J., Ha, J. Y., Reinfelder, J. R., Barkay, T., & Yee, N. (2013). Anaerobic oxidation of Hg(0) and methylmercury formation by *Desulfovibrio desulfuricans* ND132. *Geochimica et Cosmochimica Acta, 112*, 166–177. http://dx.doi.org/10.1016/j.gca.2013.03.001.

Gregoire, D. S., & Poulain, A. J. (2016). A physiological role for Hg(II) during phototrophic growth. *Nature Geoscience, 9*, 121–125. http://dx.doi.org/10.1038/ngeo2629.

Grimme, S., Antony, J., Ehrlich, S., & Krieg, H. (2010). A consistent and accurate ab initio parametrization of density functional dispersion correction (DFT-D) for the 94 elements H-Pu. *The Journal of Chemical Physics, 132*, 154104.

Grimme, S., Ehrlich, S., & Goerigk, L. (2011). Effect of the damping function in dispersion corrected density functional theory. *Journal of Computational Chemistry, 32*(7), 1456–1465. http://dx.doi.org/10.1002/jcc.21759.

Hong, L., Sharp, M. A., Poblete, S., Bieh, R., Zamponi, M., Szekely, N., ... Smith, J. C. (2014). Structure and dynamics of a compact state of a multidomain protein, the mercuric ion reductase. *Biophysical Journal*, *107*(2), 393–400. http://dx.doi.org/10.1016/j.bpj.2014.06.013.

Johs, A., Harwood, I. M., Parks, J. M., Nauss, R. E., Smith, J. C., Liang, L. Y., & Miller, S. M. (2011). Structural characterization of intramolecular Hg (2+) transfer between flexibly linked domains of mercuric ion reductase. *Journal of Molecular Biology*, *413*(3), 639–656. http://dx.doi.org/10.1016/J.Jmb.2011.08.042.

Lafrance-Vanasse, J., Lefebvre, M., Di Lello, P., Sygusch, J., & Omichinski, J. G. (2009). Crystal structures of the organomercurial lyase MerB in its free and mercury-bound forms insights into the mechanism of methylmercury degradation. *The Journal of Biological Chemistry*, *284*(2), 938–944. http://dx.doi.org/10.1074/jbc.M807143200.

Lian, P., Guo, H. B., Riccardi, D., Dong, A. P., Parks, J. M., Xu, Q., ... Guo, H. (2014). X-ray structure of a Hg^{2+} complex of mercuric reductase (MerA) and quantum mechanical/molecular mechanical study of Hg^{2+} transfer between the C-Terminal and buried catalytic site cysteine pairs. *Biochemistry*, *53*(46), 7211–7222. http://dx.doi.org/10.1021/bi500608u.

Marenich, A. V., Cramer, C. J., & Truhlar, D. G. (2009). Universal solvation model based on solute electron density and on a continuum model of the solvent defined by the bulk dielectric constant and atomic surface tensions. *The Journal of Physical Chemistry B*, *113*(18), 6378–6396. http://dx.doi.org/10.1021/jp810292n.

Parks, J. M., Guo, H., Momany, C., Liang, L. Y., Miller, S. M., Summers, A. O., & Smith, J. C. (2009). Mechanism of Hg–C protonolysis in the organomercurial lyase MerB. *Journal of the American Chemical Society*, *131*(37), 13278–13285. http://dx.doi.org/10.1021/Ja9016123.

Parks, J. M., Johs, A., Podar, M., Bridou, R., Hurt, R. A., Jr., Smith, S. D., ... Liang, L. (2013). The genetic basis for bacterial mercury methylation. *Science*, *339*(6125), 1332–1335. http://dx.doi.org/10.1126/science.1230667.

Perdew, J. P., & Wang, Y. (1992). Accurate and simple analytic representation of the electron-gas correlation-energy. *Physical Review B*, *45*(23), 13244–13249. http://dx.doi.org/10.1103/Physrevb.45.13244.

Perdew, J. P., & Yue, W. (1986). Accurate and simple density functional for the electronic exchange energy: Generalized gradient approximation. *Physical Review B*, *33*(12), 8800–8802. http://dx.doi.org/10.1103/Physrevb.33.8800.

Peterson, K. A., & Puzzarini, C. (2005). Systematically convergent basis sets for transition metals. II. Pseudopotential-based correlation consistent basis sets for the group 11 (Cu, Ag, Au) and 12 (Zn, Cd, Hg) elements. *Theoretical Chemistry Accounts*, *114*(4–5), 283–296. http://dx.doi.org/10.1007/s00214-005-0681-9.

Pitts, K. E., & Summers, A. O. (2002). The roles of thiols in the bacterial organomercurial lyase (MerB). *Biochemistry*, *41*(32), 10287–10296. http://dx.doi.org/10.1021/bi0259148.

Ragsdale, S. W., & Pierce, E. (2008). Acetogenesis and the Wood-Ljungdahl pathway of CO_2 fixation. *Biochimica et Biophysica Acta*, *1784*(12), 1873–1898. http://dx.doi.org/10.1016/j.bbapap.2008.08.012.

Riccardi, D., Guo, H. B., Parks, J. M., Gu, B. H., Liang, L. Y., & Smith, J. C. (2013). Cluster-continuum calculations of hydration free energies of anions and group 12 divalent cations. *Journal of Chemical Theory and Computation*, *9*(1), 555–569. http://dx.doi.org/10.1021/Ct300296k.

Riccardi, D., Guo, H.-B., Parks, J. M., Gu, B., Summers, A. O., Miller, S. M., ... Smith, J. C. (2013). Why mercury prefers soft ligands. *Journal of Physical Chemistry Letters*, *4*(14), 2317–2322. http://dx.doi.org/10.1021/jz401075b.

Schaefer, J. K., & Morel, F. M. M. (2009). High methylation rates of mercury bound to cysteine by Geobacter sulfurreducens. *Nature Geoscience*, *2*(2), 123–126. http://dx.doi.org/10.1038/NGEO412.

Schaefer, J. K., Rocks, S. S., Zheng, W., Liang, L. Y., Gu, B. H., & Morel, F. M. M. (2011). Active transport, substrate specificity, and methylation of Hg(II) in anaerobic bacteria. *Proceedings of the National Academy of Sciences of the United States of America, 108*(21), 8714–8719. http://dx.doi.org/10.1073/pnas.1105781108.

Senn, H. M., & Thiel, W. (2009). QM/MM methods for biomolecular systems. *Angewandte Chemie (International Edition in English), 48*(7), 1198–1229. http://dx.doi.org/10.1002/anie.200802019.

Siegbahn, P. E. M., & Himo, F. (2011). The quantum chemical cluster approach for modeling enzyme reactions. *WIREs Computational Molecular Science, 1*(3), 323–336. http://dx.doi.org/10.1002/wcms.13.

Smith, S. D., Bridou, R., Johs, A., Parks, J. M., Elias, D. A., Hurt, R. A., … Wall, J. D. (2015). Site-directed mutagenesis of HgcA and HgcB reveals amino acid residues important for mercury methylation. *Applied and Environmental Microbiology, 81*(9), 3205–3217. http://dx.doi.org/10.1128/AEM.00217-15.

Stricks, W., & Kolthoff, I. M. (1953). Reactions between mercuric mercury and cysteine and glutathione. Apparent dissociation constants, heats and entropies of formation of various forms of mercuric mercapto-cysteine and mercapto-glutathione. *Journal of the American Chemical Society, 75*(22), 5673–5681. http://dx.doi.org/10.1021/Ja01118a060.

Svetlitchnaia, T., Svetlitchnyi, V., Meyer, O., & Dobbek, H. (2006). Structural insights into methyltransfer reactions of a corrinoid iron-sulfur protein involved in acetyl-CoA synthesis. *Proceedings of the National Academy of Sciences of the United States of America, 103*(39), 14331–14336. http://dx.doi.org/10.1073/Pnas.0601420103.

Warshel, A., & Levitt, M. (1976). Theoretical studies of enzymic reactions: Dielectric, electrostatic and steric stabilization of the carbonium ion in the reaction of lysozyme. *Journal of Molecular Biology, 103*(2), 227–249.

Zhou, J., Riccardi, D., Beste, A., Smith, J. C., & Parks, J. M. (2014). Mercury methylation by HgcA: Theory supports carbanion transfer to Hg(II). *Inorganic Chemistry, 53*(2), 772–777. http://dx.doi.org/10.1021/Ic401992y.

CHAPTER SIX

Steered Molecular Dynamics Methods Applied to Enzyme Mechanism and Energetics

C.L. Ramírez*, M.A. Martí*, A.E. Roitberg[†,1]
*FCEN, UBA, Buenos Aires, Argentina
[†]University of Florida, Gainesville, FL, United States
[1]Corresponding author: e-mail address: roitberg@ufl.edu

Contents

Abstract

One of the main goals of chemistry is to understand the underlying principles of chemical reactions, in terms of both its reaction mechanism and the thermodynamics that govern it. Using hybrid quantum mechanics/molecular mechanics (QM/MM)-based methods in combination with a biased sampling scheme, it is possible to simulate chemical reactions occurring inside complex environments such as an enzyme, or aqueous solution, and determining the corresponding free energy profile, which provides direct comparison with experimental determined kinetic and equilibrium parameters. Among the most promising biasing schemes is the multiple steered molecular dynamics method, which in combination with Jarzynski's Relationship (JR) allows obtaining the equilibrium free energy profile, from a finite set of nonequilibrium reactive trajectories by exponentially averaging the individual work profiles. However, obtaining

Methods in Enzymology, Volume 578
ISSN 0076-6879
http://dx.doi.org/10.1016/bs.mie.2016.05.029

statistically converged and accurate profiles is far from easy and may result in increased computational cost if the selected steering speed and number of trajectories are inappropriately chosen. In this small review, using the extensively studied chorismate to prephenate conversion reaction, we first present a systematic study of how key parameters such as pulling speed, number of trajectories, and reaction progress are related to the resulting work distributions and in turn the accuracy of the free energy obtained with JR. Second, and in the context of QM/MM strategies, we introduce the Hybrid Differential Relaxation Algorithm, and show how it allows obtaining more accurate free energy profiles using faster pulling speeds and smaller number of trajectories and thus smaller computational cost.

1. INTRODUCTION

1.1 Computer Simulation of Chemical Reactions: The Birth of QM/MM Methods

Understanding the underlying principles of chemical reactions is one of the main goals of chemistry. At the atomic and molecular level a reaction is understood in terms of the reaction pathways that describe it, and the thermodynamics that govern it. In other words, a chemical reaction is described in terms of the atomic motions that drive the system from the reactant to the product state, and the free energy profile along the reaction path. Experimentally, chemical dynamics is inferred from chemical kinetic studies that help in defining a reaction mechanism; ie, a scheme showing all bond-breaking/making steps, including proposals for the transition states, and relative free energy values of the key involved states. To move beyond the mechanistic proposal and really understand the process at an atomic level, one has to inevitably employ quantum mechanics (QM)-based molecular dynamics (MD) simulations.

QM methods in chemistry were a natural extension of early 20th century atomic physics moving to larger molecules, and gained momentum with the developments of Hartree–Fock (HF), post HF, and density functional theory (DFT)-based methods, as well as the birth of modern electronics. However, although computer power has been ever increasing, they are still computationally demanding and are restricted—in their pure form—to simulations of a few hundred atoms and MD runs for few picoseconds. This drawback presents itself as a barrier to true understanding, since most interesting reactions occur in condensed phases, like aqueous solvent, or inside enzymes, which requires simulations of systems of thousands of atoms.

Enzyme reactions are particularly interesting, since enzymes are not only exceptional catalysts, achieving rate enhancement factors of up to 1×10^{15} times but also are highly substrate and reactant/product specific. To really understand an enzyme mechanism, we need to describe not only how the free energy changes as the atoms move along the reaction (reaction mechanism), but also how the protein framework lowers the transition state (TS) free energy, with respect to the same reaction mechanism occurring in solution. In other words, how—structurally and energetically speaking—do enzymes achieve their rate enhancement.

To overcome the computational cost problem, while nonetheless being able to simulate a reaction inside an enzyme or in bulk solvent, recent nobel laureates Warshel, Levitt, and Karplus (Warshel & Karplus, 1972; Warshel & Levitt, 1976) introduced in the 1970s the concept of multiscale modeling, in particular the hybrid quantum mechanical/molecular mechanical (QM/MM) method. QM/MM methods allow an accurate description of the enzyme active site (or the solute) where the chemical reaction takes place, which is described at the QM level of theory, while considering the environment using a less expensive MM—or classical type of—force field level of theory. Key to these methods is the coupling between the QM and MM regions, which must properly describe the environment electrostatic as well as the steric effects of the environment on the reactive subsystem.

Among the most successful QM/MM coupling schemes are the additive ones, which as shown by Eqs. (1) and (2), add an additional coupling term ($\mathrm{H}_{\mathrm{QM/MM}}$) to the Hamiltonian. This term is composed by an electrostatic term that computes the interaction between the QM electrostatic potential and the classical point charges, and a Van der Waals term that computes steric interaction between QM and MM atoms, usually using a Lennard–Jones type of potential. Most important is that in the QM Hamiltonian (H_{QM}), the potential exerted by the classical point charges is self-consistently added to the nuclei potential an thus the solved electronic structure corresponds to the actual hybrid QM/MM potential.

$$\mathrm{H}_{\mathrm{TOT}} = \mathrm{H}_{\mathrm{QM}} + \mathrm{H}_{\mathrm{MM}} + \mathrm{H}_{\mathrm{QM-MM}} \tag{1}$$

$$\mathrm{H}_{\mathrm{QM-MM}} = \sum_{i=1}^{C} q_i \int \frac{\rho(r)}{|r-\tau_i|} \mathrm{d}r + \sum_{i=1}^{C} \sum_{j=1}^{Q} \frac{q_i Z_j}{|R_j - \tau_i|} + E_{\mathrm{QM-MM}}^{\mathrm{LJ}} \tag{2}$$

where the sum in i is for all the classical particles (C) and in j is for the quantum particles (Q), q_i are the punctual charges in the MM region, $\rho(r)$ is the charge density of the QM region, τ_i the position of the classical nuclei, and R_j the position of the quantum nuclei.

Although QM/MM methods allow a description of the dynamics of large molecular systems and the study of chemical reactions, in plain MD simulations at room temperature, the system is unable to cross even moderately high barriers ($>kT$), such as those presented by enzyme reactions, thus remaining trapped in the initial (reactant) state, unless driven up the energy hill. Therefore, to simulate chemical dynamics process the system must be driven along the reaction.

1.2 Reaction Coordinate, Biased Sampling, and Jarzynski's Relationship

To drive any chemical system along a selected pathway, first a proper reaction coordinate (RC) needs to be defined. The RC consists of a combination of structural parameters, usually distances (or angles) between atoms, that provide a description of the reaction progress and that therefore can be used to force the system along. Proper sampling of the system along the desired reaction coordinate allows obtaining the corresponding free energy (FE) profile—also called potential of mean force—which describes how the free energy changes along the reaction, providing the corresponding barrier ($\Delta G^{\ddagger}$) and reaction free energy (ΔG°), both of which can be directly compared with experimental data. Moreover, by comparing the FE profiles obtained with different RC or varying conditions, alternative mechanistic proposals can be analyzed.

The key for obtaining accurate, and thus meaningful, FE profiles is to properly sample configurational space. The first, and most commonly used method, to achieve this is umbrella sampling (US) (Kumar, Rosenberg, Bouzida, Swendsen, & Kollman, 1992). In US, a harmonic potential is added to the system hamiltonian, and its equilibrium position is varied along the RC in discrete windows, until the whole RC width is evenly sampled. The resulting FE profile is obtained using Boltzmann weighting of the sampled configurations with respect to the RC. More recently, other strategies, such as metadynamics (Laio & Parrinello, 2002), adaptive biasing force (Hénin & Chipot, 2004), free energy perturbation (Zwanzig, 1954), and orthogonal space random walk (Zheng, Chen, & Yang, 2008), have been developed and applied with a varying degree of success.

Jarzynski (1997) published an outstanding work, where he showed that the free energy—an equilibrium property—could be obtained from nonequilibrium dynamics, according to what has been called since then, Jarzynski's relationship (JR), shown in Eq. (3).

$$G(\lambda) = -\beta^{-1} \ln \langle e^{-\beta W_i(\lambda)} \rangle_0 \tag{3}$$

where $G(\lambda)$ represents the free energy as a function of the reaction coordinate $\lambda, \beta^{-1} = k_B T$, where k_B is Boltzmann's constant, and T is the system temperature, W_i represents the (nonreversible) work performed to drive the system along the RC, and brackets denote configurational average, where the subscript zero denotes that the initial structures of the ensemble are equilibrated for $\lambda = 0$.

In other words, the JR shows that for any system driven externally from an initial equilibrium state to a final state, at any speed, the free energy between them can nonetheless be obtained if the works of many individual trajectories are exponentially averaged (Crooks, 1998, 2000). A clear bottleneck—as will be shown later—is that the exponential average needs to be properly converged.

The JR provides a very simple strategy to obtain FE profile from steered molecular dynamics (SMD) simulations. First, an external force is added to the system, usually, in the form of a harmonic potential whose equilibrium position moves along the chosen RC at a given velocity (v), according to Eq. (4). The external force thus steers the system along the RC moving it from reactants to products.

$$F(t) = -k(\lambda - \lambda_0 - vt) \tag{4}$$

$$w(\lambda) = \int_0^{\lambda} F(\lambda) \mathrm{d}\lambda \tag{5}$$

Individual steered trajectories are started from equilibrium configurations, and for each, the corresponding work profile along the RC is computed by simply integrating the external force. Finally, multiple works are exponentially averaged using JR to obtain the corresponding FE profile. The method, which is also called multiple steered molecular dynamics (MSMD), was initially used by Shulten and coworkers to study alanine decapeptide folding and compared with US (Park, Khalili-Araghi, Tajkhorshid, & Schulten, 2003), showing similar accuracy and potential for a reduction in computational cost. Moreover, in over a decade, it has been applied to a variety of systems and received multiple computational and experimental validations (Collin et al., 2005; Liphardt, Dumont,

Smith, Tinoco, & Bustamante, 2002; Saira et al., 2012; Xiong, Crespo, Marti, Estrin, & Roitberg, 2006).

Our group was among the first to use the MSMD approach and JR in QM/MM simulations to study an enzymatic reaction, in this case the chorismate to prephenate conversion as performed by *Bacillus subtilis* chorismate mutase (BsCM) (Crespo, Marti, Estrin, & Roitberg, 2005).

This particular reaction is involved in the biosynthesis of aromatic amino acids in bacteria, fungi, and plants, and there is extensive experimental and theoretical data in the literature. The enzyme catalyzes an intramolecular Claisen rearrangement. Kast et al. (1997) reported the experimental activation parameters for the enzyme-catalyzed reaction, $\Delta G^{\ddagger} = 15.4\,\text{kcal/mol}$, $\Delta H^{\ddagger} = 12.7\,\text{kcal/mol}$ and $\Delta S^{\ddagger} = -9.1\,\text{kcal/mol}$. Mulholland's group (Ranaghan et al., 2004), Lee Woodcock and coworkers (Lee Woodcock et al., 2003; Woodcock, Hodoscek, & Brooks, 2007), and Crespo (Crespo et al., 2005) have reported theoretical activation parameters for chorismate mutase, at different theory levels.

The reaction can be nicely described and driven forward using as RC a combination of the distance between the carbons dC–C that create the new bond and the distance of the bond that is broken dO–C (Scheme 1). We had previously computed the same reaction profile using minimum energy path, which allows obtaining the potential energy instead of free energy, using the level of theory DFT with the PBD Exchange Correlation functional. Although the obtained value of 5.3 kcal/mol is below the experimental determination of 12.7 kcal/mol, we correctly predicted the energy difference related to the catalytic enhancement, since the barrier in aqueous solvent was 8.5 kcal/mol higher (Crespo et al., 2003). With the MSMD method we were able to obtain a well-converged FE profile performing 30 runs, with a 2 Å/ps speed and a RC span of ca. 4 Å, totaling 60 ps of simulation time. The results were the same as those

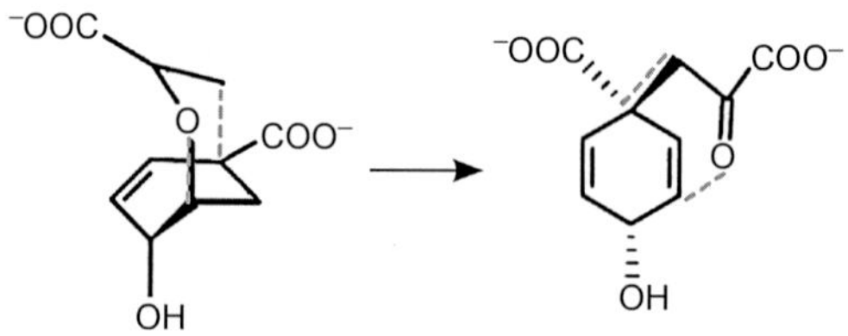

Scheme 1 Reaction from chorismate to prephenate. In *gray dotted lines* are the distances dC–C and dO–C that conform the reaction coordinate.

obtained with US using 12 windows of 5 ps each, thus the same overall computational time (Crespo et al., 2005).

Over the last decade there has been an intense debate as to whether the MSMD strategy, combined with JR, really provides a significant reduction in computational cost—yielding nonetheless accurate FE profiles—beyond the simple fact that each nonequilibrium trajectory can be run in an independent computer and thus the method is easily parallelized. Also it is not clear which of the available biased sampling method provides the best FE profile at the lowest cost. We have decided to calculate a reference set which can be considered as converged in terms of FE profile, so we can then benchmark changes in several parameters affecting the JR convergence. For this purpose, we have used the density functional tight binding (DFTB) level of theory as the QM method to determine the chorismate to prephenate conversion reaction in aqueous solution.

1.3 Quasi Reversible FE Profile of the Chorismate to Prephenate Conversion Reaction in Solution

Prior to the analysis of the impact on the FE profile of the pulling speed and the number of SMD employed, we created a reference FE profile, in order to have a comparison between the different profiles. This was achieved by performing three SMD at a pulling speed of 0.02 Å/ps. Both values were chosen because they were the ones that showed a difference, between work profiles, in less than 1 kcal/mol, which points to a quasi-static path. With this result we were able to obtain a FE profile, employing JR, that was used as the reference profile for the reaction in solution (bold line in Fig. 1A and B).

1.4 FE Profiles for the Chorismate to Prephenate Conversion in Solution Using MSMD and Jarzynski's Relationship

Fig. 1A shows the typical behavior of the work profiles obtained for a finite set of MSMD trajectories, as well as their relation with the corresponding FE profile obtained with JR and the reference FE profile. The figure nicely shows how the width of the distribution of the work values increases as the reaction moves toward products. Also the JR estimate of the FE profile increasingly overestimates the reference FE profile. Both these effects are a consequence of the fact that the further the system is driven forward by the external force, the further it moves from equilibrium. The distribution of work values is wider as one moves away from equilibrium and as a consequence, the exponential average is harder to converge and is shifted to larger values, thus overestimating the free energy. This is a consequence of the

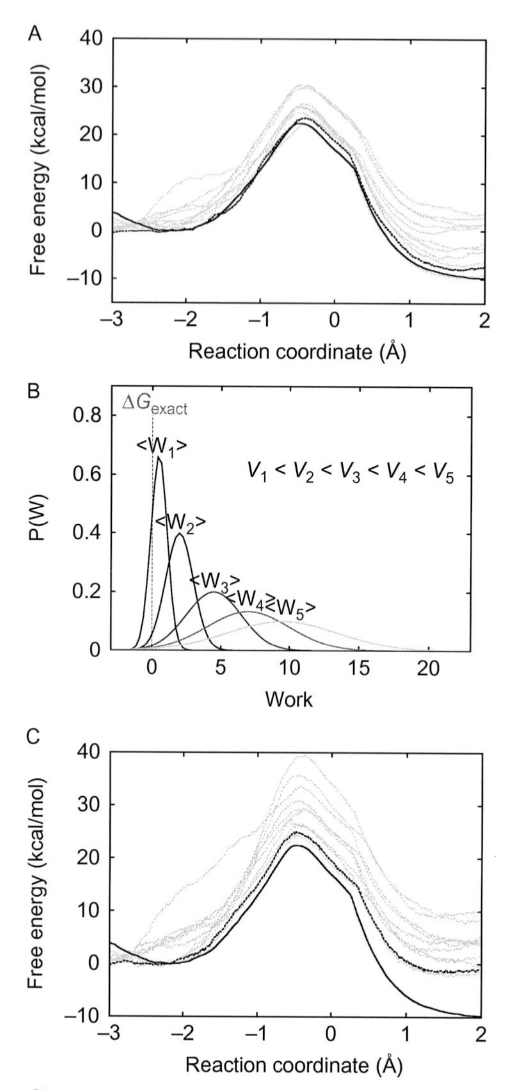

Fig. 1 (A) Work profiles (*dotted gray lines*) used to obtain a FE profile with JR (*dotted black line*), and a comparison with the reference profile employed (*continuous black line*) for the reaction of chorismate to prephenate in solution. All *dotted lines* correspond to a pulling speed of 0.5 Å/ps, while the reference has a puling speed of 0.02 Å/ps. (B) Gaussian distribution of work for different pulling speeds shown in a schematic form. It can be seen that the slower the pulling speed, the narrower the work distribution. (C) Same notation as in panel (A), but for a pulling speed of 1 Å/ps.

mathematical nature of the exponential average as can easily be seen for a Gaussian like work distribution, as shown schematically in Fig. 1B. Fig. 1B also shows why it is very important to have at least one trajectory whose work profile is close to the reference FE profile, since otherwise the exponential average will be overestimated. Indeed, Pohorille, Jarzynski, and Chipot (2010) estimated that if the standard deviation of the work distribution is around 1 $k_B T$, about one in six trajectories samples values close to the real free energy. The value decreases to 1 in 40 trajectories for $\sigma = 2k_B T$ and to 1/300,000 for $\sigma = 5k_B T$, strongly biasing the obtained FE profile if only a limited number of trajectories is performed. In other words, to have a well-converged exponential average, the works that need to be sampled are those in the lower tail of the work distribution which therefore have low occurrence probabilities.

To understand the relationship between the width of the work distribution and the pulling speed, it is interesting to analyze the data from Fig. 1C, where the work profiles, at twice the pulling speed as those shown in Fig. 1A, are shown together with the reference and obtained FE profile. The results clearly show how faster pulling results in wider distributions, which are also shifted to higher values (see also Fig. 1B) and therefore a significant overestimation of the free energy. On the contrary, extremely slow pulling speed would result in a distribution of work values which resemble a delta function, and thus the exponential average, the average, and the external work all display the same values (within the 1 $k_B T$) and thus equal the free energy, as required by the second law of thermodynamics.

Having analyzed the role of MSMD key parameter, which is the pulling speed, we will now analyze how it translates into the accuracy of the obtained FE profile, in relation to the number of trajectories and therefore the overall relative computational cost. The accuracy will be measured in terms of the overestimation of both the barrier and reaction free energy, and the total computational cost in relation to that required to obtain the reference FE profile. The results are presented in Table 1.

The results show that, for this particular system, the number of trajectories have less impact in the final FE profile that the pulling speed. This can be appreciated by the fact that the same computational cost is needed to obtain a FE profile with 20 trajectories at a pulling speed of 1 Å/ps as for one obtained using 10 trajectories at a pulling speed of 0.5 Å/ps, but the second one is more accurate.

In summary, when using MSMD combined with JR to obtain FE profile profiles, it is important to consider that since the exponential average is

Table 1 Relevant Values of the FE (in kcal/mol) Profile for the Reaction from Chorismate to Prephenate in Solution for Different Pulling Speeds (in Å/ps) and Number of Trajectories

Pulling Speed	Trajectories	$\Delta G^{\ddagger}$	ΔG^{o}	Total time	Relative Time
0.02	3	21.0	−10.0	787.5	100
0.5	10	24.6	−7.0	105.0	13.3
1.0	5	25.1	−0.6	26.3	3.3
1.0	10	25.9	−0.9	52.5	6.6
1.0	15	25.6	−0.9	78.8	10.0
1.0	20	25.2	−1.6	105.0	13.3
2.0	10	33.8	8.4	26.3	3.3

The total time (in ps) is calculated considering the time performed per simulation, multiplied for the number of trajectories performed. The relative time (in %) is calculated considering that the total time employed in performing the reference profile is 100%.

computed over a work values distribution, the more narrow (smaller SD) the more accurate the FE profile. The width of the distribution is directly proportional to the pulling speed and increases also as the system moves forward and thus further from equilibrium. For wider distributions an exponentially increasing number of trajectories is needed to sample values in the lower tail of the distribution, which are the ones required to obtain accurate FE profile. Therefore, the main drawback with JR is that in order to obtain a well-converged average, and thus accurate FE profile, either very large number of trajectories and/or very low pulling speeds are needed. These facts results in a high computational cost (only marginally smaller than that required for quasi-static transitions) and even sometimes in an insurmountable problem that prevents accurate convergence of the FE profile.

1.5 The Hybrid Differential Relaxation Algorithm

While the SMD strategy is formally correct, there are some serious convergence problems that can hinder its wider applicability. Several methods have been presented to deal with this these issues, mostly in the context of classical force fields (MM) (Ozer, Valeev, Quirk, & Hernandez, 2010). The differential relaxation strategy was inspired by the multiple time step ideas proposed by Tuckerman, Berne, and Martyna (1992) and Tuckerman and Parrinello (1994) which use different masses and time steps for the QM and MM regions. These early studies showed that while similar trajectories as those in standard MD simulations are obtained, faster (about two times) convergence of the average force can be achieved. Taken this idea forward we implemented a similar multiple time step scheme, without mass scaling, that in the context of MSMD and JR allows obtaining accurate FE profiles at a significant lower computational cost.

The method works as follows (Scheme 2), after a joint QM/MM step where both QM and MM atoms are moved a time step forward, the QM region is frozen while the MM region is allowed to move (relax) for a given number of pure classical steps, after which a new joint QM/MM step is performed. Key to hybrid differential relaxation algorithm (HyDRA) success is that the work from the external force is only accumulated in the QM/MM steps, since in the pure MM steps the reaction coordinate—which should be part only of the QM region—remains frozen and thus no force is applied. The pure classical steps allow the MM region to relax to the perturbed QM region, and therefore during the joint steps—which are the ones that matter for the FE profile—closer to equilibrium work values are obtained. In the language of the JR, only the steps where work is accumulated

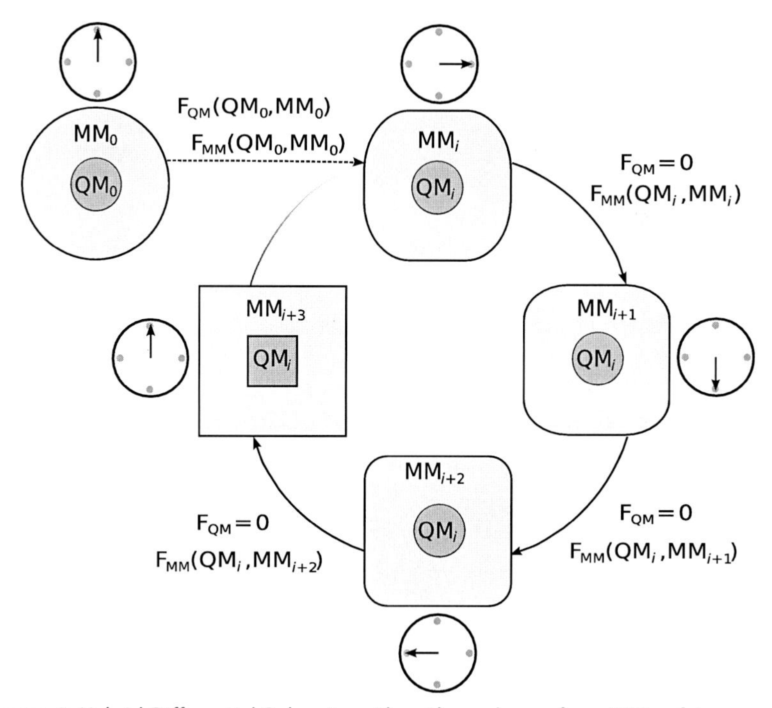

Scheme 2 Hybrid Differential Relaxation Algorithm scheme for a DRAr of 4.

(the QM/MM steps) count toward the free energy estimator. The steps where the QM region is frozen are heat (not work), and hence, they are irrelevant to the free energy calculation. The ratio between the MM and the joint QM/MM steps is called the differential relaxation algorithm ratio (DRAr). A DRAr ratio of 1 is equivalent to conventional QM/MM simulations, while with a DRAr of 10, for each joint step 10 pure classical relaxation steps are performed.

Technically implementing HyDRA is straight forward for any additive QM/MM Hamiltonian (Eq. 1) since it relies on the adequate calculation of the forces and atom coordinates for each type of step. When the simulation starts (step 0) both regions are synchronized and force contributions are computed as usual, thus forces acting on the QM atoms stem from the quantum contribution in the step 0 configuration and the QM/MM interaction contribution derived from the Lennard–Jones potential (third term in Eq. 2), which we will refer as $F_{QM}(QM_0)$ and $F_{QM}(MM_0)$. Similar reasoning applies to the forces acting for this step on the MM atoms, $F_{MM}(QM_0)$ and $F_{MM}(MM_0)$. With these forces the system is able to perform its first dynamic step, which leads to conformation *i* for both QM and MM regions. Now the DRA cycle starts. The pure classical and QM/MM forces acting on the MM atoms are computed as usual for conformation *i*, and now only the MM region moves a time step forward achieving conformation $i+1$. The QM region remains frozen in conformation *i*—which is achieved by zeroing the forces acting on the QM atoms. $F_{QM}(MM)$ forces are nonetheless stored since they will be used later. With new MM coordinates (configuration $i+1$) a new set of forces $F_{MM}(QM_i, MM_{i+1})$ acting on them is computed and the MM region is updated. After three of these steps the systems arrives at configuration $[QM_i, MM_{i+3}]$. Now the systems performs a new joint step to reach conformation $[QM_{i+1}, MM_{i+4}]$. The forces for this step are determined using the $i+3$ conformation for the MM contributions, and the *i* conformation for the quantum contributions. Considering the structures, it is important to note that in our implementation, only the conformations obtained after a joint step are meaningful and equivalent to those obtained in a standard QM/MM molecular dynamics simulations.

The method has been implemented in the AMBER (Case et al., 2012) computer simulation package and is available upon request to the author. HyDRA works with any of the available QM methods and was thoroughly tested with DFTB (Cui, Elstner, Kaxiras, Frauenheim, & Karplus, 2001; de M Seabra, Walker, Elstner, Case, & Roitberg, 2007) and pure DFT implementation of the Perdew, Burke, and Ernzerhof (Perdew, Burke, & Ernzerhof, 1996) exchange correlation functional (DFP-PBE) (Nitsche, Ferreria, Mocskos, & González Lebrero, 2014) available under request to the author.

1.6 HyDRA Scheme Improves Both Barrier and Reaction Free Energies Accuracy Determinations in MSMD–JR Simulations

To test the potential of the HyDRA we first analyzed how increasing the DRAr affects the shape and accuracy of the obtained FE profiles for the chorismate to prephenate conversion at a fixed pulling speed (in this case 100 times faster than that required to obtain the reference FE profile) both in solution and in BsCM using 10 trajectories. The results are presented in Fig. 2 and summarized in Table 2.

The results show that the standard MSMD (DRAr = 1) significantly overestimates both the barrier and reaction free energy. In particular, in

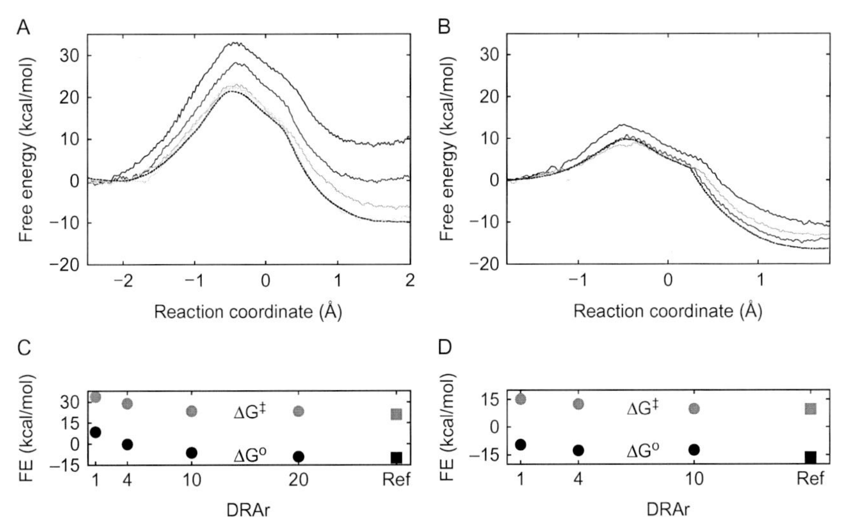

Fig. 2 FE profiles for the chorismate to prephenate reaction, obtained through 10 non-equilibrium dynamics with a pulling speed of 2 Å/ps, and reference FE profile. (A) Free energy profiles for the reaction in solution with a DRA ratio of 1, 4, 10, and 20 shown in scale of *gray*, decreasing color when the DRAr increases. (B) FE profiles in the BsCM enzyme, employing a DRAr of 1, 4, and 10, with the same color code as in (A). (C) $\Delta G^{\ddagger}$ (*gray*) and ΔG^{o} (*black*) for the FE profile shown in panel (A). *Circles* represent values for FE profile with different DRAr, and the *squares* are the values for the reference profile. (D) $\Delta G^{\ddagger}$ (*gray*) and ΔG^{o} (*black*) for the reaction in the enzyme.

Table 2 Relevant Values of the FE Profile (in kcal/mol) for the Reaction from Chorismate to Prephenate in Solution and in the Enzyme BsCM

	FE Profile	DRAr	$\Delta G^{\ddagger}$	ΔG^{o}
Solution	Reference	1	21	−10
	MSMD–JR	1	33.8	8.4
		4	29	−0.2
		10	23.6	−6.3
		20	23.4	−9.1
BsCM	Reference	1	9.6	−16.5
	MSMD–JR	1	15.2	−9.5
		4	12.5	−12.6
		10	9.9	−12.4

The pulling speed in all the cases is 2 Å/ps, and the number of trajectories is 10.

solution, plain MSMD yields a positive reaction free energy, when the reference value is −10 kcal/mol and overestimates the barrier in more than 10 kcal/mol. The HyDRA scheme significantly improves the FE profile accuracy, and the higher the DRAr, the closer the estimate to the reference FE profile. For a DRAr of 10, the obtained values for the barriers are overestimated in less than 10% of the barrier magnitude, and the free energy yields the correct trend. In general, the effect of HyDRA is more notorious for the reaction in solution than in the enzyme. This is not unexpected since the enzyme's active site is less flexible compared to the solvent and thus requires less relaxation along the reaction.

As mentioned earlier, one of the key aspects of enzyme reactions is to determine the catalytic efficiency, which corresponds to the change in the reaction free energy barrier for the reaction in the enzyme with respect to the observed in aqueous solution, ie, the $\Delta\Delta G^{\ddagger}$. For BsCM the experimental determined $\Delta\Delta G^{\ddagger}$ value is 9.1 kcal/mol, making BsCM a quite efficient enzyme. Using the reference FE profiles we obtain an efficiency of 11.4 kcal/mol for BsCM which thus is close to the experimental value. Plain MSMD yields a significantly overestimated $\Delta\Delta G^{\ddagger}$ value (18.6 kcal/mol) which HyDRA partially corrects, yielding a $\Delta\Delta G^{\ddagger}$ value of 13.5 kcal/mol.

To understand the effect of the HyDRA scheme in the context of MSMD–JR, and the opening discussion, it is interesting to analyze the work distributions obtained for increasing values of DRAr. Fig. 3 shows the work

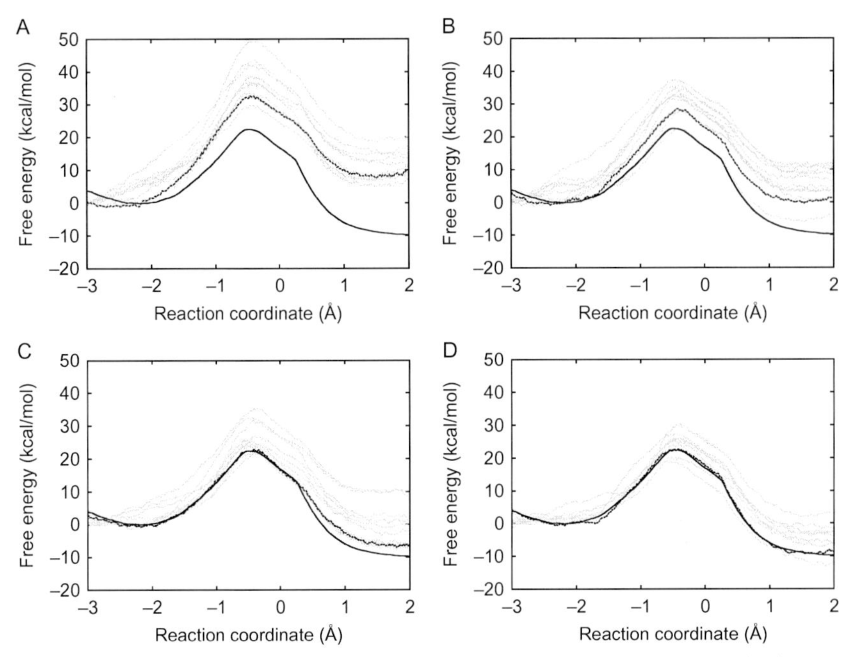

Fig. 3 Work distributions (in *gray*), Jarzynski's estimator (*dotted black line*), and reference FE profile (*solid black line*) for the chorismate to prephenate reaction in solution. In all cases the pulling speed is 2 Å/ps, and the number of trajectories is 10. Panels (A)–(D) correspond to a DRAr of 1, 4, 10, and 20, respectively.

distributions (in gray) for 10 trajectories with DRAr of 1,4, 10, and 20 for the reaction in solution, together with the Jarzynski's estimator (dotted black line) and the reference FE profile (solid black line). The results show clearly how with increasing DRAr the width of the work distribution becomes more narrow, and therefore, the Jarzynski estimate is closer to the reference profiles. Indeed for DRAr of 1 and 4 sampling is clearly inadequate since there are not any work profiles that sample the correct free energy value in the barrier and later regions. For DRAr of 10 there are already several profiles close to the reference FE profile in the transition state region, while for DARr of 20 there is even one case that goes below it. As already discussed, being able to sample these trajectories is essential for getting accurate FE profiles.

Another interesting point of analysis is related to how the HyDRA scheme impacts in the pulling speed and number of runs required to obtain accurate FEP. Table 3 shows how the obtained barrier and reaction free energy values depend on the pulling speed and number of trajectories using

Table 3 Relevant Values of the FE Profile Obtained Through JR for the Reaction of Chorismate to Prephenate in Solution

	Pulling Speed	Number of Trajectories	$\Delta G^{\ddagger}$		ΔG^{o}	
Reference FE profile	0.02	3	21		−10	
DRAr			1	10	1	10
Effect of pulling speed	0.5	10	24	22	−7	−8.1
	1	10	25.9	21.6	−0.9	−9.5
	2	10	33.8	23.6	8.4	−6.3
Effect of the number of trajectories	1	5	25.1	21.1	−0.6	−9.9
	1	10	25.9	21.6	−0.9	−9.5
	1	15	25.6	21.1	−0.9	−9.9
	1	20	25.2	21.1	−1.6	−9.9

the HyDRA scheme (using DRAr of 10). The results show that HyDRA effect on both parameters is dramatic, without HyDRA only using the slowest tests pulling speed (25 times faster than that required for the reference FEP) reasonable value for the reaction free energy and the barrier is slightly overestimated, HyDRA not only shields more accurate results at these slow speed, but it yields reasonable results even for a pulling speed which is 100 times faster than that used to obtain the reference FEP, where standard MSMD FEP is completely at odds. Concerning the number of trajectories, little difference is observed for all tested values. The results from standard MSMD are not accurate enough (especially for the reaction free energy) which possibly means that many more (in the order of hundred) simulations are required to obtain better FEPs. For HyDRA, however, the results with only five trajectories already yield accurate results, which are within 1 kcal/mol with respect to the reference FEP for both the barrier and reaction free energy.

In summary, the results presented earlier clearly show that the HyDRA scheme allows, in the context of MSMD–JR free energy profile determinations, to obtain more accurate profiles using significantly faster pulling speeds and smaller number of trajectories compared to standard QM/MM MD simulations. Moreover, the results show by comparing FE profile of the reactions in solution and enzyme accurate estimates of the enzyme catalytic power can be determined.

2. DISCUSSION

Since the introduction of Jarzynki's relationship in 1996 many theoretical studies (Crooks, 1998, 2000; Ozer et al., 2010) have taken advantage of its simplicity and easy parallelization to obtain free energy profiles of different processes at the molecular level. However, the declaimed reduction in computational cost in comparison to other schemes, such as US, and the accuracy of the obtained FE profile was more a matter of faith, than a fact. This is specially so in QM/MM simulations of enzymatic reactions where the complexity of the systems usually prevents the application of different schemes for comparative purposes, and where often no experimental accurate value for direct comparison with the FE profile is available.

The presented HyDRA scheme not only provides a method which has the potential of delivering more accurate FE profiles at significant lower computational cost but also highlight some interesting facts about the relation between the pulling speed, the number of trajectories, the resulting work profiles, and the accuracy of the resulting FE profiles. The present data add up to our previous experience in the subject and clearly show that it is very difficult to obtain accurate FE profile when the pulling speed is too fast. More important, too fast speed cannot be corrected by performing more trajectories (unless possible huge numbers are considered) and that usually small set—in the order of 10—is enough to have a good estimate of the work distribution. Therefore, when performing standard MSMD–JR FE profile estimates, the work value distribution should display deviations in the order of $k_B T$ to obtain meaningful results. HyDRA allows the system to remain closer to the equilibrium during the pulling steps—which are the joint QM/MM steps—and thus more narrow work distributions and more accurate FE profiles are obtained.

Although the method was only recently developed, during the last year we applied it to study the reaction of two essential Zn hydrolases of *Mycobacterium tuberculosis*, namely, MshB (Rv1170) and MA-amidase (Rv3717) (Romero, Martin, Ramirez, Dumas, & Marti, 2015). Both enzymes hydrolyze the difficult C–N bond of an amide. Our results allowed obtaining of accurate FE profiles for both reaction steps, the first involving Zn coordinated hydroxide nucleophilic attack to the scissile amide carbonyl that leads to meta-estable tetrahedral intermediate. And the second which involves breaking of the C–N bond and is promoted by active site base to amide nitrogen proton transfer. The obtained barriers and mechanism were similar to those reported for other Zn hydrolases like carboxypeptidase-A (Xu & Guo, 2009), thermolysis (Blumberger, Lamoureux, & Klein, 2007), angiotensin converting enzyme (Zhang, Wu, & Xu, 2013), and the anthrax lethal factor peptidase (Smith, Smith, Yang, Xu, & Guo, 2010) pointing toward a conserved (or converged) mechanism for these type of enzymes which belong to different protein families. Most important, analysis of the required QM/MM steps for obtaining the corresponding FE profiles in each case shows that HyDRA scheme resulted in 5–20 times potential reduction in the overall computational cost, highlighting the tremendous power of this HyDRA strategy.

3. METHODOLOGY

3.1 Simulation Details

For pure classical equilibration simulations and the MM part in the QM–MM dynamics all classical parameters were taken from the Amber force field, ff99SB (Hornak et al., 2006), and TIP3P model was used for the water molecules (Jorgensen, Chandrasekhar, Madura, Impey, & Klein, 1983). Parameters for the chorismate were taken from previous works from our group (Crespo et al., 2003, 2005). Simulations were performed at constant temperature using the Berendsen thermostat as implemented in SANDER, using periodic boundary conditions and either constant volume or constant pressure. Ewald Sums were used to treat long-range electrostatics.

For the QM–MM simulations the DFTB (Cui et al., 2001; de M Seabra et al., 2007) as implemented in the sander module of AMBER (Case et al., 2012) was used.

REFERENCES

Blumberger, J., Lamoureux, G., & Klein, M. L. (2007). Peptide hydrolysis in thermolysin: Ab initio QM/MM investigation of the Glu143-assisted water addition mechanism. *Journal of Chemical Theory and Computation*, *3*(5), 1837–1850. http://dx.doi.org/10.1021/ct7000792.

Case, D. A., Darden, T. A., Cheatham, T. E. I. I. I., Simmerling, C. L., Wang, J., Duke, R. E., … Kollman, P. A. (2012). *AMBER 12*. San Francisco: University of California.

Collin, D., Ritort, F., Jarzynski, C., Smith, S. B., Tinoco, I., & Bustamante, C. (2005). Verification of the Crooks fluctuation theorem and recovery of RNA folding free energies. *Nature*, *437*(7056), 231–234. http://dx.doi.org/10.1038/nature04061.

Crespo, Alejandro, Marti, M. A., Estrin, D. A., & Roitberg, A. E. (2005). Multiple-steering QM-MM calculation of the free energy profile in chorismate mutase. *Journal of the American Chemical Society*, *127*(19), 6940–6941. http://dx.doi.org/10.1021/ja0452830.

Crespo, A., Scherlis, D. A., Martí, M. A., Ordejón, P., Roitberg, A. E., & Estrin, D. A. (2003). A DFT-based QM-MM approach designed for the treatment of large molecular systems: Application to chorismate mutase. *Journal of Physical Chemistry. B*, *107*(49), 13728–13736.

Crooks, G. E. (1998). Nonequilibrium measurements of free energy differences for microscopically reversible markovian systems. *Journal of Statistical Physics*, *90*(5–6), 1481–1487. http://dx.doi.org/10.1023/A:1023208217925.

Crooks, G. E. (2000). Path-ensemble averages in systems driven far from equilibrium. *Physical Review E*, *61*(3), 2361–2366. http://dx.doi.org/10.1103/PhysRevE.61.2361.

Cui, Q., Elstner, M., Kaxiras, E., Frauenheim, T., & Karplus, M. (2001). A QM/MM implementation of the self-consistent charge density functional tight binding (SCC-DFTB) method. *The Journal of Physical Chemistry. B*, *105*(2), 569–585. http://dx.doi.org/10.1021/jp0029109.

de M Seabra, G., Walker, R. C., Elstner, M., Case, D. A., & Roitberg, A. E. (2007). Implementation of the SCC-DFTB method for hybrid QM/MM simulations within the amber molecular dynamics package. *The Journal of Physical Chemistry. A*, *111*(26), 5655–5664. http://dx.doi.org/10.1021/jp070071l.

Hénin, J., & Chipot, C. (2004). Overcoming free energy barriers using unconstrained molecular dynamics simulations. *The Journal of Chemical Physics*, *121*(7), 2904–2914. http://dx.doi.org/10.1063/1.1773132.

Hornak, V., Abel, R., Okur, A., Strockbine, B., Roitberg, A., & Simmerling, C. (2006). Comparison of multiple Amber force fields and development of improved protein backbone parameters. *Proteins*, *65*(3), 712–725. http://dx.doi.org/10.1002/prot.21123.

Jarzynski, C. (1997). Nonequilibrium equality for free energy differences. *Physical Review Letters*, *78*(14), 2690–2693. http://dx.doi.org/10.1103/PhysRevLett.78.2690.

Jorgensen, W. L., Chandrasekhar, J., Madura, J. D., Impey, R. W., & Klein, M. L. (1983). Comparison of simple potential functions for simulating liquid water. *The Journal of Chemical Physics*, *79*(2), 926–935. Retrieved from, http://www.scopus.com/inward/record.url?eid=2-s2.0-0004016501&partnerID=40&md5=7af48df275648024bbd5981b9583c129.

Kast, P., Tewari, Y. B., Wiest, O., Hilvert, D., Houk, K. N., & Goldberg, R. N. (1997). Thermodynamics of the conversion of chorismate to prephenate: Experimental results and theoretical predictions. *The Journal of Physical Chemistry. B*, *101*(50), 10976–10982. http://dx.doi.org/10.1021/jp972501l.

Kumar, S., Rosenberg, J. M., Bouzida, D., Swendsen, R. H., & Kollman, P. A. (1992). The weighted histogram analysis method for free-energy calculations on biomolecules. I. The method. *Journal of Computational Chemistry*, *13*(8), 1011–1021. http://dx.doi.org/10.1002/jcc.540130812.

Laio, A., & Parrinello, M. (2002). Escaping free-energy minima. *Proceedings of the National Academy of Sciences of the United States of America*, *99*(20), 12562–12566. http://dx.doi.org/10.1073/pnas.202427399.

Lee Woodcock, H., Hodošček, M., Sherwood, P., Lee, Y. S., Schaefer, H. F., III, & Brooks, B. R. (2003). Exploring the quantum mechanical/molecular mechanical replica path method: A pathway optimization of the chorismate to prephenate Claisen rearrangement catalyzed by chorismate mutase. *Theoretical Chemistry Accounts*, *109*(3), 140–148. http://dx.doi.org/10.1007/s00214-002-0421-3.

Liphardt, J., Dumont, S., Smith, S. B., Tinoco, I., & Bustamante, C. (2002). Equilibrium information from nonequilibrium measurements in an experimental test of Jarzynski's equality. *Science (New York, N.Y.)*, *296*(5574), 1832–1835. http://dx.doi.org/10.1126/science.1071152.

Nitsche, M. A., Ferreria, M., Mocskos, E. E., & González Lebrero, M. C. (2014). GPU accelerated implementation of density functional theory for hybrid QM/MM simulations. *Journal of Chemical Theory and Computation*, *10*, 959–967.

Ozer, G., Valeev, E. F., Quirk, S., & Hernandez, R. (2010). Adaptive steered molecular dynamics of the long-distance unfolding of neuropeptide Y. *Journal of Chemical Theory and Computation*, *6*(10), 3026–3038. http://dx.doi.org/10.1021/ct100320g.

Park, S., Khalili-Araghi, F., Tajkhorshid, E., & Schulten, K. (2003). Free energy calculation from steered molecular dynamics simulations using Jarzynski's equality. *The Journal of Chemical Physics*, *119*(6), 3559. http://dx.doi.org/10.1063/1.1590311.

Perdew, J., Burke, K., & Ernzerhof, M. (1996). Generalized gradient approximation made simple. *Physical Review Letters*, 77(18), 3865.

Pohorille, A., Jarzynski, C., & Chipot, C. (2010). Good practices in free-energy calculations. *The Journal of Physical Chemistry. B*, *114*(32), 10235–10253. http://dx.doi.org/10.1021/jp102971x.

Ranaghan, K. E., Ridder, L., Szefczyk, B., Sokalski, W. A., Hermann, J. C., & Mulholland, A. J. (2004). Transition state stabilization and substrate strain in enzyme catalysis: Ab initio QM/MM modelling of the chorismate mutase reaction. *Organic & Biomolecular Chemistry*, *2*(7), 968–980. http://dx.doi.org/10.1039/b313759g.

Romero, J. M., Martin, M., Ramirez, C. L., Dumas, V. G., & Marti, M. A. (2015). Efficient calculation of enzyme reaction free energy profiles using a hybrid differential relaxation algorithm: Application to mycobacterial zinc hydrolases. *Advances in Protein Chemistry and Structural Biology*, *100*, 33–65. http://dx.doi.org/10.1016/bs.apcsb.2015.06.006.

Saira, O.-P., Yoon, Y., Tanttu, T., Möttönen, M., Averin, D. V., & Pekola, J. P. (2012). Test of the Jarzynski and Crooks fluctuation relations in an electronic system. *Physical Review Letters*, *109*(18), 180601. http://dx.doi.org/10.1103/PhysRevLett.109.180601.

Smith, C. R., Smith, G. K., Yang, Z., Xu, D., & Guo, H. (2010). Quantum mechanical/molecular mechanical study of anthrax lethal factor catalysis. *Theoretical Chemistry Accounts*, *128*(1), 83–90. http://dx.doi.org/10.1007/s00214-010-0765-z.

Tuckerman, M., Berne, B. J., & Martyna, G. J. (1992). Reversible multiple time scale molecular dynamics. *The Journal of Chemical Physics*, *97*(3), 1990. http://dx.doi.org/10.1063/1.463137.

Tuckerman, M. E., & Parrinello, M. (1994). Integrating the Car–Parrinello equations. II. Multiple time scale techniques. *The Journal of Chemical Physics*, *101*(2), 1316. http://dx.doi.org/10.1063/1.467824.

Warshel, A., & Karplus, M. (1972). Calculation of ground and excited state potential surfaces of conjugated molecules. I. Formulation and parametrization. *Journal of the American Chemical Society*, *94*(16), 5612–5625. http://dx.doi.org/10.1021/ja00771a014.

Warshel, A., & Levitt, M. (1976). Theoretical studies of enzymic reactions: Dielectric, electrostatic and steric stabilization of the carbonium ion in the reaction of lysozyme. *Journal of Molecular Biology*, *103*, 227–249.

Woodcock, H. L., Hodoscek, M., & Brooks, B. R. (2007). Exploring SCC-DFTB paths for mapping QM/MM reaction mechanisms. *The Journal of Physical Chemistry. A*, *111*(26), 5720–5728. http://dx.doi.org/10.1021/jp0714217.

Xiong, H., Crespo, A., Marti, M., Estrin, D., & Roitberg, A. E. (2006). Free energy calculations with non-equilibrium methods: Applications of the Jarzynski relationship. *Theoretical Chemistry Accounts*, *116*(1–3), 338–346. http://dx.doi.org/10.1007/s00214-005-0072-2.

Xu, D., & Guo, H. (2009). Quantum mechanical/molecular mechanical and density functional theory studies of a prototypical zinc peptidase (carboxypeptidase A) suggest a general acid-general base mechanism. *Journal of the American Chemical Society*, *131*(28), 9780–9788. http://dx.doi.org/10.1021/ja9027988.

Zhang, C., Wu, S., & Xu, D. (2013). Catalytic mechanism of angiotensin-converting enzyme and effects of the chloride ion. *The Journal of Physical Chemistry. B*, *117*(22), 6635–6645. http://dx.doi.org/10.1021/jp400974n.

Zheng, L., Chen, M., & Yang, W. (2008). Random walk in orthogonal space to achieve efficient free-energy simulation of complex systems. *Proceedings of the National Academy of Sciences of the United States of America, 105*(51), 20227–20232. http://dx.doi.org/10.1073/pnas.0810631106.

Zwanzig, R. W. (1954). High-temperature equation of state by a perturbation method. I. Nonpolar gases. *The Journal of Chemical Physics, 22*(8), 1420. http://dx.doi.org/10.1063/1.1740409.

New Algorithms for Global Optimization and Reaction Path Determination

D. Weber, D. Bellinger, B. Engels[1]
Julius-Maximilians-Universität Würzburg, Institut für Physikalische und Theoretische Chemie, Würzburg, Germany
[1]Corresponding author: e-mail address: bernd.engels@uni-wuerzburg.de

Contents

Abstract

We present new schemes to improve the convergence of an important global optimization problem and to determine reaction pathways (RPs) between identified minima. Those methods have been implemented into the CAST program (Conformational Analysis and Search Tool).

The first part of this chapter shows how to improve convergence of the Monte Carlo with minimization (MCM, also known as Basin Hopping) method when applied to optimize water clusters or aqueous solvation shells using a simple model. Since the random movement on the potential energy surface (PES) is an integral part of MCM, we propose

Methods in Enzymology, Volume 578
ISSN 0076-6879
http://dx.doi.org/10.1016/bs.mie.2016.05.030

to employ a hydrogen bonding-based algorithm for its improvement. We show comparisons of the results obtained for random dihedral and for the proposed random, rigid-body water molecule movement, giving evidence that a specific adaption of the distortion process greatly improves the convergence of the method.

The second part is about the determination of RPs in clusters between conformational arrangements and for reactions. Besides standard approaches like the nudged elastic band method, we want to focus on a new algorithm developed especially for global reaction path search called Pathopt. We started with argon clusters, a typical benchmark system, which possess a flat PES, then stepwise increase the magnitude and directionality of interactions. Therefore, we calculated pathways for a water cluster and characterize them by frequency calculations. Within our calculations, we were able to show that beneath local pathways also additional pathways can be found which possess additional features.

1. INTRODUCTION

The construction and exploration of energy landscapes is integrant to establish understanding for many molecular phenomena (Wales, 2003; Wales, Miller, & Walsh, 1998). A great variety of methods exist to create, analyze, or transform energy surfaces. Global optimization algorithms (Wales, 2003; Wales & Scheraga, 1999), transition state (TS) search schemes (Schramm, 1998; Truhlar, Garrett, & Klippenstein, 1996), and coordinate transformations (Pappu, Hart, & Ponder, 1998) being only a small selection among the available and important techniques. Such theoretical studies provide insight and understanding of molecular properties which can be enhanced by increasing methodological quality. Within our study, we want to focus on global optimization techniques on the one hand and on the other hand on reaction path (RP) determination in order to obtain reaction profiles and barriers.

The exploration of the minima for a potential energy surface (PES) of water clusters is in the focus of the first part of this investigation. In former studies, we were able to present a new procedure for the setup of water shells around proteins (Grebner, Niebling, Schmuck, Schlucker, & Engels, 2013) based on the Gradient Only Tabu-Search (GOTS) (Grebner, Becker, Stepanenko, & Engels, 2011; Grebner, Becker, Weber, & Engels, 2012) global optimization algorithm. It was shown that this scheme can provide an improved description of the important solute–solvent interface in comparison to the standard molecular dynamics equilibration. The used strategy implied a mixture of methods where interposed diversification steps improved the convergence behavior of pure GOTS. The diversification was accomplished by using the Monte Carlo with minimization (MCM, Li & Scheraga, 1987; Basin Hopping, Wales & Doye, 1997) method to escape local regions of the potential surface. Here we present how the modification of the random jump step in the MCM procedure further improves the convergence of a global optimization of water clusters.

The second part is about global reaction path (RP) searches. Many of the established RP approaches make local predictions or deliver the TS by using a local guess. The group of local methods starts from one well-defined structure and explores the landscape by using the gradient and Hessian information (Banerjee, Adams, Simons, & Shepard, 1985; Cerjan & Miller, 1981; Henkelman & Jonsson, 1999). Other methods make use of two starting structures (chain of states methods) which are sometimes also called interpolation-based approaches (Ayala & Schlegel, 1997; Elber & Karplus, 1987; Fischer & Karplus, 1992; Maragakis, Andreev, Brumer, Reichman, & Kaxiras, 2002). Comparing both methods, the chain of states methods are superior in the manner that they might deliver further TSs on the pathway which cannot be found by completely local acting methods. Nevertheless, the complete exploration of the landscape between two structures can only be achieved using global search strategies. By exploring the energy landscape, the dynamics of the system would be more and more included and be understood. Therefore, we present a global search strategy called Pathopt (PO) (Grebner, Pason, & Engels, 2013).

2. GLOBAL OPTIMIZATION OF AQUEOUS SYSTEMS

Water is—hands down—one of the most important molecules and considered to be not "a" but "the" prerequisite for life. Not only, does it enable many other biomolecules to be in their physiologically active form in the first place, but it also participates in important biological processes (Chaplin, 2001; Cheung, García, & Onuchic, 2002). The key feature of aqueous solutions is hydrogen bonding. Those interactions are weaker than covalent bonds but much stronger than van der Waals interactions and larger than kT at room temperature (Brady & Schmidt, 1993). Therefore, they dominate the noncovalent domain of a molecular energy surface in water. Sometimes, it is sufficient to include implicit solvation using continuum models but for many applications, water has to be considered explicitly because hydrogen bonds are directional, they moderate interactions and water molecules are often involved in reaction mechanisms (Straeter & Lipscomb, 1995; Zhao, Guengerich, Voehler, & Waterman, 2005).

The most common procedure for building up solvation shells is comprised of placing the solute in a premade water cluster by cutting the interfering water atoms and equilibrating the resultant geometry through long molecular dynamics simulation runs (Geerke, Thiel, Thiel, & van Gunsteren, 2008; Lee, Luckner, Kisker, Tonge, & Engels, 2011). Experience shows that this does not necessarily provide energetically favored structures because MD is usually performed at room temperature where many conformations are accessible which are higher in terms of energy. Although giving a qualitatively correct description of the bulk, such simulations may still miss crucial conformations of the important inner water shell. Low lying minima, especially for the first solvation shell, are desirable however, due to several reasons. Besides providing access to ice structures, they are to be preferred over less favorable conformations, as starting structures for MD sampling and they provide a better description of microsolvation (Grebner, Niebling, et al., 2013).

The hydration method of Grebner, Niebling, et al. (2013) was shown to be substantially superior to the standard procedure in order to find stable water structures with a better characterization of the solvent–solute interface. Although this scheme incorporates the Metropolis MCM (also known as Basin Hopping) approach as a diversification step, the applied search was still quite local and thus demanding for large systems. The random move step was applied in the torsional rotation space which either only allows minimal changes if a to-be-changed dihedral rotates a large subgroup of the system or it does not change the system very much if only a small number of atoms is affected. Although, such purely random moves in the dihedral phase space have proven to be suitable for the global optimization of proteins, they do not take care of hydrogen bonding. We will refer to this scheme as DA-MCM (random dihedral angle MCM).

For better convergence, we suggest a random movement scheme that is considerably less local and enforces the formation of hydrogen bonds. The concept is to move a random number of water molecules, which are not buried, to a new, random location with the constraint that they must undergo hydrogen bonding at their new position.

The move step of the MCM procedure using this concept is:

(1) Find all possible destinations, where placing a water molecule would form a new hydrogen bond.

(2) Select a random number of water molecules which are not buried in the current conformation.

(3) Try to move each selected water molecule to a randomly selected destination.

This essentially is a rigid-body move under constraint and will be referred to as WM-MCM (random water movement MCM). In resemblance to the reference method, the move is accepted if the Metropolis criterion is met for the minimized structure.

2.1 Determination of Sites, Potentially Forming a Hydrogen Bond with a Water Molecule

Positions in space, where placing an appropriately oriented water molecule would form a hydrogen bond will be denoted "***targets***." They consist of an atom and an orientation vector. The atoms are selected by finding all accessible hydrogen bond donors and acceptors of the system and the orientation vector is either the covalent bond vector in case of HB donors or determined by applying the valence shell electron pair repulsion theory (Gillespie & Hargittai, 2013) using the AXE (Lindmark, 2010) method.

2.2 Identification of Nonburied Water Molecules for Movement

Checking whether a given water molecule is either surrounded by other molecules or accessible and therefore moveable in terms of this algorithm, is done via a full spherical scan around the geometric center (p^w) of the water molecule. The spherical scan is performed on a two-angle (θ, φ) grid in a certain distance to p^w. If no atoms, but the ones which are part of the examined water, are within a given distance of the resulting scan position $x^w(\theta, \varphi)$ and if none of the atoms is fixed by user request, that water is viable to be moved (Fig. 1). Water molecules which have been identified to possibly undergo random movement will be called "***candidates***".

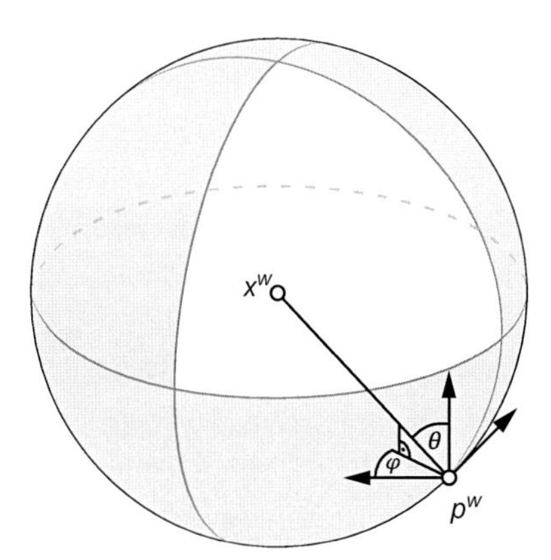

Fig. 1 Scan position $x^w(\theta, \varphi)$ indicating the sphere that has to be free of other (=not part of current water molecule) atoms: originating geometric center of the water molecule p^w as well as the angles θ and φ are also shown.

2.3 Random Water Selection and Movement

The overall movement selection scheme involves

1. Get a random number *r* from a geometric distribution.
2. For at max *r* + 1 randomly selected elements (*c*), of the ***candidates*** set, do:
   a. Move element *c*, if sterically possible, to first viable nontabu ***target*** (tries in random order) under the formation of a hydrogen bond and set that ***target*** tabu.

3. OPTIMIZATION OF WATER CLUSTERS $(H_2O)_{20}$, $(H_2O)_{30}$, AND $(H_2O)_{40}$

3.1 Algorithm Assessment Criteria

In order to analyze the quality of the proposed move technique, we chose to perform a set of MC simulations on three different water clusters, $(H_2O)_{20}$, $(H_2O)_{30}$, and $(H_2O)_{40}$, using the random water movement scheme (WM/WM-MCM) as well as a set of runs using random dihedral angle changes (DA/DA-MCM) for comparison.

The capability of a global optimization algorithm to find the lowest minimum of an energy surface of a molecular system can be judged by the absolute criterion A:

(A) Can the combined algorithm (random move + local optimization)—in theory—sample the full phase space of all minima and therefore the global minimum or will any region be left out? Or more mathematically: Is the graph connected, which is formed by the entirety of local minimum conformations as vertices with the movement algorithm defining the edges?

A qualitative appraisal about the algorithm's suitability for global optimization is to be made. The most interesting, observable characteristic is the lowest minimum energy that is found within a certain number of steps. Since Metropolis MC is a stochastic method, it is sensible to define criteria for the judging based on multiple applications of the rated techniques to the same problem (sample) with the mean of the lowest minima energies being the most significant feature of the sample. This leads to a second criterion (B).

(B) How fast does the lowest energy converge toward the global minimum on average? Or analog: What is the average lowest minimum energy value (AV) after a given time or a given number of steps?

Given a set of random walk runs, applied to the same problem, it is desirable to have a low variance and standard deviation (SD) of the lowest minimum energies in the end, because it is undesirable to accidentally obtain any high energy runaway values. Furthermore, the ideal edge case is that all calculations converge to the global minimum resulting in a final variance and final SD of zero. This leads to a further criterion C:

(C) Does every application of the procedure yield comparably low minima within the same number of steps or within the same time? Or in mathematical terms: What is the magnitude of the final SD of the lowest energies?

3.2 Calculations and Methodology

Fifty random walk MCM calculations were performed for each water cluster $(H_2O)_n$ size ($n=20$, 30, 40) and each random move scheme (DA, WM). The number of iterations was limited to 2000. Each local minimization step was performed using a low-memory Broyden–Fletcher–Goldfarb–Shanno (BFGS) implementation. Energies and gradients were obtained from force field calculations (TIP3P model; OPLS-AA, (Damm, Frontera, Tirado-Rives, Jorgensen, & Rives, 1997; Jorgensen, Maxwell, & Tirado-Rives, 1996; Kony, Damm, Stoll, & Van Gunsteren, 2002; Tirado-Rives & Jorgensen, 1992), parameters).

3.3 Computational Results

The qualitative characteristics of the graphs in Fig. 2 are quite identical for the three benchmark systems. The curves from WM-MCM calculations are, with a few exceptions, shifted toward lower energies, compared to the ones for DA-MCM. Table 1 shows that the final WM-MCM sample has a lower minimal energy, a lower mean energy, and a smaller SD for every system after 2000 steps, compared to DA-MCM.

3.4 Summary and Quality Appraisement

We discuss the quality of the proposed scheme with respect to the earlier mentioned criteria.

Criterion A Given that the algorithm is limited in several ways (ie, valid destinations must expose a HB donor or a HB acceptor), the phase space is not guaranteed to be fully sampled using the proposed scheme, even in case of an unlimited number of steps. Therefore, the approach is—in theory—inherently inferior to real random moves. In practice, however, this problem usually does not affect the quality of the results since the global minimum is very likely to exhibit at least one hydrogen bond for every water molecule, anyway.

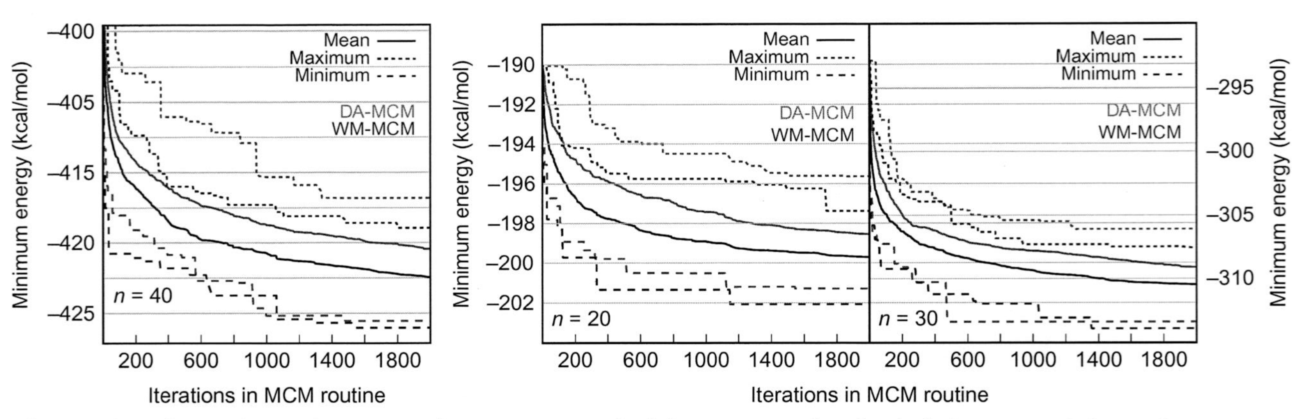

Fig. 2 Progression of mean lowest energy, maximum energy, and minimum energy for all calculation sets and cluster sizes.

Table 1 Lowest Sample Energy As Well As Mean and SD of Lowest Energies for $(H_2O)_{20}$, $(H_2O)_{30}$, and $(H_2O)_{40}$

kcal/mol	WM-MCM	DA-MCM
$(H_2O)_{20}$		
Lowest energy in final sample	−202.07	−201.289
Mean final lowest energy of sample	−199.70	−198.54
SD of final sample	1.10	1.40
$(H_2O)_{30}$		
Lowest energy in final sample	−313.89	−313.37
mean of final sample	−310.47	−309.11
SD of final sample	1.28	1.50
$(H_2O)_{40}$		
Lowest energy in final sample	−426.02	−425.54
mean of final sample	−422.45	−420.44
SD of final sample	1.68	1.87

The results show that WM-MCM is able to find lower minima for all test systems than does the DA-MCM.

Criterion B For all three test systems, it is favorable in terms of convergence to randomly move water molecules from one place to another in order to form new hydrogen bonds. WM-MCM found a $(H_2O)_{20}$ conformation, lower than −198 kcal/mol on average within around 500 steps. The average number of steps for the dihedral-randomizing calculations to get below that threshold was about 1299. Similar observations can be made with respect to the other test systems. The average $n = 30$ cluster calculation using dihedral rotations takes 1889 iterations to get below −309 kcal/mol whereas for water moves the average of the lowest energy is below that value from iteration 814 onwards. All calculations show an increasing horizontal distance between the lower mean energy curve of the random WM and the mean minimum energy curve of the DA-MCM, clearly indicating better convergence of WM-MCM. Fig. 3 visualizes the superior performance behavior of the water moving method and indicates a linear correlation.

Presuming direct proportionality between the numbers of steps required for both methods in order to converge to the same energy level, simple linear regression can be applied to find a factor relating both methods for each test

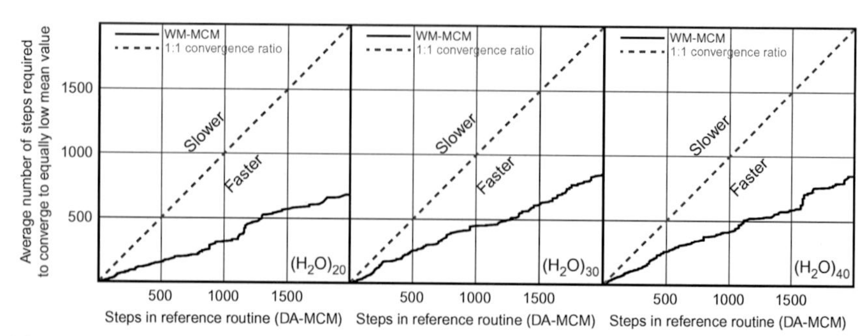

Fig. 3 Average number of WM-MCM iterations required until the mean minimum energy is converged to the same value as for the given step in the DA-MCM process; the diagonal marks the 1:1 ratio of equal minimum energy convergence behavior.

Table 2 Proportionality Factors Between the Mean Energy Convergence of SA-MCM and WM-MCM with Correlation Coefficients

System	Fitted Inclination	Correlation Coefficient
$(H_2O)_{20}$	2.85	0.995
$(H_2O)_{30}$	2.33	0.998
$(H_2O)_{40}$	2.32	0.998

system via a linear fit of the values used in Fig. 3. It turns out that random dihedral movement requires two to three times the number of steps in order to reach the same low mean energy level as the procedure moving waters around. The obtained values can be found in Table 2.

Criterion C The SD of the minimum energies after 2000 steps in WM-MCM is about 10–20% smaller than those obtained after 2000 iterations of dihedral movement and subsequent local optimization. Thus it is less likely to find a runaway value with higher energy if only a single calculation is performed.

4. ILLUSTRATING THE BASIC CONCEPTS OF PO

4.1 Reaction Pathways

Within this part of the review, we want to introduce the PO algorithm for reaction path determination which is included in the CAST (Conformational Analysis and Search Tool) (Grebner et al., 2014) program. Especially, we want to focus on the newest implementation.

Many algorithms have been developed within this field and each group of methods has its advantages and disadvantages. Therefore, to choose which method you want to use often depends on the system investigated. For example, if you are interested only in the nature of a specific TS and not interested, eg, in byproducts and intermediates than a local approach like eigenvector following (Banerjee et al., 1985) might be quite sufficient. However, the prerequisites for using this methodology are set by the fact that one has already enough information and a good clue of the location of the TS. If one wants to explore a reaction from scratch than the nudged elastic band (NEB) (Henkelman & Jonsson, 2000; Sheppard, Terrell, & Henkelman, 2008) approach is a good alternative, because it estimates a local minimum energy path, which includes additional TSs if they are present and can locate them without previous knowledge of them. Nevertheless, if the reaction of interest is more complex and different pathways exist, like in enzymatic reactions, then other search strategies might be preferred. Therefore, global search strategies are acknowledged. One of these methods is the transition path sampling (TPS) approach by Dellago (Bolhuis, Chandler, Dellago, & Geissler, 2002) or another one the discrete path sampling (DPS) approach from Wales (2002). For both approaches, it was shown that they are able to predict, eg, protein folding in a suitable way (Bolhuis & Dellago, 2015; Carr & Wales, 2008). These two approaches reach their aim in a different manner. The TPS approach does it by using a Markov Chain MC (Hastings, 1970) procedure, which connects two structures by shooting dynamical unbiased trajectories and the DPS method by building up a large database of minima and TSs and connect them.

The PO approach based upon a constraint global search strategy and the connection of pathways is directly done for the two starting structures (educt/product). The basis of PO is an initially built pathway based on the NEB pathway. The implementation of the NEB method within our program is founded on the scheme developed by Henkelman et al. (Henkelman & Jonsson, 2000; Sheppard et al., 2008), especially we use the climbing image variant (Henkelman, Uberuaga, & Jonsson, 2000).

4.2 The PO Scheme

In the first step of PO, the given two minimum structures are connected within a linear interpolation procedure by using a complete number of N structures. Between the obtained structures connecting tangent vectors are calculated. These connecting vectors are atom wise calculated. The

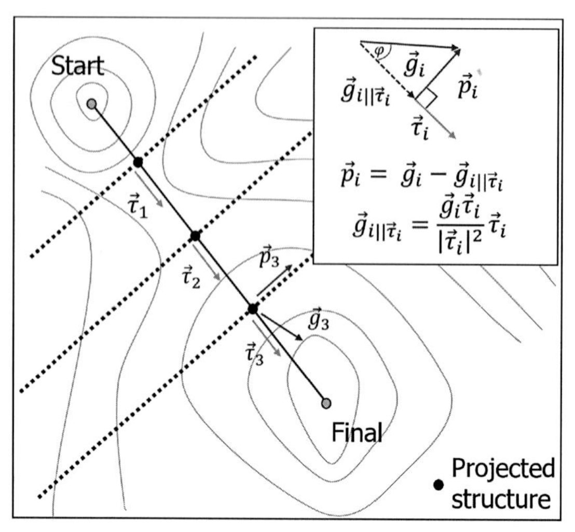

Fig. 4 The schematic representation of the PO algorithm is depicted. Three example hyperplanes and the linear pathway are shown. The projected perpendicular gradients $\vec{p}_i$ with respect to the pathway and the connecting tangent vectors $\vec{t}_i$ are illustrated which are used for the constraint optimization.

calculation of the tangents is based on the proposed method by Henkelman. A graphical illustration of this concept is depicted in Fig. 4. In the next step, each projected structure n from $n=0$ up to $n=N$ serves as a starting point of a constraint global optimization, which means it is carried out on a $k-1$ dimensional PES (hyperplane). For this purpose, we apply a modified MCM scheme (Grebner, Pason, et al., 2013). The local optimization steps are restricted to be perpendicular to the tangent vectors. To realize the constraint two different approaches are included. The first approach uses a simple projection of gradients. In consequence, the resulting perpendicular gradient $\vec{p}_i$ (first derivative of atom i) is calculated as follows:

$$\vec{p}_i = \vec{g}_i - \vec{g}_{i||\vec{\tau}_i} \quad (1)$$

with

$$\vec{g}_{i||\vec{\tau}_i} = \frac{\vec{g}_i \vec{\tau}_i}{\left|\vec{\tau}_i\right|^2} \vec{\tau}_i \quad (2)$$

The component $\vec{g}_{i||\vec{\tau}_i}$ is the gradient component of atom i of structure n along the tangent of the concerning atom pointing to the next atom of structure $n+1$ along the band. The random steps in the Monte Carlo simulation are modified as well. Only move steps perpendicular to the initial tangent vectors are carried out. In the second approach, the projection scheme is replaced by applying a quadratic bias potential acting perpendicular to the connecting band built up by the NEB tangent vectors. For the first scheme, no additional force constant has to be chosen, but for the latter method acting forces have to be carefully assigned. Therefore, the bias constant has to be adapted to the specific system which should be investigated.

The found minima obtained with the modified MCM scheme may serve as traces of possible reaction pathways and can be used to build up complete pathways. In addition, the found maximum energy points on the path can serve as guesses for possible TSs and can be further refined or characterized. For getting these pathways different connection strategies were developed. The first applied connection strategy (see Fig. 5) based upon a constraint optimization on only one hyperplane located at the midst of the initial linear pathway (Grebner, Pason, et al., 2013). Starting from the obtained minima on the hyperplane an uphill optimization by the improved dimer (Kastner & Sherwood, 2008) method is used to find the directly connected TSs. In a next step, a minimization is carried out along the first imaginary mode along both directions to find the connected minima. The obtained path segments are further proofed for their connectivity with the final and starting structures, eg, with NEB. If necessary, further cycles are applied to obtain

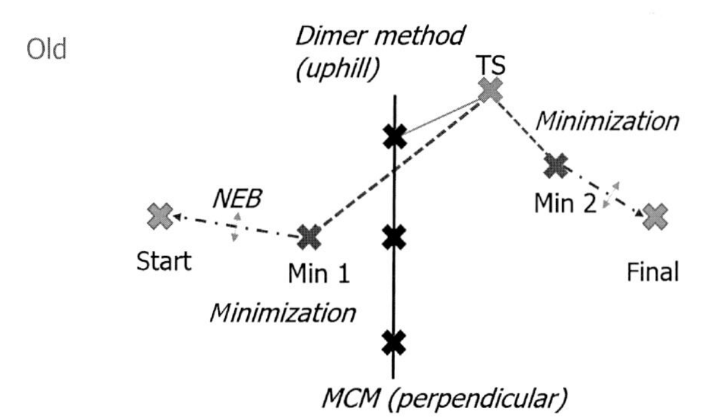

Fig. 5 The "old" connection strategy within the PO procedure is shown (Grebner, Pason, et al., 2013). Within this procedure, the constraint global optimization is carried out on only one intervening hyperplane. The found minima are further optimized to the TS and connection with start and final structure is reached by applying further PO cycles or by using NEB.

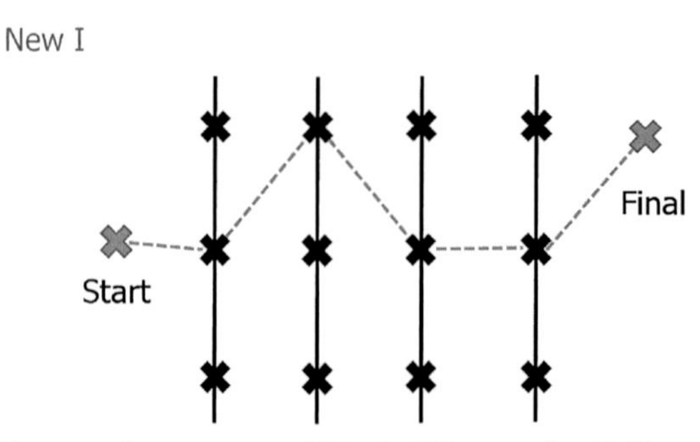

Fig. 6 The "new I" connection strategy within the PO procedure is illustrated. Within this approach, the number of simulations on intervening hyperplanes is increased. The found minima on these hyperplanes are directly connected by searching the next nearest neighbors, measured in RMSD values, between the concerning hyperplanes.

complete pathways. However, this method is very time consuming and depends on the accuracy of the uphill optimization to get complete pathways. Within the newest approach the number of hyperplanes is increased to the number of $N-2$. The found minima are directly used as reaction path determining structures, to accelerate the computation. Therefore, the found structures are aligned and connected via a minimum RMSD (root mean square deviation) criterion to obtain complete pathways. This means that next nearest neighbors on the next hyperplane are searched and connected ("new 1" connection scheme). In addition, not only nearest neighbors also second- and third-nearest neighbors are connected (Fig. 6).

Beneath this connection strategy, also a third possibility exists. In that case, the presented connection strategy is slightly modified and the connection between the minimum structures is carried out by applying NEB simulations. This strategy is called "new 2."

4.3 Ar_{38} and $(H_2O)_{20}$ Cluster Calculations

The calculations on the cluster systems should reveal that the PO algorithm is an appropriate method to describe the transition between different arrangements in these clusters. The NEB and PO calculations were carried out with the CAST program. Vibrational frequencies were obtained with the Tinker 6.2 (Kundrot, Ponder, & Richards, 1991; Pappu et al., 1998; Ponder & Richards, 1987; Ponder et al., 2010; Ren, Wu, & Ponder, 2011; Ren & Ponder, 2003; Shi et al., 2013) program. The Ar_{38} cluster was calculated by using the OPLS-AA (Damm et al., 1997; Jorgensen et al., 1996; Tirado-Rives & Jorgensen, 1992) force field. The water cluster was parameterized by using TIP3P water model within the OPLS-AA force field description.

In the case of the Argon cluster calculations both starting structures were generated by carrying out a global optimization. For the global optimizations Tabu-Search as well as MCM simulations were carried out. We sampled 500 steps at 298.15 K and applied a step size of 1.4 Å for the random MC steps. For this study, we applied the global search to generate reasonable starting structures for the RP search. For the water cluster $(H_2O)_{20}$, we used 5000 MCM steps and the same temperature and steps size settings, as for the argon clusters. In both calculations, no cutoff radius for the nonbonded interactions was applied. The general procedure for all calculations was to align the starting structures beforehand. For this we used the Kabsch (1978) alignment procedure implemented in VMD 1.9.1 (Humphrey, Dalke, & Schulten, 1996) program.

4.4 Ar_{38} Results

The argon cluster should serve as a test system for the PO procedure, because of its relatively flat and complex PES with a multitude of minima. Caused by this fact the Ar_{38} cluster is an ideal test system and hence often used in studies within the field of global optimization. The general applicability of the algorithm was already shown for smaller Lennard–Jones (LJ) clusters (Grebner, Pason, et al., 2013). In consequence, we want to focus on how good the new presented connection procedure performs and how to process and visualize the obtained data. The starting structures are two randomly chosen local minima of the Ar_{38} cluster. In Fig. 7, the results of a PO optimization carried out on 20 hyperplanes are depicted. For the simulation, 2000 MCM steps were performed per hyperplane. In these calculations the gradient projection scheme was used. The results exemplify that the obtained minima are wide spread as well as in RMSD as in energy values. Therefore, the question might arise which minima are reasonable to choose and which pathways are meaningful to include. In this manner, different strategies can be used. The RMSD criterion is already included within the built up procedure for pathways. In addition, an energy criterion for the found minima can be applied. Such a criterion is offered by the Arrhenius equation might using the relative barrier heights with respect to the NEB barrier and calculating the effect on rate constants. In accordance to this, we chose only minima from the PO simulation, which have a peak value of maximal 2 kcal/mol

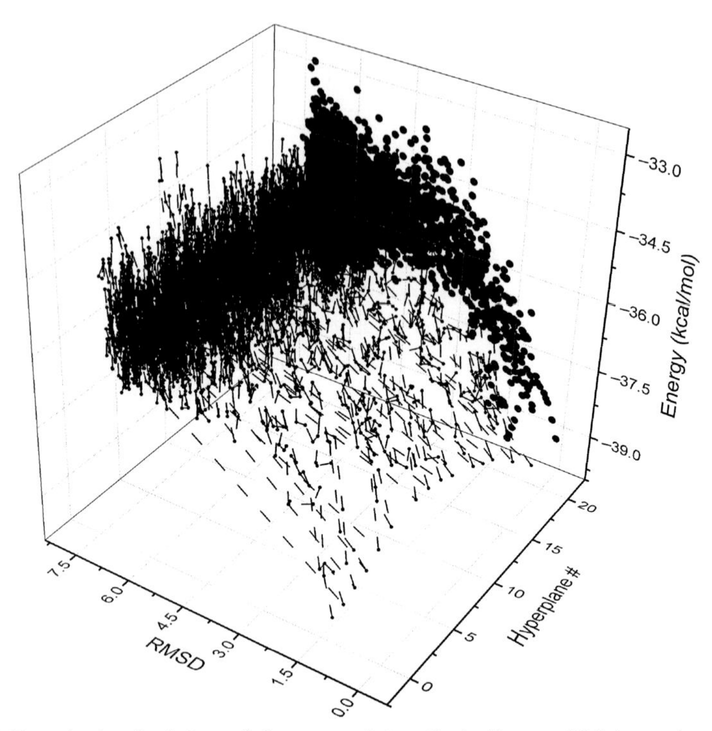

Fig. 7 The obtained minima of the constraint optimization on 20 intervening hyperplanes are shown by their energy in kcal/mol and with respect to the RMSD values, which are referenced to the NEB pathway starting structures. (See the color plate.)

higher compared with the NEB TS. The influence of the difference in activation energies within the Arrhenius formula results in a reduction of the rate constant up to 7.4 relatively to the barrier found with NEB. In consequence, it can be stated that pathways built up by minima within a relatively small energy range contribute significantly while other pathways are less important to include. By looking on the results in a per plane view, which is illustrated in Fig. 8, shows that also structures which are lower in energy compared with NEB can be found. By restricting the energy of accepted minima, the number of structures for the connection procedure can be significantly reduced.

In a next step, we want to have a closer look on the pathways, which are built up and are automatically generated within PO. Therefore, the results from a simulation on 10 hyperplanes are depicted, whereas the focus lays on the minima, which are the lowest in energy and RMSD. The excerpt is

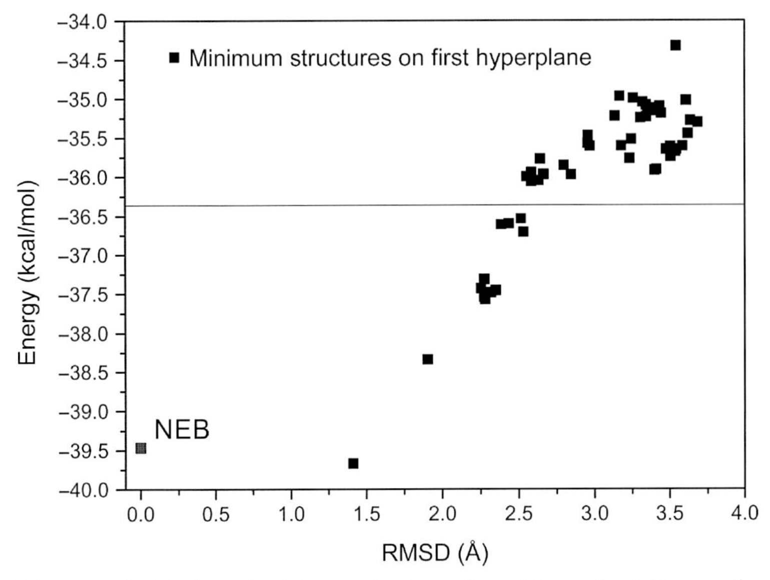

Fig. 8 Energy distribution of minimum structures found for a simulation on the first hyperplane is given. The simulation was carried out on 10 hyperplanes. The energies are given in kcal/mol and the concerning RMSD value in Å. The NEB reference (first structure of the path) is given in *green* (*dark gray* in the print version). The *black line* indicates the 2 kcal/mol energy difference relatively to the NEB barrier.

depicted in Fig. 9 where two possible pathways are highlighted in *purple* (*dark gray* in the print version) and *black*. The black pathway is the PO pathway, which is defined along the minimum RMSD change and delivers a pathway that is only for a few points significantly different with respect to the NEB pathway. For a better comparison, the NEB and the PO pathway are presented in Fig. 10. The comparison between PO path 1 and the NEB pathway shows that for structures 4–11 the pathways are very similar. The difference occurs for structure 3. Also by looking on the structural change, which is depicted in Fig. 11, it can be seen that NEB predicts a very smooth pathway which is also obtained by PO represented in path 1. There exists only one real outlier, which is related to the third structure. For path 2 much larger displacements are obtained between the structures which can also be seen in the RMSD value changes. The results indicate that PO is able to predict additional pathways besides the local pathway. Some of the pathway features can be quite similar with respect to a local NEB pathway, but might exhibit also new features.

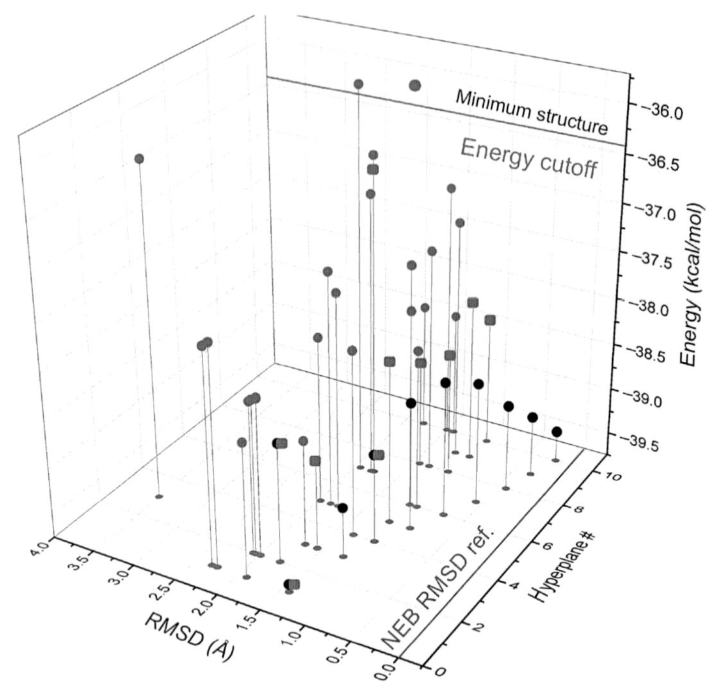

Fig. 9 The excerpt of a PO simulation on 10 hyperplanes is depicted. The minimum structures, which are the lowest in energy and RMSD are depicted as *red* (*dark gray* in the print version) points. The RMSD values are calculated with respect to the NEB structures. The 2 kcal/mol cutoff is indicated by the *orange* (*gray* in the print version) *line*. *Purple* (*dark gray* in the print version) *squares* and *black dots* mark two possible pathways obtained from a PO run.

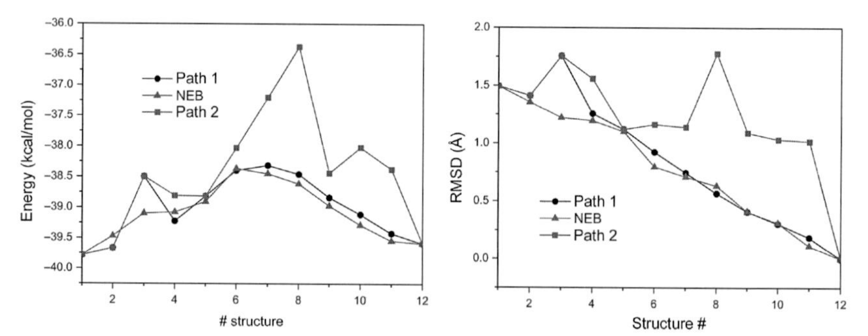

Fig. 10 On the left side, the energy profile of the NEB pathway (*red* (*dark gray* in the print version)) and two PO pathways (*purple* (*gray* in the print version)/*black*) are depicted. On the right side, the RMSD change is presented. The pathways are built up by 10 structures. Structures 1 and 12 are the start and the final structures.

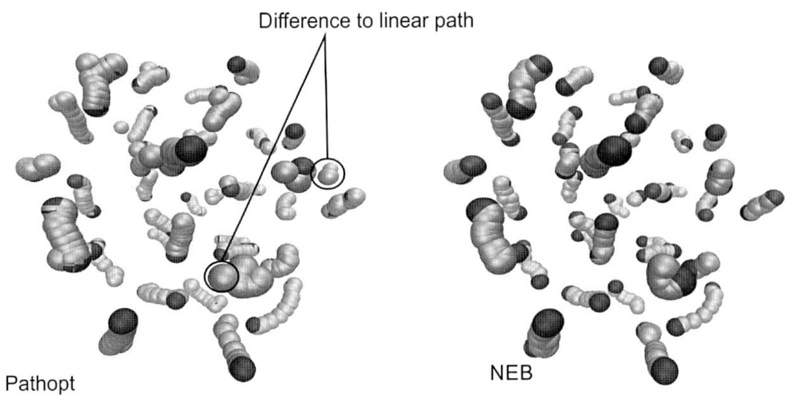

Fig. 11 On the left side, the structures building the PO path 1 are shown. In *green* and *red* the start and final structures of the simulation are presented. On the right side, the NEB pathway structures are depicted. Two differences between the PO and the NEB pathway are highlighted. (See the color plate.)

4.5 $(H_2O)_{20}$ Cluster Results

The water cluster system represents a second test case for the PO algorithm. With respect to the LJ clusters previously calculated the nonbonded interactions are stronger and have a bigger directionality due to the hydrogen bonds. Nevertheless, the same simulation strategies as for the LJ clusters were applied and yield satisfying results. We were able to characterize additional RPs. To verify the obtained pathways we also calculated the frequencies and especially looking on the first mode. Therefore, we present a pathway in Fig. 12 that we directly obtained from a PO calculation. For the discrete pathway two TSs can be located. The first one occurs for the 9th image and the second one for the 11th image. The frequencies of the first mode are $-21.17\ \text{cm}^{-1}$ (9th) and $-6.03\ \text{cm}^{-1}$ (11th), whereas the frequencies of the second and third modes are zero. In consequence, we can state that we might have found real TSs without further refinement, which might not be the case every time. Compared with the found NEB TS the PO one is of about 1 kcal/mol lower in energy.

5. SUMMARY

We have shown that a problem-specific WM algorithm is able to improve the convergence of the Metropolis Monte Carlo algorithm significantly, compared to the random deviation of DAs. The final energies after

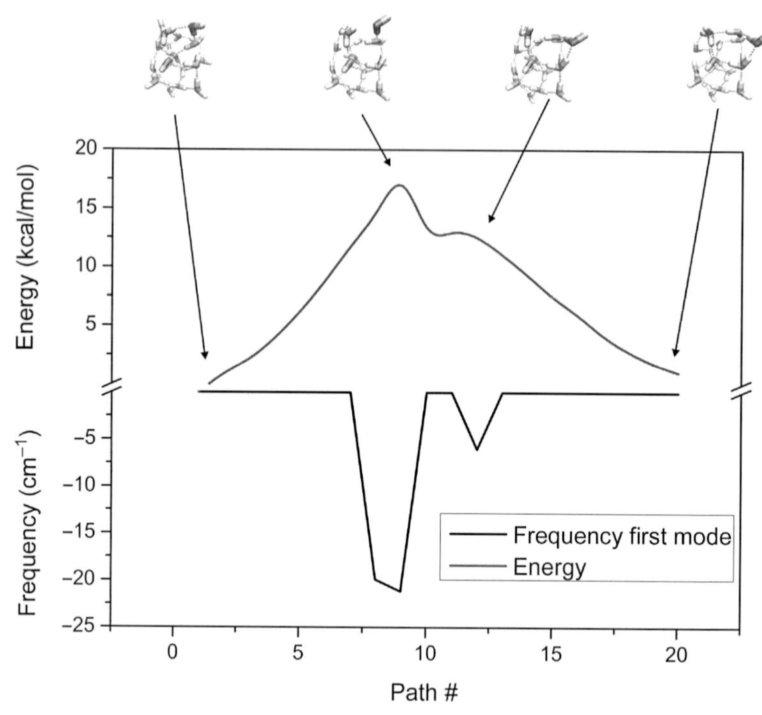

Fig. 12 The minimum energy pathway for a transition between two minimum structures of the $(H_2O)_{20}$ water cluster is shown. The energies are given in kcal/mol and the frequencies of the first vibrational mode are given in cm^{-1}.

the same number of steps are lower and their variance is smaller. The hydrogen bonding-based approach appears to be a reasonable choice in order to optimize aqueous systems.

Within our studies concerning the field of reaction path search, we were able to show that our presented PO strategy is able to find multiple reaction pathways within one simulation. For cluster systems, the Cartesian move strategy in combination with the gradient projection scheme also predicts good TS without further refinement. Nevertheless, a classification and calculation of the vibrational frequencies might be recommended as a convergence criterion.

Within our studies, we were able to show that the proper choice of the global optimization strategy can influence the quality of the obtained results. Therefore, the choice of coordinates representation and move strategies are important. In addition to get further insight, a proper sampling in combination with proper analyzation strategies might be improve the significance of the obtained results.

REFERENCES

Ayala, P. Y., & Schlegel, H. B. (1997). A combined method for determining reaction paths, minima, and transition state geometries. *Journal of Chemical Physics*, *107*(2), 375–384. http://dx.doi.org/10.1063/1.474398.

Banerjee, A., Adams, N., Simons, J., & Shepard, R. (1985). Search for stationary-points on surface. *Journal of Physical Chemistry*, *89*(1), 52–57. http://dx.doi.org/10.1021/j100247a015.

Bolhuis, P. G., Chandler, D., Dellago, C., & Geissler, P. L. (2002). Transition path sampling: Throwing ropes over rough mountain passes, in the dark. *Annual Review of Physical Chemistry*, *53*, 291–318. http://dx.doi.org/10.1146/annurev.physchem.53.082301.113146.

Bolhuis, P. G., & Dellago, C. (2015). Practical and conceptual path sampling issues. *European Physical Journal Special Topics*, *224*(12), 2409–2427. http://dx.doi.org/10.1140/epjst/e2015-02419-6.

Brady, J. W., & Schmidt, R. K. (1993). The role of hydrogen bonding in carbohydrates: Molecular dynamics simulations of maltose in aqueous solution. *The Journal of Physical Chemistry*, *97*(4), 958–966. http://dx.doi.org/10.1021/j100106a024.

Carr, J. M., & Wales, D. J. (2008). Folding pathways and rates for the three-stranded beta-sheet peptide Beta3s using discrete path sampling. *Journal of Physical Chemistry B*, *112*(29), 8760–8769. http://dx.doi.org/10.1021/jp801777p.

Cerjan, C. J., & Miller, W. H. (1981). On finding transition-states. *Journal of Chemical Physics*, *75*(6), 2800–2806. http://dx.doi.org/10.1063/1.442352.

Chaplin, M. F. (2001). Water: Its importance to life. *Biochemistry and Molecular Biology Education*, *29*(2), 54–59. http://dx.doi.org/10.1111/j.1539-3429.2001.tb00070.x.

Cheung, M. S., García, A. E., & Onuchic, J. N. (2002). Protein folding mediated by solvation: Water expulsion and formation of the hydrophobic core occur after the structural collapse. *Proceedings of the National Academy of Sciences of the United States of America*, *99*(2), 685–690. http://dx.doi.org/10.1073/pnas.022387699.

Damm, W., Frontera, A., Tirado-Rives, J., Jorgensen, W. L., & Rives, J. T. (1997). OPLS all-atom force field for carbohydrates. *Journal of Computational Chemistry*, *18*, 1955–1970. http://dx.doi.org/10.1002/(SICI)1096-987X(199712)18:16<1955::AID-JCC1>3.3.CO;2-A.

Elber, R., & Karplus, M. (1987). A method for determining reaction paths in large molecules—Application to myoglobin. *Chemical Physics Letters*, *139*(5), 375–380. http://dx.doi.org/10.1016/0009-2614(87)80576-6.

Fischer, S., & Karplus, M. (1992). Conjugate peak refinement—An algorithm for finding reaction paths and accurate transition-states in systems with many degrees of freedom. *Chemical Physics Letters*, *194*(3), 252–261. http://dx.doi.org/10.1016/0009-2614(92)85543-j.

Geerke, D. P., Thiel, S., Thiel, W., & van Gunsteren, W. F. (2008). QM-MM interactions in simulations of liquid water using combined semi-empirical/classical Hamiltonians. *Physical Chemistry Chemical Physics*, *10*(2), 297–302. http://dx.doi.org/10.1039/B713197F.

Gillespie, R. J., & Hargittai, I. (2013). *The VSEPR model of molecular geometry*. Mineola, NY: Dover Publications.

Grebner, C., Becker, J., Stepanenko, S., & Engels, B. (2011). Efficiency of Tabu-search-based conformational search algorithms. *Journal of Computational Chemistry*, *32*(10), 2245–2253. http://dx.doi.org/10.1002/Jcc.21807.

Grebner, C., Becker, J., Weber, D., Bellinger, D., Tafipolski, M., Brückner, C., & Engels, B. (2014). CAST: A new program package for the accurate characterization of large and flexible molecular systems. *Journal of Computational Chemistry*, *35*(24), 1801–1807. http://dx.doi.org/10.1002/jcc.23687.

Grebner, C., Becker, J., Weber, D., & Engels, B. (2012). Tabu search based global optimization algorithms for problems in computational chemistry. *Journal of Cheminformatics*, *4*(Suppl. 1), P10.

Grebner, C., Niebling, S., Schmuck, C., Schlucker, S., & Engels, B. (2013). Force field-based conformational searches: Efficiency and performance for peptide receptor complexes. *Molecular Physics*, *111*(16–17), 2489–2500. http://dx.doi.org/10.1080/00268976.2013.826392.

Grebner, C., Pason, L. P., & Engels, B. (2013). PathOpt—A global transition state search approach: Outline of algorithm. *Journal of Computational Chemistry*, *34*(21), 1810–1818. http://dx.doi.org/10.1002/jcc.23307.

Hastings, W. K. (1970). Monte-carlo sampling methods using markov chains and their applications. *Biometrika*, *57*(1), 97–109. http://dx.doi.org/10.2307/2334940.

Henkelman, G., & Jonsson, H. (1999). A dimer method for finding saddle points on high dimensional potential surfaces using only first derivatives. *Journal of Chemical Physics*, *111*(15), 7010–7022. http://dx.doi.org/10.1063/1.480097.

Henkelman, G., & Jonsson, H. (2000). Improved tangent estimate in the nudged elastic band method for finding minimum energy paths and saddle points. *Journal of Chemical Physics*, *113*(22), 9978–9985. http://dx.doi.org/10.1063/1.1323224.

Henkelman, G., Uberuaga, B. P., & Jonsson, H. (2000). A climbing image nudged elastic band method for finding saddle points and minimum energy paths. *Journal of Chemical Physics*, *113*(22), 9901–9904. http://dx.doi.org/10.1063/1.1329672.

Humphrey, W., Dalke, A., & Schulten, K. (1996). VMD: Visual molecular dynamics. *Journal of Molecular Graphics & Modelling*, *14*(1), 33–38. http://dx.doi.org/10.1016/0263-7855(96)00018-5.

Jorgensen, W. L., Maxwell, D. S., & Tirado-Rives, J. (1996). Development and testing of the OPLS all-atom force field on conformational energetics and properties of organic liquids. *Journal of the American Chemical Society*, *118*(45), 11225–11236. http://dx.doi.org/10.1021/ja9621760.

Kabsch, W. (1978). A discussion of the solution for the best rotation to relate two sets of vectors. *Acta Crystallographica Section A*, *34*(5), 827–828. http://dx.doi.org/10.1107/S0567739478001680.

Kastner, J., & Sherwood, P. (2008). Superlinearly converging dimer method for transition state search. *The Journal of Chemical Physics*, *128*(1), 014106. http://dx.doi.org/10.1063/1.2815812.

Kony, D., Damm, W., Stoll, S., & Van Gunsteren, W. F. (2002). An improved OPLS–AA force field for carbohydrates. *Journal of Computational Chemistry*, *23*, 1416–1429. http://dx.doi.org/10.1002/jcc.10139.

Kundrot, C. E., Ponder, J. W., & Richards, F. M. (1991). Algorithms for calculating excluded volume and its derivatives as a function of molecular-conformation and their use in energy minimization. *Journal of Computational Chemistry*, *12*(3), 402–409. http://dx.doi.org/10.1002/jcc.540120314.

Lee, W., Luckner, S. R., Kisker, C., Tonge, P. J., & Engels, B. (2011). Elucidation of the protonation states of the catalytic residues in mtKasA: Implications for inhibitor design. *Biochemistry*, *50*(25), 5743–5756. http://dx.doi.org/10.1021/bi200006t.

Li, Z., & Scheraga, H. A. (1987). Monte Carlo-minimization approach to the multiple-minima problem in protein folding. *Proceedings of the National Academy of Sciences of the United States of America*, *84*(19), 6611–6615. Retrieved from, http://www.ncbi.nlm.nih.gov/pmc/articles/PMC299132/.

Lindmark, A. F. (2010). Who needs Lewis structures to get VSEPR geometries? *Journal of Chemical Education*, *87*(5), 487–491. http://dx.doi.org/10.1021/ed800145e.

Maragakis, P., Andreev, S. A., Brumer, Y., Reichman, D. R., & Kaxiras, E. (2002). Adaptive nudged elastic band approach for transition state calculation. *Journal of Chemical Physics*, *117*(10), 4651–4658. http://dx.doi.org/10.1063/1.1495401.

Pappu, R. V., Hart, R. K., & Ponder, J. W. (1998). Analysis and application of potential energy smoothing and search methods for global optimization. *Journal of Physical Chemistry B*, *102*(48), 9725–9742. http://dx.doi.org/10.1021/jp982255t.

Ponder, J. W., & Richards, F. M. (1987). An efficient Newton-like method for molecular mechanics energy minimization of large molecules. *Journal of Computational Chemistry, 8*(7), 1016–1024. http://dx.doi.org/10.1002/jcc.540080710.

Ponder, J. W., Wu, C., Ren, P., Pande, V. S., Chodera, J. D., Schnieders, M. J., ... Head-Gordon, T. (2010). Current status of the AMOEBA polarizable force field. *Journal of Physical Chemistry B, 114*, 2549–2564. http://dx.doi.org/10.1021/jp910674d.

Ren, P. Y., & Ponder, J. W. (2003). Polarizable atomic multipole water model for molecular mechanics simulation. *Journal of Physical Chemistry B, 107*(24), 5933–5947. http://dx.doi.org/10.1021/jp027815+.

Ren, P., Wu, C., & Ponder, J. W. (2011). Polarizable atomic multipole-based molecular mechanics for organic molecules. *Journal of Chemical Theory and Computation*, 7(10), 3143–3161. http://dx.doi.org/10.1021/ct200304d.

Schramm, V. L. (1998). Enzymatic transition states and transition state analog design. *Annual Review of Biochemistry, 67*(1), 693–720. http://dx.doi.org/10.1146/annurev.biochem.67.1.693.

Sheppard, D., Terrell, R., & Henkelman, G. (2008). Optimization methods for finding minimum energy paths. *Journal of Chemical Physics. 128*(13). http://dx.doi.org/10.1063/1.2841941.

Shi, Y., Xia, Z., Zhang, J., Best, R., Wu, C., Ponder, J. W., & Ren, P. (2013). Polarizable atomic multipole-based AMOEBA force field for proteins. *Journal of Chemical Theory and Computation, 9*(9), 4046–4063. http://dx.doi.org/10.1021/ct4003702.

Straeter, N., & Lipscomb, W. N. (1995). Two-metal ion mechanism of bovine lens leucine aminopeptidase: Active site solvent structure and binding mode of L-leucinal, a gem-diolate transition state analog, by X-ray crystallography. *Biochemistry, 34*(45), 14792–14800. http://dx.doi.org/10.1021/bi00045a021.

Tirado-Rives, J., & Jorgensen, W. L. (1992). The OPLS force-field for organic and biomolecular systems. *Abstracts of Papers of the American Chemical Society, 204*, 43.

Truhlar, D. G., Garrett, B. C., & Klippenstein, S. J. (1996). Current status of transition-state theory. *The Journal of Physical Chemistry, 100*(31), 12771–12800. http://dx.doi.org/10.1021/jp953748q.

Wales, D. J. (2002). Discrete path sampling. *Molecular Physics, 100*(20), 3285–3305. http://dx.doi.org/10.1080/00268970210162691.

Wales, D. J. (2003). *Energy landscapes.* UK: Cambridge University Press.

Wales, D. J., & Doye, J. P. K. (1997). Global optimization by basin-hopping and the lowest energy structures of Lennard–Jones clusters containing up to 110 atoms. *The Journal of Physical Chemistry A, 101*(28), 5111–5116. http://dx.doi.org/10.1021/jp970984n.

Wales, D. J., Miller, M. A., & Walsh, T. R. (1998). Archetypal energy landscapes. *Nature, 394*(6695), 758–760. Retrieved from, http://dx.doi.org/10.1038/29487.

Wales, D. J., & Scheraga, H. A. (1999). Global optimization of clusters, crystals, and biomolecules. *Science, 285*(5432), 1368–1372. http://dx.doi.org/10.1126/science.285.5432.1368.

Zhao, B., Guengerich, F. P., Voehler, M., & Waterman, M. R. (2005). Role of active site water molecules and substrate hydroxyl groups in oxygen activation by cytochrome P450 158A2: A new mechanism of proton transfer. *Journal of Biological Chemistry, 280*(51), 42188–42197. http://dx.doi.org/10.1074/jbc.M509220200.

CHAPTER EIGHT

Simulation Studies of Protein and Small Molecule Interactions and Reaction

L. Yang*,†, J. Zhang*,†, X. Che*,†, Y.Q. Gao*,†,1
*College of Chemistry and Molecular Engineering, Beijing National Laboratory for Molecular Sciences, Peking University, Beijing, PR China
†Biodynamic Optical Imaging Center (BIOPIC), Peking University, Beijing, PR China
[1]Corresponding author: e-mail address: gaoyq@pku.edu.cn

Contents

Abstract

Computational studies of protein and small molecule (protein–ligand/enzyme–substrate) interactions become more and more important in biological science and drug discovery. Computer modeling can provide molecular details of the processes such as conformational change, binding, and transportation of small molecules/proteins, which are not easily to be captured in experiments. In this chapter, we discussed simulation studies of both protein and small molecules from three aspects: conformation sampling, transportations of small molecules in enzymes, and enzymatic reactions involving small molecules. Both methodology developments and examples of simulation studies in this field were presented.

Studies of protein and small molecule (protein–ligand/enzyme–substrate) interactions are among the most active areas in biological science. Structural biology lays foundations for studies of interactions between proteins and small molecules. Insights about protein–ligand binding, enzyme–substrate

Methods in Enzymology, Volume 578
ISSN 0076-6879
http://dx.doi.org/10.1016/bs.mie.2016.05.031

reaction mechanisms, and protein functions can be translated from knowledge obtained in structural biology that aims to determine the static structures of biomolecules. However, additional information is needed on the dynamics and structural flexibility that are crucial for functionality of proteins. In recent years, with the developments of computational methods and hardware, computer modeling has become a powerful complementary tool for static structural studies, by showing "the rest of the story" (ie, how motions and ensembles of alternative conformers that cannot be seen in single molecular structures also contribute to bindings and interactions). Especially, computer modeling becomes more and more important in modern drug discovery. It is now very common for the drug discovery process to involve various levels of computation (Guvench & MacKerell, 2009; Jorgensen, 2004).

Many different processes involve in enzyme catalyzed reactions, which include the binding of ligand to the protein, their conformational changes, the chemical bond making and breaking, and migration of ligand on the protein surface or inside proteins. It is often difficult for experiments to provide molecular details of these events. Molecular dynamics (MD) simulation can serve as a good tool, although in many cases, different techniques are needed. In this chapter, we will discuss our simulation studies of both protein and small molecules from three aspects: conformation sampling, transportations of small molecules in enzymes, and enzymatic reactions involving small molecules. We mainly focus on the methodology development that we hope to find usage in simulations of enzymatic reactions.

1. CONFORMATION SAMPLING OF SMALL MOLECULES AND PROTEINS

Upon binding to a protein, a ligand can bind in multiple orientations and the protein or ligand can also be deformed by the binding event. Computer modeling is helping to elucidate these factors. In this section, we first discuss enhanced sampling methods that enable us to sample efficiently the configuration space. Examples of the conformation sampling for both small molecules and proteins are presented.

1.1 Methods for Conformation Sampling

In recent years, great progress has been made in computer simulations of complex molecules. However, conformation sampling is still a major challenge for simulations of biomolecules because of multiple time scales presented in these systems. The characteristic motions in biological systems include both the fast vibrations of covalent bonds (typically femtosecond time scale) and the slow conformation transitions. The fast motions limit the integration time step of all-atom MD simulation to femtosecond, and as a result, these MD simulations can only explore typically microsecond time scale motions and their applicability is thus limited. A large number of methods have been proposed to improve MD's capabilities of handling the coexisting slow and fast degrees of freedom (DOF). In one class of methods, the fast motions are frozen, such as in the programs SHAKE (Ryckaert, Ciccotti, & Berendsen, 1977) and its variant RATTLE (Andersen, 1983), to permit a larger integration time step; or by using multiple time steps (Izaguirre, Reich, & Skeel, 1999; Minary, Tuckerman, & Martyna, 2004; Swindoll & Haile, 1984; Tuckerman, Berne, & Martyna, 1992; Tuckerman, Martyna, & Berne, 1990), to compute the slow varying interactions less frequently (eg, RESPA method) (Tuckerman et al., 1992, 1990). Alternately, to approach a longer simulation time, one can speed up slow events so that they take place fast enough to be accessed by an MD simulation. For example, the self-guided MD (SGMD) (Wu & Wang, 1998, 1999) introduces a guiding force estimated from an average of the instantaneous forces over a period of normal MD simulations to accelerate slow systematic motions. In 2007, Yang and Gao introduced a multitime scale molecular dynamics (MTS) method with the hope to extend the time scale reachable by MD simulations through slowing down the fast motions and speeding up the slow ones at the same time (Yang & Gao, 2007). One similarity between this method and SGMD is that in addition to the instantaneous forces and velocities, average forces, and velocities obtained from trajectories of normal MD simulations for a nonequilibrium spontaneous process are introduced into the Newtonian equations of motions to facilitate the computation of the slow protein conformational change. However, MTS slows down the fast DOF in addition to speeding up the motions along the slow DOF so that a relatively large time step can be used in simulations.

Another major factor, which severely restricts the sampling efficiency of molecular simulations, is the complexity and roughness of biomolecules' free energy landscape. In the sampling of the conformational space of complex molecules one is very likely to be trapped in local minima. A number of methods have been developed over the last several decades to overcome such a computational bottleneck in molecular conformation sampling. These enhanced sampling methods are mainly designed to allow an accelerated search in the configuration and/or trajectory space and thus fast thermodynamics and/or kinetics calculations. Generally, enhanced sampling methods fall into two main categories. One category contains bias potential methods which enhance the sampling of rare events by systematically modifying the potential energy surface on certain collective variables (CVs); ie, decreasing the energy barrier along a chosen reaction coordinate (RC) to enhance the sampling over the transition regions. Such methods include but are not limited to the widely used umbrella sampling (Torrie & Valleau, 1977), J-walking (Frantz, Freeman, & Doll, 1990), adaptive umbrella sampling (Bartels & Karplus, 1998), metadynamics (Bussi, Laio, & Parrinello, 2006), accelerated MD (AMD) simulation (Gao & Yang, 2006; Hamelberg, Mongan, & McCammon, 2004), conformational flooding (Grubmuller, 1995), conformational space annealing (Lee, Scheraga, & Rackovsky, 1997), hyperdynamics simulation (Voter, 1997), potential smoothing methods (Piela, Kostrowicki, & Scheraga, 1989), and adaptive biasing force methods (Darve & Pohorille, 2001). The other category is composed of generalized ensemble methods in which a more uniform and broad distribution of energy (or temperature) in a molecular simulation is generated. This latter category of methods is especially useful when it is difficult to identify suitable collective coordinates. The popular generalized ensemble methods include replica exchange molecular dynamics (REMD) (Sugita & Okamoto, 1999), parallel tempering (Hansmann, 1997), multicanonical simulation (Berg & Neuhaus, 1991), Wang–Landau algorithm (Wang & Landau, 2001), statistical temperature sampling (Kim, Straub, & Keyes, 2006), simulated tempering (Marinari & Parisi, 1992), and envelop distribution sampling method (Lin & van Gunsteren, 2013).

In order to study biological systems efficiently, Gao proposed the integrated tempering enhanced sampling (ITS) method (Gao, 2008a, 2008b; Gao, Yang, Fan, & Shao, 2008; Yang & Gao, 2009; Yang, Liu, Shao, Zhang, & Gao, 2015; Yang, Shao, & Gao, 2009). Instead of running parallel simulations at many different temperatures, ITS method generates a generalized non-Boltzmann distribution covering a broad energy range. By allowing efficient transition over energy barriers and at the same time providing the desired configuration distribution, this method allows fast sampling of the configuration space and efficient calculations of thermodynamic properties. In the implementation of ITS, an extra potential term is introduced in energy evaluations and no predefined RCs are needed. In this sense, this method combines the merits of both bias potential methods and generalized ensemble methods.

In the following, the two enhanced sampling methods, MTS (dealing with the problem of broad range of time scales) and ITS (overcoming the free energy barriers), will be discussed in detail.

1.1.1 Multitime Scale Molecular Dynamics

As mentioned earlier, one of the central questions in molecular simulations is how to capture the desired slow motions from a complexity of motions spanned in very different time scales. MTS was designed for separating the motions with respect to different time scales based on the hypothesis that most of the short-range (or local) motions, such as bond stretching and angle bending, reach local equilibrium much faster than the large conformational changes, so that there exists a time scale t_1 during which only the fast but not the slow DOF have reached "pseudo-equilibrium." Therefore, the slow DOF can be treated as evolving in a potential of mean–force resulted from the motions of the fast DOF. In MTS, the average velocities (over a time period of t_1) are used to propagate motions. As a result, one can access a much longer time scale and speed up the slow nonequilibrium processes.

First, assuming that we can separate the motions of a system into collections of slow and fast DOF, x_1 and x_2, respectively, we are interested in following the motion along x_1 (to obtain a distribution function $f_1(x_1,t)$ along the coordinates x_1) under the influence of an almost equilibrium condition of the local environment x_2.

$$f_1(x_1, t) = \int_{x_2} f(x_1, x_2, t)\, dx_2. \tag{1}$$

The earlier equation can be further written as

$$f_1(x_1, t) = \int_{x_2} f_2(x_2, t) f'(x_1; x_2, t)\, dx_2. \tag{2}$$

where $f_2(x_2,t)$ is the probability distribution function for x_2 and $f'(x_1;x_2,t)$ is a conditional probability function. Under the assumption that the motion in x_2 reaches equilibrium quickly and this equilibrium only weakly depends on x_1, $f_2(x_2,t)$ can be obtained from a normal MD simulation for a relative short period of time, during which the fast degree of freedom approaches a local equilibrium. This distribution is then used for the DOF in the propagation of dynamics to obtain $f_1(x_1,t)$.

In the implementation of this method, the dynamic quantities of the very fast DOF average out over time much faster than the slow but directional motions do. Therefore, separations of fast and slow motions are achieved automatically. It is noticeable that although the average motion of the fast DOF is much slower than their real motions, the major characteristics of the motion are kept. Due to the slowed motion in the fast DOF, the problem caused by the large separation of time scales is reduced and the integrations time step can be significantly prolonged. In MTS, the energy of the system is closely monitored. When the energy increases to a certain limit, the MTS simulation will be stopped and a new normal MD simulation will be performed using the new protein conformation as the starting point, which is followed again by another MTS simulation. These procedures are repeated, until satisfactory result is obtained, judged either by the structure or by the energy. The algorithm is summarized as:

(1) N steps of normal MD simulations are performed for a time length of t_1 with a step size of δt. During the normal MD simulation, coordinates and velocities of each atom in the molecular system are recorded at each n step.

(2) The average velocity is calculated as:

$$\langle \dot{r}_i(t) \rangle_{t_1} = \frac{[r_i(t) - r_i(t - t_1)]}{t_1} \tag{3}$$

from which, an effective guiding force is evaluated:

$$g_i(t) = m_i \frac{\langle \dot{r}_i(t) \rangle_{t_1} - \dot{r}_i(t - t_1)}{t_1}. \tag{4}$$

(3) Following the normal MD simulation and starting from time t, a larger time step $\Delta t = n\delta t$ is used to propagate the motions. And the acceleration at this time is calculated from the combination of the instantaneous force and the guiding force as:

$$\tilde{\ddot{r}}_i(t) = \frac{\lambda_1 f_i(t)}{m_i} + \frac{\lambda_2 g_i(t)}{m_i}, \tag{5}$$

where $\lambda_1 = -2N/n^2$ and $\lambda_2 = 2/n^2$, and the position is propagated as:

$$r_i(t + \Delta t) = r_i(t) + \dot{r}_i(t)\Delta t + \tilde{\ddot{r}}(t)\Delta t^2/2. \tag{6}$$

The new time step Δt and equations of motions (3)–(6) are used for N/n steps until all the positions and velocities recorded in step 1 have been used once.

(4) Repeat steps 1–3 until the end of the simulation.

1.1.2 Integrated Tempering Sampling (ITS)

Now we switch gears to a different approach of treating the multiscale motions of complex systems: ITS. In ITS, the sampled energy range is broadened through the introduction of a sum-over-temperature non-Boltzmann distribution factor. We define the non-Boltzmann factor $p(U)$ as

$$p(U) = \sum_{k=1}^{N} n_k e^{-\beta_k U}. \tag{7}$$

In Eq. (7), $\beta_k = 1/k_B T_k$, k_B is the Boltzmann constant and T_k is the kth temperature of a series of temperatures (T_1 to T_N) that allows a broad energy range to be effectively sampled in a short simulation (eg, submicroseconds). The desired system temperature lies in the range covered by T_k's and the highest value of T_k is chosen so that, on the simulation time scale, transitions over a preestimated barrier have a relatively high probability of occurrence. In MD simulations, a distribution function of Eq. (7) can be obtained by running simulations on an effective potential U_{eff}, which is a function of U, at the desired temperature (thus β_0):

$$U_{eff} = -\frac{1}{\beta_0} \ln \sum_k n_k e^{-\beta_k U}, \tag{8}$$

The biased force F_{eff} used in the simulations is

$$F_{eff} = -\frac{\partial U_{eff}}{\partial r} = \frac{\sum_k \beta_k n_k e^{-\beta_k U}}{\beta_0 \sum_k n_k e^{-\beta_k U}} F, \tag{9}$$

where F is the force in the original system.

The remaining problem is the determination or estimation of the parameters n_k which can lead to a smooth distribution function in the form of Eq. (7) by conducting simulations. In ITS, the parameters n_k are determined by the requirement that each term in the summation contributes a desired fraction to the total distribution. In other words, if we define

$$P_k = n_k \int_{\mathbf{r}} e^{-\beta_k U(\mathbf{r})} d\mathbf{r}, \tag{10}$$

integrating over the entire configuration space, one can preselect the ratios between P_k values for all k between 1 and N. Therefore, one can define a set of fixed expectation values $\{p_k^0\}$ for the normalized quantities p_k:

$$p_k = \frac{P_k}{\sum_{k=1}^{N} P_k}. \tag{11}$$

For example, if equal contribution from each P_k is desired, $p_k^0 = 1/N$ for all k. Generally speaking, $\{p_k^0\}$ needs not to be uniform and can be chosen to emphasize the sampling in a subtemperature (energy) range.

n_k can be efficiently calculated by realizing that the overlap of sampling is maximal between nearest neighboring temperatures. In this approach, n_k is updated according not to the absolute values of p_k but to the ratios between neighboring p_k values. Let's first define a series of numbers m_k,

$$m_k = \begin{cases} 1 & \text{if } k = 1 \\ n_{k+1}/n_k & \text{if } 1 < k \le N \end{cases}. \tag{12}$$

The original series of n_k is then simply related to m_k by

$$n_k = n_1 \prod_{j=1}^{k} m_j \tag{13}$$

After setting either $n_1 = 1$ or $\sum_{k=1}^{N} n_k = 1$, one can determine n_k by knowing m_k. The simulations and iterations can then be performed as follows:

(1) A sequence of $\{\beta_k,\ k = 1, N\}$ is chosen to cover the desired range of temperatures.

(2) A set of initial guesses, $m_k(0)$, for $\{m_k,\ k = 2, N\}$ is chosen and used to calculate $n_k(0)$, which are used to run the MD (or MC) simulations using the effective potential.

(3) From the trajectories, P_k and thus p_k are calculated ($p_k(0)$), and a new set of $\{m_k(1),\ k = 2, N\}$ is obtained from $m_k(1) = m_k(0)\dfrac{p_k(0)}{p_{k+1}(0)}$, to fulfill the requirement (in the next step of the calculation) that $\dfrac{p_k(1)}{p_{k+1}(1)} = 1$, for an equal contribution from all temperatures. New n_k are then calculated from $m_k(1)$ and used in the next round of simulation.

(4) The process returns to step 2 and repeats steps 2 and 3 to update $\{m_k,\ k = 1, N\}$. Starting from the second iteration step, information from earlier steps on $\{m_k,\ k = 1, N\}$ is integrated together. For example, for the jth step, m_k is calculated as

$$m_k(j) = \left[f(p_k(j-1), p_{k+1}(j-1)) \frac{p_k(j-1)}{p_{k+1}(j-1)} + \sum_{l=0}^{j-2} f(p_k(l), p_{k+1}(l)) \right] \times \frac{m_k(j-1)}{W(j)}, \tag{14}$$

where $f(p_k(j-1), p_{k+1}(j-1))$ is a weighting function that takes a larger value when the sampling in the kth and $(k+1)$th temperature ranges is more adequate. For example, the weighting function can simply be a product of p_k and p_{k+1}, or the square root of this product. W is a normalizing factor, $W = \sum_{l=0}^{j-2} f(p_k(l), p_{k+1}(l))$. Once m_k is updated, new $n_k(j)$ are calculated. The simulation continues until $\{p_k,\ k=1, N\}$ converges.

In a few tested examples, ITS was shown to be highly efficient. For example, ITS was applied to study the conformational transition of ALA-PRO peptide that has a high free energy barrier of ~20 kcal/mol for its transition between *trans* and *cis* conformations. As comparisons, AMD and REMD were also applied to the same problem. It is clear that the potential of mean force (PMF) obtained from the AMD simulations converges very slowly owing to the under-sampling of low-energy states (1.22 kcal/mol) within 800-ns trajectories (Table 1). REMD and ITS both obtain better convergence of thermodynamics calculations. In particular, ITS simulations give the best convergence with the least computational time (similar to that of AMD).

In addition to the advantages of less computational resource requirement and being predefined RC free, it is convenient in the implementation of ITS to a divided system (consisting of several subspaces) and to enhance the sampling for a preselected subsystem (selective integrated tempering

Table 1 Convergence and Computational Costs for ALA-PRO Peptide

Method	Trajectory Length (ns)	Wall Clock Time (h)	CPU Time (h)	RMSD[a] (kcal/mol)
AMD	800	12.4	99	1.22
REMD	2400	25	600	0.48
ITS	800	12.5	100	0.21

[a]RMSD: The root-mean-square deviation of free energies that is used to characterize the convergence of the methods.

sampling (SITS); Yang & Gao, 2009). For example, in explicit solvent simulations of protein folding, the protein atoms can be targeted for enhanced sampling and a large variety of protein configurations (eg, both folded and unfolded) can be sampled while the solvent is kept at the near-room-temperature configurations. The differentiated sampling of such a system is conveniently achieved by introducing an effective potential in the following form:

$$U_{eff} = E_w - \frac{1}{\beta_0} \ln \sum_k n_k e^{-\beta_k \left(E_p + E_{pw}\right)}. \tag{15}$$

In the earlier equation, the total energy U is divided into multiple components:

$$U = E_p + E_w + E_{pw}, \tag{16}$$

where E_p, E_w, and E_{pw} are, respectively, the internal energy of the subsystem of interest (protein), the internal energy of all water molecules, and the energy of interaction between the protein and water. Following the energy division strategy mentioned earlier, a bias potential can be generated in such a way that the molecule of interest would be sampled in a wide temperature and energy range, with the solvent staying at the target temperature. The application of SITS thus allows efficient sampling of the solute configurations in a well-controlled solvent environment (or enzyme surroundings).

From ITS or SITS simulations, a converged calculation yields a biased distribution function in the configuration space, $\rho_{eff}(r) \propto e^{-\beta_0 U_{eff}(r)}$. The desired distribution function at the temperature T_0 is easily calculated as

$$\rho(r) = \rho_{eff}(r) e^{-\beta_0 \left(U(r) - U_{eff}(r)\right)}, \tag{17}$$

with $\rho_{eff}(r)$ calculated from the enhanced sampling simulation and $U_{eff}(r)$ defined as in Eq. (8) or (15). Other thermodynamics properties can be readily obtained once $\rho(r)$ is known.

SITS can also be naturally introduced to QM/MM calculations, in which only a small part of the simulation system is treated using quantum mechanics (QM) and the rest by classical molecular mechanics (MM). For such calculations, since the events of interest (such as chemical reactions) normally occur within the quantum region, it is desirable to explore the molecular configurations of the quantum mechanically treated subsystem. One can easily make use of the SITS scheme to enhance the sampling over the QM region while keeping the MM part less perturbed by introducing the effective potential

$$U_{eff} = E_{MM} - \frac{1}{\beta_0} \ln \sum_k n_k e^{-\beta_k \left(E_{QM} + E_{QM/MM}\right)} \tag{18}$$

where E_{MM} is the self-energy of the MM region (eg, the solvent), E_{QM} is the self-energy of the QM region (eg, the reacting molecule), and $E_{QM/MM}$ is the energy of interaction between the QM and MM regions.

1.2 Solution Conformation and Collective Conformation/Configuration of Biomolecules

One of the important factors in enzymatic reactions is the binding of the ligand to the enzyme. In the study of protein–ligand interactions, many efforts have been paid to the conformational changes of proteins. However, investigating conformational changes of proteins in solutions using computers is still challenging, considering that macromolecules usually contain many DOF among which multiple slow motions can be presented at the same time. One approach is to reduce the complex and redundant DOFs directly in the model setup, giving rise to the coarse-grained models. But details on the atomic levels are frequently found indispensable for an ideal description of a biosystem. Therefore, more efficient sampling methods are in need to push the all-atom computation of biomolecules to larger spatial scale and longer time scale. Both ITS/SITS and MTS methods may find applications in sampling of polypeptides/proteins' conformations.

The complete conformational sampling of peptide plays an important role in peptide–protein interaction studies. In this part, we introduce a simulation protocol to deal with the challenges faced in the sampling of polypeptide conformations. The solution conformations of proline-rich peptide bradykinin (BK, RPPGFSPFR) were studied using this protocol as an example. The simulation was performed via a combination of metadynamics and SITS. In this method, before applying SITS to the simulation system, we first locally flatten the deep potential well(s) by depositing history-dependent Gaussian bias potential along chosen CVs ($s(r)$), as in metadynamics:

$$U^{\text{bias}}[s(\mathbf{r}), t] = \omega \sum_{t'<t} \exp - \frac{\{s(\mathbf{r}) - s[\mathbf{r}(t')]\}^2}{2\sigma^2} \tag{19}$$

Next, the potential depositing process is stopped once reversible transitions along the collective coordinate are observed, and the SITS method is then employed on the new rescaled PES (U_1^{meta}), which incorporates previously accumulated bias potentials. Thus the new effective PES is in the form as follows:

$$U_{\text{eff}} = U_0^{\text{meta}} - \frac{1}{\beta_0} \ln \sum_k n_k e^{-\beta_k U_1^{\text{meta}}} \tag{20}$$

By this means, the simulation system can be rapidly driven across the dominant barrier(s) along selected collective coordinate(s), meanwhile the thermodynamic calculations of which quickly converge with the aid of SITS. In the example of bradykinin, metadynamics is first applied to accelerate the sampling of the *trans–cis* isomerization of the three prolines. The biasing potentials (Eq. 19) are added sequentially to the ω dihedral angles of the three prolines. Metadynamics in this combination protocol facilitates fast thermodynamics convergence of SITS through the efficient sampling of proline isomerization (Fig. 1A). Interestingly, the calculation showed that the *trans–cis* preference is different for the three prolines in $[\text{BK}+3\text{H}]^{3+}$ (Fig. 1B). Compared to Pro^2 and Pro^3, Pro^7 has a larger tendency to adopt the *cis-*ω conformation. In contrast, the *trans-*ω conformation largely dominates in Pro^2.

The free energy profiles along the ω dihedral angles of the three prolines are shown in Fig. 2. The *top row* shows the "real" free energy surfaces (after reweighting using Eq. 17) and the *bottom row* shows the effective free energy surfaces sampled in the enhanced sampling simulations. Comparison between Fig. 2A and B shows that the combined enhanced sampling protocol decreases the free energy barrier to a large extent ($\sim 30 k_B T$), and the modified free energy surfaces are smooth which guarantee the fast convergence of thermodynamic properties. Therefore, the metadynamics–SITS protocol has achieved a complete sampling of the structures with all combinations of proline ω *trans–cis* states (Fig. 3). In Fig. 3, T represents the *trans* state of ω dihedral angle and C represents the *cis* state of ω dihedral angle. The three letters represent the states of Pro^2, Pro^3, and Pro^7, respectively, for example, CCC stands for the state with Pro^2, Pro^3, and Pro^7 all taking the *cis* conformations. In agreement with the proline isomerization studies, the probability of the TTT conformation is the largest among the eight types of states while the TTC conformation is also populated.

Although bradykinin is only a small peptide, its thorough thermodynamics properties are hard to be obtained due to the high free energy barriers associated with the isomerization of the three prolines, the motion of which is also coupled to a large number of other (faster) DOF. We have shown that

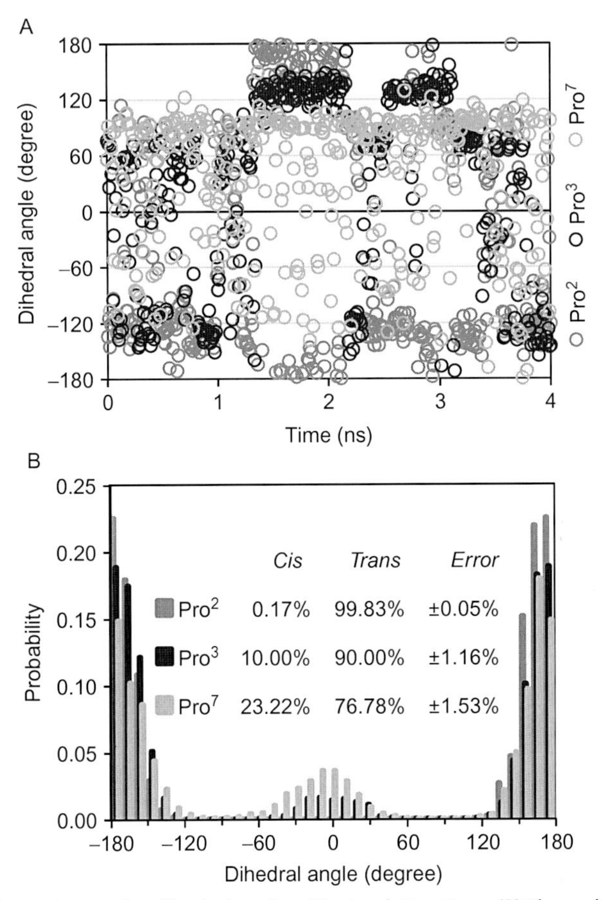

Fig. 1 (A) The variance of ω dihedral angle with simulation time. (B) The ω dihedral angle probability distribution of Pro^2, Pro^3, and Pro^7, respectively. *Reprinted with permission from Yang, I. Y., Zhang, J., Che, X., Yang, L., & Gao, Y.Q. (2016). Efficient sampling over rough energy landscapes with high barriers: A combination of metadynamics with integrated tempering sampling.* The Journal of Chemical Physics, *144, 094105. Copyright (2016) AIP Publishing LLC.*

the metadynamics–SITS protocol is efficient in sampling the configurations of such a system which contains both slow and fast DOF, but its applicability to more complex systems, especially enzymatic processes involving large protein conformation changes, is subject to further tests.

Unlike the sampling of an ensemble of structures at equilibrium, for the binding of a ligand to a large protein or enzyme one is often interested in the conformational change induced by the small molecule binding. In such cases, achieving a thorough sampling of the protein structure can be overwhelmingly demanding. But accelerated methods can be applied to follow the slow structure changes. In the following, we summarize the calmodulin

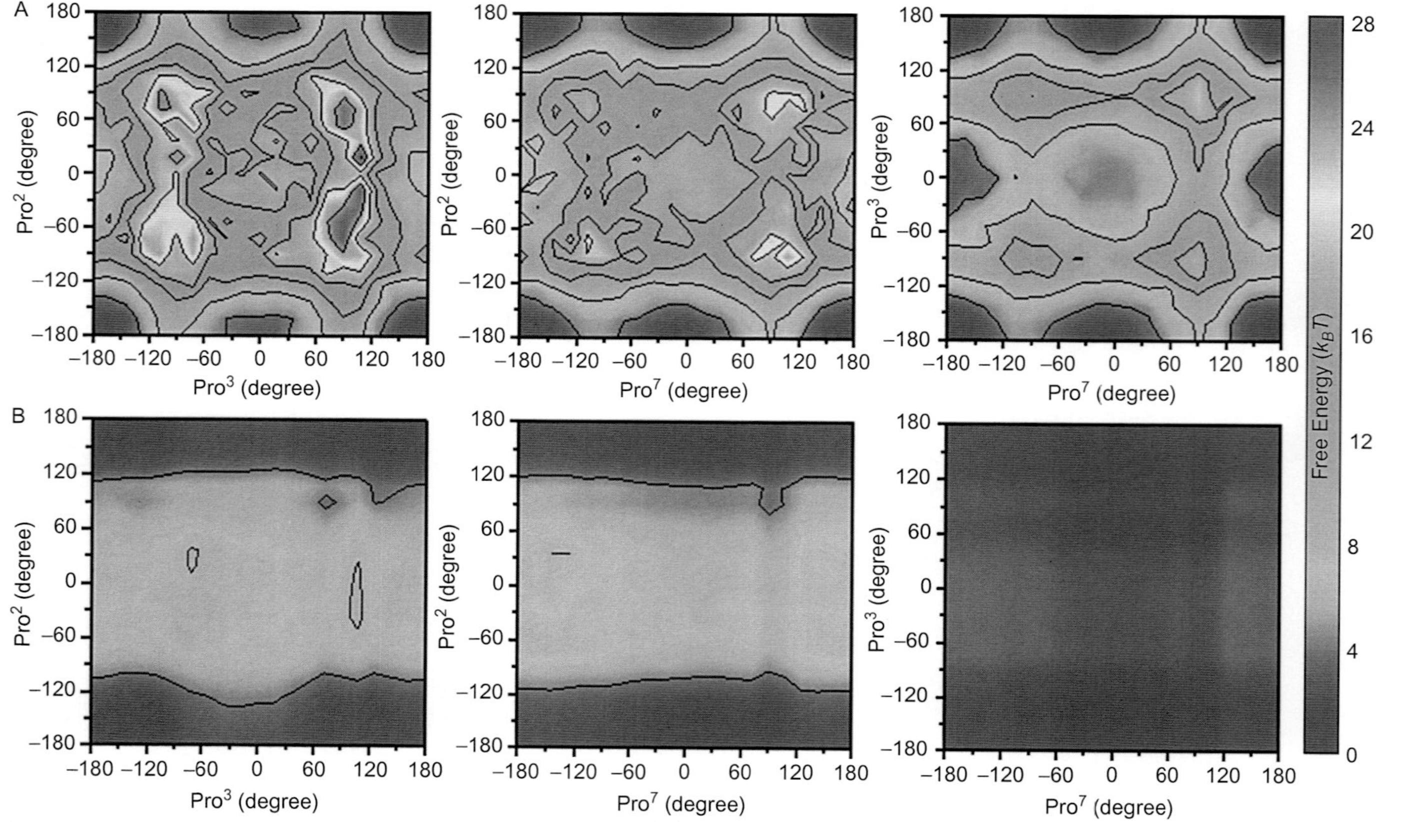

Fig. 2 Free energy landscapes of ω dihedral angle of prolines in BK. (A) The free energy surfaces after reweighting. (B) The effective free energy surfaces without reweighting. *Reprinted with permission from Yang, I. Y., Zhang, J., Che, X., Yang, L., & Gao, Y. Q. (2016). Efficient sampling over rough energy landscapes with high barriers: A combination of metadynamics with integrated tempering sampling*. The Journal of Chemical Physics, *144*, 094105. *Copyright (2016) AIP Publishing LLC*. (See the color plate.)

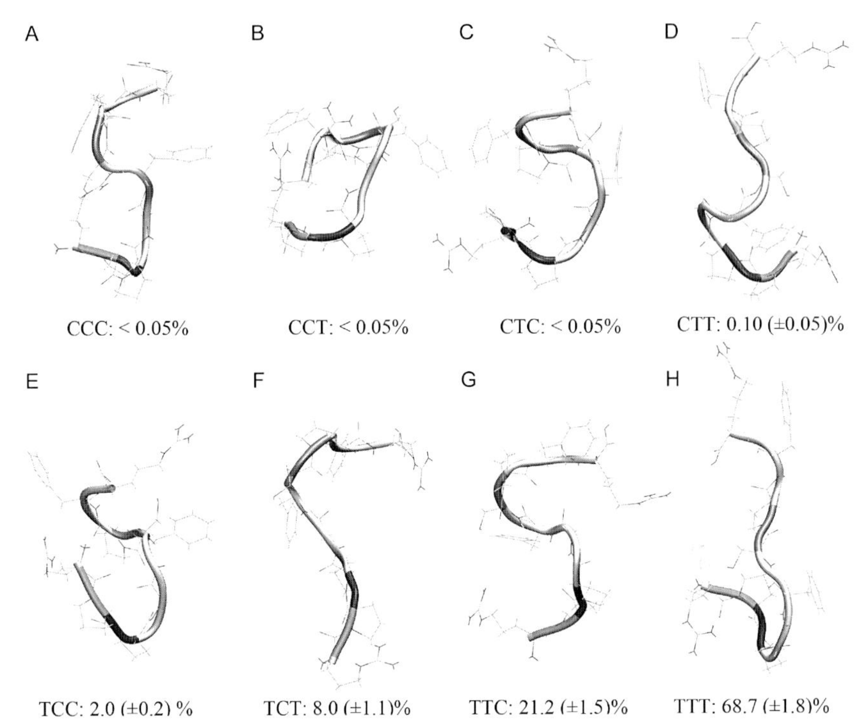

Fig. 3 The probabilities of eight types of structures defined by states of proline ω dihedral angle. *T* represents *trans* state of ω dihedral angle; *C* represents *cis* state of ω dihedral angle. The three letters represent the states of Pro^2, Pro^3, and Pro^7, respectively. *Reprinted with permission from Yang, I. Y., Zhang, J., Che, X., Yang, L., & Gao, Y. Q. (2016). Efficient sampling over rough energy landscapes with high barriers: A combination of metadynamics with integrated tempering sampling.* The Journal of Chemical Physics, *144, 094105. Copyright (2016) AIP Publishing LLC.*

conformational change in response to the release of Ca^{2+} studied by using MTS method introduced earlier. Calmodulin is a calcium binding protein (Chattopadhyaya, Meador, Means, & Quiocho, 1992), which mediates many crucial processes by binding different target proteins. Calcium binding exposes nonpolar surfaces of calmodulin, which then bind to nonpolar regions on the target proteins. Oppositely, large conformational changes of calmodulin will be induced upon the removal of the calcium ion that was originally bound to calmodulin. Such that nonpolar surfaces of calmodulin will be buried inside and the activation function of calmodulin is thus turned off.

Our simulations start from the X-ray crystal structure of Ca^{2+}-bound calmodulin 1CLL (only the N-terminal domain without Ca^{2+}). Fig. 4A shows the backbone RMSD between simulated structures and the NMR apo structure 1F70 for calmodulin N-terminal domain. In this figure, the *black solid line* segments are obtained from the normal MD simulation (time

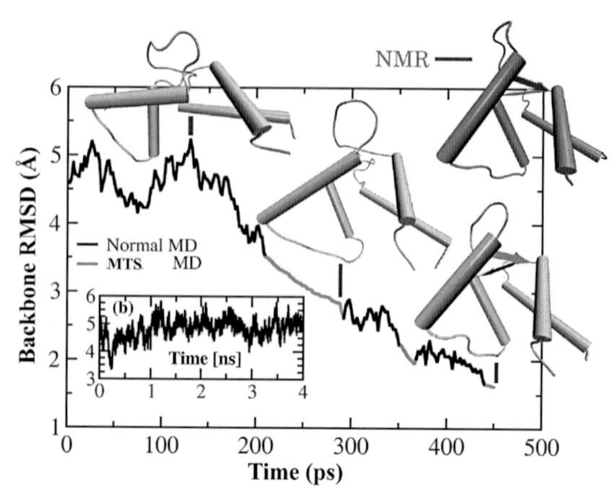

Fig. 4 (A) Backbone RMSD between simulated structures and the NMR apo structure 1F70 for Calmodulin N-terminal domain. *Black* denotes the normal simulation stage which is used to accumulate information for calculating the average. *Red* (*dark gray* in the print version) denotes the multitime scale MD simulation stage. (B) Backbone RMSD of a 4 ns normal MD simulation using the same initial structure as that used to obtain (A). *Adapted with permission from Yang, L. J., & Gao, Y. Q. (2007). An approximate method in using molecular mechanics simulations to study slow protein conformational changes.* Journal of Physical Chemistry B, 111*(11), 2969–2975. Copyright (2007) American Chemical Society.*

step equals 2 fs), which is used to calculate average forces and velocities. The *red* (*gray* in the print version) *solid line* segments are for MTS simulations (using a 20 times larger time step, ie, 40 fs). It is apparently that the large conformational transition from Ca^{2+} bound to apo structure is efficiently captured within only 500 ps. As a comparison, the change of the same backbone RMSD remains large during a normal MD simulation of 4 ns (Fig. 4B). Since it allows a quick search for the intermediate conformations, MTS is expected to be helpful in obtaining thermodynamic as well as dynamic information, when it is combined with other methods, such as umbrella sampling or transition path sampling,

1.3 Solution Conformation of Free Ligand/Substrate

In the study of protein–ligand interactions, the flexibility of a ligand could also play an important rule. However, only a limited number of conformations are usually considered in studies of ligand/protein interactions. On the other hand, a growing number of examples show that a thorough study of the ligand conformation ensemble is vital to investigate and predict ligand/substrate–protein interactions (Chen & Foloppe, 2011; Stockwell & Thornton, 2006). Moreover, in structure-based drug design, the tested conformations of ligands are the foundation for subsequent calculations. Particularly, a ligand can adopt quite different conformations before and after it binds to the effector protein, and evidence exists to show that many ligands do not bind to the proteins through their local minimum energy conformation (Perola & Charifson, 2004). In such cases, inadequate analysis of the free ligand conformation may result in wrong binding free energy prediction and could be potentially problematic when further enzymatic reaction is involved. Therefore, a thorough sampling of solution structures of the free ligands becomes necessary, which can be realized cost effectively by MD simulations combined with proper enhanced sampling methods. We expect SITS to serve as a powerful tool, as it is able to explore the conformational space of the ligand efficiently in the presence of varied solvent conditions.

In addressing the importance of free ligand conformations, we use the binding of cyclic di-nucleotides (CDNs) to stimulator of interferon genes (STING) protein as an example. CDNs, as STING agonists, remain as a hot spot in anticancer drug development, given that STING is pivotal in immune system. Upon activation, STING induces the production of various interferons and cytokines (Fu et al., 2015). CDNs contain two purine nucleotides (adenine or guanine) joined by two phosphodiester bonds and contain a cyclic backbone ring. The two phosphodiester bonds can independently adopt either the 2′5′- or the 3′5′-form, and give rise to several CDN isomers (Fig. 5). However, with subtle changes in phosphodiester linkages and their combinations in the phosphodiester ring, CDNs exhibit binding affinities to STING varied in orders of magnitude (2′3′-cGAMP, 2′2′-cGAMP $\gg$ c-di-GMP, 3′3′-cGAMP) (Wu et al., 2013). Especially, it is elusive that the binding affinity of asymmetric 2′3′-cGAMP to symmetric dimer of STING is three orders of magnitude higher than the symmetric 3′3′-cGAMP and c-di-GMP.

To unravel the intriguing problem, it is important to have a thorough knowledge of structural properties of the compounds themselves. SITS simulations of 2′2′-cGAMP, 2′3′-CGAMP, 3′3′-cGAMP, and c-di-GMP revealed several clear rules to understand and predict different preferential conformations of CDNs. It was found that the ribose with 2′-5′ phosphodiester linkage prefers the C2′-endo-like (SOUTH) conformation and the 3′-5′ linkage ribose prefers the C3′-endo-like (NORTH) conformation (Fig. 6A and B). The two types of puckers permit different spaces for the purine base rotation about the glycosidic bond. The probability to adopt *syn*-χ conformation is higher for the base linked to a C2′-endo-like pucker than that linked to a C3′-endo-like pucker

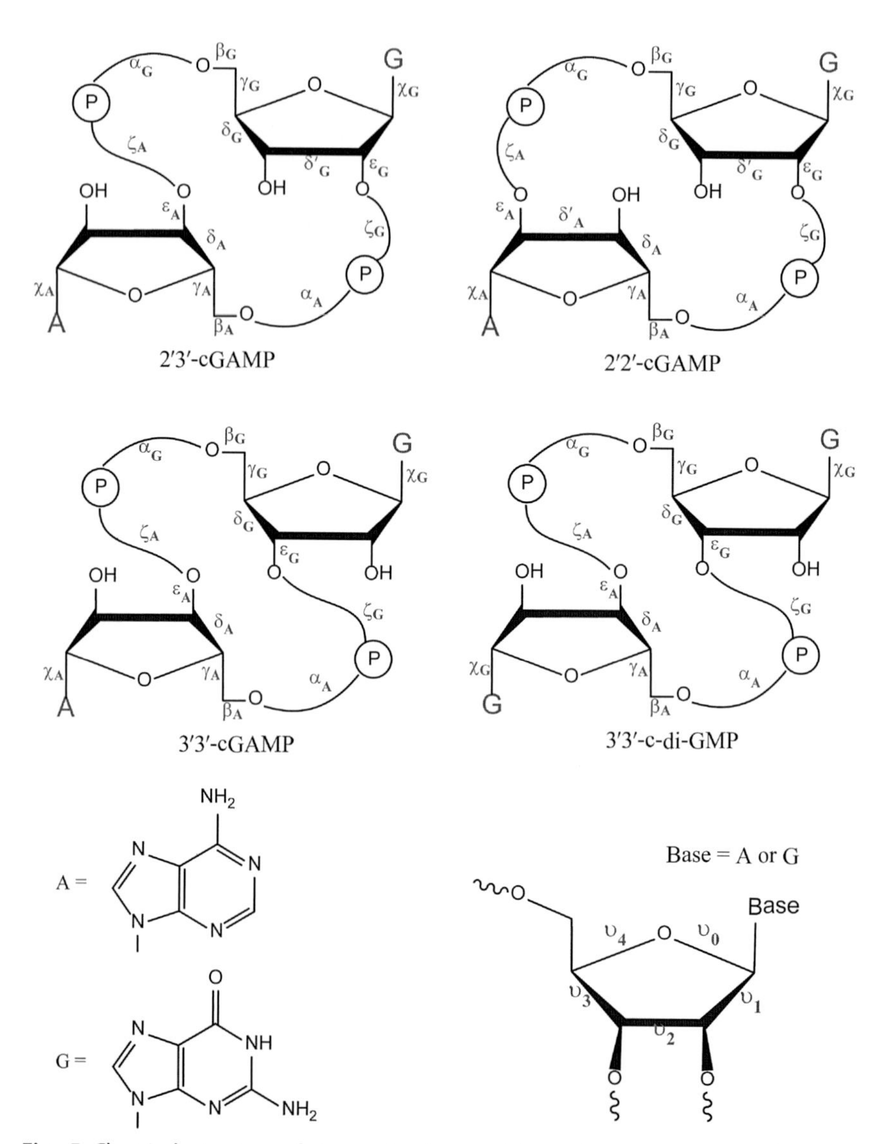

Fig. 5 Chemical structure of the cyclic dinucleotides. Dihedral angles are labeled in *magenta* (*gray* in the print version): α, O3′-P-O5′-C5′ (or O2′-P-O5′-C5′); β, P-O5′-C5′-C4′; γ, O5′-C5′-C4′-C3′; δ, C5′-C4′-C3′-C2′; δ′, C4′-C3′-C2′-O2′; ε, C4′-C3′-O3′-P (or C3′-C2′-O2′-P); ξ, C3′-O3′-P-O5′ (C2′-O2′-P-O5′); χ, O4′-C1′-N9-C4; ν_0, C4′-O4′-C1′-C2′; ν_1, O4′-C1′-C2′-C3′; ν_2, C1′-C2′-C3′-C4′; ν_3, C2′-C3′-C4′-O4′; ν_4, C3′-C4′-O4′-C1′. Subscript "A" and "G" in c-di-GMP are used to distinguish the two sets of dihedral angles. Notations in the parenthesis are the corresponding definition of dihedral angle in 2′-5′ linked nucleotide. *Reprinted with permission from Che, X., Zhang, J., Zhu, Y., Yang, L., Quan, H., & Gao, Y.Q. (2016). Structural flexibility and conformation features of cyclic dinucleotides in aqueous solutions.* The Journal of Physical Chemistry B, 120*(10), 2670–2680. Copyright (2016) American Chemical Society.*

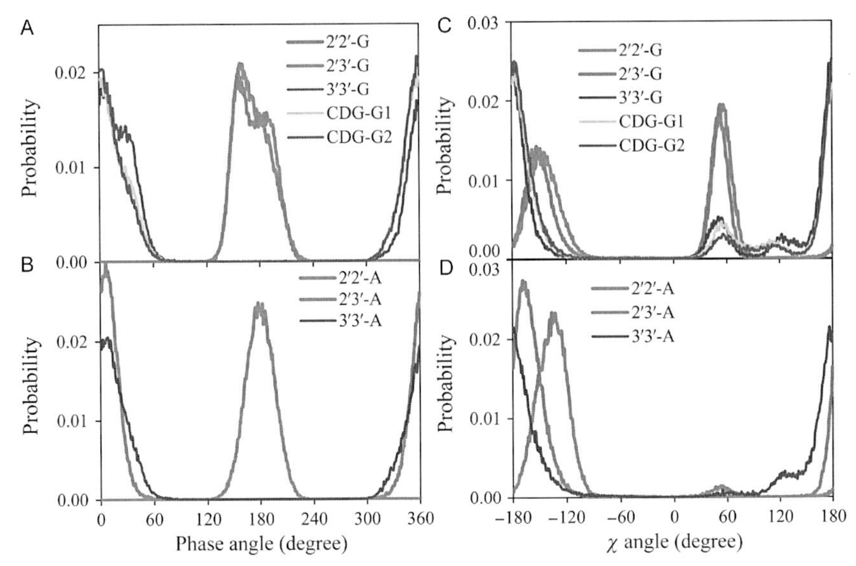

Fig. 6 Phase angle distribution (A and B) and χ angle distribution (C and D) of CDNs. In panels (A) and (C), 2′2, 2′3′, 3′3′-G denotes the angle distribution of ribose in guanosine of 2′2′, 2′3′, 3′3′-cGAMP, respectively. CDG-G1/G2 denotes the angle distribution of ribose in guanosine of c-di-GMP. In panels (B) and (D), 2′2′, 2′3′, 3′3′-A denotes the phase angle distribution of ribose in adenosine of 2′2′, 2′3′, 3′3′-cGAMP, respectively. *Adapted with permission from Che, X., Zhang, J., Zhu, Y., Yang, L., Quan, H., & Gao, Y.Q. (2016). Structural flexibility and conformation features of cyclic dinucleotides in aqueous solutions.* The Journal of Physical Chemistry B, 120*(10), 2670–2680. Copyright (2016) American Chemical Society.*

(Fig. 6C and D). In the C2′-endo-like conformation, the base orients equatorially, with much less steric clash for it to rotate from *anti*- to *syn*-χ, compared to the case of the C3′-endo-like ribose. In addition, guanosine is intrinsically more inclined than adenine to adopt the *syn*-χ conformation. Moreover, the ring size also contributes to the flexibility of CDNs. When the ring size is small, the negatively charged backbone is subject to a more severe intramolecular electrostatic repulsion. According to these findings, 2′3′-cGAMP and 2′2′-cGAMP are less flexible in aqueous solution than 3′3′-cGAMP or c-di-GMP and adopt more bound-like conformations in the free state (Fig. 7 and Table 2). It is instructive to observe that the structural flexibility of CDNs correlates with their binding affinity measured by ITC.

As another example, we discuss the concept of restricted substrate conformation when it is subject to catalysis by enzyme. A large category of enzymes catalyze the target reaction by restricting the substrate in a proper molecular conformation, in which usually the bond forming sites are brought closer and the bond breaking sites exposed to nearby active residues. Such restricted conformation of the substrate, sometimes called "near attack" conformation

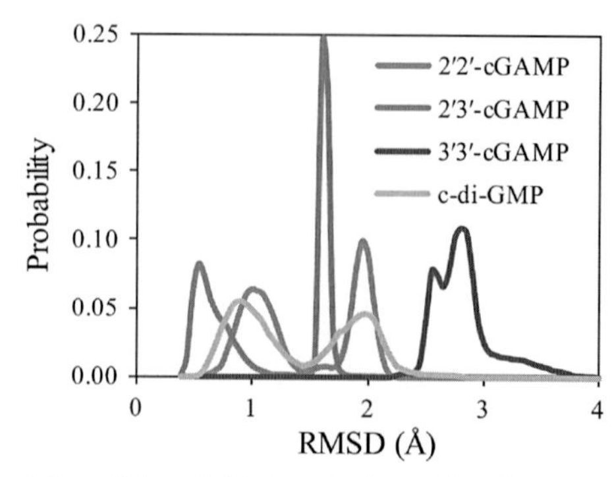

Fig. 7 Structural deviation of free CDNs in solution with reference to the protein-bound crystal structure. All heavy atoms of CDNs are included in the RMSD calculations. *Reprinted with permission from Che, X., Zhang, J., Zhu, Y., Yang, L., Quan, H., & Gao, Y.Q. (2016). Structural flexibility and conformation features of cyclic dinucleotides in aqueous solutions.* The Journal of Physical Chemistry B, 120*(10), 2670–2680. Copyright (2016) American Chemical Society.* (See the color plate.)

Table 2 Thermodynamic Properties of CDNs

	2′2′-cGAMP	2′3′-cGAMP	3′3′-cGAMP	c-di-GMP
Bound-like probability (%)	32.1 ± 0.5	38.1 ± 0.7	13.1 ± 0.7	12.4 ± 0.6
ΔS (cal/mol K)	3.4 ± 0.2	3.8 ± 0.1	−2.8 ± 0.3	−6.9 ± 0.1
ΔH (kcal/mol)	1.45 ± 0.01	1.45 ± 0.03	0.29 ± 0.02	−0.92 ± 0.02
ΔG (kcal/mol)	0.42 ± 0.03	0.30 ± 0.04	1.12 ± 0.04	1.19 ± 0.05

Reprinted with permission from Che, X., Zhang, J., Zhu, Y., Yang, L., Quan, H., & Gao, Y.Q. (2016). Structural flexibility and conformation features of cyclic dinucleotides in aqueous solutions. *The Journal of Physical Chemistry B, 120*(10), 2670–2680. Copyright (2016) American Chemical Society.

(Hur & Bruice, 2003), accounts for a small proportion of the ligand conformational ensemble, even if without the enzymatic niche. Changes in the conformation equilibrium of free substrates will directly alter the enzyme–substrate recognition and the binding free energy. Furthermore, since the population of the "near attacking" conformation is sensitive to the temperature and solvents, it is important to evaluate the intrinsic conformational changes of free ligands/substrates before one tries to explain varied rates of enzymatic reactions in different temperatures or solvent mixtures. We expect that acknowledging and understanding how the enzyme "locks" its substrate in the reactive configuration could in turn help understand the solvent effects and nonenzymatic catalysis mechanism for chemical reactions.

The chorismate acid turning to the prephenate is part of biosynthesis of phenylalanine and tyrosine, which is fulfilled by the chorismate mutase enzyme found in bacteria, fungi, and plants (Haslam, 1993). This reaction

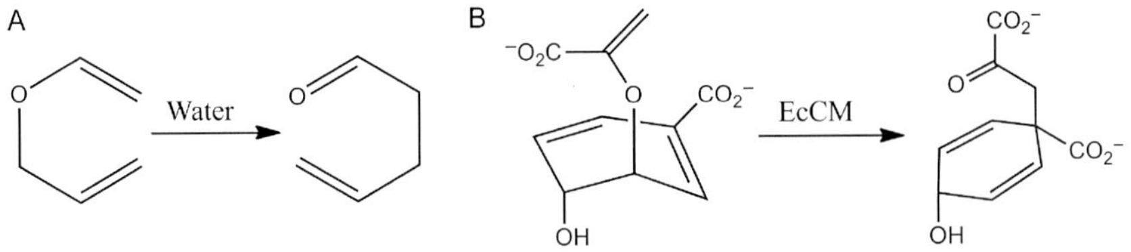

Scheme 1 Aliphatic Claisen rearrangement reaction and the indices of heavy atoms of the reactant (allyl vinyl ether in (A), chorismate ion in (B)) and the product (4-pentenal in (A) and prephenate ion in (B)).

is known to organic chemists as Claisen rearrangement (Scheme 1). Similar to many other chemical reactions, Claisen rearrangement is subjected to a wide range of solvent and catalyst options to enhance its rate (Coates, Rogers, Hobbs, Curran, & Peck, 1987). The water acceleration (Chanda & Fokin, 2009) of it is more than intriguing, given its potential application in terms of green chemistry. In seek of how water facilitates this reaction, nature's design provides vital clues. Enzyme structures of two chorismate mutase (EcCM and BsCM) (Lee, Stewart, Clardy, & Ganem, 1995), show common features that: (i) the substrate is locked in the requisite chair conformer for the rearrangement and that (ii) active site residues form hydrogen bonds to the bond breaking site oxygen. Follow this line, many groups reported both experimentally and computationally that there exists an equilibrium between pseudoequatorial and pseudoaxial conformations of chorismate acid in solutions, which moreover shows strong dependence on the solvent constituents (Guimarães, Repasky, Chandrasekhar, Tirado-Rives, & Jorgensen, 2003). These findings indeed pose the study of enzyme reactions into sharper perspectives: influence of diverse solution conformations of substrates can no longer be "averaged out," and the binding orientation or the binding mode of substrates in the active pocket is extremely important.

Unlike protein–ligand binding, enzyme–substrate interactions entail more detailed information on chemical properties of the substrate, including both molecular configurations and electronic configurations. Therefore, in order to study the precatalysis conformation of free substrates, QM/MM simulations are better options. SITS–QM/MM method mentioned in Section 1.1.2 is suitable to serve as a reliable and efficient tool to explore the conformational details of enzyme substrates in solutions. A SITS–QM/MM MD study demonstrated that for ally vinyl ether (AVE), the compact and reactive conformers can be significantly enriched in the aqueous solution in contrast to common organic solvents, mainly due to the large cohesive energy density of water (Fig. 8). The hydrophobic effect which stabilizes the compact conformers plays an important role in the water

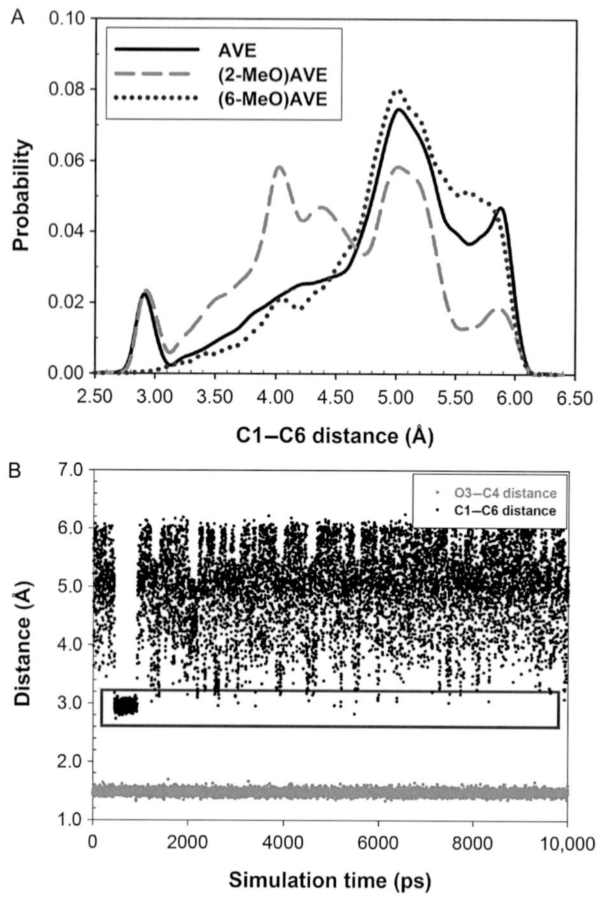

Fig. 8 (A) Distribution functions of the distance between C1 and C6 atoms of different reactants in aqueous solution. The *black solid*, *red dashed* (*gray* in the print version), and *blue dotted* (*black* in the print version) *lines* correspond to the parent, 2-methoxyl substituted, and 6-methoxyl substituted AVEs, respectively. Conformations with C1–C6 distance shorter than 3.25 Å (where can be reached by the right-bottoms of the Gaussian-like peaks) were defined to be compact, and the others were defined to be extended. (B) Exemplary profiles of C1–C6 distance (*black dots*) and O3–C4 distance (*red* (*gray* in the print version) *dots*) along a 10 ns trajectory. A wide range of C1–C6 distance can be adopted, the *blue-colored* (*dark gray* in the print version) *inset box circles* out the frames containing compact conformers. O3 and C4 atoms are separated by a normal C–O single bond length with tiny fluctuation. On average, there were tens of transitions between different conformations during per nanosecond driven by SITS. *Reprinted with permission from Zhang, J., Yang, Y. I., Yang, L., & Gao, Y. Q. (2015a). Conformational pre-adjustment in aqueous claisen rearrangement revealed by SITS-QM/MM MD simulations.* The Journal of Physical Chemistry B, 119*(17), 5518–5530. doi: 10.1021/jp511057f. Copyright (2015) American Chemical Society.*

acceleration effect on Claisen rearrangement (Zhang, Yang, Yang, & Gao, 2015a). Besides, different substituents may alter the inclination of AVE molecule to adopt the compact conformation, thus leading to the various substituent effects on the reaction rate (Fig. 8A). Such studies showed that SITS–QM/MM MD is capable of yielding information on reaction rate and reaction mechanism in a complex environment without predefined RC or preassumed mechanism. We are applying such a protocol to molecular simulations of enzymatic reactions.

2. TRANSPORTATION OF SMALL MOLECULES IN PROTEINS

Transportation of small molecules in proteins, which can be categorized as a specific mode of interaction between proteins and small molecules, is common and important process in biological systems. Intramolecular protein tunnels have been found in a number of enzymes where it is necessary to protect unstable and reactive intermediates from bulk solution. For example, protein tunnels for the migration of ammonia have been found in all members of the glutamine amidotransferase family of enzymes (Douangamath et al., 2002; Krahn et al., 1997; LaRonde-LeBlanc, Resto, & Gerratana, 2009; Larsen et al., 1999; Mouilleron, Badet-Denisot, & Golinelli-Pimpaneau, 2006; Raushel, Thoden, & Holden, 2003; Schmitt, Panvert, Blanquet, & Mechulam, 2005; Teplyakov, Obmolova, Badet, & Badet-Denisot, 2001; Thoden, Holden, Wesenberg, Raushel, & Rayment, 1997; van den Heuvel et al., 2002; Wojcik, Seidle, Bieganowski, & Brenner, 2006). In addition to the transport of ammonia, tunnels have been identified for the translocation of other reactive species. For example, tryptophan synthase (Barends, Dunn, & Schlichting, 2008; Dunn, Niks, Ngo, Barends, & Schlichting, 2008; Hyde, Ahmed, Padlan, Miles, & Davies, 1988), acetyl–coenzyme A synthase/carbon monoxide dehydrogenase (Maynard & Lindahl, 2001; Tan, Loke, Fitch, & Lindahl, 2005; Tan, Volbeda, Fontecilla-Camps, & Lindahl, 2006), 4-hydroxy-2-ketovalerate aldolase/acylating acetaldehyde dehydrogenase (Manjasetty, Powlowski, & Vrielink, 2003), and carbamoyl phosphate synthetase (CPS) use molecular tunnels to transport indole, carbon monoxide, acetaldehyde, and carbamate, respectively.

Although structural biology experiments can help us identify the shape and length of the tunnel inside the protein, there are significant difficulties in experimentally tracking the transportation of small molecules from one active site to the next. MD simulations can serve as a useful tool to investigate the kinetics and dynamics details of the mechanism for small molecule migration at the atomic level. Especially, the RCs in thermodynamic calculations are usually clearly defined for transportation problems (along the protein tunnel), so that, the enhanced sampling method such as umbrella sampling and steered MD can be easily adopted. For example, the transfer of NH_3 in imidazole glycerol phosphate synthase and glucosamine-6-phosphate synthase has been studied using steered MD (Amaro & Luthey-Schulten, 2004; Amaro, Myers, Davisson, & Luthey-Schulten, 2005; Floquet et al., 2007). The free energy profile for NH_3 transport in AmtB was calculated by an umbrella sampling technique (Lin, Cao, & Mo, 2006) and it was also investigated at an elevated temperature of 323 K utilizing conventional MD simulations (Yang, Xu, Zhu, Chen, & Jiang, 2007). These simulations have led to a better understanding of the transport mechanisms for ammonia and helped to explain the conformational changes of key residues in these proteins.

As a good example, CPS, was studied by MD simulation (Fan, Lund, Shao, Gao, & Raushel, 2009; Fan, Lund, Yang, Raushel, & Gao, 2008; Lund, Fan, Shao, Gao, & Raushel, 2010). CPS from *Escherichia coli* catalyzes one of the most complex reactions in biological chemistry. The product of this enzymatic transformation, carbamoyl phosphate, is utilized in the biosynthesis of arginine and pyrimidine nucleotides (Anderson, 1977; Anderson & Meister, 1966; Pierard & Wiame, 1964). Carbamoyl phosphate is formed from glutamine, bicarbonate, and two molecules of MgATP through a series of four separate reactions. The generally accepted reaction mechanism is summarized in Scheme 2. In the initial reaction, the first molecule of ATP is used to phosphorylate bicarbonate to generate the reactive intermediate, carboxy phosphate. Glutamine is hydrolyzed to glutamate and ammonia, followed by the reaction of ammonia with the carboxy phosphate

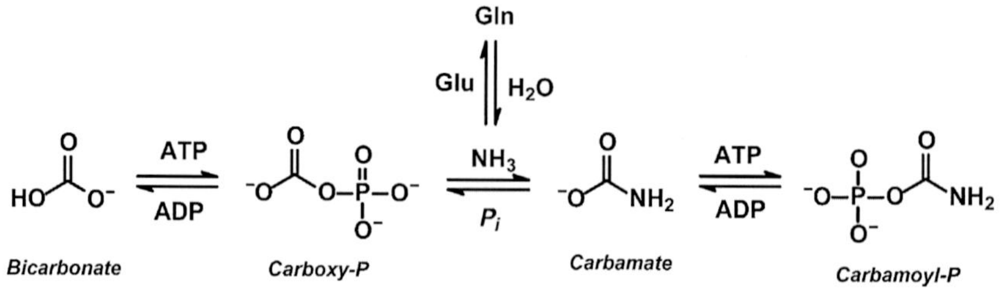

Gln
Glu H_2O
ATP ADP NH_3 P_i ATP ADP
Bicarbonate *Carboxy-P* *Carbamate* *Carbamoyl-P*

Scheme 2 Carbamoyl phosphate is formed from glutamine, bicarbonate, and two molecules of MgATP through a series of four separate reactions.

intermediate to form carbamate. Finally, a second molecule of ATP is used to phosphorylate carbamate to form the ultimate product, carbamoyl phosphate. There are thus four separate reactions and three discrete, reactive, and unstable intermediates involved in this reaction mechanism: carboxy phosphate, ammonia, and carbamate.

X-ray crystal structure shows that CPS is a heterodimeric protein that is composed of two subunits of molecular weight ~40 and ~118 kDa, respectively (Matthews & Anderson, 1972). There are three spatially distinct active sites in CPS that are connected by two molecular tunnels that extend nearly 100 Å from one end of the protein to the other as illustrated in Fig. 9 (Thoden et al., 1997). The small subunit contains the active site for the hydrolysis of glutamine, which is hydrolyzed to glutamate and ammonia via a thioester intermediate. The ammonia derived from this reaction travels ~45 Å to the active site for the synthesis of the carboxy phosphate intermediate, located in the N-terminal half of the large subunit. The carbamate intermediate, formed by the reaction of ammonia with carboxy phosphate,

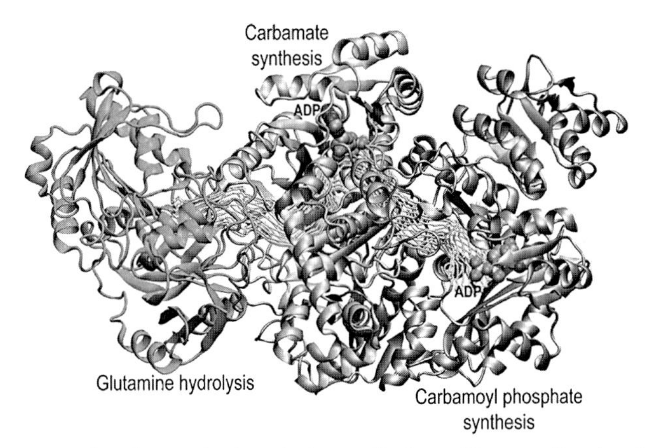

Fig. 9 Structure of carbamoyl phosphate synthetase. The small subunit that contains the active site for the hydrolysis of glutamine is shown in *green*. The N-terminal domain of the large subunit that contains the active site for the synthesis of carboxy phosphate and carbamate is shown in *red*. The C-terminal domain of the large subunit that contains the active site for the synthesis of carbamoyl phosphate is shown in *blue*. The two molecular tunnels for the translocation of ammonia and carbamate are shown in *yellow dotted lines*. The image was constructed from PDB file: 1c30. *Reprinted with permission from Fan, Y. B., Lund, L., Shao, Q., Gao, Y. Q., & Raushel, F. M. (2009). A combined theoretical and experimental study of the ammonia tunnel in carbamoyl phosphate synthetase.* Journal of the American Chemical Society, 131*(29), 10211–10219. doi: 10.1021/ja902557r. Copyright (2009) American Chemical Society.* (See the color plate.)

is subsequently transported ~45 Å to the active site for the synthesis of carbamoyl phosphate. This active site is located within the C-terminus of the large subunit where carbamate is phosphorylated by the second molecule of MgATP to form the ultimate product, carbamoyl phosphate. CPS thus contains two long molecular tunnels for the transport of ammonia and carbamate, respectively.

To study carbamate transport in CPS, the large subunit from the X-ray crystal structure of CPS (PDB codes: 1c30 and 1bxr) was taken as the initial structure for molecular modeling. Three structural complexes were created for the simulations (Lund et al., 2010) and these are denoted as complexes A, B, and C in Table 3.

Based on the X-ray crystal structure and MD simulations, the entire carbamate tunnel can be roughly separated into three continuous water-filled pockets connected by two relatively narrow regions near Ala-23 and Gly-575. The first water pocket locates in the region near the first nucleotide-binding site of the large subunit to Ala-23. There are 21 waters in the first pocket, which is partly connected to the water surrounding ADP and water outside the tunnel. As a result, water can exchange between this pocket and the exterior. The middle water pocket is located between Ala-23 and Gly-575, which is filled with nine water molecules. The third water pocket occupies the space near Gly-575 and extends to the ATP-binding site for carbamoyl phosphate synthesis. There are also nine water molecules in this last pocket. In contrast to the first water cluster, the latter two pockets are tightly sealed and no exchange of water with the exterior was observed during the simulations. The major difference between the second and third

Table 3 Definition of the Complexes with Different Ligands Bound

		Substrate(s)		
Name	**PDB Code**	**Site 1[a]**	**Site 2[b]**	**Arg-306[c]**
A	1bxr	ADP	ATP	Open
B	1bxr	ADP + Pi	ATP	Closed
C	1c30	ADP + Pi	ADP	Open

[a]Site for carbamate synthesis.
[b]Site for carbamoyl phosphate synthesis.
[c]When Arg-306 is hydrogen bonded to the phosphate site 1 the tunnel is closed and it is open when this residue is hydrogen bonded to Glu-25, Glu-383, and Glu-604.

water clusters is that the former is more compact while the latter is more extended in one dimension along the tunnel.

Three 40 ns conventional MD simulations were performed with carbamate positioned in the middle of the three water pockets using complex A. Fig. 10 showed the probability distribution for the position of carbamate in the tunnel. Three distinct narrow distributions verify the findings obtained from the X-ray structures that the entire tunnel is separated into three segments. The *blue curve* (*black* in the print version) in Fig. 10 shows that carbamate is localized at a distance of 10 Å and at a position above the center of Glu-25, Glu-383, and Glu-604, corresponding to the first segment of the tunnel. The *red* (*gray* in the print version) *curve* centered at a distance of 23 Å shows the second segment, which extends from Ala-23 to Gly-575. When carbamate approaches the active site for carbamoyl phosphate synthesis, it is solvated in the third water pocket with a distance distribution that centers at 33 Å. As illustrated in Fig. 10, Ala-23 and Gly-575 separate the entire tunnel into three segments, at a distance of 17 and 29 Å, respectively. These conventional MD simulations indicate the existence of well-structured water pockets within the tunnel and the possibility of molecular

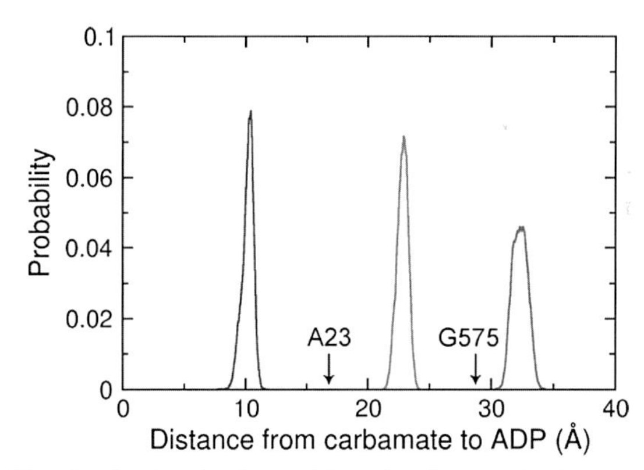

Fig. 10 Probability distribution for the position of carbamate in the tunnel. The distance is defined from the carbon atom of carbamate to the phosphate of ADP bound at the active site for the carbamate formation. The distributions in the first (near Glu-25, Glu-383, and Glu-604), second (in the middle of the tunnel), and third water pockets (near Glu-577 and Glu-916) are shown in *blue* (*black* in the print version), *red* (*gray* in the print version), and *green* (*gray* in the print version), respectively. *Reprinted with permission from Lund, L., Fan, Y. B., Shao, Q., Gao, Y. Q., & Raushel, F. M. (2010). Carbamate transport in carbamoyl phosphate synthetase: A theoretical and experimental investigation.* Journal of the American Chemical Society, *132(11), 3870–3878. doi: 10.1021/ja910441v. Copyright (2010) American Chemical Society.*

valves at the narrow parts of the tunnel near Ala-23 or Gly-575 that may control the transport of carbamate.

Next, the free energy profile for the migration of carbamate along the tunnel was calculated using the umbrella sampling technique. In the umbrella sampling, a biasing harmonic potential was applied with a force constant of 40 kcal mol^{-1} Å^{-2}. A total of 113 windows were used for each of the three systems (listed in Table 3) with a step size of 0.25 Å starting from 5 Å and ending at 33 Å. An equilibration simulation for each window was performed and the corresponding PMF along the RC was computed every 500 ps using the weighted histogram analysis method (WHAM) (Kumar, Bouzida, Swendsen, Kollman, & Rosenberg, 1992; Kumar, Rosenberg, Bouzida, Swendsen, & Kollman, 1995) until convergence was observed. The last four 500-ps trajectories, without yielding significantly different PMF when used separately, were combined as a 2-ns simulation for WHAM analysis. Because of the repulsion between carbamate and phosphate in complexes B and C when the RC is shorter than 10 Å, the PMFs for these two systems were calculated from 10 to 33 Å. The calculated PMFs along the tunnel for the three complexes are shown in Fig. 11.

In Fig. 11, the *red solid* (*gray* in the print version) *line* represents the free energy profile calculated for the complex A. It is clear that three minima can be found at distances of 10, 23, and 31 Å from ADP and two maxima located at 17 and 26 Å. The free energies differences of the first and third minima, relative to the middle minimum at 23 Å, are 5.8 and 1.6 kcal/mol, respectively, whereas the two barriers have heights of 6.7 and 8.4 kcal/mol, respectively. In the initial complex prepared for the umbrella sampling calculation, carbamate was positioned in the middle of the tunnel. Within 1-ns relaxation, the side chain of Arg-306 rotates away from the ADP-binding site and forms multiple hydrogen bonds to Glu-25, Glu-383, and Glu-604. Apparently, the release of phosphate into solution frees Arg-306 and the negative charges of the three glutamates attract the side chain of Arg-306.

The *solid blue* (*black* in the print version) *line* in Fig. 11 shows the results obtained for complex B. Compared to complex A, the PMF along the first segment of the tunnel for the complex B is significantly higher. Through the analyses of dynamics and structures obtained in umbrella sampling calculations, it is observed that the reaction product prevents Arg-306 from rotating toward Glu-25, Glu-383, and Glu-604 and thus effectively blocks the migration of carbamate into the tunnel. With phosphate present, there is

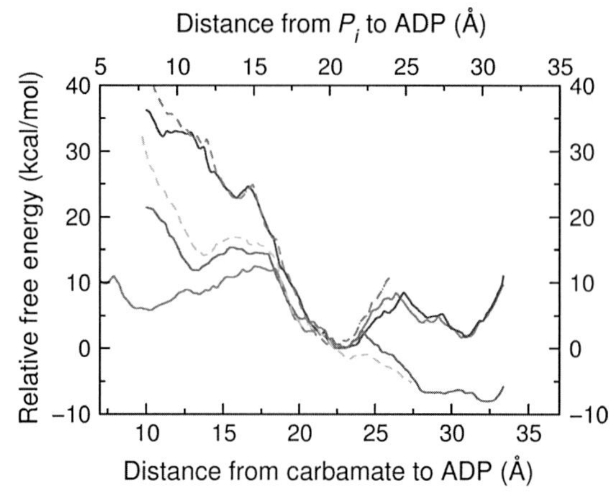

Fig. 11 Potentials of mean force from the site of carbamate formation to the site of utilization. The reaction coordinate, labeled on the bottom, is defined as the distance from the carbon atom of carbamate to the phosphate of ADP bound at the active site for carbamate formation. The PMFs, with respect to this reaction coordinate, are presented as the *solid lines* for the complexes A, B, and C in *red* (*gray* in the print version), *blue* (*black* in the print version), and *green* (*dark gray* in the print version), respectively. The other reaction coordinate, labeled on the top, is defined as the distance from the carbon atom of carbamate to the phosphate product. The PMFs, with respect to the second reaction coordinate, are shown as the *dashed curves* for the complexes B and C in *red* (*gray* in the print version) and *yellow* (*light gray* in the print version), respectively. *Reprinted with permission from Lund, L., Fan, Y. B., Shao, Q., Gao, Y. Q., & Raushel, F. M. (2010). Carbamate transport in carbamoyl phosphate synthetase: A theoretical and experimental investigation.* Journal of the American Chemical Society, 132*(11), 3870–3878. doi: 10.1021/ja910441v. Copyright (2010) American Chemical Society.*

a strong repulsion between carbamate and the closed gate formed by phosphate and the side chain of Arg-306. Carbamate is thus trapped in the active site. Therefore, a high free energy barrier between 10 and 15 Å is pronounced as shown on the *blue* (*black* in the print version) *curve*. The separation between the first and second water pockets near Ala-23 is also observed, which is corresponding to the sharp peak at 17 Å resulted from the narrow passage. The negative charges on Glu-25, Glu-383, and Glu-604 prohibit the entrance of carbamate. However, in complex A, the rotation of Arg-306, after the release of phosphate, shields these negative charges from carbamate, thus allowing it to pass through. The free energy profile from 19 to 33 Å is nearly identical to that for complex A. Therefore, the conformational change of Arg-306 only affects carbamate transport over short distances.

In the simulation for complex C, the side chain of Arg-306 was manually rotated toward the side chains of Glu-25, Glu-383, and Glu-604 before the umbrella sampling calculation was initiated. The free energy (*green solid* (*dark gray* in the print version) *curve* in Fig. 11) is significantly higher than that of complex A at distances shorter than 13 Å. Such a difference is induced by the strong anion–anion repulsion between phosphate and carbamate. The free energy curve is similar to that for the complex A in the range from 13 to 23 Å, although the PMF between 13 and 18 Å is 1.5–2.5 kcal/mol higher than that for complex A. This increase of PMF is also a result of the repulsion between phosphate and carbamate. The free energy profile after the narrow passage near Ala-23 is overwhelmingly downhill. Such a PMF indicates that the transport of carbamate needs to overcome a significant barrier at 15 Å. Once this barrier is passed, the NH_3 translocation is spontaneous and fast.

The strong repulsion between carbamate and phosphate makes estimation of the PMF difficult when carbamate is close to its site of formation. Thus, the distance between phosphorus (in phosphate) and carbon (in carbamate) was defined as a new RC in umbrella sampling simulations for complexes B and C. As shown in Fig. 11, using the newly defined RC the details of the PMF were reproduced (*red dashed* (*dark gray* in the print version) *line* to *blue solid* (*black* in the print version) *line* and *yellow dashed* (*gray* in the print version) *line* to *green solid* (*dark gray* in the print version) *line*). As a result, the simulations based on the second RC further confirm that the PMFs calculated on the previous RC are reliable for the first and second segments of the tunnel.

Furthermore, the thermodynamic influences of phosphate and the conformational change of Arg-306 can be observed when comparing the PMFs for each of the three systems in the range of 10–17 Å (Fig. 11). At these positions, the PMFs of A and C are significantly lower than that of B, indicating that the rotation of Arg-306 effectively shields the charge–charge repulsion of Glu-25, Glu-383, and Gu-604 with carbamate. The repulsion between phosphate and carbamate results in the PMFs of B and C to steeply increase at RC shorter than 13 Å. The PMF segments of A and C are nearly parallel to each other in the range from 13 to 17 Å, illustrating the structural similarity for these two systems. However, a small difference does exist. The repulsion from phosphate also lifts the PMF of C by about 2 kcal/mol, relative to that of A.

In order to understand the influence of phosphate on the residue Arg-306 conformational change, two more umbrella sampling simulations were also performed with complexes A and B. The distance between the guanidinium carbon of Arg-306 and the carboxylate carbon of Glu-604

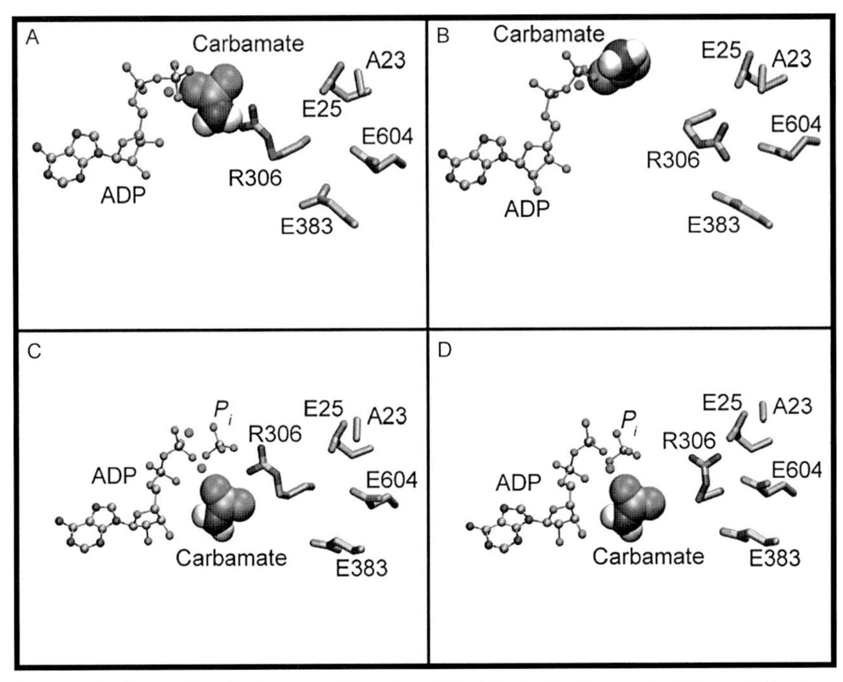

Fig. 12 Conformational change of the Arg-306 side chain. In panels (A) and (B), phosphate is released, but in panels (C) and (D), it remains bound. The side chain of Arg-306 orients toward ADP (or phosphate) in panels (A) and (C) while it forms a salt bridge with Glu-25, Glu-383, and Glu-604 in panels (B) and (D). *Reprinted with permission from Lund, L., Fan, Y. B., Shao, Q., Gao, Y. Q., & Raushel, F. M. (2010). Carbamate transport in carbamoyl phosphate synthetase: A theoretical and experimental investigation.* Journal of the American Chemical Society, 132*(11), 3870–3878. doi: 10.1021/ja910441v. Copyright (2010) American Chemical Society.*

was defined as the RC and separated into 20 windows. The side chain of Arg-306 forms hydrogen bonds with Glu-604 when the distance is less than 5 Å (Fig. 12B and D) while the side chain points toward ADP when the distance greater than 7 Å (Fig. 12A and C). These two conformations are defined as the open and closed forms, respectively. Carbamate was positioned near the phosphorus of ADP. In the open form, the passage leading to the tunnel entrance is filled with water while in the closed form the transport of carbamate is blocked by the side chain of Arg-306. Relaxation was performed until equilibrium was reached which was checked by comparing the PMFs for two continuous 500-ps trajectories. The two 500-ps trajectories were followed by a 1-ns trajectory and collected for data analysis. The free energy profiles from the 2-ns umbrella sampling are shown in Fig. 13.

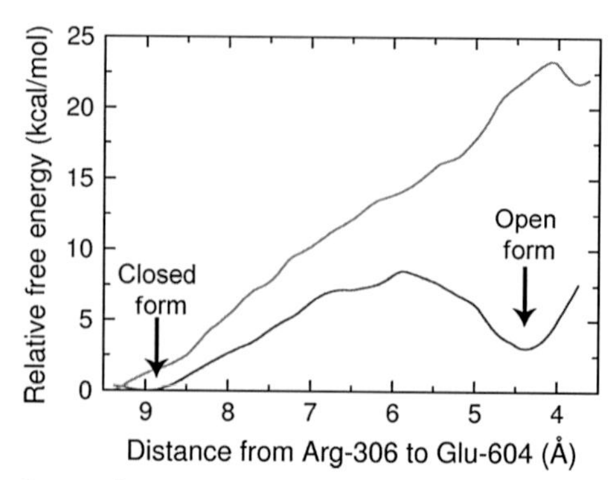

Fig. 13 Potentials of mean force for the rotation of the Arg-306 side chain. The reaction coordinate is defined as the distance between the carbon atom in the guanidinium group of Arg-306 and the carbon atom of the side chain carboxylate group of Glu-604. The PMFs for complexes A and B are plotted as the *blue* (*black* in the print version) and *red* (*dark gray* in the print version) *curves*, respectively. *Reprinted with permission from Lund, L., Fan, Y. B., Shao, Q., Gao, Y. Q., & Raushel, F. M. (2010). Carbamate transport in carbamoyl phosphate synthetase: A theoretical and experimental investigation.* Journal of the American Chemical Society, 132*(11), 3870–3878. doi: 10.1021/ja910441v. Copyright (2010) American Chemical Society.*

After the release of phosphate into solution, carbamate partially fills the void and forms a hydrogen bond with the side chain of Arg-306 (Fig. 12A). This hydrogen bond is the major obstacle to the rotation of Arg-306. The overall free energy barrier is 8.0 kcal/mol and the free energy increases by 3.0 kcal/mol after the rotation relative to the closed conformation (*blue* (*black* in the print version) *curve* in Fig. 13). When phosphate is bound, the electrostatic interactions between phosphate and Arg-306 prohibit this conformational change. As shown in Fig. 12D, when Arg-306 forms hydrogen bonds with Glu-25, Glu-383, and Glu-604, the free energy is 22 kcal/mol higher than that for the closed conformation, in which phosphate and Arg-306 bind to one another (*red* (*dark gray* in the print version) *curve* in Fig. 13), with a barrier of 24 kcal/mol at a distance of 4.1 Å. The conformational change from the closed to the open form is favored when phosphate is released. Based on the transition state theory, the barrier height of 24 kcal/mol corresponds to a reaction rate of approximately one turnover every few hours. This rate is more than a 1000-fold slower than the synthesis of carbamoyl phosphate. The importance of Arg-306 and the trio of glutamate residues within the tunnel is confirmed by the fact that neither of the mutants, R306A nor E25Q/E383Q/E604Q, are able to synthesize carbamoyl phosphate (Kim & Raushel, 2004).

Similarly, the ammonia transportation tunnel was also studied using both MD and umbrella sampling MD (Fan et al., 2009, 2008). The MD simulations provided strong evidence for the existence and function of a sealed molecular tunnel for the transport of ammonia from the site of formation to the site of utilization. Normal MD simulations for each section of the nearly 60 Å tunnel indicate that the passageway is tightly sealed since leakage of ammonia was never observed. The transfer of ammonia through this tunnel is a rapid process with a free energy barrier of 7.2 kcal/mol at a gate that is formed by the convergence of Cys-232, Ala-251, and Ala-314 (Fig. 14). The initial injection of the ammonia into the tunnel is favored with a large free energy drop of at least 7 kcal/mol. When ammonia passes the tunnel exit from the small subunit near Thr-37, it enters an area filled with approximately 27 waters, which associate together by forming hydrogen bonds. Tyr-261 from the large subunit splits these water molecules into two clusters. The transfer of the ammonia is favored by a free energy potential from Thr-37 to a gate enclosed by the side chains of Cys-232, Ala-251, and Ala-314.

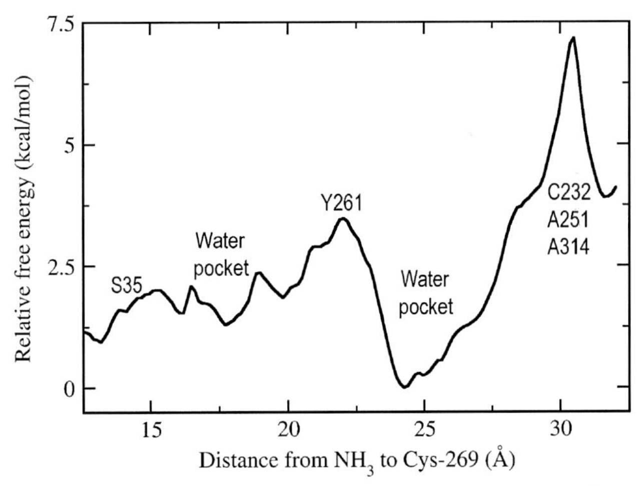

Fig. 14 Potentials of mean force from Thr-37 in the small subunit to the narrow turn surrounded by Cys-232, Ala-251, and Ala-314. The reaction coordinate is defined as the distance from the nitrogen atom in the ammonia to the carbonyl carbon of the thioester intermediate in Cys-269. *Reprinted with permission from Fan, Y. B., Lund, L., Shao, Q., Gao, Y. Q., & Raushel, F. M. (2009). A combined theoretical and experimental study of the ammonia tunnel in carbamoyl phosphate synthetase.* Journal of the American Chemical Society, 131*(29), 10211–10219. doi: 10.1021/ja902557r. Copyright (2009) American Chemical Society.*

From the earlier examples of carbamate and ammonia migrations in CPS, it is clear that computer simulations can be very helpful for studying the mechanism of transportations of small molecules in proteins. Since the transportation tunnels are usually buried inside the large proteins, the detailed transportation processes are very difficult to be captured directly in experiments. Utilizing MD simulations and free energy calculation methods (such as umbrella sampling) not only thermodynamics but also dynamics of transportation can be obtained at the atomic level. Furthermore, insights gained from simulations can provide us hints of modifying the interactions between small molecules and proteins. Therefore, mutations along the protein tunnels can be designed to facilitate or block the migrations of small molecules.

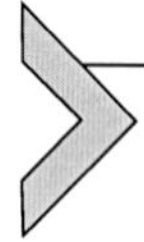

3. QM/MM MD SIMULATIONS OF ENZYMATIC REACTIONS

In the following discussion, we turn from physical to the chemical processes. Experiments that resolve the enzyme–substrate (or analogue) structures lay down the ground for the theoretical study of the catalysis process. Though, most structural experiments involve active pocket mutations (loss-of-function) or replacement of the natural substrate for nontarget analogues to prevent the substrate converting and releasing. Adequate relaxation of the enzyme structure (especially the active pocket) and optimization of the substrate binding mode hence appears necessary in the preparation for the subsequent mechanism study, as discussed earlier. According to the well-assessed active pocket structure as well as the binding mode, one can propose (or guess) possible mechanisms. As is often the case, various possible schemes exist for one same reaction, which are largely dependent on chemical intuition or experience. Sometimes single residue mutation study will rule out many of the hypotheses and narrow down the range of options. A common practice is to evaluate several most "promising" and "reasonable" mechanisms in a trial-and-error manner. Specifically, each mechanism is characterized by a unique set of artificially chosen RCs, with the energy (or free energy) profile along these RCs obtained through various CV-based sampling methods, then the mechanism showing smallest overall barrier is accepted as the correct mechanism. Metadynamics (Laio & Parrinello, 2002) and umbrella sampling (Torrie & Valleau, 1977) are very popular methods implemented in this framework. Both of the methods entail predefined but low-dimensional RCs or CVs and assume that the slowest dynamic motions lie exactly along the chosen RCs or CVs.

Both metadynamics and umbrella sampling prove their usefulness in mechanistic studies of enzymatic reactions given a set of reasonable predefined RCs. A number of enzymology applications of metadynamics and umbrella can be seen from specialized reviews (Ensing, De Vivo, Liu, Moore, & Klein, 2006; Kästner, 2011). Most recently, it was used to examine the catalysis process of NEIL1 protein, which excises the oxidation-damaged pyrimidines in DNA duplex (Yeo, Goodman, Schirle, David, & Beal, 2010). Human NEIL1 bears intriguing but mysterious RNA editing effect (Yeo et al., 2010), namely the processing of NEIL1 pre-mRNA before translation causes a lysine to arginine change in the lesion recognition loop of the protein and hence slows down the excising rate of certain substrates. In order to address the RNA editing effect, umbrella sampling on QM (SCC-DFTB; Gaus, Cui, & Elstner, 2014)/MM (AMBER) MD trajectories was adopted, to validate possible reaction mechanisms (details to be published). According to the free energy profiles calculated by umbrella sampling along different RCs, the base-activated pathway was ruled out due to the very high barrier at the initial step (Fig. 15B). In contrast, the ribose-protonated pathway seems viable for a Tg tautomer (Tg2H) given that the overall free energy barrier is moderate for an enzymatic reaction. This feasible ribose-protonated mechanism involves residue 242 (altered by RNA editing) directly in the catalysis chemistry, where 242R/K plays as a proton donor to the cleaved Tg product (Fig. 15). Along this line, the turnover rate is expected to vary for enzymes with either R or K at 242 position, because despite that both R and K are protonated basic residues, the basicity (pK_a) is yet different. Arginine is generally more basic than lysine, and more geometrically rigid, hence would likely be more reluctant if it were to donate a proton to Tg. As a consequence, the Tg-excising process executed by NEIL1-242R will be slower than NEIL1-242K to some extent (c.30-fold at room temperature).

Albeit many current enzymatic reaction mechanisms are proposed based on the CV-reliant sampling methods, these approaches, indeed, are suitable for ruling out various options. It remains interesting to further develop methods to give more conclusive and predictive answers. In some cases, debates on the mechanisms for the very same enzymatic reaction arise from the different CV choices. In addition, transition pathways and transition states for each reaction can be varied, but conform to certain statistics. Besides the energy-minimum pathway, which dominates in terms of the

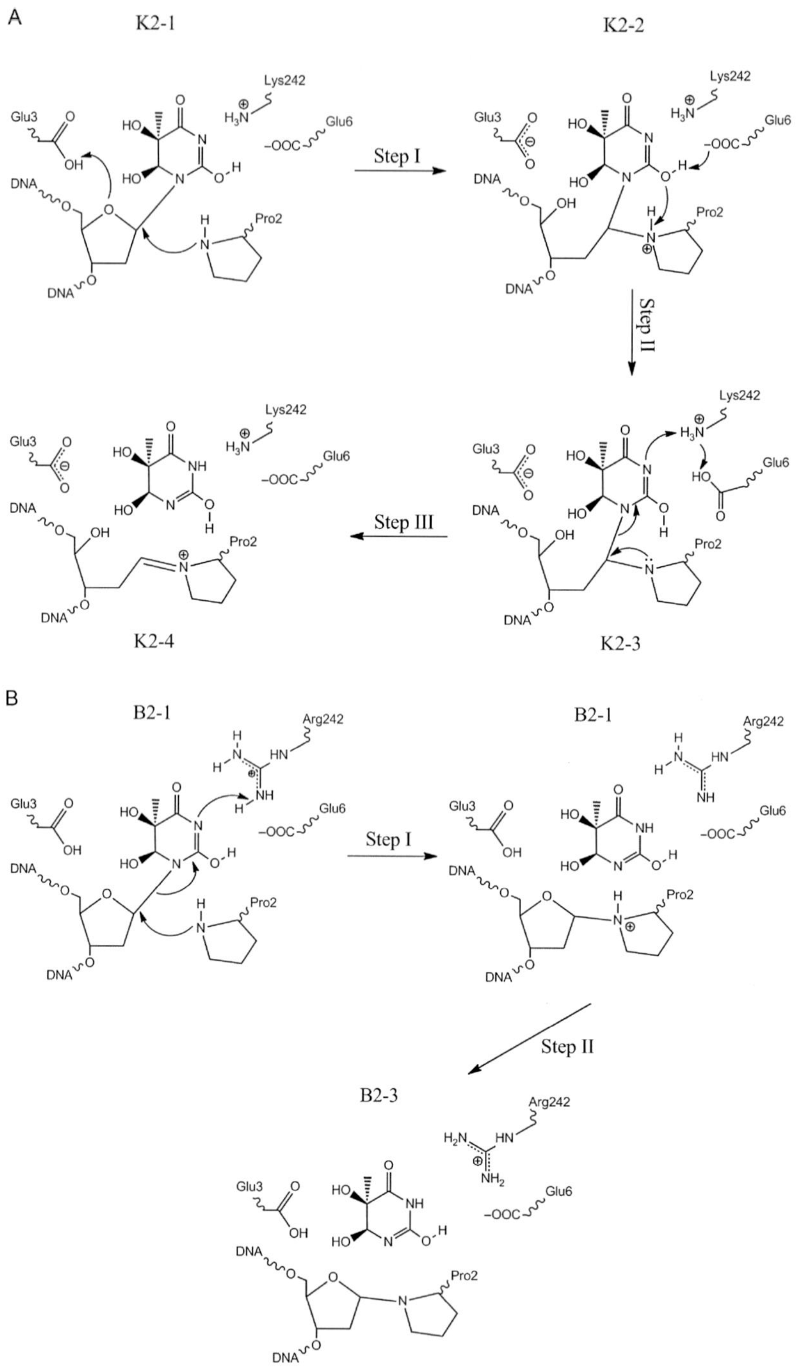

Fig. 15 (A) Ribose-protonated pathway initiated with NEIL(K242)-Tg2H structure; the reactant, intermediate, and product states are termed as K2-1, K2-2, and K2-3, respectively. (B) Base-activated pathway initiated with NEIL(R242)-Tg2H structure; the reactant, intermediate, and product states are termed as B2-1, B2-2, and B2-3, respectively.

statistical weight, pathways slightly deviated from the exact energy-minimum pathway defined by CVs may also account for a large proportion of real enzyme reactions. Moreover, alternative mechanisms, if exist in an enzyme reaction, are difficult to be captured by one single set of CVs, thus may lead to incomplete understanding of how enzyme works. Therefore, CV-free sampling methods that can realize spontaneous enzyme catalyzed transformations are needed. Transition path sampling (and its variations) is a representative method belonging to this category (Bolhuis, Chandler, Dellago, & Geissler, 2002; Moroni, Bolhuis, & van Erp, 2004). The procedure of TPS ensures the authenticity of the dynamics of the system, and it in principle requires no a priori knowledge of RCs. Despite that TPS effectively yields an ensemble of transition paths, the rate constant cannot be directly obtained until another separately operated procedure, usually the flux calculation, is done. When applied to calculate the rate constant of a rare event, a conventional combination of flux calculations and TPS sometimes may not perform so efficiently as anticipated due to the slow convergence.

Enhanced sampling of reactive trajectories (ESoRT) was inspired by TPS, in which we take advantages of both ITS/SITS-MD and TPS (shooting) methods (Bolhuis et al., 2002; Fu, Yang & Gao, 2007; Gaus et al., 2014; Moroni et al., 2004). The prerequisite for the ESoRT is to generate trajectories which connect the reactant and product without or with few predefined RCs. Successfully reacted trajectories from ITS/SITS-MD simulations are collected and registered as "reactive trajectories," with atomic coordinates, velocities, charges as well as the effective weighting factors recorded. Then, along each reactive trajectory, a series of configurations, which contain both reactant-like structures and configurations near the transition states but exclude product-like ones, are selected as the initial structures for the following transition trajectory shooting. Finally, after a number of short NVE simulations, the rate constant is retrieved as the ratio between the probability of reactive and that of nonreactive trajectories, each reweighted to reflect their probability distribution in the unbiased ensemble.

Contrasting to traditional TPS, in the framework of ESoRT, we first perform a thorough search over phase space of the reactant and identify the reactive regions by SITS simulations. Transition path sampling is then performed starting from these phase space points. Since the initial configurations for trajectory shootings are selected from the ensemble of phase space points biased toward successful transitions, this trajectory sampling method dramatically increases the efficiency of rate constant calculations and the reaction pathway searching. By all accounts, the ESoRT method enjoys advantages from three major aspects: (1) the ensemble of trajectories (defined as reactive trajectories) are automatically generated through SITS instead of a nontrivial procedure as in TPS; (2) ESoRT samples the phase space more thoroughly hence enables the search of multiple pathways separated by high barriers, because it avoids the entrapment of the trajectories in certain pathway(s); and (3) the calculation of the reaction rate is free of predetermined RCs and quickly converges.

ESoRT was recently validated in the mechanistic study of chemical reactions in complex environment (eg, in solution), for example, dynamics and kinetics study of in-water Claisen rearrangement (Scheme 1) (Zhang, Yang, Yang, & Gao, 2015b). Reaction pathway information can be obtained by projecting the calculated visiting probabilities to the postselected coordinate. For instance, by conventionally choosing the atomic coordination as the RC, the activated transition structure was observed to prefer a chair-like conformation and exhibited a concerted bond forming/rupturing feature. In addition to a calculated absolute reaction rate constant in agreement with experiments, the results highlight the water as a hydrogen donor which stabilizes the transition state and emphasize as well the dipole–dipole interactions between bulk water and the transforming chemical compound.

4. SUMMARY

In this chapter, we discussed in details three key modes of interactions between small molecules and proteins: binding, transportation, and reaction. Due to the large computational demand, we chose different approaches, ranging from normal MD simulations to globally enhanced sampling, to umbrella sampling and trajectory sampling to deal with the different aspects that involve in ligand/protein interactions and reaction. To investigate the binding mode and affinity of small molecules, we put emphasis on the conformation sampling of free ligands/substrates as well as proteins. The more efficient sampling method, MTS and ITS/SITS, were introduced. Then the migration of small molecules (carbamate and ammonia) in molecular tunnels of proteins was studied by using conventional MD and umbrella sampling. The complete free energy profile along the tunnel was calculated from traditional umbrella sampling. MD simulations of small molecules starting from different locations of the tunnel gave us more hints about the key residues along the tunnel. Finally, enzymatic reactions and associated computation techniques were discussed. Both the method requiring predefined RC/CV (umbrella sampling and metadynamics) and the RC/CV-independent method SITS/ESoRT were introduced.

The interactions between small molecules and proteins can be explored and illuminated from different perspectives, and the understanding of which is important for the advancement of basic science and drug development. Computational tools can help provide insight into the binding mode, transportation pathway, and reaction kinetics. Here, we only survey specifically some of the computational strategies to demonstrate the value of these approaches in answering important chemical and biological questions. The application of various methods allowed a faster sampling of the protein and small molecular conformations and their interactions, but in many cases with the loss of real dynamics. It remains interesting to explore the relation between the protein structure fluctuation/conformational change and ligand binding/reaction. The coupling of protein dynamics with these more local events, eg, binding through formation of ligand–protein hydrogen bonds and breaking/making chemical bonds of the ligand, could be very important when the efficient conversion of chemical energy to mechanical work is needed (such as that happens in the hydrolysis of ATP and protein motors). In these cases, long range structure changes including allosteric effects may exist and direct investigation of the coupling between protein structure change and enzymatic reactions becomes necessary, although challenging. Readers should be also aware that there are many other useful methods (such as docking, MMPB/GBSA, etc.) to study interactions of small molecules and proteins that are not covered in this chapter.

REFERENCES

Amaro, R., & Luthey-Schulten, Z. (2004). Molecular dynamics simulations of substrate channeling through an alpha-beta barrel protein. *Chemical Physics*, *307*(2-3), 147–155. http://dx.doi.org/10.1016/j.chemphys.2004.05.019.

Amaro, R. E., Myers, R. S., Davisson, V. J., & Luthey-Schulten, Z. A. (2005). Structural elements in IGP synthase exclude water to optimize ammonia transfer. *Biophysical Journal*, *89*(1), 475–487. http://dx.doi.org/10.1529/biophysj.104.058651.

Andersen, H. C. (1983). Rattle—A velocity version of the shake algorithm for molecular-dynamics calculations. *Journal of Computational Physics*, *52*(1), 24–34.

Anderson, P. M. (1977). Binding of allosteric effectors to carbamyl-phosphate synthetase from Escherichia-coli. *Biochemistry*, *16*(4), 587–593. http://dx.doi.org/10.1021/Bi00623a005.

Anderson, P. M., & Meister, A. (1966). Control of Escherichia coli carbamyl phosphate synthetase by purine and pyrimidine nucleotides. *Biochemistry*, *5*(10), 3164–3169. http://dx.doi.org/10.1021/Bi00874a013.

Barends, T. R. M., Dunn, M. F., & Schlichting, I. (2008). Tryptophan synthase, an allosteric molecular factory. *Current Opinion in Chemical Biology*, *12*(5), 593–600. http://dx.doi.org/10.1016/j.cbpa.2008.07.011.

Bartels, C., & Karplus, M. (1998). Probability distributions for complex systems: Adaptive umbrella sampling of the potential energy. *Journal of Physical Chemistry B*, *102*(5), 865–880.
Berg, B. A., & Neuhaus, T. (1991). Multicanonical algorithms for 1st order phase-transitions. *Physics Letters B*, *267*(2), 249–253.
Bolhuis, P. G., Chandler, D., Dellago, C., & Geissler, P. L. (2002). Transition path sampling: Throwing ropes over rough mountain passes, in the dark. *Annual Review of Physical Chemistry*, *53*(1), 291–318.
Bussi, G., Laio, A., & Parrinello, M. (2006). Equilibrium free energies from nonequilibrium metadynamics. *Physical Review Letters*, *96*(9), 090601.
Chanda, A., & Fokin, V. V. (2009). Organic synthesis "on water" *Chemical Reviews*, *109*(2), 725–748.
Chattopadhyaya, R., Meador, W. E., Means, A. R., & Quiocho, F. A. (1992). Calmodulin structure refined at 1.7 Angstrom resolution. *Journal of Molecular Biology*, *228*(4), 1177–1192. http://dx.doi.org/10.1016/0022-2836(92)90324-D.
Chen, I. J., & Foloppe, N. (2011). Is conformational sampling of drug-like molecules a solved problem? *Drug Development Research*, *72*(1), 85–94. http://dx.doi.org/10.1002/ddr.20405.
Coates, R. M., Rogers, B. D., Hobbs, S. J., Curran, D. P., & Peck, D. R. (1987). Synthesis and Claisen rearrangement of alkoxyallyl enol ethers. Evidence for a dipolar transition state. *Journal of the American Chemical Society*, *109*(4), 1160–1170.
Darve, E., & Pohorille, A. (2001). Calculating free energies using average force. *Journal of Chemical Physics*, *115*(20), 9169–9183.
Douangamath, A., Walker, M., Beismann-Driemeyer, S., Vega-Fernandez, M. C., Sterner, R., & Wilmanns, M. (2002). Structural evidence for ammonia tunneling across the (beta alpha)(8) barrel of the imidazole glycerol phosphate synthase bienzyme complex. *Structure*, *10*(2), 185–193. http://dx.doi.org/10.1016/S0969-2126(02)00702-5.
Dunn, M. F., Niks, D., Ngo, H., Barends, T. R. M., & Schlichting, I. (2008). Tryptophan synthase: The workings of a channeling nanomachine. *Trends in Biochemical Sciences*, *33*(6), 254–264. http://dx.doi.org/10.1016/j.tibs.2008.04.008.
Ensing, B., De Vivo, M., Liu, Z., Moore, P., & Klein, M. L. (2006). Metadynamics as a tool for exploring free energy landscapes of chemical reactions. *Accounts of Chemical Research*, *39*(2), 73–81.
Fan, Y. B., Lund, L., Shao, Q., Gao, Y. Q., & Raushel, F. M. (2009). A combined theoretical and experimental study of the ammonia tunnel in carbamoyl phosphate synthetase. *Journal of the American Chemical Society*, *131*(29), 10211–10219. http://dx.doi.org/10.1021/ja902557r.
Fan, Y., Lund, L., Yang, L. J., Raushel, F. M., & Gao, Y. Q. (2008). Mechanism for the transport of ammonia within carbamoyl phosphate synthetase determined by molecular dynamics Simulations. *Biochemistry*, *47*(9), 2935–2944. http://dx.doi.org/10.1021/bi701572h.
Floquet, N., Mouilleron, S., Daher, R., Maigret, B., Badet, B., & Badet-Denisot, M. A. (2007). Ammonia channeling in bacterial glucosamine-6-phosphate synthase (Glms): Molecular dynamics simulations and kinetic studies of protein mutants. *FEBS Letters*, *581*(16), 2981–2987. http://dx.doi.org/10.1016/j.febslet.2007.05.068.
Frantz, D. D., Freeman, D. L., & Doll, J. D. (1990). Reducing quasi-ergodic behavior in Monte-Carlo simulations by J-walking—Applications to atomic clusters. *Journal of Chemical Physics*, *93*(4), 2769–2784.
Fu, J., Kanne, D. B., Leong, M., Glickman, L. H., McWhirter, S. M., Lemmens, E., … Kim, Y. (2015). STING agonist formulated cancer vaccines can cure established tumors resistant to PD-1 blockade. *Science Translational Medicine*, 7(283). http://dx.doi.org/10.1126/scitranslmed.aaa4306. ARTN 283ra52.
Fu, X. B., Yang, L. J., & Gao, Y. Q. (2007). Selective sampling of transition paths. *Journal of Chemical Physics*, *127*(15), 154106.

Gao, Y. Q. (2008a). An integrate-over-temperature approach for enhanced sampling. *Journal of Chemical Physics*, *128*(6), 064105.
Gao, Y. Q. (2008b). Self-adaptive enhanced sampling in the energy and trajectory spaces: Accelerated thermodynamics and kinetic calculations. *Journal of Chemical Physics*, *128*(13), 134111.
Gao, Y. Q., & Yang, L. J. (2006). On the enhanced sampling over energy barriers in molecular dynamics simulations. *Journal of Chemical Physics*, *125*(11), 114103.
Gao, Y. Q., Yang, L. J., Fan, Y. B., & Shao, Q. (2008). Thermodynamics and kinetics simulations of multi-time-scale processes for complex systems. *International Reviews in Physical Chemistry*, *27*(2), 201–227.
Gaus, M., Cui, Q., & Elstner, M. (2014). Density functional tight binding: Application to organic and biological molecules. *Wiley Interdisciplinary Reviews: Computational Molecular Science*, *4*(1), 49–61.
Grubmuller, H. (1995). Predicting slow structural transitions in macromolecular systems—Conformational flooding. *Physical Review E*, *52*(3), 2893–2906.
Guimarães, C. R. W., Repasky, M. P., Chandrasekhar, J., Tirado-Rives, J., & Jorgensen, W. L. (2003). Contributions of conformational compression and preferential transition state stabilization to the rate enhancement by chorismate mutase. *Journal of the American Chemical Society*, *125*(23), 6892–6899.
Guvench, O., & MacKerell, A. D. (2009). Computational evaluation of protein-small molecule binding. *Current Opinion in Structural Biology*, *19*(1), 56–61. http://dx.doi.org/10.1016/j.sbi.2008.11.009.
Hamelberg, D., Mongan, J., & McCammon, J. A. (2004). Accelerated molecular dynamics: A promising and efficient simulation method for biomolecules. *Journal of Chemical Physics*, *120*(24), 11919–11929.
Hansmann, U. H. E. (1997). Parallel tempering algorithm for conformational studies of biological molecules. *Chemical Physics Letters*, *281*(1–3), 140–150.
Haslam, E. (1993). *Shikimic acid: Metabolism and metabolites*. Chichester, New York: Wiley.
Hur, S., & Bruice, T. C. (2003). The near attack conformation approach to the study of the chorismate to prephenate reaction. *Proceedings of the National Academy of Sciences of the United States of America*, *100*(21), 12015–12020. http://dx.doi.org/10.1073/pnas.1534873100.
Hyde, C. C., Ahmed, S. A., Padlan, E. A., Miles, E. W., & Davies, D. R. (1988). 3-Dimensional structure of the tryptophan synthase alpha-2-beta-2 multienzyme complex from Salmonella-Typhimurium. *Journal of Biological Chemistry*, *263*(33), 17857–17871.
Izaguirre, J. A., Reich, S., & Skeel, R. D. (1999). Longer time steps for molecular dynamics. *Journal of Chemical Physics*, *110*(20), 9853–9864.
Jorgensen, W. L. (2004). The many roles of computation in drug discovery. *Science*, *303*(5665), 1813–1818. http://dx.doi.org/10.1126/Science.1096361.
Kästner, J. (2011). Umbrella sampling. *Wiley Interdisciplinary Reviews: Computational Molecular Science*, *1*(6), 932–942.
Kim, J., & Raushel, F. M. (2004). Access to the carbamate tunnel of carbamoyl phosphate synthetase. *Archives of Biochemistry and Biophysics*, *425*(1), 33–41. http://dx.doi.org/10.1016/j.abb.2004.02.031.
Kim, J., Straub, J. E., & Keyes, T. (2006). Statistical-temperature Monte Carlo and molecular dynamics algorithms. *Physical Review Letters*, *97*(5), 050601.
Krahn, J. M., Kim, J. H., Burns, M. R., Parry, R. J., Zalkin, H., & Smith, J. L. (1997). Coupled formation of an amidotransferase interdomain ammonia channel and a phosphoribosyltransferase active site. *Biochemistry*, *36*(37), 11061–11068. doi: 10.1021/Bi9714114.
Kumar, S., Bouzida, D., Swendsen, R. H., Kollman, P. A., & Rosenberg, J. M. (1992). The weighted histogram analysis method for free-energy calculations on biomolecules. 1. The method. *Journal of Computational Chemistry*, *13*(8), 1011–1021.

Kumar, S., Rosenberg, J. M., Bouzida, D., Swendsen, R. H., & Kollman, P. A. (1995). Multidimensional free-energy calculations using the weighted histogram analysis method. *Journal of Computational Chemistry*, *16*(11), 1339–1350. http://dx.doi.org/10.1002/Jcc.540161104.

Laio, A., & Parrinello, M. (2002). Escaping free-energy minima. *Proceedings of the National Academy of Sciences*, *99*(20), 12562–12566. http://dx.doi.org/10.1073/pnas.202427399.

LaRonde-LeBlanc, N., Resto, M., & Gerratana, B. (2009). Regulation of active site coupling in glutamine-dependent NAD(+) synthetase. *Nature Structural & Molecular Biology*, *16*(4), 421–429. http://dx.doi.org/10.1038/nsmb.1567.

Larsen, T. M., Boehlein, S. K., Schuster, S. M., Richards, N. G. J., Thoden, J. B., Holden, H. M., & Rayment, I. (1999). Three-dimensional structure of Escherichia coli asparagine synthetase B: A short journey from substrate to product. *Biochemistry*, *38*(49), 16146–16157. http://dx.doi.org/10.1021/Bi9915768.

Lee, J., Scheraga, H. A., & Rackovsky, S. (1997). New optimization method for conformational energy calculations on polypeptides: Conformational space annealing. *Journal of Computational Chemistry*, *18*(9), 1222–1232.

Lee, A., Stewart, J. D., Clardy, J., & Ganem, B. (1995). New insight into the catalytic mechanism of chorismate mutases from structural studies. *Chemistry & Biology*, *2*(4), 195–203.

Lin, Y. C., Cao, Z. X., & Mo, Y. R. (2006). Molecular dynamics simulations on the Escherichia coli ammonia channel protein AmtB: Mechanism of ammonia/ammonium transport. *Journal of the American Chemical Society*, *128*(33), 10876–10884. http://dx.doi.org/10.1021/ja0631549.

Lin, Z. X., & van Gunsteren, W. F. (2013). Enhanced conformational sampling using enveloping distribution sampling. *Journal of Chemical Physics*, *139*(14), 144105–144107.

Lund, L., Fan, Y. B., Shao, Q., Gao, Y. Q., & Raushel, F. M. (2010). Carbamate transport in carbamoyl phosphate synthetase: A theoretical and experimental investigation. *Journal of the American Chemical Society*, *132*(11), 3870–3878. http://dx.doi.org/10.1021/ja910441v.

Manjasetty, B. A., Powlowski, J., & Vrielink, A. (2003). Crvstal structure of a bifunctional aldolase-dehydrogenase: Sequestering a reactive and volatile intermediate. *Proceedings of the National Academy of Sciences of the United States of America*, *100*(12), 6992–6997. http://dx.doi.org/10.1073/pnas.1236794100.

Marinari, E., & Parisi, G. (1992). Simulated tempering—A new Monte-Carlo scheme. *Europhysics Letters*, *19*(6), 451–458.

Matthews, S. L., & Anderson, P. M. (1972). Evidence for presence of 2 nonidentical subunits in carbamyl phosphate synthetase of Escherichia-coli. *Biochemistry*, *11*(7), 1176–1183. http://dx.doi.org/10.1021/Bi00757a010.

Maynard, E. L., & Lindahl, P. A. (2001). Catalytic coupling of the active sites in Acetyl–CoA synthase, a bifunctional CO-channeling enzyme. *Biochemistry*, *40*(44), 13262–13267. http://dx.doi.org/10.1021/Bi015604+.

Minary, P., Tuckerman, M. E., & Martyna, G. J. (2004). Long time molecular dynamics for enhanced conformational sampling in biomolecular systems. *Physical Review Letters*, *93*(15). http://dx.doi.org/10.1103/Physrevlett.93.150201. Artn 150201.

Moroni, D., Bolhuis, P. G., & van Erp, T. S. (2004). Rate constants for diffusive processes by partial path sampling. *The Journal of Chemical Physics*, *120*(9), 4055–4065.

Mouilleron, S., Badet-Denisot, M. A., & Golinelli-Pimpaneau, B. (2006). Glutamine binding opens the ammonia channel and activates glucosamine-6P synthase. *Journal of Biological Chemistry*, *281*(7), 4404–4412. http://dx.doi.org/10.1074/jbc.M511689200.

Perola, E., & Charifson, P. S. (2004). Conformational analysis of drug-like molecules bound to proteins: An extensive study of ligand reorganization upon binding. *Journal of Medicinal Chemistry*, *47*(10), 2499–2510. http://dx.doi.org/10.1021/jm030563w.

Piela, L., Kostrowicki, J., & Scheraga, H. A. (1989). The multiple-minima problem in the conformational-analysis of molecules—Deformation of the potential-energy hypersurface by the diffusion equation method. *Journal of Physical Chemistry, 93*(8), 3339–3346.

Pierard, A., & Wiame, J. M. (1964). Regulation + mutation affecting glutamine dependent formation of carbamyl phosphate in Escherichia coli. *Biochemical and Biophysical Research Communications, 15*(1), 76–81. http://dx.doi.org/10.1016/0006-291x(64)90106-8.

Raushel, F. M., Thoden, J. B., & Holden, H. M. (2003). Enzymes with molecular tunnels. *Accounts of Chemical Research, 36*(7), 539–548. http://dx.doi.org/10.1021/ar020047k.

Ryckaert, J. P., Ciccotti, G., & Berendsen, H. J. C. (1977). Numerical-integration of Cartesian equations of motion of a system with constraints—Molecular-dynamics of N-alkanes. *Journal of Computational Physics, 23*(3), 327–341.

Schmitt, E., Panvert, M., Blanquet, S., & Mechulam, Y. (2005). Structural basis for tRNA-dependent amidotransferase function. *Structure, 13*(10), 1421–1433. http://dx.doi.org/10.1016/j.str.2005.06.016.

Stockwell, G. R., & Thornton, J. M. (2006). Conformational diversity of ligands bound to proteins. *Journal of Molecular Biology, 356*(4), 928–944. http://dx.doi.org/10.1016/j.jmb.2005.12.012.

Sugita, Y., & Okamoto, Y. (1999). Replica-exchange molecular dynamics method for protein folding. *Chemical Physics Letters, 314*(1–2), 141–151.

Swindoll, R. D., & Haile, J. M. (1984). A multiple time-step method for molecular-dynamics simulations of fluids of chain molecules. *Journal of Computational Physics, 53*(2), 289–298.

Tan, X. S., Loke, H. K., Fitch, S., & Lindahl, P. A. (2005). The tunnel of acetyl-coenzyme a synthase/carbon monoxide dehydrogenase regulates delivery of CO to the active site. *Journal of the American Chemical Society, 127*(16), 5833–5839. http://dx.doi.org/10.1021/ja043701v.

Tan, X. S., Volbeda, A., Fontecilla-Camps, J. C., & Lindahl, P. A. (2006). Function of the tunnel in acetylcoenzyme A synthase/carbon monoxide dehydrogenase. *Journal of Biological Inorganic Chemistry, 11*(3), 371–378. http://dx.doi.org/10.1007/s00775-006-0086-9.

Teplyakov, A., Obmolova, G., Badet, B., & Badet-Denisot, M. A. (2001). Channeling of ammonia in glucosamine-6-phosphate synthase. *Journal of Molecular Biology, 313*(5), 1093–1102. http://dx.doi.org/10.1006/Jmbi.2001.5094.

Thoden, J. B., Holden, H. M., Wesenberg, G., Raushel, F. M., & Rayment, I. (1997). Structure of carbamoyl phosphate synthetase: A journey of 96 angstrom from substrate to product. *Biochemistry, 36*(21), 6305–6316. http://dx.doi.org/10.1021/Bi970503q.

Torrie, G. M., & Valleau, J. P. (1977). Non-physical sampling distributions in Monte-Carlo free-energy estimation—Umbrella sampling. *Journal of Computational Physics, 23*(2), 187–199.

Tuckerman, M., Berne, B. J., & Martyna, G. J. (1992). Reversible multiple time scale molecular-dynamics. *Journal of Chemical Physics, 97*(3), 1990–2001. http://dx.doi.org/10.1063/1.463137.

Tuckerman, M. E., Martyna, G. J., & Berne, B. J. (1990). Molecular-dynamics algorithm for condensed systems with multiple time scales. *Journal of Chemical Physics, 93*(2), 1287–1291.

van den Heuvel, R. H. H., Ferrari, D., Bossi, R. T., Ravasio, S., Curti, B., Vanoni, M. A., ... Mattevi, A. (2002). Structural studies on the synchronization of catalytic centers in glutamate synthase. *Journal of Biological Chemistry, 277*(27), 24579–24583. http://dx.doi.org/10.1074/jbc.M202541200.

Voter, A. F. (1997). Hyperdynamics: Accelerated molecular dynamics of infrequent events. *Physical Review Letters, 78*(20), 3908–3911.

Wang, F. G., & Landau, D. P. (2001). Efficient, multiple-range random walk algorithm to calculate the density of states. *Physical Review Letters*, *86*(10), 2050–2053.
Wojcik, M., Seidle, H. F., Bieganowski, P., & Brenner, C. (2006). Glutamine-dependent NAD (+) synthetase—How a two-domain, three-substrate enzyme avoids waste. *Journal of Biological Chemistry*, *281*(44), 33395–33402. http://dx.doi.org/10.1074/jbc.M607111200.
Wu, J. X., Sun, L. J., Chen, X., Du, F. H., Shi, H. P., Chen, C., & Chen, Z. J. J. (2013). Cyclic GMP-AMP is an endogenous second messenger in innate immune signaling by cytosolic DNA. *Science*, *339*(6121), 826–830. http://dx.doi.org/10.1126/science.1229963.
Wu, X. W., & Wang, S. M. (1998). Self-guided molecular dynamics simulation for efficient conformational search. *Journal of Physical Chemistry B*, *102*(37), 7238–7250. http://dx.doi.org/10.1021/Jp9817372.
Wu, X. W., & Wang, S. M. (1999). Enhancing systematic motion in molecular dynamics simulation. *Journal of Chemical Physics*, *110*(19), 9401–9410.
Yang, L. J., & Gao, Y. Q. (2007). An approximate method in using molecular mechanics simulations to study slow protein conformational changes. *Journal of Physical Chemistry B*, *111*(11), 2969–2975.
Yang, L. J., & Gao, Y. Q. (2009). A selective integrated tempering method. *Journal of Chemical Physics*, *131*(21). http://dx.doi.org/10.1063/1.3266563. Artn 214109.
Yang, L. J., Liu, C. W., Shao, Q., Zhang, J., & Gao, Y. Q. (2015). From thermodynamics to kinetics: Enhanced sampling of rare events. *Accounts of Chemical Research*, *48*(4), 947–955.
Yang, L. J., Shao, Q., & Gao, Y. Q. (2009). Comparison between integrated and parallel tempering methods in enhanced sampling simulations. *Journal of Chemical Physics*, *130*(12), 124111.
Yang, H. Y., Xu, Y. C., Zhu, W. L., Chen, K. X., & Jiang, H. L. (2007). Detailed mechanism for AmtB conducting NH4+/NH3: Molecular dynamics simulations. *Biophysical Journal*, *92*(3), 877–885. http://dx.doi.org/10.1529/biophysj.106.090191.
Yeo, J., Goodman, R. A., Schirle, N. T., David, S. S., & Beal, P. A. (2010). RNA editing changes the lesion specificity for the DNA repair enzyme NEIL1. *Proceedings of the National Academy of Sciences*, *107*(48), 20715–20719.
Zhang, J., Yang, Y. I., Yang, L., & Gao, Y. Q. (2015a). Conformational pre-adjustment in aqueous Claisen rearrangement revealed by SITS-QM/MM MD simulations. *The Journal of Physical Chemistry. B*, *119*(17), 5518–5530. http://dx.doi.org/10.1021/jp511057f.
Zhang, J., Yang, Y. I., Yang, L., & Gao, Y. Q. (2015b). Dynamics and kinetics study of "In-Water" chemical reactions by enhanced sampling of reactive trajectories. *The Journal of Physical Chemistry. B*, *119*(45), 14505–14514.

CHAPTER NINE

How to Run FAST Simulations

M.I. Zimmerman*, G.R. Bowman*,†,1

*Washington University School of Medicine, St. Louis, MO, United States
†Center for Biological Systems Engineering, Washington University School of Medicine, St. Louis, MO, United States
[1]Corresponding author: e-mail address: bowman@biochem.wustl.edu

Contents

Abstract

Molecular dynamics (MD) simulations are a powerful tool for understanding enzymes' structures and functions with full atomistic detail. These physics-based simulations model the dynamics of a protein in solution and store snapshots of its atomic coordinates at discrete time intervals. Analysis of the snapshots from these trajectories provides thermodynamic and kinetic properties such as conformational free energies, binding free energies, and transition times. Unfortunately, simulating biologically relevant timescales with brute force MD simulations requires enormous computing resources. In this chapter we detail a goal-oriented sampling algorithm, called fluctuation amplification of specific traits, that quickly generates pertinent thermodynamic and kinetic information by using an iterative series of short MD simulations to explore the vast depths of conformational space.

1. INTRODUCTION

One of the largest challenges in using molecular dynamics (MD) simulations to study enzymes is achieving adequate sampling to accurately

Methods in Enzymology, Volume 578
ISSN 0076-6879
http://dx.doi.org/10.1016/bs.mie.2016.05.032

represent its equilibrium structural ensemble and conformational transitions (Dror, Dirks, Grossman, Xu, & Shaw, 2012; Zwier & Chong, 2010). In other words, the conformational space of a protein is extraordinarily large and transitions between two given conformations may be separated by numerous kinetically slow steps that require a great deal of simulation-time to observe. To put this into perspective, many enzymatic reactions/conformational transitions occur on the millisecond–second timescale, but a typical desktop computer may only be able to simulate a few nanoseconds of dynamics per day. Therefore, it could take a desktop computer hundreds to millions of years to simulate a particular event.

One approach to overcome the limitations of MD simulations is to build specialized supercomputers. For example, the development of powerful purpose-built hardware for MD simulations, such as the ANTON supercomputer, allows for much longer timescale simulations (Shaw et al., 2008). However, this approach is typically too expensive for the average researcher.

An alternative approach is to run many short timescale simulations on different computers. A single, long simulation will eventually generate multiple independent samples of rare events. Running multiple simulations can capture the same independent events in parallel. Running simulations in parallel maximizes the use of commodity hardware since obtaining larger aggregate simulation times can be easily achieved through the addition of processors rather than increasing a processor's speed. For these reasons, massively parallelized distributed computing projects, such as Folding@home, have been very successful in using MD to capture long timescale conformational transitions of proteins such as folding and allostery (Bowman, Bolin, Hart, Maguire, & Marqusee, 2015; Bowman & Geissler, 2012; Bowman & Pande, 2010; Pande et al., 2003; Plattner & Noé, 2015; Shirts & Pande, 2000; Voelz, Bowman, Beauchamp, & Pande, 2010).

Markov state models (MSMs) provide an elegant framework for analyzing protein simulation data, whether it is generated by a single simulation or many of them (Bowman & Pande, 2014; Chodera & Noé, 2014; Pande, Beauchamp, & Bowman, 2010). An MSM is essentially a map of the different conformations a protein adopts. The basic construction of an MSM consists of the following steps: (1) cluster all of the simulation data into discrete "microstates," for example, with a k-centers clustering algorithm based on the protein backbone, (2) generate a transition count matrix, an $N \times N$ matrix of all the observed transitions from microstate i to j for a specified lag time (ie, observation interval), and (3) generate a transition probability matrix, an $N \times N$ matrix created from the transition count matrix detailing the probability of transitioning from state i to state j. The transition probability matrix contains a wealth of information. For example, the first eigenvector of this matrix specifies the equilibrium probabilities of all the states. Other eigenvalues and eigenvectors specify the rates of transitioning between different sets of states and which states are involved. One powerful attribute of MSMs is that it is equally valid to build them from a single long simulation or a set of simulations, even if these simulations are not initiated from a Boltzmann distribution. Furthermore, freely available software packages, such as MSMBuilder and pyEMMA, provide a readily accessible means to construct and analyze these models (Beauchamp et al., 2011; Bowman, Huang, & Pande, 2009; Scherer et al., 2015).

MSMs' ability to extract the equilibrium thermodynamics and kinetics of a system irrespective of the distribution of starting conformations for a set of simulations opens the doors to interactively sample desired regions of conformational space. For example, adaptive sampling algorithms iteratively run a set of simulations, build an MSM, and then select starting points for new simulations that will help to reduce statistical uncertainty in the model (Adhikari, Freed, & Sosnick, 2013; Bacci, Vitalis, & Caflisch, 2015; Bowman, Ensign, & Pande, 2010; Doerr & De Fabritiis, 2014; Hinrichs & Pande, 2007; Voelz, Elman, Razavi, & Zhou, 2014; Weber & Pande, 2011). Various adaptive sampling schemes have been developed to enhance and automate the construction of MSMs. For example, Hinrichs and Pande have developed an adaptive sampling scheme that spawns new simulations from the states that contribute most to the statistical uncertainty in an MSM's principle eigenvectors and eigenvalues (Hinrichs & Pande, 2007). Other methods spawn simulations from states based on the number of times they have been observed or the number of neighbors they are connected to in order to discover new states more quickly (Weber & Pande, 2011). These methods will generally explore conformational space more efficiently than brute force simulations. However, they will not necessarily sample specific events of interest to a researcher before thoroughly exploring other, less relevant regions of conformational space.

We have developed a goal-oriented sampling algorithm, called fluctuation amplification of specific traits (FAST), which draws inspiration from adaptive sampling and the multiarmed bandit problem to efficiently identify structures with a desired physical property (Zimmerman & Bowman, 2015). For example, FAST can be used to identify the preferred pathways between active and inactive states of an enzyme or it can be used to identify potentially druggable pockets that are not apparent from existing crystal structures. The FAST algorithm achieves this by balancing between exploiting promising structures (ie, searching around promising solutions for even better ones) and broad exploration (ie, searching unexplored regions of conformational space for entirely new solutions). The following sections of this chapter provide details on the algorithm and the parameters relevant to setting up FAST simulations.

2. FAST ALGORITHM

The FAST algorithm can be used to find structures that optimize any geometric function (ϕ) of protein conformations. At the heart of the FAST-ϕ algorithm is the reward function used to decide which states to simulate for future runs of sampling. The FAST-ϕ reward function is modeled after a simple solution to the multiarmed bandit problem,

$$r_{\phi}(i) = \overline{\phi}(i) + \alpha\overline{\psi}(i)$$

where the reward (r_{ϕ}) for state i is the sum of a directed component, $\overline{\phi}(i)$, and an undirected component, $\overline{\psi}(i)$, with scaling parameter α. The set of directed components corresponds to a feature-scaled list of traits that one wishes to exploit (such as the RMSD to a target structure) and the set of undirected components corresponds to a feature-scaled list of some statistical function that facilitates state-space exploration (such as the number of observations per state). Feature scaling transforms a list of values to range from 0 to 1. Directed and undirected components to the FAST ranking can be either positively feature-scaled to favor large values

$$\overline{\phi}(i) = \frac{\phi(i) - \phi_{\min}}{\phi_{\max} - \phi_{\min}}$$

or negatively feature-scaled to favor small values

$$\overline{\phi}(i) = \frac{\phi_{\max} - \phi(i)}{\phi_{\max} - \phi_{\min}}$$

The variables $\phi_{\min}$ and $\phi_{\max}$ are the minimum and maximum values of $\phi(i)$ observed in a simulation dataset, respectively.

Although the reward function may change for the specific type of FAST-ϕ sampling, the basic algorithm remains the same:

(1) Start a swarm of N simulations from a structure or set of structures, such as one or more known crystal structures,

(2) Cluster all the simulation data collected so far into discrete conformational states. This can be accomplished by using a k-centers algorithm on the RMSD between select protein atoms (such as backbone heavy-atoms), with a specified distance cutoff. The distance cutoff will specify the maximum distance between structures in a cluster to the cluster center and will dictate the total number of clusters generated.

(3) Rank all of the states discovered using the FAST-ϕ reward function.

(4) Start a new swarm of simulations from the top N structures that maximize the FAST-ϕ reward function.

(5) Repeat steps 2–4 until some convergence criterion is met or a predetermined amount of simulation has been conducted.

(6) Build an MSM from the final data set to capture the proper thermodynamics and kinetics, thereby correcting for any bias introduced by selecting starting conformations from each swarm of simulations according to our reward function instead of a Boltzmann distribution.

As mentioned, the directed and undirected components to the ranking can vary depending on the specific problem at hand. Specific traits for the directed component of FAST sampling will be discussed in a later section. In early applications of the FAST-ϕ reward function, the undirected component was chosen to be the negatively feature-scaled number of observations of each state

$$\overline{\psi}(i) = \frac{C_{\max} - C(i)}{C_{\max} - C_{\min}}$$

where $C_{\min}$ and $C_{\max}$ are the minimum and maximum number of observations of any state, respectively. This version of the undirected component was selected based on a simple Bayesian model that suggests it should maximize the discovery of new states. One could also use alternative statistical measures, such as existing adaptive sampling methods, in place of our counts-based, undirected component.

3. FAST SAMPLING PARAMETERS

The FAST algorithm contains many parameters, in the form of input and output, which can be reduced to those that are relevant for running MD simulations, building MSMs, or propagating a run of goal-oriented sampling. With the large number of parameters required for running FAST, it can be a daunting task to set up expensive simulations without a good feel for reasonable values. In this section, we will detail some of the main parameters that are used in FAST simulations, how to determine reasonable values, and how they may interact with one another. For the sake of brevity and clarity, parameters relevant to running individual MD simulations will not be discussed; there are many software packages that can perform these simulations that provide extensive tutorials and user manuals, such as Amber, CHARMM, Gromacs, and NAMD (Brooks et al., 2009; Case et al., 2005; Phillips et al., 2005; Van Der Spoel et al., 2005).

3.1 Number of Runs

Typically one will run FAST sampling until some convergence criterion is achieved. In some circumstances this is very straightforward, although a convergence criterion is not always easy to deduce. Running FAST simulations from a starting structure to a specified target (eg, using FAST-RMSD between known conformations) will produce a simple convergence criterion since there is a single end state; simulations can be terminated once the end state is discovered. On the other hand, there may not be obvious criteria for terminating a set of FAST simulations for more open-ended problems, such as searching for conformations with large solvent accessible surface areas (SASAs) using FAST-SASA as a heuristic for discovering unknown druggable pockets. In the case of FAST-SASA, one might want to stop simulations when the solvent-accessible surface area ceases to increase as rounds continue, but we have shown that in this scenario the undirected component to the FAST-SASA reward function will aid in the discovery of multiple pathways to large SASA states, which is desirable because a diversity of potential druggable sites can be discovered. In practice, it is convenient to run simulations for a specified number of runs and continue the runs if sampling is deemed insufficient, since the algorithm is easy to restart from a previous run or preexisting set of data.

3.2 The α Scaling Parameter

The scaling parameter, α, is used in the FAST-ϕ reward function to weight the relative importance of exploiting physical traits and increasing state exploration. Large α values will increase the exploration of state space by favoring states that have not been observed as frequently, whereas smaller values will place more emphasis on exploiting structures with promising traits. Emphasizing the trait-based component of the reward function will increase the likelihood FAST tries to hop over larger energy barriers rather than try new solutions. Through the analysis of synthetic trajectories generated with existing MSMs, we have seen that sampling results are largely insensitive to values of α between 0.5 and 1.5. However, it is possible that the energy landscape of the protein being simulated, as well as the gradient of conformational space that one is attempting to follow, may change this observation. If traits are very monotonically increasing/decreasing, it is expected that smaller values of α will optimize FAST-ϕ's performance, whereas if traits require significant backtracking, larger values of α will optimize FAST-ϕ's performance. For these reasons, unless one has special insight into the nature of a particular protein's energy landscape, an $\alpha = 1$ is a safe choice.

3.3 Number of Simulations per Run

The number of simulations to perform during each run of FAST sampling is an important decision to maximize computational resources, ensure a good swath of conformations, and accelerate the observation of rare-events. A main advantage to running simulations in parallel over generating a single trajectory is that many parallel jobs on multiple processors can be efficiently used to generate sizeable datasets; thus, the biggest factor in selecting the number of simulations per run is attempting to generate the largest aggregate simulation time with the resources available. Additionally, more simulations per run allow for a greater spread of starting states that will identify a diversity of potential pathways to explore, which will better allow for the circumvention of dead-ends. Despite this improvement, the number of simulations should be balanced with the individual simulation lengths so that there is a reasonable amount of aggregate simulation time per run; having too much aggregate simulation per run means that there will be less total runs and less FAST enhancement. As an example, if one wishes to observe a process that takes 1 μs of simulation to observe, using 40 simulations per run with 10 ns lengths would generate 400 ns of aggregate simulation per run, meaning that after three runs the aggregate simulation is much larger than the expected mean first passage time. Alternatively, 10 simulations per run of 10 ns lengths would take 10 runs to total 1 μs, which will provide more FAST enhancement by offering extra chances to adaptively explore conformational space.

3.4 Simulation Length

Individual simulations must be longer than the Markov time for the final MSM. If simulations are shorter than this timescale, then the final model will violate the Markov assumption and be of little utility. However, one also wants simulations to be as short as possible to maximize the number of different runs that can be performed and to prevent simulations from wandering far from the region of conformational space one hopes to explore. In practice, we have often found that simulation lengths between 10 and 20 ns satisfy these constraints, in large part because models with a Markov time much greater than this would often be insufficient for the applications we have pursued.

3.5 Atom Indices Used for Clustering

The atomic indices that are used to cluster simulation data with a specified method into discrete microstates are the core of how states are defined. When clustering simulation data in-between runs of FAST sampling, it is important to recognize that the structures within a state are similar only in the atomic indices specified for clustering. Usually it is beneficial to cluster conformations in a holistic fashion, based on the backbone heavy atoms (C_α, C_β, CO, N, and O), so that different clusters represent global changes in a protein's conformation. However, situations arise where one is interested in an aspect of a protein structure, and that using the entire protein backbone for clustering would include unnecessary detail that drowns out the relevant structural motions. For example, we used FAST-RMSD to study the transition between apo and holo conformations of TEM-1 β-lactamase to understand how a surprising cryptic pocket opens up. These conformations have a global RMSD of ~0.26 Å, so an extremely high-resolution model would be required if the data were clustered based on a global RMSD (Horn & Shoichet, 2004; Jelsch, Mourey, Masson, & Samama, 1993). To avoid an unnecessarily large number of clusters, we chose to instead cluster the data based on just the atoms of the two helices surrounding the pocket we were interested in.

3.6 Resolution of Clustering

One must balance between having enough clusters to resolve valuable differences but not so many as to make the statistical component of the reward function ineffective. For example, we have previously used a *k*-centers clustering algorithm that continues to divide conformational space into smaller groups until the maximum distance from any structure to its cluster center is less than a predetermined distance cutoff. This distance cutoff controls the level of structural similarity within and between clusters as well as the total number of clusters created during each run of sampling. Large distance cutoffs will generate fewer clusters with many structures per cluster, whereas a small distance cutoff will generate numerous clusters with few structures per cluster. A good distance-cutoff value will be small enough that a structure pulled from a cluster will be an accurate representation of other structures in that cluster, but also large enough that the number of observations of each state reflects the sampling for that region in conformational space. If a distance cutoff is particularly small, there may be many clusters with only a single conformation, which the FAST undirected component will rank extraordinarily highly. This is not desirable if it is because extremely similar states are falsely considered separate. On the other hand, if the clusters are too coarse, then one may miss a valuable region of conformational space that FAST's reward function would otherwise discover.

4. APPLICATIONS

FAST-ϕ can be tailored to provide pertinent thermodynamic and kinetic information for any region in conformational space that can be identified with a calculable order parameter. The central hypothesis of FAST is that gradients exist in conformational space with respect to specific traits and that they can be exploited through sampling states with large or small values of some trait that one wishes to maximize or minimize, respectively. Although there are limitless possibilities for the directed component to the FAST-ϕ ranking, we will discuss a few that have been used to study enzyme function and structural ensembles.

FAST-SASA aims to uncover structural states of enzymes with large SASA under the assumption that a large-SASA state will likely have large pocket openings that can be used to discover or design novel therapeutics. While enzymatic function is generally critical for cellular and biological processes, enzymatic reactions can be detrimental to human health. As an example, the enzyme TEM-1 β-lactamase is produced in certain bacteria as a means of hydrolyzing β-lactam antibiotics to confer antibiotic resistance (Bush, 2013). Antibiotic resistant bacteria are swiftly becoming a global health concern due to the overuse of antibiotic treatments, so it is desired to find molecular ways to inhibit the antibiotic resistant nature of β-lactamase (Laxminarayan et al., 2013). If complete atomistic structures exist where the enzyme has a large pocket opening, computational docking of small molecules to this region can aid in the discovery of ligands that will inhibit its function (Jones, Willett, Glen, Leach, & Taylor, 1997). While the crystal structure of β-lactamase has a single large pocket (its active site), there is little diversity in locations to dock small molecules against; multiple pocket openings will increase the likelihood of successful docking. Fortunately, proteins are not static and pockets will emerge during the course of an MD simulation (Bowman et al., 2015). FAST-SASA will accelerate the observation of large pocket openings by favoring states with an already large SASA, as we have previously shown for the enzyme β-lactamase (Bowman & Geissler, 2012; Zimmerman & Bowman, 2015). We foresee that FAST-SASA, or related FAST-ϕ algorithms that more quantitatively detail global or specific pocket volumes, will be an invaluable tool for discovering druggable sites on enzymes that do not display obvious pockets in crystal structures.

FAST-RMSD is intended to reveal the equilibrium conformational transition pathway between two known enzyme structures with accurate thermodynamics and kinetics. Enzymes are dynamic proteins that often transition between many conformations that are relevant to their biological function (Benkovic & Hammes-Schiffer, 2003; Elber & Karplus, 1987; Garcia-Viloca, Gao, Karplus, & Truhlar, 2004). Knowledge of their transition pathway, along with their kinetic rates and conformational free energies, can provide significant insight into their mechanisms of action and intrinsic regulation. It is often the case that structural studies of enzymes will identify multiple conformational populations, although it will be unable to detail the structural intermediates between them. For example, nuclear magnetic resonance spectra may identify conformational heterogeneity through the analysis of chemical shifts, although intermediates between populations are too short-lived or have too small a population to observe (Kleckner & Foster, 2011). Additionally, crystallographic studies may detail enzyme structures in various substrate-binding conformations, but they will not reveal the relative populations of these conformations. Given two atomically detailed structures as input, a start and a target, FAST-RMSD can efficiently identify the equilibrium transition pathway between them by biasing the starting structures of simulations originally spawned from the start toward the target. As mentioned earlier, if the conformational change that one is attempting to observe takes place for a portion of the total protein, it is beneficial to define states based solely on the atom indices of that region. Additionally, all RMSDs that are used in the reward function should be confined to this region of the protein. By doing this, the global changes to the protein will not wash away the observation of (in terms of RMSD values) relevant conformational transitions.

Clever use of the FAST-ϕ reward function can also provide valuable structural information in cases where experiments suggest a conformational transition but are unable to produce an atomistic description of the relevant structures. For example, a FRET experiment could provide a low-resolution view of a conformational change that occurs in some enzyme. Without an all-atom representation of a target structure, FAST-RMSD would be unable to elucidate the nature of the conformational change. Despite this, directed components to the reward function can be deduced that will explain these data. For example, a FAST-distance algorithm can be devised to favor transitions from some known structure, say where the dyes in the FRET study would be far apart, to new structures where the dyes would be brought together:

$$\overline{\phi}(i) = \frac{d_{\max} - d(i)}{d_{\max} - d_{\min}}$$

where d is the distance between the dyes. The resulting model could then be used to help explain the origins of the experimental observation and to plan new experiments.

ACKNOWLEDGMENT

G.R.B. holds a Career Award at the Scientific Interface from the Burroughs Wellcome Fund.

REFERENCES

Adhikari, A. N., Freed, K. F., & Sosnick, T. R. (2013). Simplified protein models: Predicting folding pathways and structure using amino acid sequences. *Physical Review Letters*, *111*(2), 028103–028104. http://dx.doi.org/10.1103/PhysRevLett.111.028103.

Bacci, M., Vitalis, A., & Caflisch, A. (2015). A molecular simulation protocol to avoid sampling redundancy and discover new states. *Biochimica et Biophysica Acta*, *1850*(5), 889–902. http://dx.doi.org/10.1016/j.bbagen.2014.08.013.

Beauchamp, K. A., Bowman, G. R., Lane, T. J., Maibaum, L., Haque, I. S., & Pande, V. S. (2011). MSMBuilder2: Modeling conformational dynamics on the picosecond to millisecond scale. *Journal of Chemical Theory and Computation*, 7(10), 3412–3419. http://dx.doi.org/10.1021/ct200463m.

Benkovic, S. J., & Hammes-Schiffer, S. (2003). A perspective on enzyme catalysis. *Science*, *301*(5637), 1196–1202. http://dx.doi.org/10.1126/science.1085515.

Bowman, G. R., Bolin, E. R., Hart, K. M., Maguire, B. C., & Marqusee, S. (2015). Discovery of multiple hidden allosteric sites by combining Markov state models and experiments. *Proceedings of the National Academy of Sciences of the United States of America*, *112*(9), 2734–2739. http://dx.doi.org/10.1073/pnas.1417811112.

Bowman, G. R., Ensign, D. L., & Pande, V. S. (2010). Enhanced modeling via network theory: Adaptive sampling of Markov state models. *Journal of Chemical Theory and Computation*, *6*(3), 787–794. http://dx.doi.org/10.1021/ct900620b.

Bowman, G. R., & Geissler, P. L. (2012). Equilibrium fluctuations of a single folded protein reveal a multitude of potential cryptic allosteric sites. *Proceedings of the National Academy of Sciences of the United States of America, 109*(29), 11681–11686. http://dx.doi.org/10.1073/pnas.1209309109.

Bowman, G. R., Huang, X., & Pande, V. S. (2009). Using generalized ensemble simulations and Markov state models to identify conformational states. *Methods, 49*(2), 197–201. http://dx.doi.org/10.1016/j.ymeth.2009.04.013.

Bowman, G. R., & Pande, V. S. (2010). Protein folded states are kinetic hubs. *Proceedings of the National Academy of Sciences of the United States of America, 107*(24), 10890–10895. http://dx.doi.org/10.1073/pnas.1003962107.

Bowman, G. R., Pande, V. S., & Noé, F. (Eds.). (2014). An introduction to Markov state models and their application to long timescale molecular simulation. *Advances in Experimental Medicine and Biology* (Vol. 797). Dordrecht, The Netherlands: Springer. http://dx.doi.org/10.1007/978-94-007-7606-7.

Brooks, B. R., Brooks, C. L., III, Mackerell, A. D., Jr., Nilsson, L., Petrella, R. J., Roux, B., et al. (2009). CHARMM: The biomolecular simulation program. *Journal of Computational Chemistry, 30*(10), 1545–1614. http://dx.doi.org/10.1002/jcc.21287.

Bush, K. (2013). Proliferation and significance of clinically relevant β-lactamases. *Annals of the New York Academy of Sciences, 1277*(1), 84–90. http://dx.doi.org/10.1111/nyas.12023.

Case, D. A., Cheatham, T. E., Darden, T., Gohlke, H., Luo, R., Merz, K. M., et al. (2005). The Amber biomolecular simulation programs. *Journal of Computational Chemistry, 26*(16), 1668–1688. http://dx.doi.org/10.1002/jcc.20290.

Chodera, J. D., & Noé, F. (2014). Markov state models of biomolecular conformational dynamics. *Current Opinion in Structural Biology, 25*, 135–144. http://dx.doi.org/10.1016/j.sbi.2014.04.002.

Doerr, S., & De Fabritiis, G. (2014). On-the-fly learning and sampling of ligand binding by high-throughput molecular simulations. *Journal of Chemical Theory and Computation, 10*(5), 2064–2069. http://dx.doi.org/10.1021/ct400919u.

Dror, R. O., Dirks, R. M., Grossman, J. P., Xu, H., & Shaw, D. E. (2012). Biomolecular simulation: A computational microscope for molecular biology. *Annual Review of Biophysics, 41*(1), 429–452. http://dx.doi.org/10.1146/annurev-biophys-042910-155245.

Elber, R., & Karplus, M. (1987). Multiple conformational states of proteins: A molecular dynamics analysis of myoglobin. *Science, 235*(4786), 318–321. http://dx.doi.org/10.1126/science.3798113.

Garcia-Viloca, M., Gao, J., Karplus, M., & Truhlar, D. G. (2004). How enzymes work: Analysis by modern rate theory and computer simulations. *Science, 303*(5655), 186–195. http://dx.doi.org/10.1126/science.1088172.

Hinrichs, N. S., & Pande, V. S. (2007). Calculation of the distribution of eigenvalues and eigenvectors in Markovian state models for molecular dynamics. *The Journal of Chemical Physics, 126*(24), 244101–244112. http://dx.doi.org/10.1063/1.2740261.

Horn, J. R., & Shoichet, B. K. (2004). Allosteric inhibition through core disruption. *Journal of Molecular Biology, 336*(5), 1283–1291. http://dx.doi.org/10.1016/j.jmb.2003.12.068.

Jelsch, C., Mourey, L., Masson, J. M., & Samama, J. P. (1993). Crystal structure of Escherichia coli TEM1 beta-lactamase at 1.8 A resolution. *Proteins: Structure, Function, and Bioinformatics, 16*(4), 364–383. http://dx.doi.org/10.1002/prot.340160406.

Jones, G., Willett, P., Glen, R. C., Leach, A. R., & Taylor, R. (1997). Development and validation of a genetic algorithm for flexible docking. *Journal of Molecular Biology, 267*(3), 727–748. http://dx.doi.org/10.1006/jmbi.1996.0897.

Kleckner, I. R., & Foster, M. P. (2011). An introduction to NMR-based approaches for measuring protein dynamics. *Biochimica et Biophysica Acta, 1814*(8), 942–968. http://dx.doi.org/10.1016/j.bbapap.2010.10.012.

Laxminarayan, R., Duse, A., Wattal, C., Zaidi, A. K. M., Wertheim, H. F. L., Sumpradit, N., et al. (2013). Antibiotic resistance-the need for global solutions. *The Lancet Infectious Diseases*, *13*(12), 1057–1098. http://dx.doi.org/10.1016/S1473-3099(13)70318-9.

Pande, V. S., Baker, I., Chapman, J., Elmer, S. P., Khaliq, S., Larson, S. M., et al. (2003). Atomistic protein folding simulations on the submillisecond time scale using worldwide distributed computing. *Biopolymers*, *68*(1), 91–109. http://dx.doi.org/10.1002/bip.10219.

Pande, V. S., Beauchamp, K., & Bowman, G. R. (2010). Everything you wanted to know about Markov state models but were afraid to ask. *Methods*, *52*(1), 99–105. http://dx.doi.org/10.1016/j.ymeth.2010.06.002.

Phillips, J. C., Braun, R., Wang, W., Gumbart, J., Tajkhorshid, E., Villa, E., et al. (2005). Scalable molecular dynamics with NAMD. *Journal of Computational Chemistry*, *26*(16), 1781–1802. http://dx.doi.org/10.1002/jcc.20289.

Plattner, N., & Noé, F. (2015). Protein conformational plasticity and complex ligand-binding kinetics explored by atomistic simulations and Markov models. *Nature Communications*, *6*, 7653. http://dx.doi.org/10.1038/ncomms8653.

Scherer, M. K., Trendelkamp-Schroer, B., Paul, F., Pérez-Hernández, G., Hoffmann, M., Plattner, N., et al. (2015). PyEMMA 2: A software package for estimation, validation, and analysis of Markov models. *Journal of Chemical Theory and Computation*, *11*(11), 5525–5542. http://dx.doi.org/10.1021/acs.jctc.5b00743.

Shaw, D. E., Chao, J. C., Eastwood, M. P., Gagliardo, J., Grossman, J. P., Ho, C. R., et al. (2008). Anton, a special-purpose machine for molecular dynamics simulation. *Communications of the ACM*, *51*(7), 91–97. http://dx.doi.org/10.1145/1364782.1364802.

Shirts, M., & Pande, V. S. (2000). COMPUTING: Screen savers of the world unite! *Science*, *290*(5498), 1903–1904. http://dx.doi.org/10.1126/science.290.5498.1903.

Van Der Spoel, D., Lindahl, E., Hess, B., Groenhof, G., Mark, A. E., & Berendsen, H. J. C. (2005). GROMACS: Fast, flexible, and free. *Journal of Computational Chemistry*, *26*(16), 1701–1718. http://dx.doi.org/10.1002/jcc.20291.

Voelz, V. A., Bowman, G. R., Beauchamp, K., & Pande, V. S. (2010). Molecular simulation of ab initio protein folding for a millisecond folder NTL9(1–39). *Journal of the American Chemical Society*, *132*(5), 1526–1528. http://dx.doi.org/10.1021/ja9090353.

Voelz, V. A., Elman, B., Razavi, A. M., & Zhou, G. (2014). Surprisal metrics for quantifying perturbed conformational dynamics in Markov state models. *Journal of Chemical Theory and Computation*, *10*(12), 5716–5728. http://dx.doi.org/10.1021/ct500827g.

Weber, J. K., & Pande, V. S. (2011). Characterization and rapid sampling of protein folding Markov state model topologies. *Journal of Chemical Theory and Computation*, 7(10), 3405–3411. http://dx.doi.org/10.1021/ct2004484.

Zimmerman, M. I., & Bowman, G. R. (2015). FAST conformational searches by balancing exploration/exploitation trade-offs. *Journal of Chemical Theory and Computation*, *11*(12), 5747–5757. http://dx.doi.org/10.1021/acs.jctc.5b00737.

Zwier, M. C., & Chong, L. T. (2010). Reaching biological timescales with all-atom molecular dynamics simulations. *Current Opinion in Pharmacology*, *10*(6), 745–752. http://dx.doi.org/10.1016/j.coph.2010.09.008.

Bridging Enzymatic Structure Function via Mechanics: A Coarse-Grain Approach

S. Sacquin-Mora[1]
Laboratoire de Biochimie Théorique, CNRS UPR9080, Institut de Biologie Physico-Chimique, Paris, France
[1]Corresponding author: e-mail address: sacquin@ibpc.fr

Contents

Abstract

Flexibility is a central aspect of protein function, and ligand binding in enzymes involves a wide range of structural changes, ranging from large-scale domain movements to small loop or side-chain rearrangements. In order to understand how the mechanical properties of enzymes, and the mechanical variations that are induced by ligand binding, relate to enzymatic activity, we carried out coarse-grain Brownian dynamics simulations on a set of enzymes whose structures in the unbound and ligand-bound forms are available in the Protein Data Bank. Our results show that enzymes are remarkably heterogeneous objects from a mechanical point of view and that the local rigidity of individual residues is tightly connected to their part in the protein's overall structure and function. The systematic comparison of the rigidity of enzymes in their unbound and bound forms highlights the fact that small conformational changes can induce large mechanical effects, leading to either more or less flexibility depending on the enzyme's architecture and the location of its ligand-biding site. These mechanical variations target a limited number of specific residues that occupy key locations for enzymatic activity, and our approach thus offers a mean to detect perturbation-sensitive sites in enzymes, where the addition or removal of a few interactions will lead to important changes in the proteins internal dynamics.

Methods in Enzymology, Volume 578
ISSN 0076-6879
http://dx.doi.org/10.1016/bs.mie.2016.05.022

1. INTRODUCTION

From nucleic acids to proteins, most biomolecular machines rely on conformational changes that are essential for their function (Alberts, 1998; Teilum, Olsen, & Kragelund, 2009, 2011). It is the case in particular for enzymes, where ligand-binding events have been shown to trigger a large variety of structural variations in the protein, such as large domain movements, or more local loops or side-chain rearrangements (Amemiya, Koike, Fuchigami, Ikeguchi, & Kidera, 2011). The Protein Database (PDB) (Berman et al., 2002) now contains a large number of entries of identical proteins, whose structures were determined under different conditions (with or without a ligand molecule), and which exhibit conformational changes that are related to their biological activity (Amemiya, Koike, Kidera, & Ota, 2012). These conformational changes also impact the proteins internal dynamics and mechanics in a way that structural data alone cannot describe properly. Experimentally, data regarding the flexibility of individual residues within proteins can be obtained via Debye–Waller temperature factors (from X-ray crystallography) or order parameters (from NMR spectroscopy). Many theoretical approaches have also been developed for the investigation of protein conformational flexibility, for example, coarse-grain representations combined with elastic network models (ENMs) have been successfully used to study protein dynamics (Atilgan, Okan, & Atilgan, 2012; Bahar, Lezon, Yang, & Eyal, 2010; Sanejouand, 2013; Tama & Sanejouand, 2001). In that perspective, the ProPHet (Probing Protein Heterogeneity) program combines a coarse-grain ENM with Brownian dynamics (BD) simulations and permits to investigate protein mechanical properties on a residue level. Early work using ProPHet on hemoproteins, and globins in particular, showed that a limited set of residues within a protein present specific mechanical properties in relation with the protein's function, for example, catalysis (Sacquin-Mora, Laforet, & Lavery, 2007), or the diffusion of small ligand along internal tunnels (Bocahut, Bernad, Sebban, & Sacquin-Mora, 2011; Oteri, Baaden, Lojou, & Sacquin-Mora, 2014). In this study, we focus our attention toward a set of enzymes with various folds and functions, which undergo structural changes upon ligand binding and whose unbound and ligand-bound forms are available in the PDB (Koike, Amemiya, Ota, & Kidera, 2008). The systematic investigation of the enzymes mechanical properties in both the unbound and bound structures shows how a protein's local flexibility is strongly connected to its structure and function, and how ligand-binding events can induce very specific mechanical changes within the proteins under study.

2. MATERIAL AND METHODS

2.1 Coarse-Grain Simulations

Coarse-Grain BD simulations were performed using the ProPHet program (Lavery & Sacquin-Mora, 2007) on a subset of 31 nonredundant enzymes listed in Table 1, whose unbound and ligand-bound forms are available in the PDB and whose structural change upon ligand binding were investigated in an earlier work by Koike et al. (2008). The proteins are classified depending on their enzymatic reaction type: hydrolase (15 cases), transferase (9 cases), ligase, isomerase, lyase, and oxidoreductase ("others" category, 7 cases).

Diverging from most common CG models, where each residue is described by a single pseudo-atom (Tozzini, 2005), ProPHet uses a more detailed representation that was originally developed by Zacharias (2003), and which enables different residues to be distinguished. The amino acids are represented by one pseudo-atom located at the Cα position, and either one or two (for larger residues) pseudo-atoms replacing the side chain (with the exception of Gly) (Zacharias, 2003), thus leading to a reduction of the number of heavy atoms (non hydrogen atoms) in the system by a factor of 3 approximately. Interactions between the pseudo-atoms are treated according to the standard ENM (Tozzini, 2005), that is, pseudo-atoms closer than the cutoff parameter, $R_c = 9$ Å, are joined by gaussian springs that all have identical spring constants of $\gamma = 0.42\ \mathrm{N\ m^{-1}}$ ($0.6\ \mathrm{kcal\ mol^{-1}\ Å^{-2}}$). The spring constant in this model has a slightly smaller value than in most one-point-per-residue coarse-grain models, where it is usually set to roughly $1.0\ \mathrm{kcal\ mol^{-1}\ Å^{-2}}$ (Atilgan et al., 2001; Doruker, Jernigan, & Bahar, 2002; Yang & Bahar, 2005). This value was chosen to offset the higher spring density of the Zacharias representation.

The springs are taken to be relaxed for the starting conformation of the protein, ie, its crystallographic structure. The simulations use an implicit solvent representation via the diffusion and random displacement term in the equation of motion (Ermak & McCammon, 1978)

$$\mathbf{r}_i = \mathbf{r}_i^0 + \sum_j \frac{\mathbf{D}_{ij}^0 \mathbf{F}_j^0}{kT} \Delta t + \mathbf{R}_i(\Delta t),$$

Table 1 Enzymes Studied in This Work

Protein Name	Category	PDB + Chain ID, Unbound Form	PDB + Chain ID, Bound Form	rmsd (Å)	rmsΔk (kcal mol^{-1} Å^{-2})
Lipase	Hydrolase	1i6wA	1r4zB	0.51	7.90
Ribonuclease MC1	Hydrolase	1bk7A	1ucdA	0.56	2.79
Ribonuclease T1	Hydrolase	1q9eB	1bviC	0.88	5.62
Ribonuclease A	Hydrolase	1izrA	1rtaE	1.07	6.90
Endoglucanase	Hydrolase	1l8fA	1oa7A	0.49	8.68
Endo-1,4-beta-xylanase	Hydrolase	1h14A	1h12A	0.17	5.17
Endo-1,4-beta-xylanase II	Hydrolase	1xyoA	1refB	0.70	8.60
Dextranase	Hydrolase	1ogmX	1ogoX	0.42	6.92
Chitinase 1	Hydrolase	1ll6A	1ll4A	0.22	2.76
T4 lysozyme	Hydrolase	180lA	148lE	0.38	4.00
Human lysozyme	Hydrolase	208lA	1lzsE	1.19	5.66
Subtilisin BPN′	Hydrolase	1gnvA	1suaA	0.23	6.80
Pepsin	Hydrolase	5pep_	1psaA	0.77	7.50
Glucoamylase	Hydrolase	1lf6B	1lf9A	0.45	8.32
Cellulosomal xylanase Z	Hydrolase	1jjfA	1jt2A	0.41	4.21
T-protein	Transferase	1wosA	1wopA	0.23	3.89
Amylomaltase	Transferase	1fp9A	1eswA	0.54	9.09

EPSP synthase	Transferase	1rf5A	1rf6A	3.73	17.54
Pyrophosphokinase	Transferase	1im6A	1f9hA	2.68	12.12
Mycolic acid synthase	Transferase	1kp9A	1kphC	3.41	18.31
Adenosine kinase	Transferase	1lioA	1liiA	3.13	13.44
Arginin kinase	Transferase	3m10A	1rl9A	3.02	14.06
DNA methyltransferase	Transferase	1nw8A	1nw5A	0.36	4.83
N-AGA-transferase 1	Transferase	1fo9A	1foaA	0.35	11.67
Alanyl-tRNA synthetase	Ligase	1ygbA	1yfrB	0.98	6.39
FK506-binding protein	Isomerase	1j4rB	1J4hA	0.73	4.42
Galactose mutarotase	Isomerase	1snzB	1so0A	0.25	9.79
Carbonic anhydrase	Lyase	1fsnA	1lgdA	0.59	8.31
Hyaluronate lyase	Lyase	1flsA	1i8qA	0.53	3.39
Shikimate dehydrogenase	Oxidoreductase	1p74A	1p77A	1.21	6.38
Heme oxygenase 1	Oxidoreductase	1s8cC	1oykA	0.76	7.64

The rmsd values are taken from Koike, R., Amemiya, T., Ota, M., & Kidera, A. (2008). Protein structural change upon ligand binding correlates with enzymatic reaction mechanism. *Journal of Molecular Biology, 379*(3), 397–401. doi: 10.1016/j.jmb.2008.04.019.

where $\mathbf{r}_i$ and $\mathbf{r}_i^0$ denote the position vector of particle i before and after the time step Δt, $\mathbf{F}_i$ is the force on particle i, $\mathbf{R}_i(\Delta t)$ is a random displacement, and hydrodynamic interactions are included through the configuration-dependent diffusion tensor $\mathbf{D}$ (Pastor, Venable, & Karplus, 1988).

The reduced representation for the proteins and the implicit solvent model of BD naturally allow a considerably larger time step than all-atom molecular dynamics. After monitoring the stability of the system for a number of simulations with different time steps and running times, we chose $\Delta t = 10$ fs, a value in agreement with the time steps classically chosen in the literature for BD simulations on flexible systems (Wade, Davis, Luty, Madura, & McCammon, 1993), and ran simulations of 200,000 BD steps at 300 K. For a 200 residues protein, a complete CG-BD simulation takes around 2 h on a standard desktop computer.

These simulations lead to deformations of roughly 1.5 Å root-mean-square deviation (rmsd) with respect to the protein starting conformation (which corresponds to the system's equilibrium state). The simulations are then analyzed in terms of the fluctuations of the mean distance between each pseudo-atom belonging to a given amino acid and the pseudo-atoms belonging to the remaining protein residues. The inverse of these fluctuations yields an effective force constant k_i that describes the ease of moving a pseudo-atom i with respect to the overall protein structure:

$$k_i = \frac{3k_B T}{\left\langle (d_i - \langle d_t \rangle)^2 \right\rangle},$$

where $\langle\rangle$ denotes an average taken over the whole simulation and $d_i = \langle d_{ij} \rangle j^*$ is the average distance from particle i to the other particles j in the protein (the sum over j^* implies the exclusion of the pseudo-atoms belonging to the same residue as i). The distances between the Cα pseudo-atom of residue k and the Cα pseudo-atoms of the adjacent residues $k-1$ and $k+1$ are excluded, since the corresponding distances are virtually constant. The force constant for each residue k in the protein is the average of the force constants for all its constituent pseudo-atoms i. We will use the term "rigidity profile" to describe the ordered set of force constants for all the residues in a given protein (as in Fig. 1A). In the case of multidomain proteins (such as chitinase, glucoamylase, or EPSP synthase), we proceed to a domain separation approach using the domains definition given by the PDBsum website (https://www.ebi.ac.uk/thornton-srv/databases/cgi-bin/pdbsum/GetPage.pl?pdbcode=index.html) (Laskowski, Chistyakov, & Thornton, 2005).

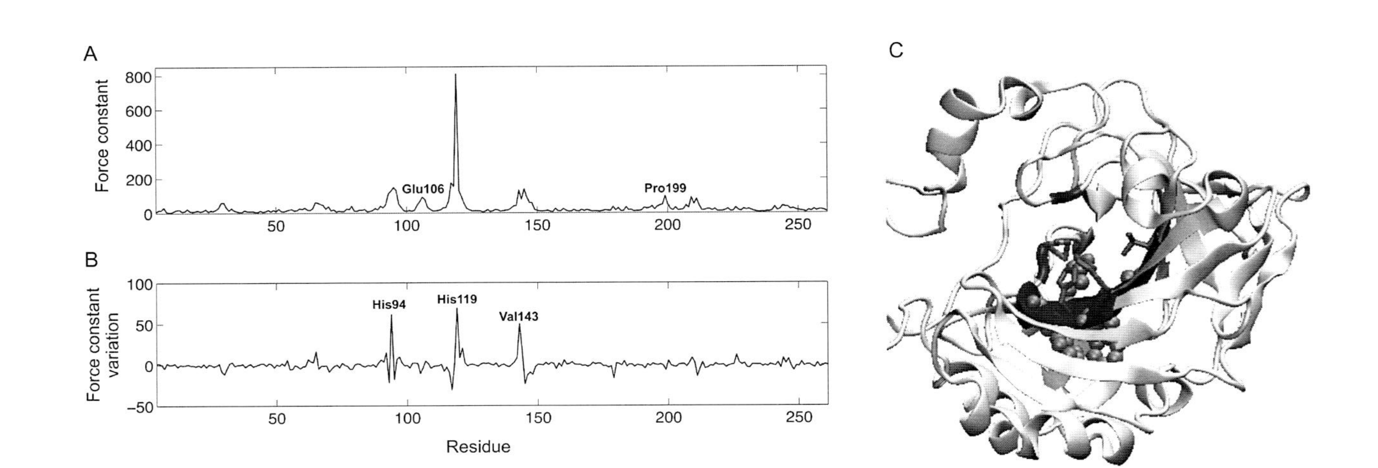

Fig. 1 Carbonic anhydrase. (A) Rigidity profile. (B) Force constant variation upon ligand binding. Force constants in Figs. 1–7 are in kcal mol^{-1} $Å^{-2}$. (C) Cartoon representation with the most rigid residues highlighted in *blue*, residues undergoing a rigidity increase upon ligand binding are shown as *red sticks* and residues undergoing a rigidity decrease as *green* van der Waals spheres. This *color code* is also valid for Figs. 2–7 which were all prepared using Visual Molecular Dynamics (Humphrey, Dalke, & Schulten, 1996). (See the color plate.)

That is, after a single BD simulation on the whole protein structure, we treat each one of the protein domains separately for the force constants calculation, ie, we only consider the distance fluctuations between residues belonging to the same domain. Note that the domain separation approach does not influence the elastic network representation used during the BD simulation, which still includes all residues from all domains.

Following earlier studies, which showed that the small ligands had little influence on calculated force constants (Sacquin-Mora et al., 2007; Sacquin-Mora & Lavery, 2006), we chose not to include ligands or prosthetics groups in the protein representations.

3. RESULTS AND DISCUSSION

3.1 General Properties of the Force Constant Spectra

We start by illustrating the results of our force constant calculations with carbonic anhydrase (PDB codes 1FSN/1LGD), a soluble lyase with an αβ-roll. Its rigidity profile and structure are shown in Fig. 1A and C, respectively. The first striking observation regarding the force constants is their high variability, since they can change sharply from one residue to the next. Here the values range from 3 to over 800 kcal mol^{-1} $Å^{-2}$ with a standard deviation of 56 around an average of 29 kcal mol^{-1} $Å^{-2}$ (note that 1 kcal $mol^{-1}Å^{-2}$=0.07 nN $Å^{-1}$). The largest force constants can be found for residues located in the protein's core, in the central β-strands, and are highlighted in *blue* in Fig. 1C. Additional rigidity peaks located on the loops surrounding the ligand correspond to catalytic residues Glu106 and Pro199.

3.1.1 *Multidomain Proteins*

In the case of the two-domains EPSP synthase (1RF5/1RF6), a transferase with two αβ-prism domains encasing a S3P-GLP ligand, the force constants for the two domains follow one another thus giving rise to a horizontally repeating pattern (see Fig. 2A). The most rigid residues are located at the junction between the two domains (see Fig. 2E), a behavior common to most multidomain proteins, which reflects the fact that domain movements leave the residues on the domain interface virtually undisturbed (Sacquin-Mora, 2014; Sacquin-Mora et al., 2007). In the ProPHet approach, this will lead to high force constants, since the mainly rotational movements of the domains do not modify the distance of other residues to these hinge points. Similar observations were made using normal mode analysis of ENMs

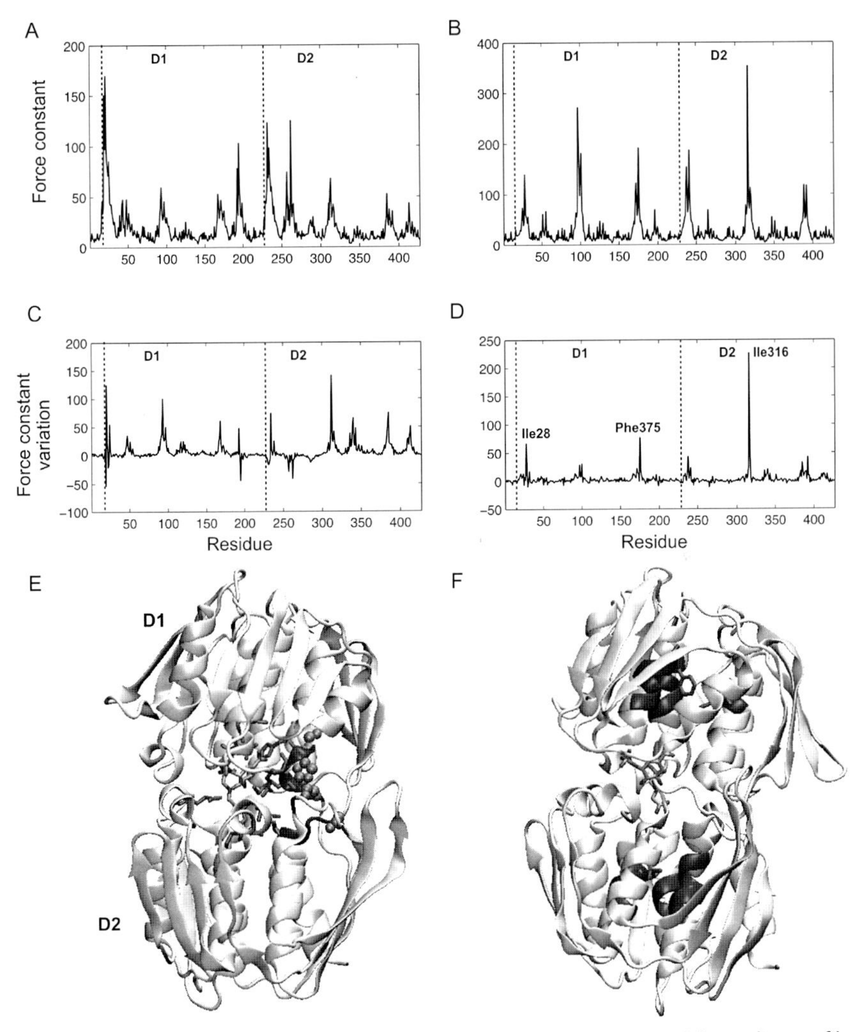

Fig. 2 EPSP synthase. (A) Rigidity profile without domain separation. (B) Rigidity profile after domain separation. (C) Force constant variation upon ligand binding without domain separation. (D) Force constant variation upon ligand binding after domain separation. The domains limits are shown as *vertical dashed lines*. (E) Cartoon representation without domain separation. (F) Cartoon representation after domain separation.

(Bahar & Jernigan, 1999; Isin, Doruker, & Bahar, 2002), and the domain separation protocol described in Section 2.1 was developed to avoid this effect dominating the force constant spectra. The result of this procedure is shown in Fig. 2B. While the force constant spectrum maintains it periodicity, we can now observe a set of new rigidity peaks indicating a shift of highly rigid residues from the domains interface toward the domains cores, ie, the three central α-helices which are highlighted in blue in Fig. 2F. A similar effect of domain separation can be seen for glucoamylase (1LF6/1LF9) and chitinase (1LL6/1LL4) in Figs. 3E and F, and 4E and F, respectively.

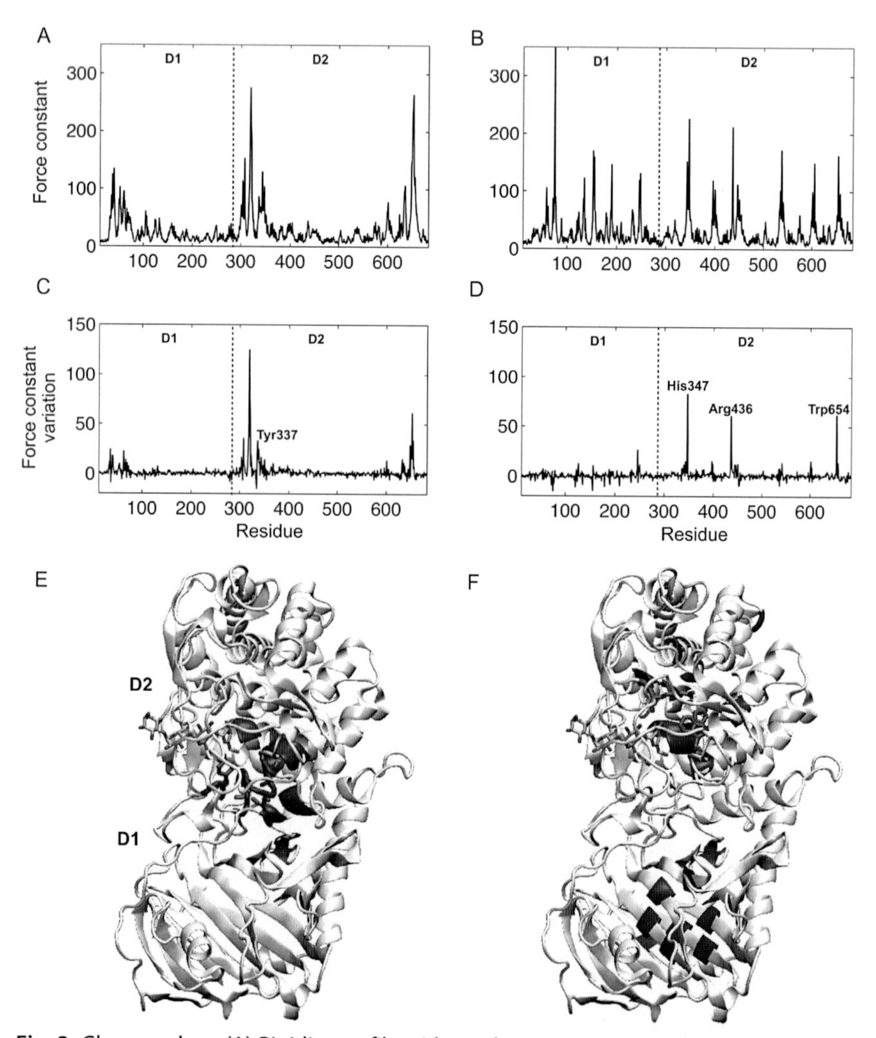

Fig. 3 Glucoamylase. (A) Rigidity profile without domain separation. (B) Rigidity profile after domain separation. (C) Force constant variation upon ligand binding without domain separation. (D) Force constant variation upon ligand binding after domain separation. The domains limits are shown as *vertical dashed lines*. (E) Cartoon representation without domain separation. (F) Cartoon representation after domain separation.

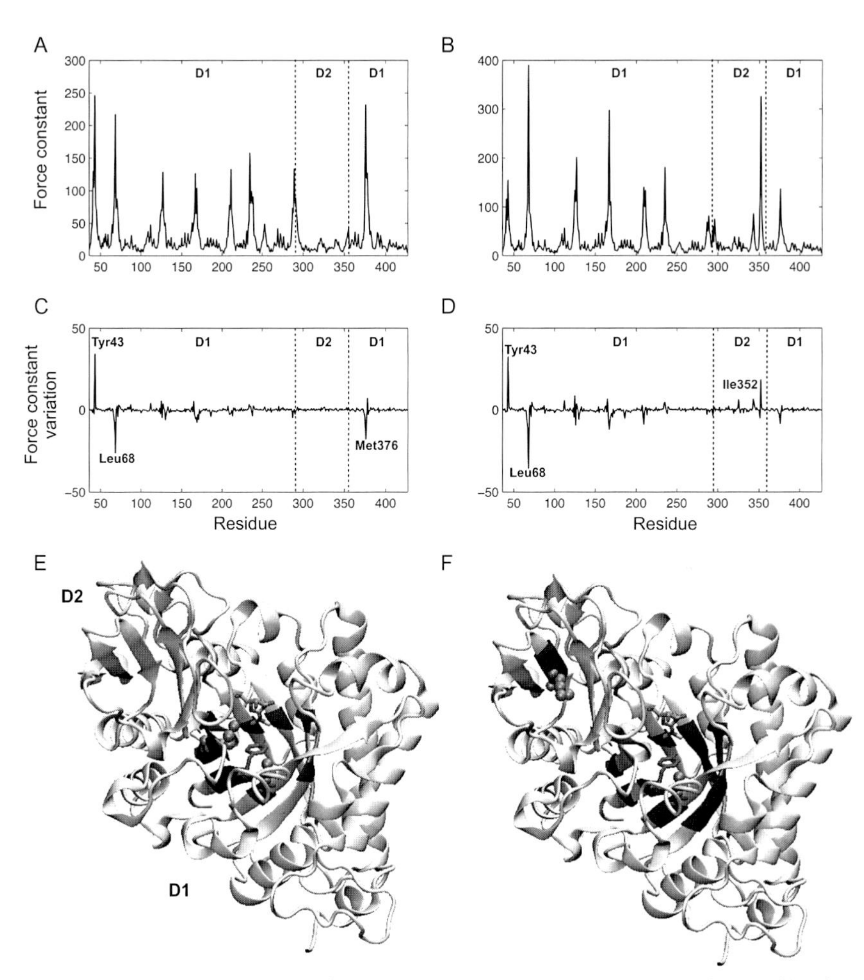

Fig. 4 Chitinase. (A) Rigidity profile without domain separation. (B) Rigidity profile after domain separation. (C) Force constant variation upon ligand binding without domain separation. (D) Force constant variation upon ligand binding after domain separation. The domains limits are shown as *vertical dashed lines*. (E) Cartoon representation without domain separation. (F) Cartoon representation after domain separation.

3.1.2 Identification of Active Site Residues

Going back to carbonic anhydrase, its rigidity profile is dominated by residue His 119, which points directly toward the ligand (see Fig. 1) and catalytic residues Glu106 and Pro199 also correspond to local rigidity peaks, despite being themselves located on flexible loops. In EPSP synthase, the rigidity peaks in Fig. 2A include catalytic residues Lys20, Thr93, Gln168,

Asp312, and Arg385. Catalytic residues Glu636 in glucoamylase, and Asp169 and Tyr239 in chitinase (see Figs. 3A and 4A, respectively) also happen to be highly rigid. These few examples turn out to be fairly representative of the general mechanical properties of enzymes. In a earlier study on a group of almost 100 enzymes from all the main enzymatic families (Sacquin-Mora et al., 2007), active site residues, as defined in the Catalytic Site Atlas database (Porter, Bartlett, & Thornton, 2004) or in an earlier elastic network study (Yang & Bahar, 2005), turned out to be amongst the most strongly fixed residues within the protein structures. In particular, residues with force constants above the average represented only 28% of the total set (with an overall distribution that was highly skewed toward the lower values), but this subset was highly enriched in active site residues, since it contained 78% of all such residues and only 25% of other residues.

3.1.3 Mechanical Signatures of Specific Protein Folds

As shown in previous mechanical studies done on members of the globin family (Bocahut et al., 2011; Stadler et al., 2012) and homologous hydrogenases (Oteri et al., 2014), proteins with similar folds will also present similar rigidity profiles. In some cases, the distribution of high force constants residues along the sequence can be a signature of the overall protein structure. For example, the regular series of rigidity peaks in the force constant spectra of glucoamylase and chitinase (after domain separation, in Figs. 3B and 4B, respectively) correspond to residues lying within the β-strands of their distorted β-sandwich (for glucoamylase) and αβ-barrel (for chitinase) domains. Another "architectural fingerprint" can be observed for heme oxygenase (1S8C/1OYK) with groups of rigidity peaks going through a maximum value in their center (see Fig. 5A and C) that are characteristic of the α-helices forming this oxidoreductase.

Proteins can also present rigid residues far from their active site, and in some cases these residues, which are often highly conserved (Ptitsyn, 1998; Ptitsyn & Ting, 1999), have been identified as playing a key role in protein folding (Sacquin-Mora & Lavery, 2006). In particular, in a recent study on 14 different protein families that combines calculations with ProPhet and sequence alignments (Sacquin-Mora, 2015), we have shown that the analysis of conserved mechanical properties within a protein family can lead to the identification of its folding nucleus.

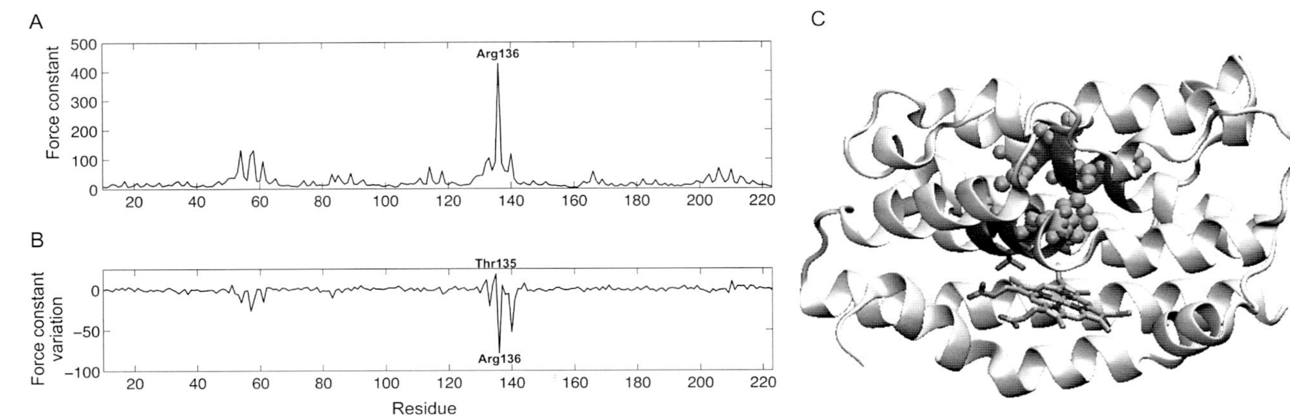

Fig. 5 Heme oxygenase. (A) Rigidity profile. (B) Force constant variation upon ligand binding. (C) Cartoon representation.

3.2 "E pur si muove": Mechanical Variations in Enzymes Undergoing Structural Changes upon Ligand Binding

All the proteins listed in Table 1 undergo structural changes upon ligand binding, with rmsd on the Cα atoms ranging from 0.1 to 4 Å. Therefore, for each enzyme, we made a systematic comparison of its rigidity profiles between the unbound and ligand-bound structures. The resulting force constant variation plots are highly heterogeneous along the protein sequences and show how the enzymes mechanical response to structural changes is closely related to their biological activity.

From a quantitative point of view, Fig. 6 shows the rmsd variation of the force constants upon ligand binding in each enzyme as a function of the rmsd between the unbound and bound structures. There is a clear positive correlation between these two values for enzymes from the transferase category (correlation coefficient = 0.87) but not for enzymes

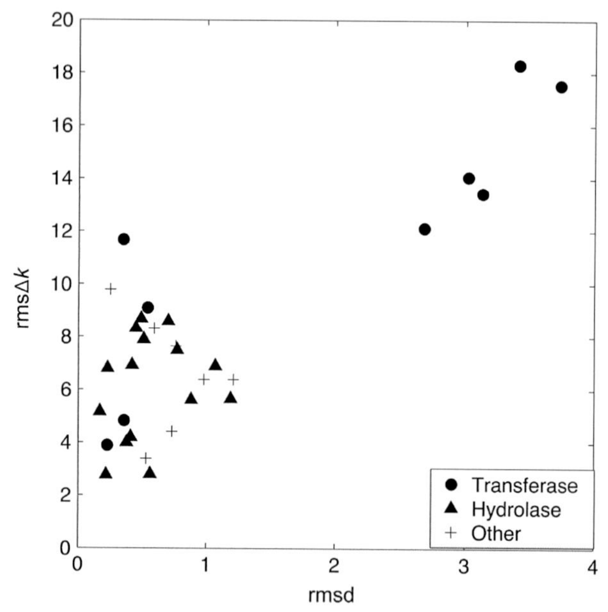

Fig. 6 Root-mean-square variations of the force constants as a function of rmsd values, upon ligand binding for 31 enzymes. Transferases are shown as *black dots*, hydrolases as *black triangles*, and enzymes from the remaining functional categories (ligase, isomerase, lyase, and oxidoreductase) as *crosses. The rmsd values are taken from Koike, R., Amemiya, T., Ota, M., & Kidera, A. (2008). Protein structural change upon ligand binding correlates with enzymatic reaction mechanism.* Journal of Molecular Biology, *379(3), 397–401. doi: 10.1016/j.jmb.2008.04.019.*

from the two remaining groups (hydrolases and others). While most proteins undergo small structural changes upon ligand binding, with rmsd values below 1.5 Å, transferase is the only category where enzymes can undergo large structural variations, with five enzymes whose rsmd values are above 2.5 Å. For transferases, the large mechanical changes upon ligand binding result from important structural variations, leading to a shielding of the ligand molecule, which is often located at the boundary between two domains, from the water environment (Koike et al., 2008). For example, in EPSP synthase, the protein will undergo a structural transition from an open to a closed conformation, so that for the enzyme in its bound form, the ligand finds itself buried between the two domains (see Fig. 2E).

From a qualitative perspective, ligand binding can produce different effects, such as a rigidity increase, or decrease or a mixed response, on the protein mechanical properties, depending on the location of the ligand-binding site relative to its core. This response only affects a very limited set of residues within the global protein structure, and interestingly, catalytic or ligand-binding residues will often undergo a noticeable rigidity increase as can be seen from the examples that follow.

3.2.1 Rigidity Increase upon Ligand Binding

The force constant variation plots, before and after domain separation, for glucoamylase are shown in Fig. 3C and D, respectively. In both cases, we can see that the small structural change (rmsd = 0.45 Å) resulting from ligand binding is sufficient to induce a remarkable rigidity increase for a few residues. When we look at the protein as the whole (without domain separation), the conformational transition leads to the rigidification of residues that are located at the domain interface and shown in red on Fig. 3E. In particular, we can notice a rigidity peak for Tyr337, which points directly toward the acarbose ligand. After domain separation, the rigidity increase now mostly concerns residues that are located in the core of the αα-barrel domain (D2), and especially Arg436 and Trp654, which surround the buried head of the acarbose ligand (see Fig. 3F).

3.2.2 Flexibility Increase upon Ligand Binding

In heme oxygenase, the insertion of the heme-ligand inside the α-bundle seems to disrupt the protein fold and leads to an important rigidity decrease of the central core residues (shown in green in Fig. 5C). Interestingly, the only residue presenting a slight increase of its force constant upon ligand binding is Thr135 (see Fig. 5B), which points toward the heme group. For endoglucanase (1L8F/1OA7), Fig. 7 shows that the ligand site lies next to a small β-barrel domain which also becomes more flexible after cellobiose binding to the protein.

3.2.3 Mixed Mechanical Responses

In carbonic anhydrase, ligand binding induces an important rigidity increase for residues His94, His119, and Val143 (see Fig. 1B and C), which lie at the center of the β-strands in the enzyme's core. Simultaneously, residues located on the borders of those same β-strands will become more flexible. If we look at the force constant variations upon ligand binding before domain separation in EPSP synthase, we can see in Fig. 2C that the few residues undergoing a flexibility increase (and shown in green in Fig. 2E) lie on the domain interface, but away from the glyphosphate ligand. On the contrary, amino acids with increased force constants (shown in red in Fig. 2E) include numerous catalytic residues (such as Lys20, Thr93, Gln168, Asp312, Glu340, and Arg385) and completely surround the ligand, which will finds itself encased in an extremely rigid cage. After domain separation (Fig. 2D), the rigidity increase upon ligand binding mostly concerns the core residues of the two αβ-prism domains, thus indicating a more stable structure for the protein in its closed conformation. In the case of chitinase, the rigidity profile is dominated by the large αβ-barrel domain, and both approaches (without and with domain separation) will show an increase of the force constant for the ligand-binding residue Tyr43.

3.2.4 Ligand Binding with No Mechanical Impact

Our last example concerns an enzyme where the ligand-binding site is located on the protein's surface, instead of at a domain boundary or into a cleft. In that case, ligand binding will have little to no effect on the protein's mechanical properties, as we can see in Fig. 8 for xylanase Z (1JJF/1JT2), where the core of the αβα-sandwich domain maintains its original rigidity after binding of the FER ligand.

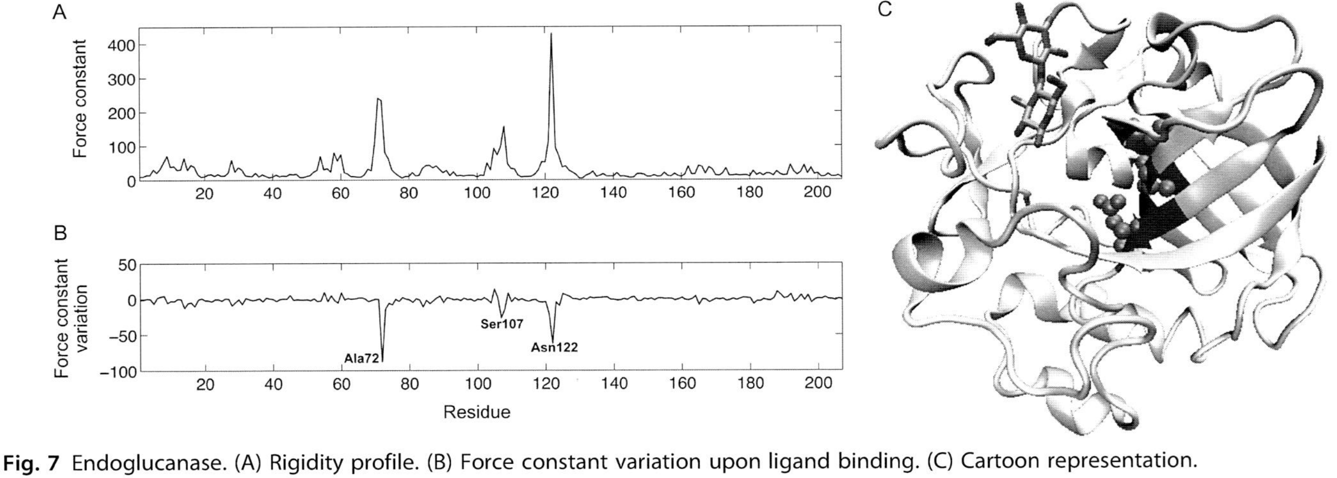

Fig. 7 Endoglucanase. (A) Rigidity profile. (B) Force constant variation upon ligand binding. (C) Cartoon representation.

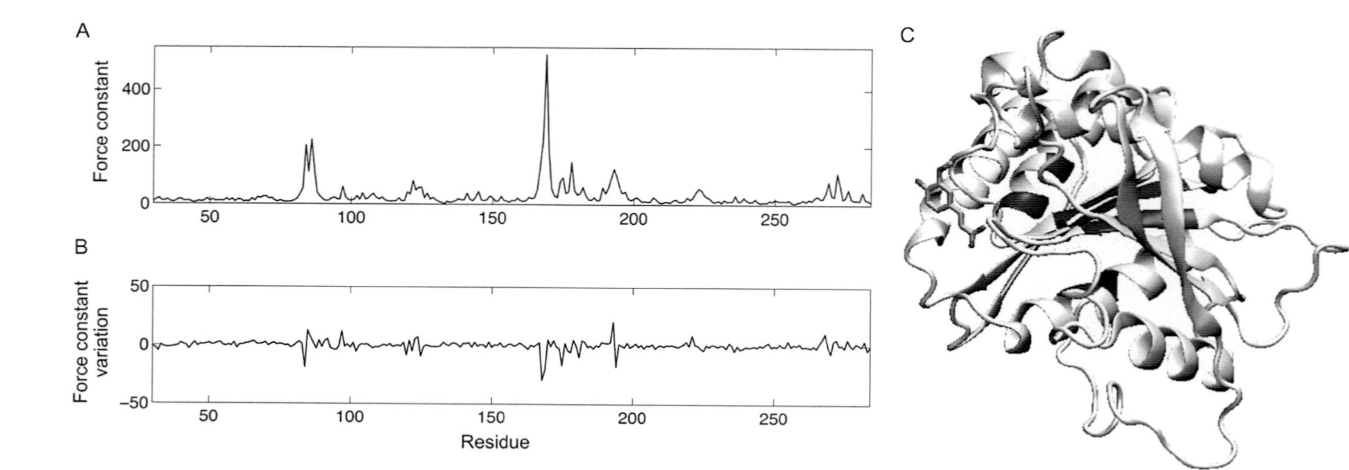

Fig. 8 Xylanase Z. (A) Rigidity profile. (B) Force constant variation upon ligand binding. (C) Cartoon representation.

4. CONCLUDING REMARKS

Combining a CG-ENM protein representation and BD simulations, we can investigate protein mechanical properties on the residue level. The rigidity profiles resulting from our calculations highlight how proteins complex structures lead to equally complex properties. The local mechanics of proteins are very diverse, with key residues, either from the structural or functional point of view, presenting important force constants. Binding site residues in particular will often be highly rigid, thus making rigidity within the overall protein structure a good guide for locating catalytic activity in enzymes. This result might seem at first sight surprising, since flexibility in the catalytic site was originally considered essential for an enzyme to carry out its activity (Daniel, Dunn, Finney, & Smith, 2003). The reverse has however already been found by analysis of temperature factors (Bartlett, Porter, Borkakoti, & Thornton, 2002; Yang & Bahar, 2005; Yuan, Zhao, & Wang, 2003) and by looking at the residue fluctuations associated with the low frequency normal nodes representing collective motions (Yang & Bahar, 2005), and our findings support these results.

In a second step, we made a systematic comparison of the force constants from enzymes in their unbound and ligand-bound forms. Early work by Gutteridge and Thornton (2005) has shown that the induced fit motions in most enzymes are less than 1 Å between the apo and substrate bound forms across the whole protein. It is also the case of the vast majority (23 out of 31) of the enzymes in our study. However, these small structural changes are still sufficient to induce a noticeable response in the enzymes mechanics. These rigidity variations are not uniformly or randomly scattered across the protein structure but are on the contrary nonmonotonous, highly heterogenous and will target very specific parts in the molecules, namely, the hinge (on the domain boundaries) and binding site residues. Quantitatively, transferases, which are the only proteins in this study that undergo large conformational changes upon binding, also seem to be the only functional category where the amplitude of the force constants variation correlates well with the amplitude of the structural change. Qualitatively, the enzymes present various patterns in their mechanical changes, that is, a global rigidity increase or decrease, a mixed response or a very weak response to conformational variations, and these patterns could not be related to a specific enzymatic reaction mechanism. However, we can observe a constant trend in our enzymes, which is the rigidity increase of ligand surrounding residues, even though the ligand molecule is not explicitly represented in our elastic model. This effect of ligand binding on protein flexibility was already observed both theoretically (via ENM calculation) and experimentally (via NMR spectroscopy) in the aminoglycoside phosphotransferase (Wieninger, Serpersu, & Ullmann, 2011) and seems to be a general property of enzymes. Overall, investigating protein mechanical variations permits to highlight a small subset of residues likely to play an essential part in enzymatic activity and which represent potentially interesting target for the inhibition, modulation, or enhancement of enzyme function in a protein engineering perspective.

REFERENCES

Alberts, B. (1998). The cell as a collection of protein machines: Preparing the next generation of molecular biologists. *Cell*, *92*(3), 291–294.

Amemiya, T., Koike, R., Fuchigami, S., Ikeguchi, M., & Kidera, A. (2011). Classification and annotation of the relationship between protein structural change and ligand binding. *Journal of Molecular Biology*, *408*(3), 568–584. http://dx.doi.org/10.1016/j.jmb.2011.02.058.

Amemiya, T., Koike, R., Kidera, A., & Ota, M. (2012). PSCDB: A database for protein structural change upon ligand binding. *Nucleic Acids Research*, *40*(Database issue), D554–D558. http://dx.doi.org/10.1093/nar/gkr966.

Atilgan, A. R., Durell, S. R., Jernigan, R. L., Demirel, M. C., Keskin, O., & Bahar, I. (2001). Anisotropy of fluctuation dynamics of proteins with an elastic network model. *Biophysical Journal*, *80*(1), 505–515.

Atilgan, C., Okan, O. B., & Atilgan, A. R. (2012). Network-based models as tools hinting at nonevident protein functionality. *Annual Review of Biophysics*, *41*, 205–225. http://dx.doi.org/10.1146/annurev-biophys-050511-102305.

Bahar, I., & Jernigan, R. L. (1999). Cooperative fluctuations and subunit communication in tryptophan synthase. *Biochemistry*, *38*(12), 3478–3490.

Bahar, I., Lezon, T. R., Yang, L. W., & Eyal, E. (2010). Global dynamics of proteins: Bridging between structure and function. *Annual Review of Biophysics*, *39*, 23–42. http://dx.doi.org/10.1146/annurev.biophys.093008.131258.

Bartlett, G. J., Porter, C. T., Borkakoti, N., & Thornton, J. M. (2002). Analysis of catalytic residues in enzyme active sites. *Journal of Molecular Biology*, *324*(1), 105–121.

Berman, H. M., Battistuz, T., Bhat, T. N., Bluhm, W. F., Bourne, P. E., Burkhardt, K., Berman, H. M., Battistuz, T., Bhat, T. N., Bluhm, W. F., Bourne, P. E., Burkhardt, K., … Zardecki, C. (2002). The Protein Data Bank. *Acta Crystallographica. Section D, Biological Crystallography*, *58*(Pt. 6 No 1), 899–907.

Bocahut, A., Bernad, S., Sebban, P., & Sacquin-Mora, S. (2011). Frontier residues lining globin internal cavities present specific mechanical properties. *Journal of the American Chemical Society*, *133*(22), 8753–8761. http://dx.doi.org/10.1021/ja202587a.

Daniel, R. M., Dunn, R. V., Finney, J. L., & Smith, J. C. (2003). The role of dynamics in enzyme activity. *Annual Review of Biophysics and Biomolecular Structure*, *32*, 69–92.

Doruker, P., Jernigan, R. L., & Bahar, I. (2002). Dynamics of large proteins through hierarchical levels of coarse-grained structures. *Journal of Computational Chemistry*, *23*(1), 119–127.

Ermak, D. L., & McCammon, J. A. (1978). Brownian dynamics with hydrodynamic interactions. *The Journal of Chemical Physics*, *69*, 1352–1360.

Gutteridge, A., & Thornton, J. (2005). Conformational changes observed in enzyme crystal structures upon substrate binding. *Journal of Molecular Biology*, *346*(1), 21–28. http://dx.doi.org/10.1016/j.jmb.2004.11.013.

Humphrey, W., Dalke, A., & Schulten, K. (1996). VMD: Visual molecular dynamics. *Journal of Molecular Graphics*, *14*(1), 33–38. 27–38.

Isin, B., Doruker, P., & Bahar, I. (2002). Functional motions of influenza virus hemagglutinin: A structure-based analytical approach. *Biophysical Journal*, *82*(2), 569–581.

Koike, R., Amemiya, T., Ota, M., & Kidera, A. (2008). Protein structural change upon ligand binding correlates with enzymatic reaction mechanism. *Journal of Molecular Biology*, *379*(3), 397–401. http://dx.doi.org/10.1016/j.jmb.2008.04.019.

Laskowski, R. A., Chistyakov, V. V., & Thornton, J. M. (2005). PDBsum more: New summaries and analyses of the known 3D structures of proteins and nucleic acids. *Nucleic Acids Research*, *33*(Database issue), D266–D268.

Lavery, R., & Sacquin-Mora, S. (2007). Protein mechanics: A route from structure to function. *Journal of Biosciences*, *32*(5), 891–898.

Oteri, F., Baaden, M., Lojou, E., & Sacquin-Mora, S. (2014). Multiscale simulations give insight into the hydrogen in and out pathways of [NiFe]-hydrogenases from Aquifex aeolicus and Desulfovibrio fructosovorans. *The Journal of Physical Chemistry B*, *118*(48), 13800–13811. http://dx.doi.org/10.1021/jp5089965.

Pastor, R. W., Venable, R., & Karplus, M. (1988). Brownian dynamics simulation of a lipid chain in a membrane bilayer. *The Journal of Chemical Physics*, *89*, 1112–1127.

Porter, C. T., Bartlett, G. J., & Thornton, J. M. (2004). The Catalytic Site Atlas: A resource of catalytic sites and residues identified in enzymes using structural data. *Nucleic Acids Research*, *32*(1), D129–D133.

Ptitsyn, O. B. (1998). Protein folding and protein evolution: Common folding nucleus in different subfamilies of c-type cytochromes? *Journal of Molecular Biology*, *278*(3), 655–666.

Ptitsyn, O. B., & Ting, K. L. (1999). Non-functional conserved residues in globins and their possible role as a folding nucleus. *Journal of Molecular Biology*, *291*(3), 671–682.

Sacquin-Mora, S. (2014). Motions and mechanics: Investigating conformational transitions in multi-domain proteins with coarse-grain simulations. *Molecular Simulation*, *40*(1–3), 229–236. http://dx.doi.org/10.1080/08927022.2013.843176.

Sacquin-Mora, S. (2015). Fold and flexibility: What can proteins' mechanical properties tell us about their folding nucleus? *Journal of the Royal Society Interface*, *12*(112), 20150876. http://dx.doi.org/10.1098/rsif.2015.0876.

Sacquin-Mora, S., Laforet, E., & Lavery, R. (2007). Locating the active sites of enzymes using mechanical properties. *Proteins*, *67*(2), 350–359. http://dx.doi.org/10.1002/prot.21353.

Sacquin-Mora, S., & Lavery, R. (2006). Investigating the local flexibility of functional residues in hemoproteins. *Biophysical Journal*, *90*(8), 2706–2717. http://dx.doi.org/10.1529/biophysj.105.074997.

Sanejouand, Y. H. (2013). Elastic network models: Theoretical and empirical foundations. *Methods in Molecular Biology*, *924*, 601–616. http://dx.doi.org/10.1007/978-1-62703-017-5_23.

Stadler, A. M., Garvey, C. J., Bocahut, A., Sacquin-Mora, S., Digel, I., Schneider, G. J., … Zaccai, G. (2012). Thermal fluctuations of haemoglobin from different species: Adaptation to temperature via conformational dynamics. *Journal of the Royal Society, Interface*, *9*(76), 2845–2855. http://dx.doi.org/10.1098/rsif.2012.0364.

Tama, F., & Sanejouand, Y. H. (2001). Conformational change of proteins arising from normal mode calculations. *Protein Engineering*, *14*(1), 1–6.

Teilum, K., Olsen, J. G., & Kragelund, B. B. (2009). Functional aspects of protein flexibility. *Cellular and Molecular Life Sciences*, *66*(14), 2231–2247. http://dx.doi.org/10.1007/s00018-009-0014-6.

Teilum, K., Olsen, J. G., & Kragelund, B. B. (2011). Protein stability, flexibility and function. *Biochimica et Biophysica Acta, 1814*(8), 969–976. http://dx.doi.org/10.1016/j.bbapap.2010.11.005.

Tozzini, V. (2005). Coarse-grained models for proteins. *Current Opinion in Structural Biology, 15*(2), 144–150.

Wade, R. C., Davis, M. E., Luty, B. A., Madura, J. D., & McCammon, J. A. (1993). Gating of the active site of triose phosphate isomerase: Brownian dynamics simulations of flexible peptide loops in the enzyme. *Biophysical Journal, 64*(1), 9–15.

Wieninger, S. A., Serpersu, E. H., & Ullmann, G. M. (2011). ATP binding enables broad antibiotic selectivity of aminoglycoside phosphotransferase(3')-IIIa: An elastic network analysis. *Journal of Molecular Biology, 409*(3), 450–465. http://dx.doi.org/10.1016/j.jmb.2011.03.061.

Yang, L. W., & Bahar, I. (2005). Coupling between catalytic site and collective dynamics: A requirement for mechanochemical activity of enzymes. *Structure, 13*(6), 893–904.

Yuan, Z., Zhao, J., & Wang, Z. X. (2003). Flexibility analysis of enzyme active sites by crystallographic temperature factors. *Protein Engineering, 16*(2), 109–114.

Zacharias, M. (2003). Protein-protein docking with a reduced protein model accounting for side-chain flexibility. *Protein Science, 12*(6), 1271–1282.

A Networks Approach to Modeling Enzymatic Reactions

P. Imhof[1]

Institute of Theoretical Physics, Free University Berlin, Berlin, Germany
[1]Corresponding author: e-mail address: petra.imhof@fu-berlin.de

Contents

Abstract

Modeling enzymatic reactions is a demanding task due to the complexity of the system, the many degrees of freedom involved and the complex, chemical, and conformational transitions associated with the reaction. Consequently, enzymatic reactions are not determined by precisely one reaction pathway. Hence, it is beneficial to obtain a comprehensive picture of possible reaction paths and competing mechanisms. By combining individually generated intermediate states and chemical transition steps a network of such pathways can be constructed. Transition networks are a discretized representation of a potential energy landscape consisting of a multitude of reaction pathways connecting the end states of the reaction. The graph structure of the network allows an easy identification of the energetically most favorable pathways as well as a number of alternative routes.

1. INTRODUCTION

A popular approach to computationally investigate enzymatic reaction mechanisms is a combined quantum mechanical/molecular mechanical (QM/MM) ansatz (Díaz & Field, 2004; Eurenius, Chatfield, Brooks, & Hodoscek, 1996; Field, 2002; Florián, Goodman, & Warshel, 1995; Fothergill, Goodman, Petruska, & Warshel, 1995; Gao & Truhlar, 2002;

Methods in Enzymology, Volume 578
ISSN 0076-6879
http://dx.doi.org/10.1016/bs.mie.2016.05.025

Garcia-Viloca, Gao, Karplus, & Truhlar, 2004; Gregersen, Lopez, & York, 2004; Lopez, York, DeJaegere, & Karplus, 2002; Warshel & Weiss, 1980) in which the active site of the enzyme and part of its surrounding are modeled by quantum chemical methods and the remainder of the protein and (some of) the solvent is treated by classical MM force fields. When applying semi-empirical methods for the QM part, and with the power of modern computers also to a limited extend for first-principle methods, even a fair amount of sampling the fluctuations of the active site and its environment along the reaction pathway is feasible. This way, the impact of the fluctuating environment can be analyzed and free energy barriers can be estimated. Enzymatic reactions usually comprise a few chemical steps with considerable free energy barriers, rendering the transition over such barriers rare events. Consequently, a large variety of "enhanced sampling" techniques have been developed most of which introduce a bias to the system to force it along a, generally low number of reaction coordinates, expressed as internal coordinates or more complicated collective variables.

A number of recent approaches use adaptive ways to modify the free energy landscape (Barducci, Bonimi, & Parrinello, 2011; Dama, Hocky, Sun, & Voth, 2015; Ensing, Laio, Parrinello, & Klein, 2005; Leines & Ensing, 2012; Zheng, Chen, & Yang, 2008) or the composition of the reaction coordinate (E, Ren, & Vanden-Eijnden, 2002; Rosta, Nowotny, Yang, & Hummer, 2011; Rosta, Woodcock, Brooks, & Hummer, 2009; Vanden-Eijnden & Venturoli, 2009).

Optimization-based methods, on the other hand, do not give direct access to free energies, but allow for a comprehensive exploration of the potential energy landscape. Chain-of-states optimizers such as nudged elastic band (Henkelman, Jóhannesson, & Jónsson, 2000; Henkelman, Uberuage, & Jonsson, 2000) or conjugate peak refinement (Fischer & Karplus, 1992) find an optimal path between given end states, ie, reactant and product state of a reaction, without a priori definition of a reaction coordinate. The optimality of the resulting minimum energy path, however, depends on the initial guess of the path. For only the two end states provided, this initial path is a linear interpolation, a rather crude approximation for end states with very different geometries.

Dividing instead the transition from reactant state to product state of the full reaction, into many subtransitions, ie, from reactant to a first intermediate, then from first to second intermediate, etc., will likely help the path minimizers to determine optimal pathways for the individual subtransition.

Finally, piecing those together, the optimal energy profile for the whole reaction can be obtained. In computational setups including the full enzyme, there is likely not only a single reaction path. For example, proton transfers play an important role in reaction mechanisms employing acid–base catalysis. The number of possible proton transfer reactions and the order of the individual steps grows rapidly with the number of possible proton donor and acceptor groups in the reactive center. Especially with such reactions that consist of multiple steps, several competing pathways going from reactant to product via a number of intermediates are conceivable. Some pathways are clearly distinct, offering fundamentally different mechanisms. Other reaction pathways can as well branch resulting in alternative routes, or merge at different intermediate steps, forming a bundle of possible connections between two stations along the mechanism. It is therefore desirable to explore the energy surface with regard to intermediate states and their connections so as to generate a comprehensive picture of possible reaction paths (Fig. 1).

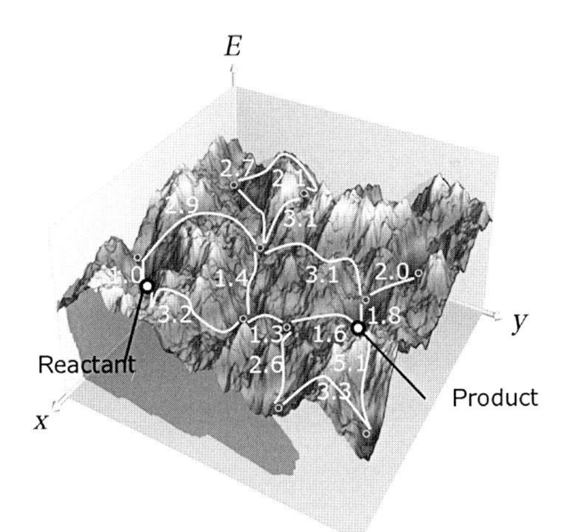

Fig. 1 Schematic energy landscape with "valleys" (*blue*) and "mountains" (*yellow/orange*). *Yellow points* mark end states of a reaction, *green dots* are intermediate states. *White connections* with transition barriers indicate a variety of possible pathways. *Adapted with permission from Noé, F., Krachtus, D., Smith, J. C., & Fischer, S. (2006). Transition networks for the comprehensive characterization of complex conformational change in proteins.* Journal of Chemical Theory and Computation, 2, *840–857. Copyright (2006) American Chemical Society.* (See the color plate.)

On the potential energy landscape, this idea is put into practice by discrete path sampling techniques (Wales, 2002, 2006) where sequences of local minima and transition states are combined to a network of connected pathways. Molecular clusters, conformational changes in peptides, and recently a proton transfer reaction of a small molecule in a water shell are among the successful applications of this method (Ball et al., 1996; Becker & Karplus, 1996; Brooks, Onuchic, & Wales, 2001; Ivchenko et al., 2014; Somani & Wales, 2013; Wales, 2006; Wales & Doye, 2003). The huge number of stationary points in a protein system prevents complete sampling. Restricting the stationary point search to the sampling region of interest, networks of pathways can also be constructed for complex conformational transitions in proteins (Noé, Krachtus, Smith, & Fischer, 2006). The path sampling or transition network approaches provide a comprehensive picture of complex transitions that are likely not captured in a single reaction pathway. Recently, network approaches that allow to explore complex chemical reaction channels for small biomolecules (Rappoport, Galvin, Zubarev, & Aspuru-Guzik, 2014) and transition metal complexes (Bergeler, Simm, Proppe, & Reiher, 2015) have been reported.

A transition networks approach is also a means for modeling enzymatic reactions, whether in active site models or in full protein environment. In hybrid QM/MM approaches, the region of interest is the active site, comprising all those atoms that are likely involved in the chemistry. Restricting the sampling of intermediate states and their connections to the active site is therefore legitimate. Other protein degrees of freedom are optimized so as to adapt to new geometries of the active site, but not explicitly sampled. The actual sampling is carried out in user selected degrees of freedom. Note that this selection of degrees of freedom (usually on the order of 5–10) also introduces some sort of bias. However, not the pathway is biased, but the way the search space is limited. Within that search space a multitude of local minima is generated that are subsequently connected so as to generate the transition network. On the network, the energetically most favorable pathway together with the next best pathways can be determined.

In this chapter, we first give examples for enzymatic reactions that have been explored by modeling individual steps between possible reaction intermediates and their subsequent connection. Then, we describe our transition networks method for the automated sampling of reaction pathways on potential energy surface. Finally, we give an example of reaction path sampling in a full-enzyme model.

2. NETWORKS OF REACTION PATHWAYS

2.1 Networks from Combination of Pathways

A QM/MM analysis of the reaction mechanism in the restriction enzyme EcoRV has been reported in Imhof, Fischer, and Smith (2009). The phosphodiester hydrolysis reaction leads to DNA backbone cleavage at a specific recognition site. From available crystallographic data, both reactant and product state of the reaction could be modeled. A comparison of the active site geometry in the two end states (see Fig. 2) suggests the reaction mechanism to consist of several steps.

The possibilities which have been explored as individual transitions are sketched in Fig. 3A. Note that already for the first proton transfer in the step Ra1, there are four possible "a1" intermediates, marked by the *differently colored hydrogen atom positions*. The associated four transitions are indicated by the *differently colored arrows*. Each of the possible intermediates in scheme Fig. 3A has been optimized individually. Then, the resulting intermediate states have been connected by minimum energy path calculation, using conjugate peak refinement (Fischer & Karplus, 1992), and the energy of the highest transition state along those optimized pathways has been determined. The resulting network of all connections is shown in Fig. 3B. Some intermediate

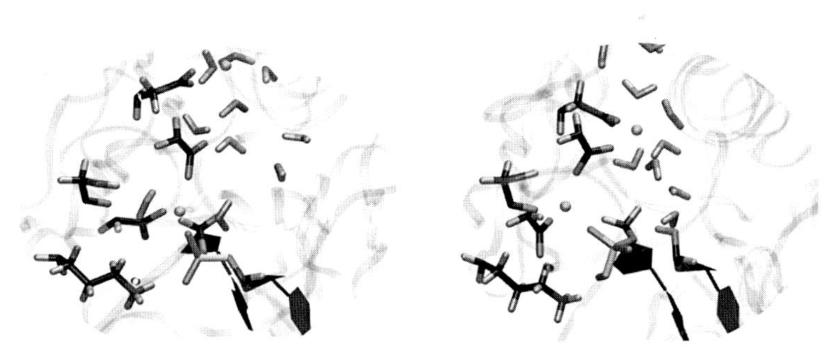

Fig. 2 Reactant (*left*) and product (*right*) state of the DNA backbone phosphodiester hydrolysis reaction in the enzyme EcoRV. Only those active site atoms treated quantum mechanically are shown in licorice for clarity. The protein is indicated by *ribbons*. The *bonds marked in yellow* (*gray* in the print version) are the scissile bond and the newly formed bond, respectively. *Adapted with permission from Imhof, P., Fischer, S., & Smith, J. C. (2009). Catalytic mechanism of DNA backbone cleavage by the restriction enzyme Eco RV: A quantum mechanical/molecular mechanical simulations analysis.* Biochemistry, 48, *9061–9075. Copyright (2009) American Chemical Society.*

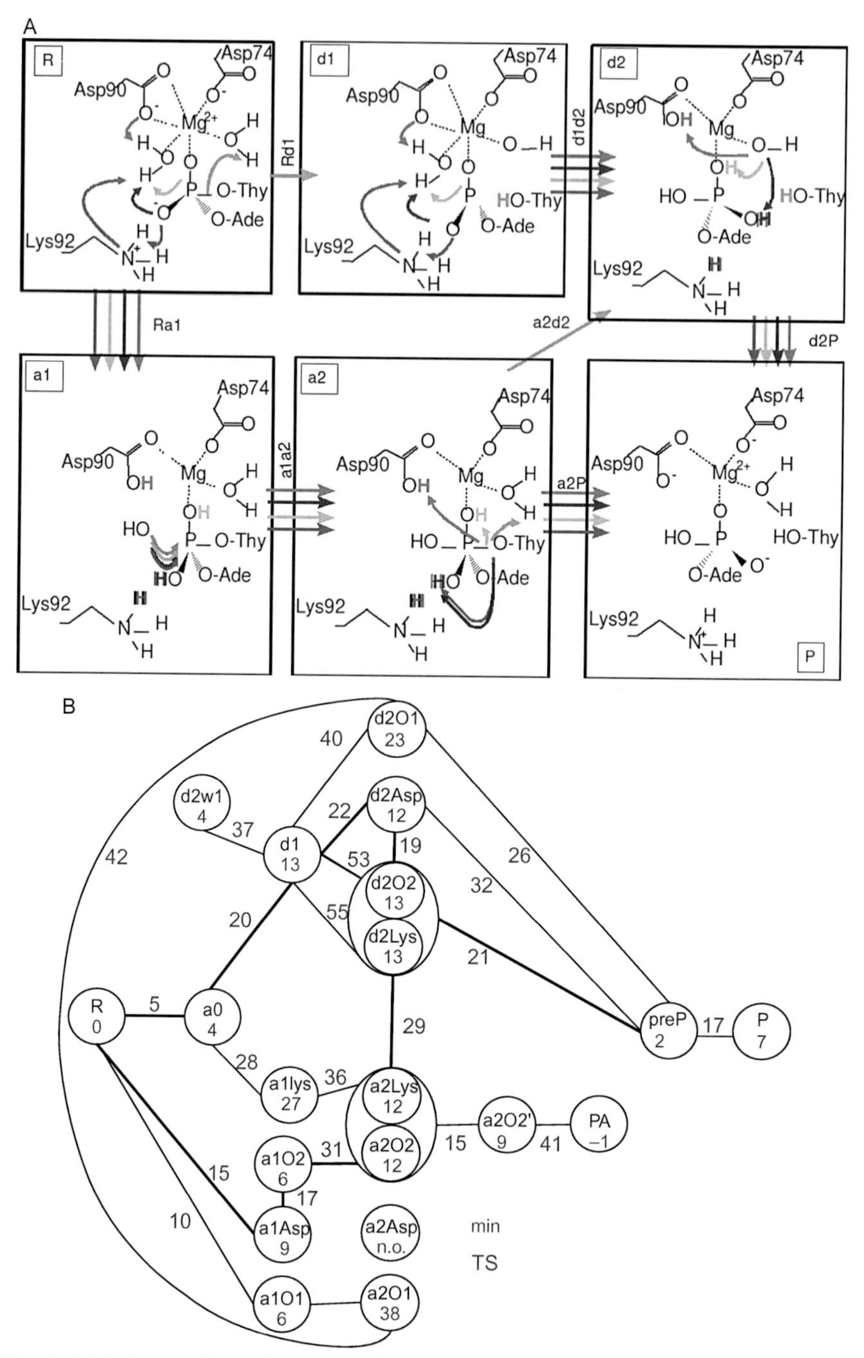

Fig. 3 (A) Scheme of possible steps of the phosphodiester hydrolysis reaction in the enzyme EcoRV. (B) Network of computed pathways. Relative energies (in kcal/mol) are given in *blue numbers inside the circles* for minima and in *red numbers next to the*

states can be reached via different routes. These states are shown as an ellipse around two circles which actually represent the same state. The two different labels mark the pathways along which the common states can be reached.

The energetically most favorable pathway, ie, the one whose rate-determining barrier is the lowest among all rate-determining barriers (indicated by *thick lines* in Fig. 3B), R–a0–d1–d2Asp–d2O2–P is a dissociative mechanism. The network representation shows also two associative mechanisms as next best pathways. All productive pathways lead through intermediate d2O2/d2Lys. Branching between dissociate and associative mechanism can occur with the first transition from the reactant state or later at state a0. Furthermore, "dead-ends," ie, states that are not intermediate to another state, are shown.

A similar strategy has been adopted for simulating the ATP hydrolysis in myosin (Kiani & Fischer, 2014).

The question whether the glycosidic bond hydrolysis, catalyzed by thymine DNA glycosylase (TDG) proceeds via an associative or dissociative mechanism could be answered by two-dimensional energy surface scans of the C1′–N1 bond dissociation/nucleophilic water attack (Kanaan, Crehuet, & Imhof, 2015). The clearly favorable dissociative mechanism, however, consists even in its simplest form of several individual steps (see Fig. 4). An alternative to this direct mechanism, a histidine-assisted mechanism, comprises the same essential steps but these are accompanied with a number of proton transfer steps between the histidine residue and the target base.

These proton transfers can in principle take place at various stages along each of the two computed mechanisms, connecting the intermediate of one mechanism to an intermediate of the other one. Fig. 5 shows the network of computed pathways: the direct and histidine-assisted pathway, and the connections between the two mechanism. As can be seen from the free energies calculated for all transitions (*arrows*) in the network, the glycosidic bond cleavage in TDG does not follow a single, dominating mechanism, but

connection lines for transition states. The states are labeled according to the scheme in (A). The labels Asp, Lys, O1, and O2 refer to the different hydrogen atom positions, *colored red, green, yellow*, and *blue*, respectively, in subfigure (A). Note that the transition from preproduct preP to product P is not shown in the scheme. *Reprinted with permission from Imhof, P., Fischer, S., & Smith, J. C. (2009). Catalytic mechanism of DNA backbone cleavage by the restriction enzyme Eco RV: A quantum mechanical/molecular mechanical simulations analysis.* Biochemistry, 48, *9061–9075. Copyright (2009) American Chemical Society.* (See the color plate.)

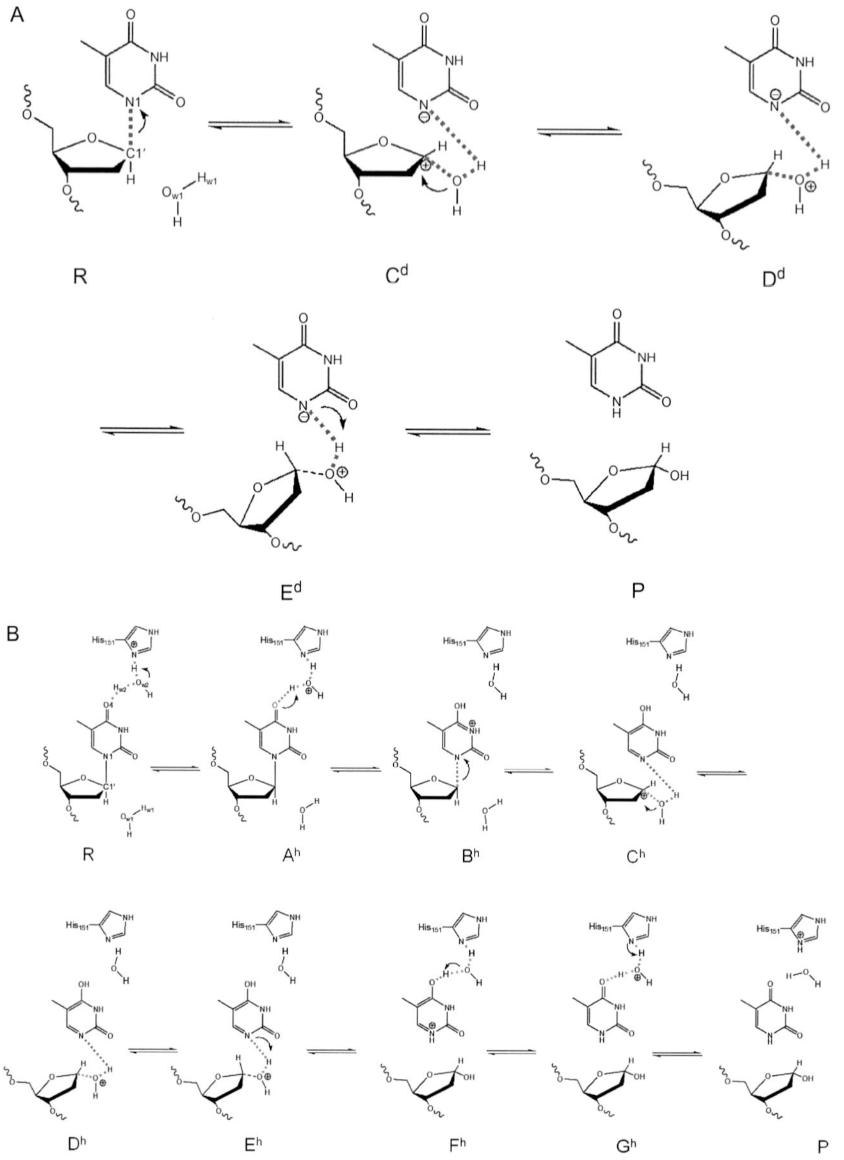

Fig. 4 Scheme of glycosidic bond hydrolysis pathways in TDG: (A) direct mechanism and (B) histidine-assisted mechanism. *Adapted with permission from Kanaan, N., Crehuet, R., & Imhof, P. (2015). Mechanism of the glycosidic bond cleavage of mismatched thymine in human thymine DNA glycosylase revealed by quantum mechanical/molecular mechanical calculations.* The Journal of Physical Chemistry B, 119, *12365–12380. Copyright (2015) American Chemical Society.*

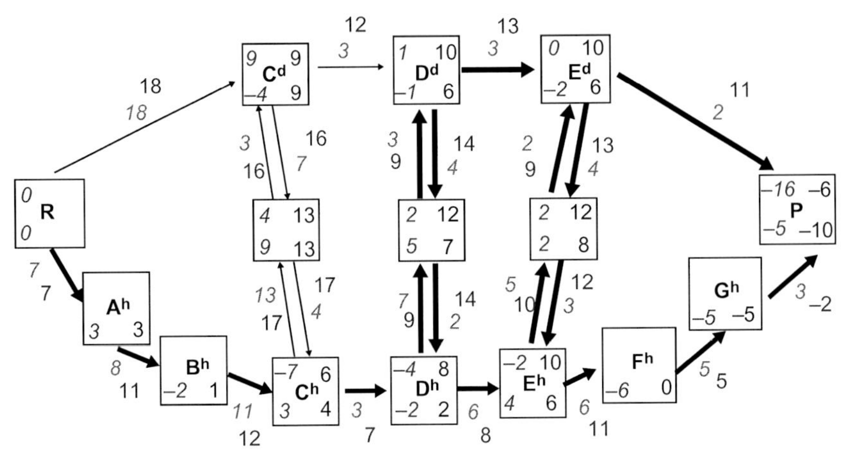

Fig. 5 Network of computed reaction pathways for glycosidic bond cleavage in TDG. *Blue* (*black* in the print version) *numbers in the boxes* are free energies of minima relative to the reactant state, *red* (*dark gray* in the print version) *numbers next to the connection lines* are relative transition free energies (in kcal/mol). *Arrows* indicate the transitions computed. *Thick arrows* highlight the most likely pathways. The "direct mechanism," R–C^d–D^d–E^d–P is shown as the upper route, the lower route R–A^h–B^h–C^h–D^h–E^h–F^h–G^h–P represents the "histidine mechanism." *Adapted with permission from Kanaan, N., Crehuet, R., & Imhof, P. (2015). Mechanism of the glycosidic bond cleavage of mismatched thymine in human thymine DNA glycosylase revealed by quantum mechanical/molecular mechanical calculations.* The Journal of Physical Chemistry B, 119, *12365–12380. Copyright (2015) American Chemical Society.*

can rather take several pathways of similar probability. The network also shows that state C^d, the dissociative intermediate with anionic base, can only be reached via high free energy barriers and C1′–N1 bond dissociation likely takes place with a protonated thymine base (B^h to C^h). Once, then the nucleophilic water molecule has attacked, all routes to product require comparable activation energies (Kanaan et al., 2015).

2.2 Comprehensive Transition Networks

Transition networks are applied in order to generate a comprehensive collection of possible reaction pathways over the energy landscape of the enzymes. Transition networks are a discrete and simplified representation of the potential energy surface of the system (eg, the enzyme-substrate complex), in which the known potential energy minima can be regarded as the nodes of a web and the connections represent the known transition states between pairs of minima. These networks have the mathematical structure of a weighted graph. In such a graph the vertices represent the minima, which are connected by edges representing the transition states. The graph weights, which are assigned to the edges, represent the energy barriers for the corresponding transition (Noé et al., 2006).

The network is generated by a uniform sampling along selected degrees of freedom (eg, bond lengths, dihedral angles, metal ion positions, and protonation states) with subsequent minimization yielding the network nodes.

Transitions between pairs of nodes are computed using the conjugate peak refinement method. Edge weights are determined by the relative energy of the highest transition state along the minimum energy path connecting two nodes. From the analysis of a transition network, the energetically most favorable pathways as well as a number of alternative pathways can be obtained.

2.2.1 Generating Network Nodes by Sampling Selected Degrees of Freedom

An initial "direct path" is computed by conjugate peak refinement from the direct connection of reactant and product state. This path ensures that the final transition network actually connects the two final end states. Moreover, the barrier computed for this transition, ie, the relative energy of the highest transition state along the "direct path" serves as a threshold for all other nodes and edges in the network to be constructed.

To generate a transition network, it is required to define a number of degrees of freedom which are allowed to vary. The most important types of degrees of freedoms for chemical reactions are described in the following.

2.2.1.1 Sampling Bond Lengths

The major degrees of freedom to vary in a substitution reaction are the dissociation of the scissile "old bond" and the formation of a new bond. These bond distances are those typically used for potential energy scans or other approaches that require predefined reaction coordinates.

The bond distances are sampled by stepwise increasing and decreasing, respectively, the distance between the atoms forming the old bond and new bond atoms, respectively, together with other atoms that are connected forming an attacking or leaving group, respectively. The atom common to both old and new bond is not displaced.

2.2.1.2 Metal Ion or Water Positions

Sampling of metal ion or water molecule positions is realized by stepwise moves of the ions or the water molecules in x, y, and z directions, thereby generating new positions on a grid. Ions positions are accepted under the condition that the distance between the ions exceeds a user-defined value, typically 3 Å, thus excluding electrostatically unfavorable configurations.

2.2.1.3 Side Chain Rotation

Side chains are rotated around selected dihedral angles with the number of possible dihedral angles per side chain depending on the type of residue. For example, aspartate can be rotated around the C_{α}–C_{β} bond or the C_{β}–C_{γ} bond (Fig. 6).

Rotations that lead to chemically equivalent conformations because of symmetry, such as a rotation of an aspartate side chain around the C_{β}–C_{γ} bond that exchanges the positions of O_{δ} and O_{ε}, ie, only relabels the carboxyl oxygen atoms, are not considered as new samples. The user chooses the residues to be included in the side chain sampling, the dihedral angles to be considered for all those residues, and the step size.

2.2.1.4 Protonation Sampling

The last sampling step varies the protonation pattern while maintaining the number of protons in the system. To this end, residues whose protonation state is to be varied need to be defined by the user. Moreover, allowed protonation pattern, ie, possible protonation sites together with a description of the protonation site have to be set. The possible proton positions can be defined by internal or cartesian displacements relative to their host atoms.

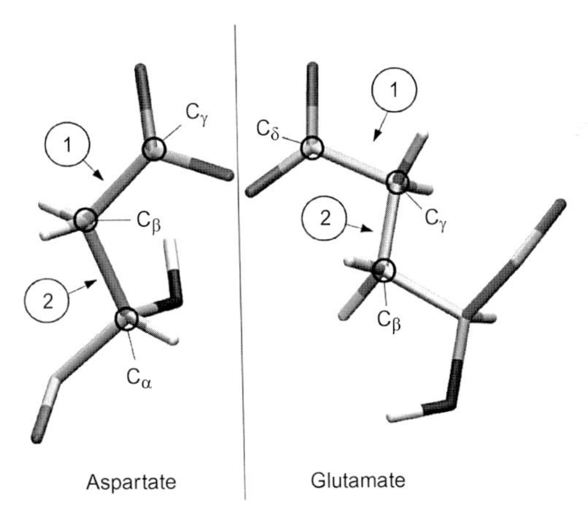

Fig. 6 Side chain conformations can be sampled by stepwise rotation around selected dihedral angles. The "outermost" dihedral angle is regarded as the first one, the second dihedral angle is the next toward the backbone, etc. Hence, chemically similar dihedral angles such as rotation of the carboxylate group in aspartate or glutamate have the same rank. Furthermore, rotation around dihedral angles of higher rank move more atoms and consequently leads to a large displacement. Note that this counting is reverse to the nomenclature of the labeling by χ_n.

2.2.1.5 Rejection of Nonmeaningful Structures

For each residue boundary conditions can be defined for the protonation sampling. These conditions ensure that only chemically meaningful protonation patterns are considered. For example an aspartate should carry at most one proton, whereas a water molecule must not be fully deprotonated while formation of OH^- and H_3O^+ ions is allowed.

The sampling steps described earlier are performed in the order of their description. Configurations in which the moved atoms show collisions to heavy atoms, ie, they are closer than 1 Å, are rejected in each sampling step. For the protonation sampling, the rejection criterion is extended to collisions with hydrogen atoms. Placement of the protons is always carried out as a last step.

2.2.2 Nodes

2.2.2.1 Geometry Optimization of Network Nodes

The configurations generated by the sampling procedures described earlier are minimized to give the network nodes. To avoid the computational cost for full minimizations of unfavorable structures, we perform stepwise minimizations. First, the generated configurations are optimized to a gradient of 0.1 kcal/mol/Å. Samples, which do not reach this gradient within 20 minimization steps are then rejected. In the second step, the gradient criterion for the geometry optimization is reduced by a factor of 10, and the maximal number of optimization steps is increase by a factor of 10. Again, configurations with higher gradients are rejected. The remaining samples undergo further minimizations and are finally energy minimized to a gradient below 0.001 kcal/mol/Å. From these, only configurations whose relative minimum energies are lower than the transition state energy of the direct, interpolated pathway are used as transition network nodes.

2.2.2.2 Node State Assignment

The local minima that fulfill gradient and energy criteria are then assigned their actual state. This is necessary since the final minimum structure can represent a different state than the initial sample. For example, a sample that has started as two steps along the new bond and three steps of rotation around a certain dihedral angle does not represent anything close to a local minimum. Instead, during the optimization, it "slides down" the energy surface into a local minimum that corresponds to a state representing three steps along the new bond and two dihedral rotation steps. All minimum energy structures are therefore compared to the reactant state. The displacement along the selected degrees of freedom defines the state of the node. Furthermore, out of the several samples that may end up in the same state after optimization, only one minimum is kept as a network node.

The state of the node is represented by a state vector with as many components as degrees of freedom used for the assignment. The value of the component is the number of steps along that coordinate which the respective node differs from the reactant state. For the reactant state vector all components equal zero. The only exception, both for the reactant state as for all other states, is the component that characterizes the protonation pattern. Here, absolute values are given in the sense that the proton pattern is encoded directly. All protonation sites are labeled "1" if the site is occupied and "0" if it is empty. By keeping the same order of protonation sites for all states, these labels represent a binary number. The corresponding decimal number is then the entry in the state vector. In a typical molecular simulation setup, each atom has an individual label, at least given by its index, or, in biochemical simulations, by segment name, residue number, and atom name. Interchanging two atoms of the same kind, eg, swapping to protons, exchanging the two oxygen atoms of a carboxyl group, or rotation of a methyl group, leads to a chemically equivalent structure. In order to avoid introducing artificial new nodes with chemically equal structures, atoms are relabeled before the state assignment. To do this in a computationally efficient manner, for relabeling protons and water molecules, we use an implementation of the "Hungarian algorithm" (Kuhn, 1955; Stern, 2012–2014).

2.2.3 Edges

The edges of the network are subsequently constructed by connecting adjacent minima. To obtain a full network, all possible connections (edges) between all vertices would have to be calculated. This is, however, not feasible and it is also not necessary for our purposes. We therefore restrict the computation of transitions to those, in which the connected states differ by a user-defied maximal number of steps, and a maximal number of protonation changes, termed neighbors. In addition, the user can set a maximal root mean square deviation between pairs of nodes so as to also restrict cartesian displacements.

For all pairs of nodes that are checked as edge candidates a minimal difference convention is applied. That is, for the putative target node of the candidate edge all relevant atoms, symmetrically equivalent atoms, protons, water molecules, or ions, are relabeled such that the sum of distances between the same atoms in the two node structures is minimal. This way, artificial transitions such as methyl group rotations can be avoided. All neighbor pairs are assigned an edge. For all edges a pathway connecting edge source and edge target is computed using the conjugate peak refinement. The edge weight is determined as the relative energy of the highest transition state (first-order saddle point) along this pathway. The energy reference is the energy of the reactant state of the total network. Since we are only interested in the highest transition state energy of the connection, we do not need to compute other saddle points along the connection or the converged minimum energy pathway, which is a substantial saving in computational cost.

2.2.4 Best Pathways

The optimized nodes and computed edges are then combined to a graph, the transition network. Only those edges whose weight is below the transition state energy of the initial "direct" path are kept in the graph. This may lead to a graph that consists of several subgraphs that are not connected to each other. In that case, only the subgraph that contains the terminal reactant and product states is considered further. The other subgraphs represent regions of the energy surface that have well-connected minima, but are separated from the relevant part by high energy barriers.

From the remaining, essential subgraph all dead-ends, ie, nodes with only one edge and the corresponding edges, are removed, unless these nodes are reactant or product state. The best pathway, that is the most probable pathway, is then determined as the shortest path on the graph, connecting reactant and product state, using (Dijkstra's algorithm in Dijkstra, 1959). Since the probability of a reaction path to be followed is determined by the highest energy point along that path, the cost of a path on the transition graph is computed from the highest edge weight along that path (Noé et al., 2006). Upon request, the k next best pathways, or all next pathways within an energy range, are computed (see also Noé et al., 2006).

2.2.5 Application to an Enzymatic Hydrolysis Reaction

To illustrate the transition network approach, we consider the peptide bond hydrolysis as catalyzed by the enzyme carboxy peptidase (CPA) (Chikvaidze, Erdel, & Imhof, submitted). For this reaction, several mechanisms have been discussed in the literature, the most popular among which are the so-called "promoted-water mechanism" and the "anhydride mechanism."

In order to perform a transition network analysis, reactant and product state of the reaction have to be modeled. Starting from the crystal structure of CPA complexed to an intact Gly–Tyr peptide, the enzymatic system has been setup in a water box and subjected to molecular dynamics simulations so as to relax and equilibrate the system. From a snapshot taken from the MD simulation the bulk water has then been removed, except for a water shell of 5 Å around the protein. Then, QM/MM optimization has been carried out using PM3 as the QM method and utilizing improved parameters for the central Zn-ion (Brothers, Suarez, Deerfield, & Merz, 2004). The QM region (104 atoms in total) comprises the Gly–Tyr peptide, the Zn-ion and its coordinating residues His196, His69, and Glu72, as well as Glu270 that may form the putative anhydride, and Arg127 that positions the substrate. Furthermore, three water molecules are treated quantum mechanically. All atoms that are further than 10 Å of the peptide substrate are kept stationary, leaving about 900 flexible atoms. From the same MD snapshot also a product structure has been modeled and energy minimized with the same QM/MM setup.

A direct connection of the two final end states results in a barrier if 48 kcal/mol. This barrier has been set as threshold for node and edge energies.

For the construction of the transition network, the following degrees of freedom have been selected for sampling: the bond length of the C–N peptide bond for two steps, the distance between the C atom of the peptide bond and the oxygen atom of one water molecule (the nucleophile) for three steps, and the distance between the peptide C atom and C_δ of Glu270, all with 0.5 Å stepsize; rotation around the C_β–C_γ bond of Glu72 and Glu270, rotation around the C_β–C_γ bond of His69 and His196, each with a step size of 120 degree; protonation of the peptide substrate, all QM-treated water molecules, and the two Glu residues. Applying boundary conditions for glutamate and water residues seven protons are allowed to relocate, resulting in 16 different protonation patterns (see also Fig. 7).

The maximal number of structures to sample with those settings is 15,552 (Fig. 8).

After minimization and state assignment, ~247 different nodes that also meet the energy requirements are kept. More than 1000 nodes are rejected because of their too high energy. For the assignment of edges, the maximal difference between two nodes is: four steps in total along any degree of freedom that is not protonation, two transferred protons, and a root mean square deviation of 2.5 Å in the heavy atoms.

The resulting neighbor list contains ~7000 entries. For all those entries, conjugate peak refinement calculations are initiated so as to find the rate-determining transition states, and hence the edge weights, between the

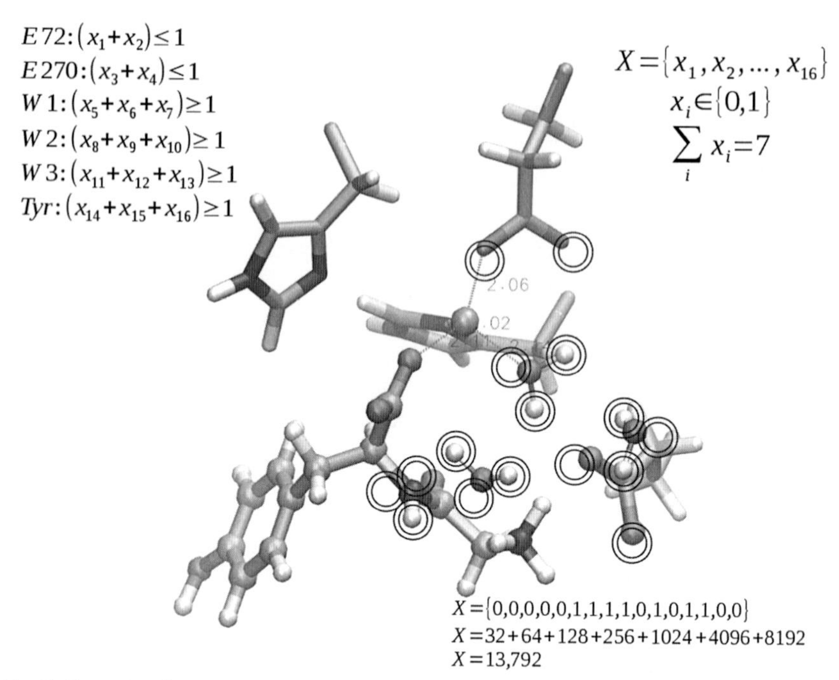

Fig. 7 Example of protonation sites, encircled positions, in the active site of carboxypeptidase. Seven protons (inside circles) are to be distributed over 16 protonation sites, $x_1, \ldots, x_{16}$. In the *upper left corner*, the boundary conditions applied for the protonation sampling of the different residues are given. The binary protonation code {0000011110101100} transformed into a decadic number (13,792) uniquely describes the protonation pattern.

respective pair of nodes. After removing those connections which fail SCF convergence after several rounds, about 5000 edges remain. Retaining only edges whose energy is below the threshold given by the "direct path" calculation, 1874 edges remain, out of which 1794 form a connected set.

The best pathway on that transition network has a highest barrier of 27 kcal/mol (marked in *dark blue* (*black* in the print version) in Fig. 9). Within further 5 kcal/mol, another 16 pathways can be found. All best pathways travel through nodes 0.0.0.1.0.0.13792 and 0.3.0.0.0.0.30400. The latter represents the tetrahedral intermediate, formed by attack of the nucleophilic water molecule. As can be seen from the change in the protonation code (last digit), this water attack is accompanied by a change in protonation pattern. This is a proton transfer from the nucleophilic water molecule to Glu270. The active site structures of the minima along the best pathway are depicted in Fig. 10. Within the energy range of 27–30 kcal/mol four possibilities of subsequent C–N bond scission are found (indicated by

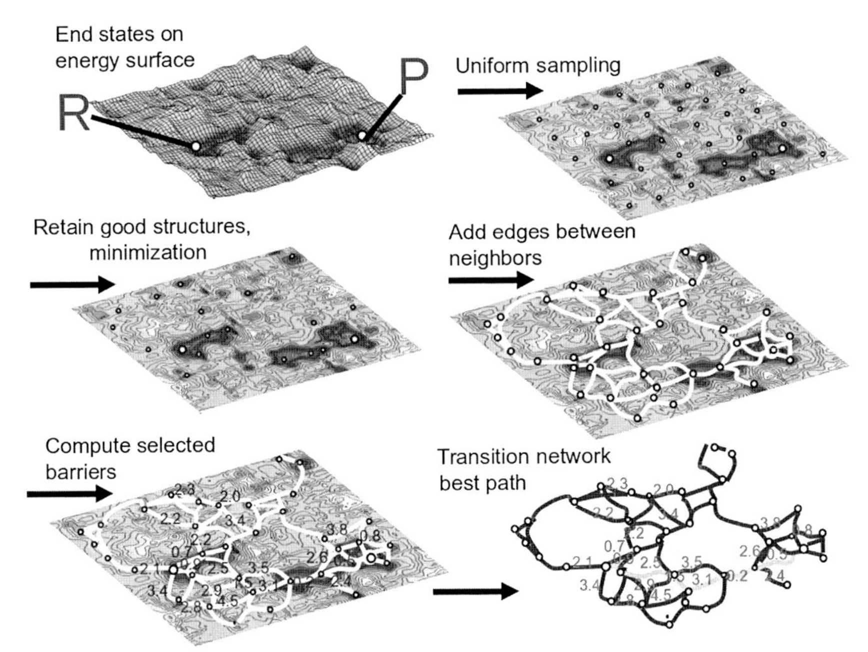

Fig. 8 Steps to generate a transition network between two end states (reactant, R; product, P) of a reaction, schematically shown in a two-dimensional energy landscape. The underlying potential energy landscape is sampled around discrete positions by uniform sampling which are subsequently minimized. Nodes identified as neighbors are connected by edges, and transitions are computed along discrete pathways between neighbors. The edge weights are given by the relative energies of the transition state. On the transition network, the best path is identified by graph theoretical algorithms. *Adapted with permission from Noé, F., Krachtus, D., Smith, J. C., & Fischer, S. (2006). Transition networks for the comprehensive characterization of complex conformational change in proteins.* Journal of Chemical Theory and Computation, 2, *840–857. Copyright (2006) American Chemical Society.* (See the color plate.)

turquoise (*dark gray* in the print version), *green* (*light gray* in the print version), *yellow* (*light gray* in the print version), and *orange* (*gray* in the print version) in Fig. 9), which all differ only in the protonation pattern of the cleaved intermediate. The final step toward product is another proton transfer that has a comparable transition state energy in all best pathways. In case of the pathway with 29 kcal/mol highest barrier (*yellow* (*light gray* in the print version) route in Fig. 9), this final proton transfer is concerted with the C–N bond dissociation. The state code of the subnetwork of best pathways (Fig. 9) moreover shows that no χH196 rotation or χE72 rotation occurs along the most favorable pathways. Both these residues are tightly coordinated to the Zn-ion throughout the entire reaction mechanism.

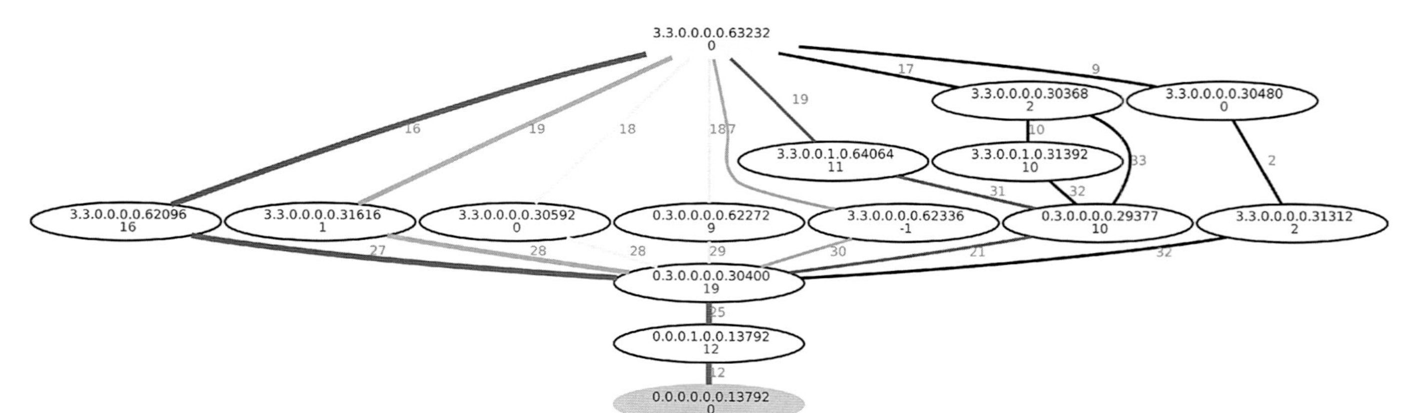

Fig. 9 Transition network of the peptide hydrolysis in carboxypeptidase, computed with QM/MM. Only the best pathway, and pathways with barriers that are at most 5 kcal/mol higher in energy are shown. *Gray* and *yellow* (*light gray* in the print version) *mark* the final reactant and product state, respectively. *Edges* are colored according to the ranking of the pathway they correspond to in following order: *blue* (*black* in the print version), *turquoise* (*dark gray* in the print version), *green* (*light gray* in the print version), *yellow* (*light gray* in the print version), *orange* (*gray* in the print version), *red* (*black* in the print version), and *black*. *Bottom blue* (*black* in the print version) *numbers within the circles* are the node energies relative to the reactant state, *red* (*dark gray* in the print version) *numbers next to the connection lines* are the transition state energies used as edge weights. All energies are in kcal/mol. The *top black numbers in the circles* encode the state of the node as displacement relative to the reactant state as: oldBond.newBond.χH196 .χH69 .χE270 .χE72 .protonation.

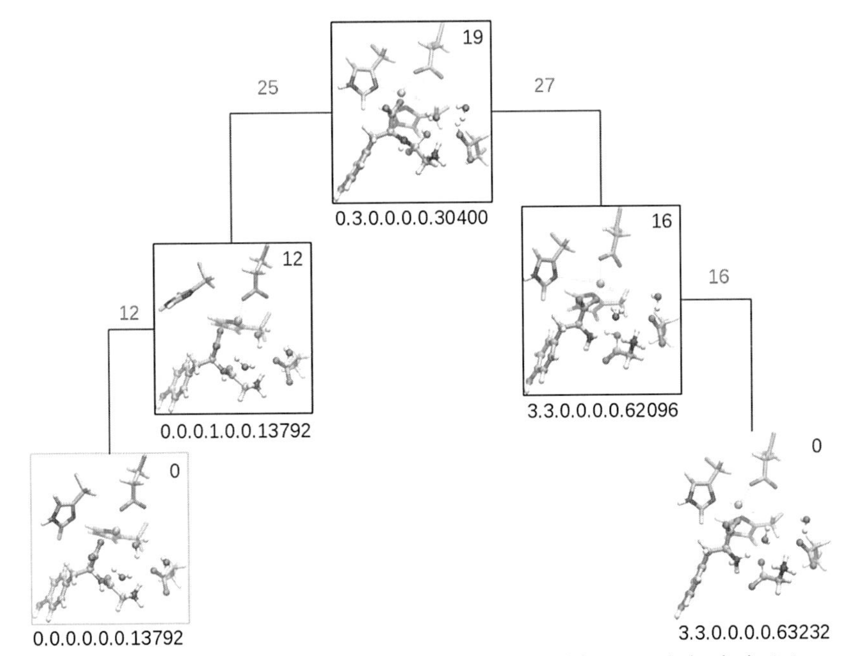

Fig. 10 Best pathway on the transition network computed for peptide hydrolysis in carboxy peptidase. *Black digits* encode the state of the nodes (minima) as displacement steps relative to the reactant state: oldBond.newBond.χH196 .χH69 .χE270 .χE72 .protonation. *Blue numbers inside the boxes* are relative minimum energies and *red numbers next to the connection lines* are relative transition state energies for the connection between to nodes, as indicated by the *lines.* All energies are in kcal/mol. (See the color plate.)

The best pathways can all be described as "promoted-water mechanism" in agreement with Xu and Guo (2009) and Alvarez-Santos, Ganzalez-Lafont, and Lluch (1994). The energetics of our model shows higher barriers than those reported in Xu and Guo (2009). Note, however, that we report only potential energies, in contrast to the free energies computed by Xu and Guo. Another difference is the usage of the semiempirical PM3 method, instead of DFTB with subsequent B3LYP refinement. Comparison of the energetics obtained in a truncated active site model by DFTB//B3LYP and AM1 calculations (Alvarez-Santos et al., 1994; Xu & Guo, 2009) shows substantially higher barriers in both cases, 32 and 38 kcal/mol, respectively, than the free energy calculations for the full enzymatic system. Moreover, the calculations presented here are carried out for the Gly–Tyr peptide which has been reported to be processed only at a low rate (Quiocho & Lipscomb, 1971).

No "anhydride pathway" was found in the entire network. No tetrahedral intermediate formed by Glu270 attack could be trapped. All minima that show Glu270 attacked to the peptide have an already broken C–N bond.

Transitions to such an intermediate require C–N bond dissociation before or in concert with Glu270 attack, which are energetically very unfavorable. Furthermore, the positively charged N-terminal group of the Gly–Tyr peptide interacts favorably with the anionic side chain of Glu270, rendering breaking this salt bridge highly unlikely.

3. CONCLUSION

Enzymatic reactions are unlikely to follow only one precise pathway. By sampling several putative mechanisms, either "manually" generated or obtained from an approach that automatically samples the energy landscape, or more comprehensive picture can be obtained. Intermediate states that are common to more than one mechanism are branching and merging points at which the reaction can take another route.

Combining different mechanisms to a network where such branching/merging points represent the nodes of the network, allows for the comparison of alternative routes, and for the identification of characteristics that are common to apparent different reaction mechanisms. Transition networks allow an automatic generation of many pathways, the identification of energetically optimal routes and important stations (intermediates as well as reaction steps) that can be decisive for a type of mechanism such as an important proton transfer step that must precede the departure of a leaving group.

Still, even with a limited set of degrees of freedom to be explored, a transition network on a potential energy landscape provides a good estimate for likely pathways, capturing the essential (rate-determining, high barrier) transitions for the chemistry to take place and the important internal coordinate involved in these steps. With such information, the minimum energy pathways calculations may be easily augmented by free energy calculations biased along the best pathways.

ACKNOWLEDGMENT

We thank Frank Noé for making the sources of his transition networks implementation for conformational transitions and other useful libraries such as MolTools available to us.

REFERENCES

Alvarez-Santos, S., Ganzalez-Lafont, A., & Lluch, J. M. (1994). On the water-promoted mechanism of peptide cleavage by carboxypeptidase A. A theoretical study. *Canadian Journal of Chemistry*, *72*, 2077–2083.

Ball, K. D., Berry, R. S., Kunz, R. E., Li, F.-Y., Proykova, A., & Wales, D. J. (1996). From topographies to dynamics on multidimensional potential energy surfaces of atomic clusters. *Science*, *271*, 963.

Barducci, A., Bonimi, M., & Parrinello, M. (2011). Metadynamics. *Wiley Interdisciplinary Reviews: Computational Molecular Science*, *1*, 826.

Becker, O. M., & Karplus, M. (1996). The topology of multidimensional potential energy surfaces: Theory and application to peptide structure and kinetics. *The Journal of Chemical Physics*, *106*, 1495.

Bergeler, M., Simm, G. N., Proppe, J., & Reiher, M. (2015). Heuristics-guided exploration of reaction mechanisms. *Journal of Chemical Theory and Computation*, *11*, 5712–5722.

Brooks, C. L., III, Onuchic, J. N., & Wales, D. J. (2001). Taking a walk on a landscape. *Science*, *293*, 612.

Brothers, E. N., Suarez, D., Deerfield, D. W., 2nd., & Merz, K. M., Jr. (2004). PM3-compatible zinc parameters optimized for metalloenzyme active sites. *Journal of Computational Chemistry*, *25*, 1677–1692.

Chikvaidze, M., Erdel, F., & Imhof, P. submitted.

Dama, J., Hocky, G. M., Sun, R., & Voth, G. (2015). Exploring valleys without climbing every peak: More efficient and forgiving metabasin metadynamics via robust on-the-fly bias domain restriction. *Journal of Chemical Theory and Computation*, *11*, 5638–5650.

Díaz, N., & Field, M. J. (2004). Insights into the phosphoryl-transfer mechanism of cAMP-dependent protein kinase from quantum chemical calculations and molecular dynamics simulations. *Journal of the American Chemical Society*, *126*, 529–542.

Dijkstra, E. A. (1959). A note on two problems in connection with graphs. *Numerische Mathematik*, *1*, 269.

E, W., Ren, W., & Vanden-Eijnden, E. (2002). String method for the study of rare events. *Physical Review B*, *66*, 052301.

Ensing, B., Laio, A., Parrinello, M., & Klein, M. L. (2005). A recipe for the computation of the free energy barrier and the lowest free energy path of concerted reactions. *The Journal of Physical Chemistry B*, *109*(14), 6676–6687.

Eurenius, K. P., Chatfield, D. C., Brooks, B. R., & Hodoscek, M. (1996). Enzyme mechanisms with hybrid quantum and molecular mechanical potentials. I. theoretical considerations. *International Journal of Quantum Chemistry*, *60*, 1189–1200.

Field, M. J. (2002). Simulating enzyme reactions: Challenges and perspectives. *Journal of Computational Chemistry*, *23*, 48–57.

Fischer, S., & Karplus, M. (1992). Conjugate peak refinement: An algorithm for finding reaction paths and accurate transition states in systems with many degrees of freedom. *Chemical Physics Letters*, *194*, 252–261.

Florián, J., Goodman, M. F., & Warshel, A. (1995). Computer simulation of the chemical catalysis of DNA polymerases: Discriminating between alternative nucleotide insertion mechanisms for T7 DNA polymerase. *Journal of the American Chemical Society*, *117*, 11619–11627.

Fothergill, M., Goodman, M. F., Petruska, J., & Warshel, A. (1995). Structure-energy analysis of the role of metal ions in phosphodiester bond hydrolysis by DNA polymerase I. *Journal of the American Chemical Society*, *117*, 11619–11627.

Gao, J., & Truhlar, D. G. (2002). Quantum mechanical methods for enzyme kinetics. *Annual Review of Physical Chemistry*, *53*, 467–505.

Garcia-Viloca, M., Gao, J., Karplus, M., & Truhlar, D. G. (2004). How enzymes work: Analysis by modern rate theory and computer simulations. *Science*, *303*, 186–195.

Gregersen, B. A., Lopez, X., & York, D. M. (2004). Hybrid QM/MM study of thio effects in transphosphorylation reactions: The role of solvation. *Journal of the American Chemical Society*, *126*, 7504–7513.

Henkelman, G., Jóhannesson, G., & Jónsson, H. (2000). Methods for finding saddle points and minimum energy paths. In *Progress on theoretical chemistry and physics* (pp. 269–300): Dordrecht: Kluwer Academic Publishers.

Henkelman, G., Uberuage, B. P., & Jonsson, H. (2000). A climbing image nudged elastic band method for finding saddle points and minimum energy paths. *The Journal of Chemical Physics*, *113*, 9901–9904.

Imhof, P., Fischer, S., & Smith, J. C. (2009). Catalytic mechanism of DNA backbone cleavage by the restriction enzyme Eco RV: A quantum mechanical/molecular mechanical simulations analysis. *Biochemistry*, *48*, 9061–9075.

Ivchenko, O., Whittleston, C. S., Carr, J. M., Imhof, P., Goerke, S., Zaiss, M., et al. (2014). Proton transfer pathways, energy landscape, and kinetics in a creatine-water system. *The Journal of Physical Chemistry B*, *118*, 1969–1975.

Kanaan, N., Crehuet, R., & Imhof, P. (2015). Mechanism of the glycosidic bond cleavage of mismatched thymine in human thymine DNA glycosylase revealed by quantum mechanical/molecular mechanical calculations. *The Journal of Physical Chemistry B*, *119*, 12365–12380.

Kiani, F. A., & Fischer, S. (2014). Catalytic strategy used by the myosin motor to hydrolyze ATP. *Proceedings of the National Academy of Sciences of the United States of America*, *111*(29), E2947–E2956.

Kuhn, H. W. (1955). The hungarian method for the assignment problem. *Naval Research Logistics Quarterly*, *2*, 83–97.

Leines, G. D., & Ensing, B. (2012). Path finding on high-dimensional free energy landscapes. *Physical Review Letters*, *109*, 020601.

Lopez, X., York, D. M., DeJaegere, A., & Karplus, M. (2002). Theoretical studies on the hydrolysis of phosphate diesters in the gas phase, solution and RNase A. *International Journal of Quantum Chemistry*, *86*, 10–16.

Noé, F., Krachtus, D., Smith, J. C., & Fischer, S. (2006). Transition networks for the comprehensive characterization of complex conformational change in proteins. *Journal of Chemical Theory and Computation*, *2*, 840–857.

Quiocho, F. A., & Lipscomb, W. N. (1971). Carboxypeptidase A: A protein and an enzyme. *Advances in Protein Chemistry*, *25*, 1–78. Academic Press.

Rappoport, D., Galvin, C. J., Zubarev, D. Y., & Aspuru-Guzik, A. (2014). Complex chemical reaction networks from heuristics-aided quantum chemistry. *Journal of Chemical Theory and Computation*, *10*(3), 897–907.

Rosta, E., Nowotny, M., Yang, W., & Hummer, G. (2011). Catalytic mechanism of RNA backbone cleavage by ribonuclease H from quantum mechanics/molecular mechanics simulations. *Journal of the American Chemical Society*, *133*(23), 8934–8941.

Rosta, E., Woodcock, H. L., Brooks, B. R., & Hummer, G. (2009). Artificial reaction coordinate "tunneling" in free-energy calculations: The catalytic reaction of RNase H. *Journal of Computational Chemistry*, *30*(11), 1634–1641.

Somani, S., & Wales, D. J. (2013). Energy landscapes and global thermodynamics for alanine peptides. *The Journal of Chemical Physics*, *139*(12), 121909.

Stern, K. L. (2012–2014). https://github.com/KevinStern/software-and-algorithms/blob/master/src/main/java/blogspot/software_and_algorithms/stern_library/optimization/HungarianAlgorithm.

Vanden-Eijnden, E., & Venturoli, M. (2009). Revisiting the finite temperature string method for the calculation of reaction tubes and free energies. *The Journal of Chemical Physics*, *130*, 194103.

Wales, D. J. (2002). Discrete path sampling. *Molecular Physics, 100*(20), 3285–3305.
Wales, D. J. (2006). Energy landscapes: Calculating pathways and rates. *International Reviews in Physical Chemistry, 25*(1–2), 237–282.
Wales, D. J., & Doye, J. P. K. (2003). Stationary points and dynamics in high-dimensional systems. *The Journal of Chemical Physics, 119*, 12409.
Warshel, A., & Weiss, R. M. (1980). An empirical valence bond approach for comparing reactions in solutions and in enzymes. *Journal of the American Chemical Society, 102*, 6218–6226.
Xu, D., & Guo, H. (2009). Quantum mechanical/molecular mechanical and density functional theory studies of a prototypical zinc peptidase (carboxypeptidase A) suggest a general acid-general base mechanism. *Journal of the American Chemical Society, 131*, 9780–9788.
Zheng, L., Chen, M., & Yang, W. (2008). Random walk in orthogonal space to achieve efficient free-energy simulation of complex systems. *Proceedings of the National Academy of Sciences of the United States of America, 105*, 20227–20232.

CHAPTER TWELVE

Conformational Sub-states and Populations in Enzyme Catalysis

P.K. Agarwal*,†,1, N. Doucet‡, C. Chennubhotla§, A. Ramanathan¶, C. Narayanan‡

*Computational Biology Institute, Oak Ridge National Laboratory, Oak Ridge, TN, United States
†University of Tennessee, Knoxville, TN, United States
‡INRS—Institut Armand-Frappier, Université du Québec, Laval, QC, Canada
§University of Pittsburgh, Pittsburgh, PA, United States
¶Oak Ridge National Laboratory, Oak Ridge, TN, United States
[1]Corresponding auhtor: e-mail address: pratul@agarwal-lab.org

Contents

Abstract

Enzyme function involves substrate and cofactor binding, precise positioning of reactants in the active site, chemical turnover, and release of products. In addition to formation of crucial structural interactions between enzyme and substrate(s), coordinated motions within the enzyme–substrate complex allow reaction to proceed at a much faster rate, compared to the reaction in solution and in the absence of enzyme. An increasing number of enzyme systems show the presence of conserved protein motions that are important for function. A wide variety of motions are naturally sampled

Methods in Enzymology, Volume 578
ISSN 0076-6879
http://dx.doi.org/10.1016/bs.mie.2016.05.023

(over femtosecond to millisecond time-scales) as the enzyme complex moves along the energetic landscape, driven by temperature and dynamical events from the surrounding environment. Areas of low energy along the landscape form *conformational sub-states*, which show higher conformational populations than surrounding areas. A small number of these protein conformational sub-states contain functionally important structural and dynamical features, which assist the enzyme mechanism along the catalytic cycle. Identification and characterization of these higher-energy (also called *excited*) sub-states and the associated populations are challenging, as these sub-states are very short-lived and therefore rarely populated. Specialized techniques based on computer simulations, theoretical modeling, and nuclear magnetic resonance have been developed for quantitative characterization of these sub-states and populations. This chapter discusses these techniques and provides examples of their applications to enzyme systems.

1. INTRODUCTION

A number of factors contribute to an enzyme's ability to accelerate a biochemical reaction by several orders of magnitude as compared to the reaction in water (Agarwal, 2006; Benkovic & Hammes-Schiffer, 2003). It is well known that enzymes achieve this acceleration by providing special structural and electrostatic environments suitable for the chemical reaction to occur (Warshel et al., 2006). The designated function of an enzyme molecule is, therefore, to provide an environment that is considerably different from the bulk solution, and which favors the reaction to proceed at a much faster rate than otherwise possible. A wealth of information is available about the role of structure in enzyme function (Knowles, 1991), and it is also widely acknowledged that enzymes provide electrostatic and structural stabilization of the transition state (TS) and other intermediates during the reaction (Benkovic, 1992). Recent investigations show that even if these two are taken into account, the catalytic efficiency cannot be fully explained for a large number of enzymes (Agarwal, 2005). Emerging evidence indicates that enzyme rates may be closely tied to the ability of enzymes to sample through a number of alternate structures (or *conformational sub-states*) such that it allows the reactive environment to achieve structural and electrostatic complementarity to the TS (and other intermediate states) along the reaction (Ramanathan, Savol, Burger, Chennubhotla, & Agarwal, 2014).

In the more familiar paradigm, the catalytic cycle involves the binding of substrate(s), and if required cofactor(s), and the positioning of these into the correct orientation into the active site. This is followed by the actual chemical step or substrate turnover. Then, the product(s) and spent cofactor are released (Boehr, McElheny, Dyson, & Wright, 2006). Inherently, the movement of various molecules in and out of the active site would require enzyme motion; such passive motions could be considered as a consequence of these binding/release events. However, evidence continues to build strongly in favor of some of the protein motions playing a much more active or *promoting* role in the catalytic cycles of enzymes. Protein motions occur over 12 orders of magnitude in time, allowing enzymes to sample vastly different conformations. Large-scale motions or conformational fluctuations at long time-scales enable the sampling of high-energy intermediates or a group of conformations (*conformational sub-state*). Experimental evidence continues to indicate that the rate of sampling of correct motions is somehow connected to the time-scale (or rate) of the various events in the enzyme cycle (Boehr, Dyson, & Wright, 2006; Henzler-Wildman et al., 2007). In particular, if the rate-limiting event of enzyme cycle requires a conformational motion of the enzyme, then the catalytic efficiency is directly tied to the sampling of the conformational sub-state and conformations transitions into this sub-state (Ramanathan et al., 2014).

Investigations indicate that sampling of these conformational sub-states is an intrinsic property of an enzyme and is observed even in the apo form of the enzyme (Eisenmesser, Bosco, Akke, & Kern, 2002). In a number of nuclear magnetic resonance (NMR) experiments, it has been observed that an enzyme does not necessarily experience a unique structural signature (corresponding to a homogeneous population) at each substep along the catalytic cycle (Boehr, McElheny, et al., 2006). Rather, it is much more likely that the enzyme samples not only the majority of conformations in a single sub-state but that it also samples the minor conformations associated with the neighboring sub-states. Overall, this is an indication of the fact that the enzyme landscape is full of conformational states, which are sampled over different time-scales (Ramanathan, Savol, Langmead, Agarwal, & Chennubhotla, 2011).

To understand the catalytic efficiency of the full enzyme cycle as well as to understand the relevance of motions and conformational sub-states along the catalytic cycle, it is important to quantitatively characterize the sub-states and the conformational transitions along the landscape (Ramanathan et al., 2014). Experimental information has been difficult to obtain because the time-scales involved can vary over several orders of magnitude (see Fig. 1) (Kleckner & Foster, 2011), while each experimental technique has an inherent limitation of time-scale(s)/resolution. It has been recently suggested that a joint effort between computational and experimental techniques can provide much more useful ways of obtaining data and deciphering the

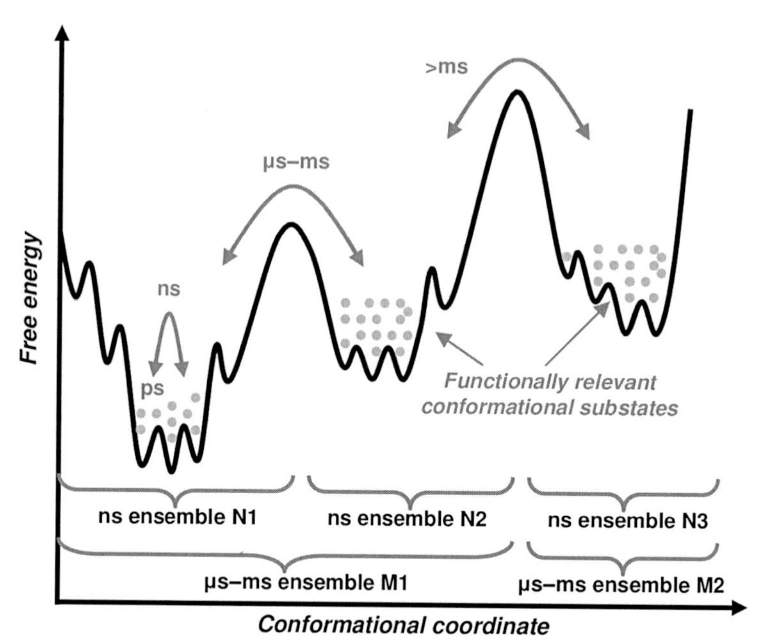

Fig. 1 Conformational sub-states in enzyme landscape. Protein motions allow an enzyme to sample a variety of conformational sub-states. Motions within sub-states occur on fast time-scales (ps–ns) and conformational fluctuations at longer time-scales (μs–ms and >ms) allow overcoming large barriers, enabling access to other sub-states in the conformation hierarchy. *Gray dots* indicate unique conformations. The conformations within each well (sub-state) form ns ensembles (N1, N2, and N3), while μs-ms ensembles would correspond to wider areas (M1 and M2). *This figure is partially based on information from Kleckner, I. R., & Foster, M. P. (2011). An introduction to NMR-based approaches for measuring protein dynamics.* Biochimica et Biophysica Acta, 1814*(8), 942–968.*

information. In this chapter, we discuss computational and theoretical techniques that allow the characterization of the dynamics and the associated conformational sub-states. Discussion is also presented regarding the current methodologies to extract information on conformational sub-states using NMR relaxation dispersion experiments. Finally, we provide two illustrative examples on how these techniques are applied to enzyme catalysis.

2. THEORETICAL CONCEPTS

Proteins, including enzymes, sample distinct conformations enabled by internal motions on a wide range of time-scales (Frauenfelder & Leeson, 1998; Kleckner & Foster, 2011). The motions within proteins range from bond vibrations that occur on femtosecond time-scales to slow conformational fluctuations of domains that occur on microsecond (or longer) time-scales. The intermediate time-scales are movements of loops and coordinated movements of beta-sheets. As depicted in Fig. 1, the various low energy regions of the protein energy landscape (valleys or energy wells) are separated by barriers associated with conformational changes on the microsecond–millisecond (μs–ms) or longer (>ms) time-scale, while within-well sampling operates at faster picosecond–nanosecond time-scales (ps–ns) (Henzler-Wildman & Kern, 2007). Thermodynamic sampling governs sub-state populations and their transitions according to temperature and energy barrier height. Some energy wells have more population (*densely populated*) while others may be less populated (*scarcely populated*). Emerging evidence suggests that protein function (such as enzyme catalysis) is facilitated by conformational sub-states that promote specific interactions between reactants at various stages in the catalytic cycle (Boehr, McElheny, et al., 2006; Eisenmesser et al., 2002; Goodey & Benkovic, 2008; Henzler-Wildman et al., 2007).

Experimental investigations and computer simulations show that protein internal motions generate a distribution, or ensemble, of structures. Depending upon the experimental technique used or time-scale of computer simulations, data on different types of ensembles can be collected. For example, in Fig. 1, N1, N2 and N3 correspond to ps–ns ensembles, while M1 and M2 correspond to μs–ms ensembles. Thermodynamical sampling allows enzymes to visit many of these alternative conformations; therefore, during analysis it would not be possible to assign catalytic relevance a priori to a single conformation and ignore functional characteristics of all others. On the contrary, these alternative conformations or conformational sub-states (and their transitions) are increasingly understood as promoting function (Fraser et al., 2009, 2011). In this new paradigm, the internal protein motions are critical to, not a byproduct of, enzyme function.

The challenge associated with characterizing the conformational sub-states partially arises from the fact that the protein conformational landscape is multidimensional and the conformational sub-states involve multiple dimensions (or variables). Therefore, techniques are required to reduce the multidimensional representation into lower dimensional space (with limited variables). In the past, several definitions of conformational sub-states or intermediates have been used where classification of conformations is based on geometrical quantities from active sites (such as important interatomic distances or angles) or based on other quantities such as orientation of a few residues or loop regions. However, this requires an assumption and/or prior knowledge of the importance of residues. In reality, the conformational sub-states are defined by thermodynamical sampling; therefore, any technique that proposes to identify correct conformational sub-states should also be able to provide clear energetic separation in the landscape. In this chapter, we use a fundamentally different definition of conformational sub-states, which is based on identification of functionally relevant conformational fluctuations and does not require prior knowledge of the mechanism or important residues. This is an automated process and one of the advantages is that it can achieve consistency with internal energy-based separation. This is discussed in more detail in Section 5.

3. ASCERTAINING CONFORMATIONAL SUB-STATES AND POPULATIONS FROM RELAXATION DISPERSION NMR

Enzyme active sites achieve TS stabilization by forming critical interactions with substrate(s). At TS, the enzyme structures themselves are under strain and a part of higher-energy conformations. Fig. 2 shows a schematic of how an enzyme in ground state *A* samples another state *B* with higher-energy conformations, enabled by conformational transitions that occur on slow time-scales (milliseconds). Sub-state *B* could contain functionally relevant conformations that stabilize the TS and therefore promote the

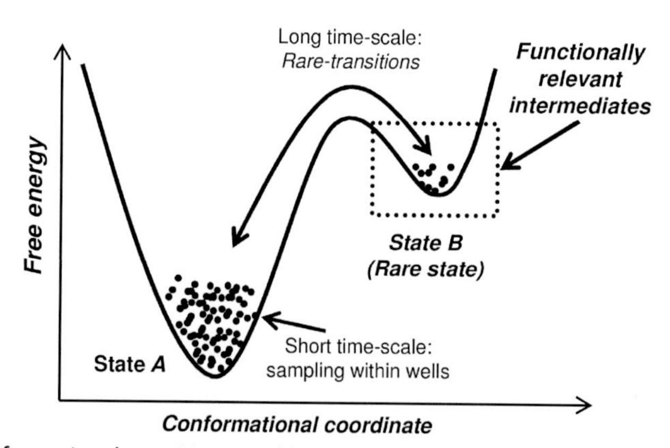

Fig. 2 Conformational transitions enable enzymes to sample higher-energy sub-states. In a hypothetical case where enzyme samples two states *A* and *B*, state *A* is in lower energy and has higher population and state *B* is higher in energy and has lower population. State *B* contains conformations that are functionally relevant; therefore, in this case the sampling of conformational transitions at long time-scale (and the rate of this sampling) that enable access to state *B* will be important for function as well.

enzyme reaction. Due to the higher energy of state B, the population in this state is lower and sampled infrequently. When enzymes are investigated using techniques such as NMR (either in apo or substrate-bound forms), the resulting dynamic equilibrium ensemble data are populated by a mix of conformational populations, with the major population represented by the ground state A and the minor excited population by state B. The minor population of state B has been referred to as *hidden* or *invisible*, as it is masked by the major population of state A.

Experimental NMR relaxation investigation allows one to decompose this NMR data into information about the two states A and B and the populations associated with them. Spin relaxation rates, especially those of transverse (R_2) magnetization, have been particularly useful to characterize the rate of conformational exchange between such states on the millisecond time-scale, further providing quantifiable information on the equilibrium site populations (p_A and p_B). In recent years, the relaxation-compensated Carr–Purcell–Meiboom–Gill (rcCPMG) and $R_{1\rho}$ experiments have been extensively used to characterize functionally relevant conformational exchange at specific local and/or global atomic sites within enzymes, specifically because they sample the millisecond time-scale of catalysis (k_{cat}) in most enzyme systems (Kleckner & Foster, 2011; Lisi & Loria, 2016; Palmer, 2015). The rcCPMG experiment has been the method of choice to probe a two-site chemical exchange experienced by $^1H–^{15}N$ backbone amide vectors (^{15}N-CPMG), providing a precise atomic-scale measure of the millisecond dynamics experienced by nearly every residue on a given enzyme. Despite considerable experimental improvements in recent years—especially with respect to higher molecular weight systems (Kay, 2015)—routine ^{15}N-CPMG analysis remains dependent on the lack of resonance overlap and overall quality of the two-dimensional $^1H–^{15}N$ heteronuclear single quantum coherence (HSQC) spectra of enzymes under study. Analyzing the line shape, width, and chemical shift of the $^1H–^{15}N$ HSQC resonances allows one to extract information on residues experiencing conformational exchange, providing quantifiable data on the *invisible* excited states rarely populated in solution (p_B). The theory and experimental implementation of the rcCPMG experiment are beyond the reach of the present review and have been treated in detail elsewhere (Ishima, 2014; Palmer, Kroenke, & Loria, 2001). Here, we provide a quick overview of the type of information obtained from multiple-field fits of the R_2 relaxation data measured as a function of the spacing τ_{CP} between successive π pulses in a CPMG echo sequence. Useful parameters extracted from such experiments include the weight of equilibrium populations (p_A and p_B), the chemical shift differences between each state ω_A and ω_B ($\Delta\omega$), and the rate of conformational exchange (k_{ex}) between each state.

NMR is particularly well suited to study subtle structural changes occurring in proteins, since distinct populations A and B arising from slow structural transitions between two or more states give rise to magnetically distinct line shapes and chemical shift resonances in NMR spectra (eg, ω_A and ω_B in a two-site exchange system) (Fig. 3). This is particularly obvious for systems where the exchange rate (k_{ex}) between the two states A and B occurs on a slow time-scale ($k_{ex} \ll \Delta\omega$), and for approximately equal population ratios (eg, 50% p_A and p_B), or slightly skewed population ratios (eg, 70% p_A and 30% p_B). Note that k_{ex} corresponds to the arrow that goes between the two states in Fig. 2. While this *ideal* situation provides structural information on populated states exchanging on a time-scale slow enough to allow the appearance of individual line shapes for each populated state, it does not provide any functional information on the rarely populated ($p_B < 5$–10%), higher-energy conformers exchanging on a much faster time-scale (eg, $k_{ex} > 1000\ s^{-1}$). This situation is typical of enzyme systems experiencing conformational exchange between highly skewed populations on the fast NMR regime ($k_{ex} \gg \Delta\omega$), where the functionally relevant, higher-energy conformer B is so rarely populated that it is often *invisible* (ie, displaying a highly broadened line shape) in the NMR spectrum, despite it being potentially involved in stabilizing a TS, positioning a specific reactant, and/or releasing a reaction product. Fig. 3 presents theoretical examples where the rcCPMG method provides the theoretical means to extract, analyze, and quantify these *hidden*, low-populated states experiencing conformational exchange on the millisecond time-scale, which often overlaps with the rate of catalysis (k_{cat}) in many enzyme systems.

4. OBTAINING CONFORMATIONAL SUB-STATES FROM SIMULATIONS

The general scheme used for identification and characterization of the conformational sub-states is described in Fig. 4. The protein conformational landscape is highly multidimensional (hundreds to thousands of dimensions correspond to the number of atoms and internal degrees of motion). It is not practical to inspect each dimension (or even two or three at a time) for characterization of the full landscape. For visualization and interpretation, methods are required to find alternate representation that reduce this

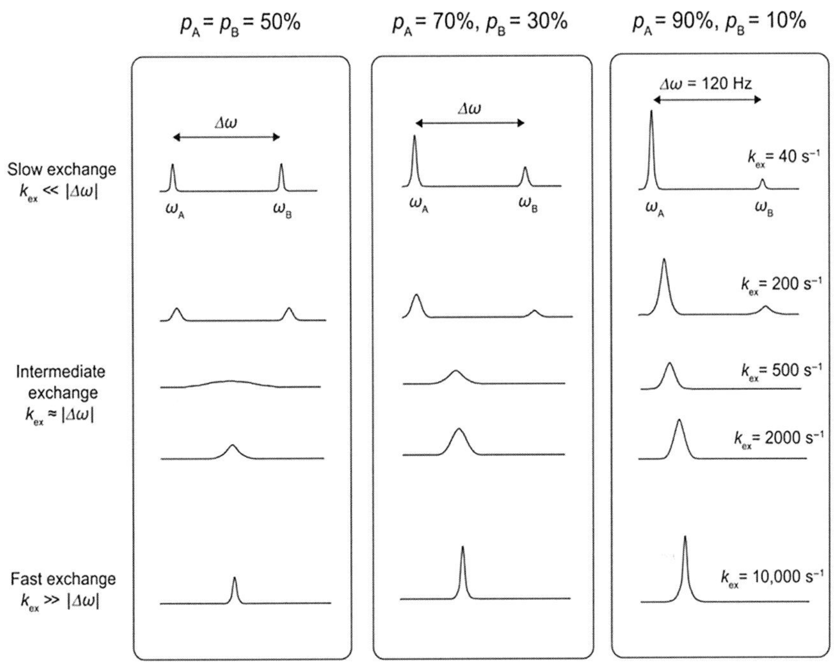

Fig. 3 Two-site conformational exchange experienced by a ^{1}H–^{15}N amide vector sampling distinct weighted populations of states *A* and *B* on various NMR time-scales. All three columns report on simulated two-state chemical exchange where the populations of each state (p_A and p_B) are skewed, with $\Delta\omega = 120$ Hz and $k_{ex} = 40$, 200, 500, 2000, or 10,000 s^{-1}. For clarity, values of $\Delta\omega$ and k_{ex} are only labeled in the right column. Slow exchanging populations sampling states *A* and *B* give rise to two distinct line shapes corresponding to magnetically distinct conformers of equal weight (column 1, ω_A and ω_B), separated by chemical shift difference $\Delta\omega$. Resonance signal broadening results from increased rates of chemical exchange between each state (local dynamics) on the intermediate and fast time-scale regimes, where a single weighted-average chemical shift population is observed ($k_{ex} > 500\ s^{-1}$ in this particular example). A single, sharper signal with distinct intensity and line width is observed on faster time-scales. A similar behavior is observed when the p_A and p_B population ratio is significantly skewed in favor of ground state *A* (columns 2 and 3). While excited state *B* is *invisible* on intermediate and fast time-scales, the shape and chemical shift of the resulting NMR signal is proportional to each population state. Relaxation dispersion NMR experiments such as the rcCPMG and $R_{1\rho}$ methods provide the theoretical means to extract, analyze, and quantify these *hidden*, low-populated states experiencing conformational exchange on the millisecond time-scale, which often overlaps with the rate of catalysis (k_{cat}) in many enzyme systems. *Adapted from Kempf, J. G., & Loria, J. P. (2004). Measurement of intermediate exchange phenomena.* Methods in Molecular Biology, 278, *185–231.*

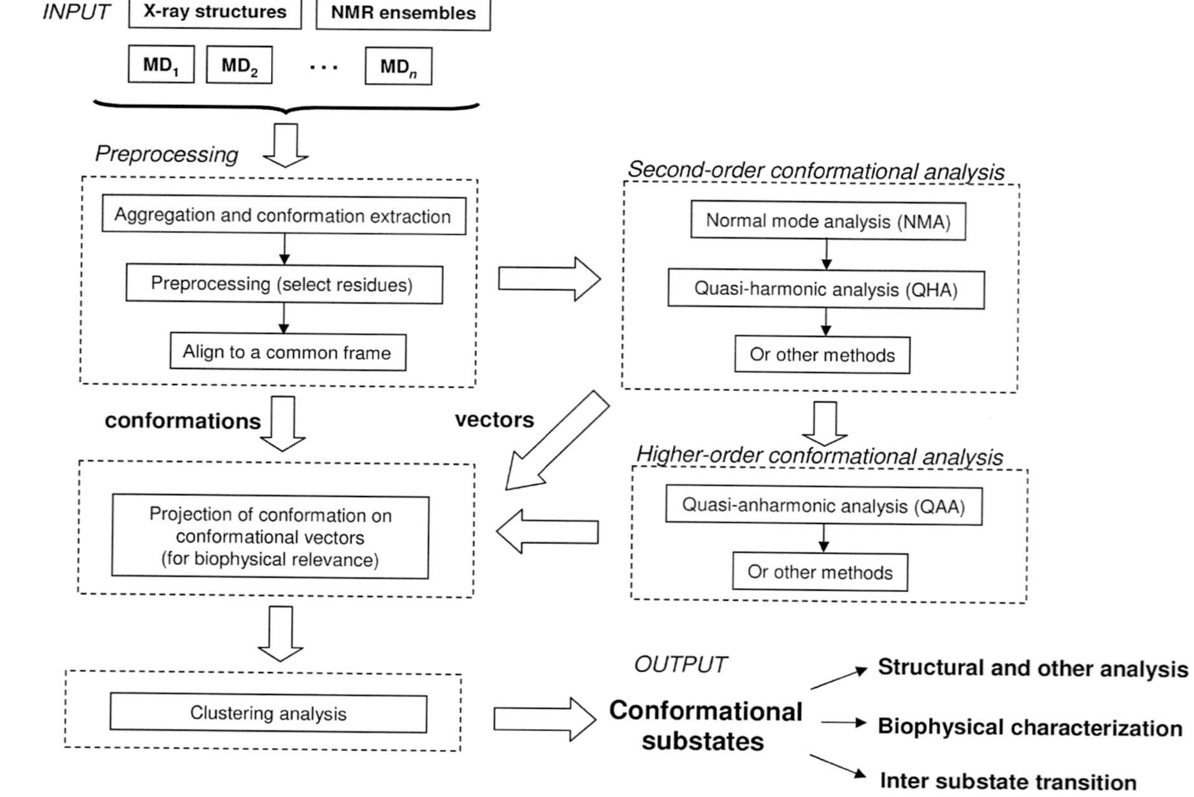

Fig. 4 Protocol for extraction of conformational sub-states and populations. The input to the method is a set of structures or conformations (from X-ray, NMR, or MD simulations), which after preprocessing are used to obtain second-order conformational vectors. For more accurate characterization, higher-order methods are used. The obtained vectors are used for projecting the conformations from an initial set of conformations. The results are analyzed by clustering method to classify the conformations into sub-states.

complex multidimensional conformational landscape to a simple representation. Typically, this is done by finding a set of a limited number of alternate variables (typically ranging between 3 and 10) where the entire conformational dataset can be visualized at a time. A number of techniques have been used to identify such variables or independent coordinates based on the use of principal component analysis (PCA) or related second-order methods. These methods provide eigenvectors corresponding to large-scale protein and enzyme motions, and the conformations can then be projected on these independent coordinates to identify the locations of various sub-states along the conformational landscape. A number of second-order methods are available to obtain the protein motions from individual protein structures (conformation) or a collection of conformations. These are briefly described later.

4.1 Normal Mode Analysis (NMA)

This method computes vibrational modes of a molecular system by diagonalization of the Hessian matrix (Leach, 2001; Qui & Bahar, 2005). Assuming the molecular system has N atoms, the Hessian matrix is a $3N \times 3N$ matrix. The elements of the matrix are the second-order energy derivatives with respect to the displacement of atomic positions in the x, y, and z directions. The elements can be computed analytically (for small systems) or computationally (for larger systems like enzymes). Once the Hessian matrix is computed, it is diagonalized to solve for the eigenvectors and eigenvalues.

$$\varepsilon_i \boldsymbol{F} = \varepsilon_i \omega_i \tag{1}$$

The time-scale of molecular vibration is determined by taking the inverse square root of the eigenvalues (ε) obtained after diagonalization. The eigenvector (ω) corresponding to the eigenvalue represents the vibrational modes, which are a set of displacement vectors for the atoms in the molecular or the protein conformation.

4.2 Time-Averaged Normal Coordinate Analysis (TANCA)

This method is similar to NMA in the sense that the vibrational modes of a protein (or protein vibrational modes) are obtained by diagonalization of the Hessian matrix (Hathorn, Sumpter, Noid, Tuzun, & Yang, 2002; Tuzun, Noid, Sumpter, & Yang, 2002). However, NMA suffers from some limitation when considering a highly flexible molecular system such as a protein.

NMA uses a reference structure and the eigenvalues and eigenvectors thus obtained are only relevant to the reference starting structure. The method therefore weights highly toward the high frequency motions and less toward the low frequency motions. Moreover, the low frequency obtained using NMA are not reliable for molecular conformations that differ considerably from the reference structure for NMA. In enzyme function, the low frequency motions are more important as they are required for overcoming the high-energy barriers. Techniques such as TANCA can partially overcome this problem by diagonalization of the Hessian matrix which has been constructed numerically by averaging the elements over time (from multiple structures). This allows the fast frequency motions to be removed by averaging, providing more accurate low frequency modes that are relevant for enzyme function (Ramanathan & Agarwal, 2009).

4.3 Quasi-Harmonic Analysis

This method computes protein vibrational modes from a set of protein conformations that are sampled using either the molecular dynamics (MD) (or Monte Carlo) type simulations (Perahia, Levy, & Karplus, 1990). Quasi-harmonic analysis (QHA) is a powerful method in obtaining protein vibrational modes that are representative of longer time-scales or the low frequency vibrations, by utilizing the information from a set of structures, which may be separated by a long time-scale—or from different parts of the protein conformational space. The vibrational modes are obtained by diagonalization of the atomic fluctuation matrix. For a protein with N atoms, the atomic fluctuation matrix, $\boldsymbol{F}$ is a symmetric $3N \times 3N$ matrix with term $F_{\alpha\beta}$ defined as:

$$\boldsymbol{F}_{\alpha\beta} = \left\langle m_\alpha^{1/2}(x_\alpha - \langle x_\alpha \rangle) m_\beta^{1/2}(x_\beta - \langle x_\beta \rangle)\right\rangle \tag{2}$$

where α, β run through the $3N$ degrees of freedom in Cartesian space and m_α is the mass of atom corresponding to the αth degree of freedom and x_α are the Cartesian coordinates of the atom corresponding to the αth degree of freedom. Quantities in $\langle\rangle$ denote an average determined from MD simulation. To obtain the eigenmodes (vibrational modes), diagonalization of the atomic fluctuation matrix is performed (see Eq. 1). The time-scale of protein vibration is determined by taking the inverse square root of the eigenvalues (ε) obtained after diagonalization. The eigenvector (ω) corresponding to the eigenvalue represents the protein vibrational modes, which are a set of displacement vectors for the atoms in the protein confirmation. Note one of the benefits of QHA is that multiple MD trajectories can be combined to construct the atomic fluctuation matrix—thus allowing vibrational modes to be computed that represent conformational changes between different areas of the protein conformational space.

4.4 Elastic Anisotropic Network Models

This type of calculation uses coarse-grained NMA to obtain protein conformational modes (Doruker, Jernigan, & Bahar, 2002; Haliloglu, Bahar, & Erman, 1997). These calculations use a simple parameter harmonic potential for the particles in the system. The eigenmodes are obtained by the diagonalization of Kirchhoff's matrix, which is similar to the Hessian matrix, but uses a reduced model of the protein and treats the protein motions as Gaussian-type motions.

For an enzyme-based system, only the coordinates of enzyme, substrate, and cofactor are used for calculating the vibrational modes. Typically nonreactive ions and solvent molecules are excluded from this analysis. Each of the methods discussed earlier provides a set of vector (or independent vectors), each $3N$ in length, corresponding to x, y, and z displacement vectors for N atoms. In addition to the identification of vectors the other requirement to obtain conformational sub-states is a set of enzyme conformations. The conformations are usually collected from MD simulations. A single trajectory or multiple trajectories can be used. In addition, it is also possible for conformations from NMR ensembles and a collection of X-ray structures to be used for projection and subsequent analysis.

Typically, a small number of vectors (3–10) are used for further analysis, as they can capture a large fraction of the protein motion (in terms of variance). The first few vectors correspond to the slowest motions that occur at slow time-scales. Using these vectors, the conformations from the ensemble are projected on each of these vectors. Once the conformations (MD and/or X-ray/NMR based) are projected, a clustering method is used for identification of the *sub-states*. Well-defined clusters represent sub-states while diffused conformations over landscape (where it is difficult to define clusters) correspond to a flat landscape with little possibility of sub-states. It should also be noted that the inability to get proper clusters (sub-states) could also be an indication of the limitation of the underlying computational methodology used for the vector extraction. Note that the same set of conformations can be used as input to obtain the independent coordinates (vectors) and then to obtain the projections on the vectors.

5. ANHARMONIC CONFORMATIONAL ANALYSIS

The techniques discussed in Section 4 are second-order methods based on the assumption that the protein sub-states and overall landscape can be described by harmonic or quasi-harmonic wells (H, Q in Fig. 5). However, a number of experiments, including the one combining neutron scattering experiments with computational modeling of the thermophilic rubredoxin protein, indicate that the protein motions are anharmonic and that the onset of anharmonic structural changes activates functionally relevant motions (Borreguero et al., 2011).

This assumption of second-order motions has an important consequence. It can lead to inaccurate characterization of protein conformational sub-states in that heterogeneous conformers are incorrectly grouped into a single conformational sub-state (Fig. 6). Neighboring conformations in a conformational sub-state are expected to have similar energy and functionally relevant geometrical/kinetic properties. When the protein motions are anharmonic with multiple wells in the conformational landscape, the use of higher-order methods is necessary for identifying them.

5.1 Quasi-Anharmonic Analysis

To quantify the anharmonic time-dependent conformational changes (A in Fig. 5), we have introduced an approach called quasi-anharmonic analysis (QAA) (Burger et al., 2012; Ramanathan, Agarwal, Kurnikova, & Langmead, 2010; Ramanathan, Savol, Agarwal, & Chennubhotla, 2012; Ramanathan et al., 2011; Savol, Burger, Agarwal, Ramanathan, & Chennubhotla, 2011). QAA uses fourth-order statistics (for analytical

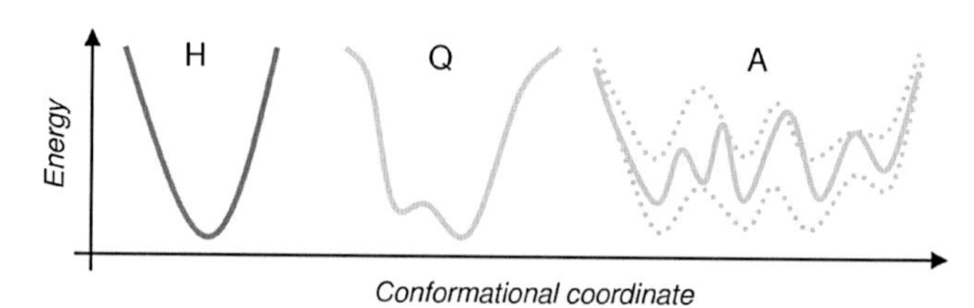

Fig. 5 Protein energy landscape (along individual conformational coordinates) can be classified as harmonic (H), quasi-harmonic (Q), or anharmonic (A). Harmonic landscapes with a single sub-state (well) can be well described by second-order methods. Quasi-harmonic landscapes can be approximated by second-order methods depending on how well a harmonic function is able to fit into the multiple sub-states. Anharmonic landscapes with multiple sub-states are poorly approximated by second-order methods and require higher-order methods for accurate characterization.

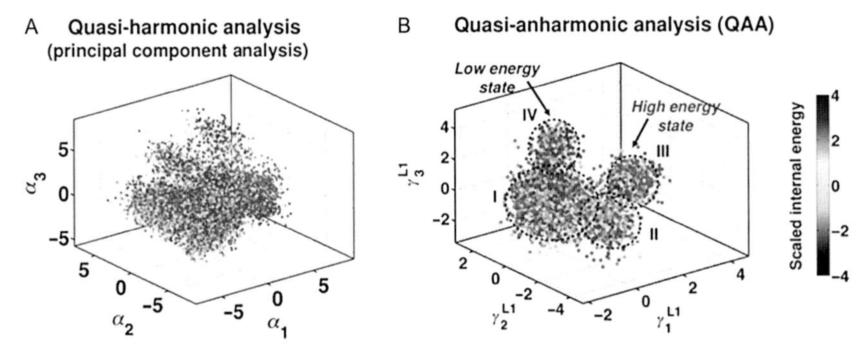

Fig. 6 The benefit of using a higher-order statistical method allows identification of conformational sub-states with homogeneous properties. Conformational sub-states identification for protein ubiquitin was performed using two different methods. (A) The conformational sub-states identified by second-order methods such as quasi-harmonic analysis and principal component analysis do not achieve clear separation and the population of the conformations shows mixed properties. (B) Using a higher-order method such as quasi-anharmonic analysis (QAA) allows identification of sub-states that are clearly separated and conformations have homogeneous properties. In both panels, *each dot* corresponds to a single conformation and the *coloring* is by scaled internal conformational energy. (See the color plate.)

convenience) to describe the atomic fluctuations and summarizes the internal motions using a small number of dominant anharmonic modes. We have successfully demonstrated this approach in the context of protein functions such as molecular recognition (ubiquitin and lysozyme) and enzyme catalysis (human cyclophilin A) (Ramanathan et al., 2011). An *emergent* property of QAA is that by characterizing anharmonicity in positional fluctuations, our method discovers energetically homogeneous conformational sub-states. Note that emergent implies that the discovered homogeneity is achieved without any prior knowledge of the internal energy of the systems.

After aligning the MD ensemble, the QAA approach is to use a small set of anharmonic basis vectors to represent the positional deviation vectors of each conformer:

$$\delta \mathbf{x} = \mathbf{A}\gamma \tag{3}$$

Here $\delta\mathbf{x}$ denotes the positional deviation vector of size $(3N \times 1)$, where N is the number of atoms, $\mathbf{A}$ is a matrix of size $(3N \times m)$ (where $m \ll 3N$) derived from an approximate diagonalization of a tensor built to hold the fourth-order statistics of positional deviations $\delta\mathbf{x}$. The anharmonic modes of motion, which are the columns of $\mathbf{A}$, are sorted in the decreasing order of their norms for convenience. Each anharmonic basis vector $\mathbf{a}_i$ in the matrix $\mathbf{A}$ has an excitation coefficient γ_i. Just like PCA, $\mathbf{A}$ fully decorrelates the input ensemble, ie, there are no second-order dependencies between the elements of γ. In addition, matrix $\mathbf{A}$ is guaranteed to reduce fourth-order dependencies. By construction, $\mathbf{A}$ can be nonorthogonal (unlike the PCA modes), meaning that exciting anharmonic mode $\mathbf{a}_i$ can also effect $\mathbf{a}_j$ because of nonorthogonal coupling between the basis vectors. By design, QAA ignores any nonlinear coupling that may exist in the fluctuations between different parts of a protein.

The various steps involved in performing QAA are (see Fig. 4):

1. Apply Gaussian-weighted root mean square deviations superposition algorithm on the MD ensemble to identify rigid and flexible residues (Ramanathan et al., 2012).
2. Use rigid residues to iteratively align the MD ensemble. Find the positional deviations from the iteratively derived mean conformer.
3. Build a low-dimensional representation for the fluctuations of the backbone or C_α atoms using PCA.
4. Choose a low-dimensional subspace m (say 50 or so) that captures 90–95% of overall variance. This initial projection onto the top m PCA bases reduces the dimensionality of the problem from $3N$ to m and helps speed up convergence of the learning algorithm for QAA.
5. Learn the QAA matrix $\mathbf{A}$, sort the anharmonic modes in decreasing order of their magnitude. For visualization, build a (dimension-reduced) anharmonic space of coefficients: γ_1, γ_2, and γ_3.
6. To gain additional insights into the conformational landscape, label (or color) each triplet in the anharmonic space by either experimental or computational features such as scaled internal energy, interdomain distance, etc.

5.2 Identifying Conformational Sub-states

Observe that neighboring conformers in QAA space have biophysically relevant coordinates (Fig. 6), and this coherence is an emergent property of the QAA representation. Based on this observation, we hypothesize that nearest neighbors in the QAA space can be hierarchically clustered to form dynamically and kinetically related metastable states.

To this end, we consider each frame in the trajectory as a node in an undirected graph constructed in the m-dimensional QAA space. We connect each node to a small number of its nearest Euclidean neighbors.

The edge weights are binary (either 0 or 1), denoting either the presence or absence of an edge. We then cluster this network using a hierarchical Markov diffusion framework that we have proposed previously (Burger & Chennubhotla, 2012; Burger et al., 2012; Chennubhotla & Bahar, 2007a, 2007b; Savol et al., 2011). First, we initiate a Markov chain to propagate on the network and identify a set of putative cluster centers. The total number of clusters is automatically determined by the algorithm, with the rule that every node in the network has some Markov probability of transitioning into at least one of the putative cluster. Second, a Markov transition matrix is built using this reduced representation based on the principle that Markov chains initiated on both the fine scale and coarse scale representations of the network should reach stationary distributions simultaneously. This principle in turn helps build a hierarchical representation of the network and promotes formation of metastable clusters in the data. We expect that fine-grained hierarchy levels will produce many small clusters containing close neighbors in the QAA space; that is, within each such cluster most members will be drawn from the same, narrow time window. As Markov diffusion progresses (fine-grained to coarse-grained), conformers that are more distant neighbors will be connected by edges in the diffused network and will therefore be assigned to the same cluster. Thus, the hierarchical clustering can highlight dynamical connections between conformers at different time-scales.

QAA is one component of a larger suite of tools that we developed, termed as anharmonic conformational analysis, to probe molecular recognition pathways (Savol et al., 2011) and to reveal intermediate states in intrinsically disordered proteins (Burger et al., 2012).

6. EXAMPLES OF CONFORMATIONAL SUB-STATES IN ENZYME CATALYSIS

6.1 Hydride Transfer Catalyzed by Dihydrofolate Reductase

The enzyme dihydrofolate reductase (DHFR) catalyzes the conversion of substrate dihydrofolate (DHF) to tetrahydrofolate (THF) and uses nicotinamide adenosine dinucleotide phosphate (NADPH) as a cofactor. The catalytic cycle of *Escherichia coli* DHFR (EcDHFR) consists of at least five sub-states corresponding to intermediates associated with the substrate (DHF) and cofactor (NADPH) binding, the chemical step of hydride transfer and the product (THF) and spent cofactor ($NADP^+$) release. Using ^{15}N spin relaxation NMR techniques (mostly the rcCPMG described earlier),

Wright and coworkers discovered that each intermediate along the catalytic cycle represents sampling of populations from multiple conformational sub-states (Fig. 7) (Bhabha, Biel, & Fraser, 2015; Boehr, Dyson, et al., 2006; Boehr, McElheny, et al., 2006).

Each intermediate samples the majority of conformational populations relevant to the current state; however, a small fraction of the population also samples the neighboring sub-states (Boehr, McElheny, et al., 2006). The conformational sub-states are stabilized by interactions with the substrate and cofactor. Therefore, the minority of conformations sampled in the present state over time become the major conformations for the subsequent intermediate in the cycle. The rate of the conversion between different intermediate states was observed to coincide with the rate of sampling of interconversion between the majority and minority of conformational states (see *dashed arrows* in Fig. 7). For example, the maximum rate of hydride transfer step (the rate-limiting step of the EcDHFR mechanism at pH > 8.5) is measured to be 950 s^{-1}. The time-scale of this event coincides

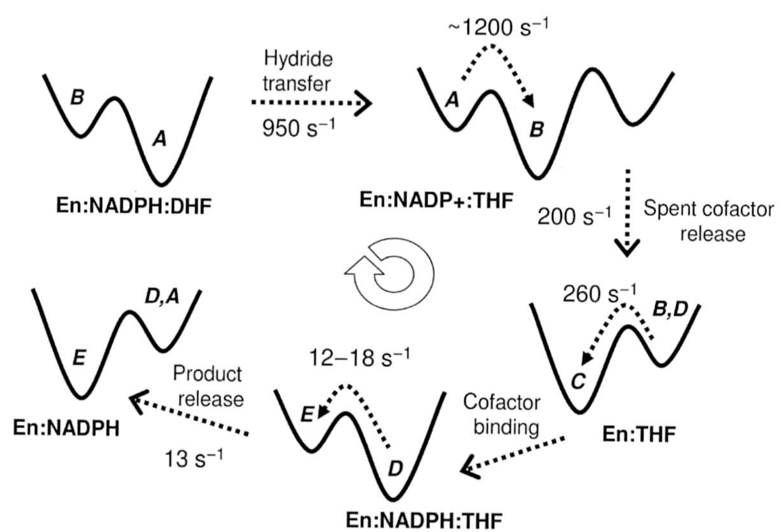

Fig. 7 Conformational sub-states associated with the catalytic activity of the enzyme *E. coli* DHFR. The DHFR (En) catalytic cycle consists of five intermediate states associated with binding and release of cofactor NAPDH, substrate DHF, spent substrate NADP+, and product THF. Each of these intermediates sample multiple enzyme conformations sub-states (*A*, *B*, *C*, *D*, or *E*). The available rates of conversion between the intermediates and the conformational exchange between sub-states are labeled. Note that for each of these intermediate states the lower energy well will have higher population and the higher-energy wells will have much lower conformational populations. *Adapted from Boehr, D. D., McElheny, D., Dyson, H. J., & Wright, P. E. (2006). The dynamic energy landscape of dihydrofolate reductase catalysis.* Science, 313*(5793), 1638–1642.*

with the rate of conformational exchange between its reactant and product conformational intermediates, which has been measured to be about 1200 s^{-1}. The explanation for this observation is that the conformational sampling allows the enzyme in the reactant ground state (protonated DHF + NADPH) to sample conformations that favor the product state (THF + $NAPD^{+}$). Once the enzyme reaches the product-favoring conformational state, the hydride transfer occurs. (There is a proton transfer that accompanies hydride transfer, which is considered to precede hydride transfer and occurs almost instantly; therefore, protonated DHF is considered as the reactive species in the ground state for the chemical step.) The overall rate of the reaction is controlled by the slowest step in the reaction. Under normal conditions the slowest event is product (THF) release, which occurs roughly at 13 s^{-1}, while the rate of conversion of the conformation sub-states is 12–18 s^{-1}.

The details about the fraction of major and minor population in the DHFR catalytic cycle are not fully available from this study. It is proposed that R_2 relaxation dispersion experiments can generally only characterize higher-energy conformations that make up at least 1–2% of the ensemble. Therefore minor state populations in EcDHFR would be of similar order. Furthermore, there may be additional excited states that are not accessible to the R_2 relaxation dispersion experiments. Alternate techniques are required to obtain more detailed information about the conformational populations.

6.2 *Cis/Trans* Isomerization Catalyzed by Cyclophilin A

The computational technique QAA has been successfully used to identify conformational sub-states associated with the *cis/trans* isomerization reaction catalyzed by the enzyme human cyclophilin A (CypA) (Ramanathan et al., 2011). This enzyme belongs to the class of enzymes known as peptidyl-prolyl isomerases and catalyzes the rotation of amide bond preceding to proline residues in a wide variety of substrates, including peptides and proteins (Agarwal, 2004; Agarwal, Geist, & Gorin, 2004). MD simulations in conjunction with reaction pathway sampling were used to sample conformations along the reaction pathway. The dihedral angle associated with the amide bond was used as a reaction coordinate and umbrella sampling was used to collect 18,500 conformations along the enzyme reaction catalyzed by CypA in a representative substrate. QAA was then used to identify multiscale hierarchy characterization of protein landscape both in time- and length-scales. Results are depicted in Fig. 8. The conformational landscape is characterized and separated into a number of conformational sub-states (marked by *ellipses*). The conformations are colored according to the reaction coordinate (amide bond dihedral angle).

The characterization of the conformational snapshots along the CypA reaction with QAA reveals the following results:

1. QAA correctly identifies the sub-state that corresponds to the TS area. In Fig. 8, this corresponds to the highest point in the free energy profile (and *green* (*gray* in the print version) *colored* conformations). The first iteration of QAA (called level I) identifies conformational sub-states associated with the reactant, product, and in the vicinity of TS (labeled as TS′) as well as a mixed energy state (labeled as cluster I). Note that QAA is a way to represent the data in lower dimensional space; therefore, it allows identification of different sub-states in multiple iterations. The mixed cluster is then further characterized (level II), which allows identification of the more accurate conformational sub-state associated with the TS (TS″). It is very important to note that the information about the reaction coordinate is *not* an input to the QAA method. Therefore, the identification of the functionally relevant reactant, product, and TS sub-states is an emergent property.
2. The conformations within the sub-states are sampled at fast time-scales and the transitions at long time-scales enable access to functionally important sub-states with higher energy. Without requiring any prior knowledge, the transition from lower to TS sub-state (*black arrows* in subpanel A) indicates the reaction promoting intrinsic protein dynamics with implications for the mechanism (panels C and D). Therefore, the theoretical foundation of QAA provides us with a unique methodology to bridge the gaps in time-scales as well as length-scales. Note that this is related to the short-lived *hidden* conformational sub-states that have been discovered using X-ray and NMR techniques, and influence enzyme catalysis in CypA (Fraser et al., 2009).
3. The conformational transitions that enable access to the TS sub-state (and other sub-states) can be identified by drawing an arrow between various centers of mixed clusters and the sub-state (see *black arrow*) and the conformations along this arrow provides detailed atomic level information about the fluctuations in various parts of the enzyme (lower panel, Fig. 8). Interestingly, this information from CypA coincides with the NMR spin relaxation experiments. Furthermore, the identification of highly flexible regions and the conserved hydrogen bonds between these regions provides information about a network of residues that promote catalysis.

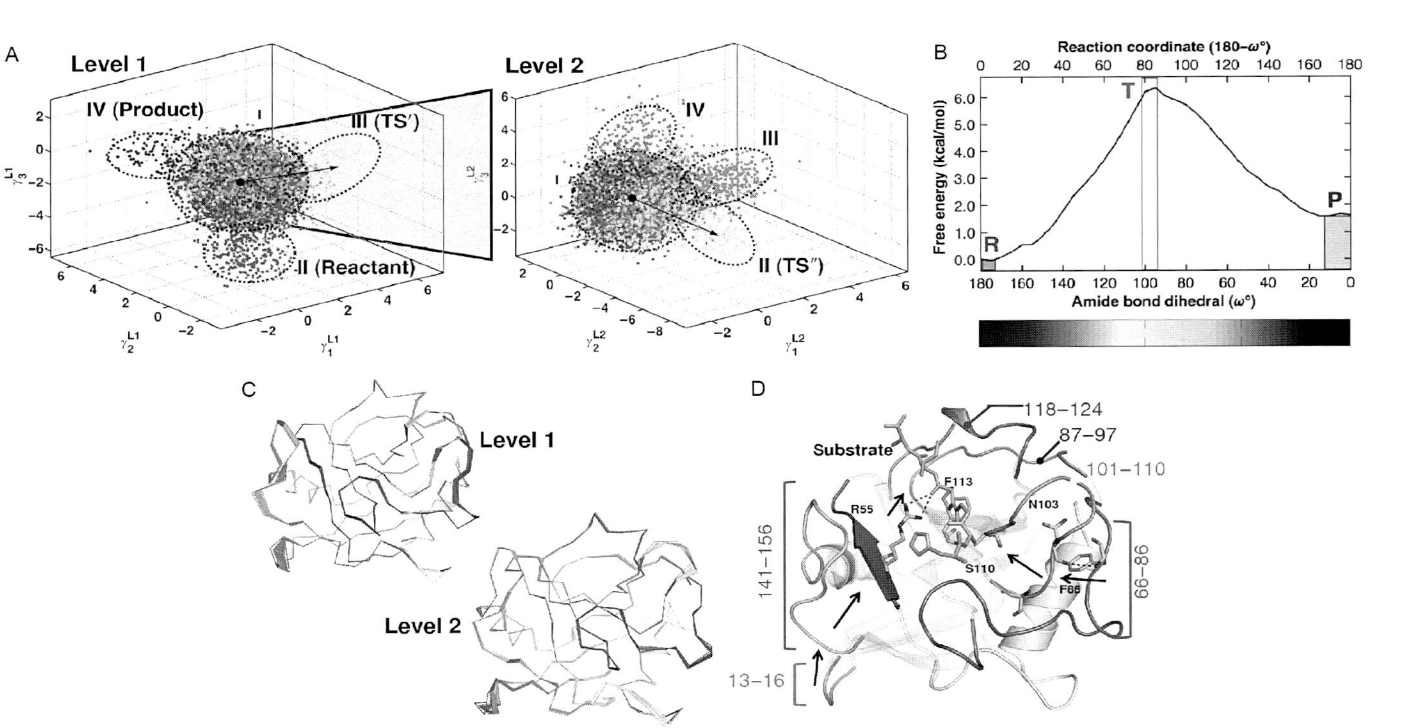

Fig. 8 Computational method QAA allows identification of multiscale hierarchy associated with catalysis by enzyme CypA. (A) Multilevel (two levels shown) hierarchy of conformational sub-states. Each *colored dot* represents a single sampled conformation; *ellipses* indicate sub-states; TS′, TS″, and T indicate transition state area. (B) The free energy profile and conformations in (A) are colored according to reaction coordinate, (C) conformational change between sub-states corresponding to *black arrow* in (A), and (D) impact of identified motions on CypA's mechanism. *Adapted from Ramanathan, A., Savol, A. J., Langmead, C. J., Agarwal, P. K., & Chennubhotla, C. S. (2011). Discovering conformational sub-states relevant to protein function.* PLoS ONE, 6*(1), e15827.* (See the color plate.)

Overall, for results for CypA indicate that the enzyme spends most of its time sampling lower energy states (ground states), including the reactant and product sub-states. Over long time-scales, the conformational fluctuations based on the topology of the enzyme allow sampling of the rare-intermediate states or the functionally relevant sub-states. These sub-states contain the correct geometric and dynamical properties for the formation of TS and thus for promoting the reaction mechanism.

QAA offers a number of advantages over alternate approaches: (1) QAA allows automatic separation of the conformations into sub-states that capture global to local motions (length-scales) and varying over fast to slow time-scales; (2) a unique advantage of QAA *is that without any prior knowledge of functionally related coordinates, it identifies intermediate sub-states with functionally important structural and dynamical features*; and (3) comparison of the functionally relevant sub-states provides insights into the reaction promoting intrinsic enzyme dynamics as well as a network of coupled motions (or enzyme residue networks) that enable solvent–enzyme thermodynamical coupling.

7. SUMMARY AND CONCLUSIONS

Enzymes sample a wide variety of conformations in solution. Only a small fraction of these conformations contain the correct structural and dynamical features that allow the stabilization of the various intermediates during the enzyme mechanism. The sampling of alternate conformations is an intrinsic property of enzymes, enabled by the topology, and driven by temperature and the surrounding environment. Identification and characterization of these conformational sub-states can be achieved though the use of experimental techniques such as NMR and computer simulations.

New techniques are required to process the experimental data and simulations data. NMR relaxation dispersion experiments such as the rcCPMG methodology allow the extraction of useful conformational exchange information related to the functional behavior of apo and ligand-bound enzymes in solution, including the weight of equilibrium populations sampled by the enzyme (p_A and p_B), the chemical shift differences between each state ω_A and ω_B ($\Delta\omega$), and the rate of conformational exchange (k_{ex}) between the ground (p_A) and excited (p_B) population states. Atomic-scale analysis of these parameters in conjunction with computer simulations provides a wide range of information on the importance of these excited sub-states in enzyme function and catalysis. For computer simulations, the conformational fluctuations are identified using second-order or more accurate higher-order methods. The projections of the conformations on the fluctuations (or vectors) allow the identification of conformational sub-states. Characterization of these states with various biophysical properties (energy, reaction coordinate(s), and other parameters) provides detailed information on the reaction mechanism.

The current challenge with these techniques remains the validation of the conformational sub-states identified, as they are rarely populated and only last for very short durations. Joint efforts between experimental and computational techniques would enable more accurate characterization of these functionally relevant conformers. For example, m in QAA can be selected such that it reproduces p_A and p_B obtained from NMR. This would allow validation of the computational methodology, and as a benefit detailed and quantitative information will then be available about conformational sub-states and populations associated with enzyme catalysis.

ACKNOWLEDGMENTS

This work was supported in part by a multi-PI grant from the National Institute of General Medical Sciences (NIGMS) of the National Institutes of Health (NIH) under award number R01GM105978 to N.D., C.S.C., and P.K.A., and a Natural Sciences and Engineering Research Council of Canada (NSERC) Discovery Grant under award number RGPIN 402623-2011 (to N.D.). N.D. holds a Fonds de Recherche Québec—Santé (FRQS) Research Scholar Junior 1 Career Award. Supercomputing time for a number of described studies conducted over the years at Oak Ridge National Laboratory was made available by peer-reviewed INCITE and ALCC allocations.

REFERENCES

Agarwal, P. K. (2004). Cis/trans isomerization in HIV-1 capsid protein catalyzed by cyclophilin A: Insights from computational and theoretical studies. *Proteins*, *56*(3), 449–463.

Agarwal, P. K. (2005). Role of protein dynamics in reaction rate enhancement by enzymes. *Journal of the American Chemical Society*, *127*(43), 15248–15256.

Agarwal, P. K. (2006). Enzymes: An integrated view of structure, dynamics and function. *Microbial Cell Factories*, *5*, 2.

Agarwal, P. K., Geist, A., & Gorin, A. (2004). Protein dynamics and enzymatic catalysis: Investigating the peptidyl-prolyl cis-trans isomerization activity of cyclophilin A. *Biochemistry*, *43*(33), 10605–10618.

Benkovic, S. J. (1992). Catalytic antibodies. *Annual Review of Biochemistry*, *61*, 29–54.

Benkovic, S. J., & Hammes-Schiffer, S. (2003). A perspective on enzyme catalysis. *Science*, *301*(5637), 1196–1202.

Bhabha, G., Biel, J. T., & Fraser, J. S. (2015). Keep on moving: Discovering and perturbing the conformational dynamics of enzymes. *Accounts of Chemical Research*, *48*(2), 423–430.

Boehr, D. D., Dyson, H. J., & Wright, P. E. (2006). An NMR perspective on enzyme dynamics. *Chemical Reviews*, *106*(8), 3055–3079.

Boehr, D. D., McElheny, D., Dyson, H. J., & Wright, P. E. (2006). The dynamic energy landscape of dihydrofolate reductase catalysis. *Science*, *313*(5793), 1638–1642.

Borreguero, J. M., He, J. H., Meilleur, F., Weiss, K. L., Brown, C. M., Myles, D. A., et al. (2011). Redox-promoting protein motions in rubredoxin. *Journal of Physical Chemistry B*, *115*(28), 8925–8936.

Burger, V., & Chennubhotla, C. (2012). Nhs: Network-based hierarchical segmentation for cryo-electron microscopy density maps. *Biopolymers*, *97*(9), 732–741.

Burger, V. M., Ramanathan, A., Savol, A. J., Stanley, C. B., Agarwal, P. K., & Chennubhotla, C. S. (2012). Quasi-anharmonic analysis reveals intermediate states in the nuclear co-activator receptor binding domain ensemble. *Pacific Symposium on Biocomputing*, 70–81.

Chennubhotla, C., & Bahar, I. (2007a). Markov methods for hierarchical coarse-graining of large protein dynamics. *Journal of Computational Biology*, *14*(6), 765–776.

Chennubhotla, C., & Bahar, I. (2007b). Signal propagation in proteins and relation to equilibrium fluctuations. *PLoS Computational Biology*, *3*(9), 1716–1726.

Doruker, P., Jernigan, R. L., & Bahar, I. (2002). Dynamics of large proteins through hierarchical levels of coarse-grained structures. *Journal of Computational Chemistry*, *23*(1), 119–127.

Eisenmesser, E. Z., Bosco, D. A., Akke, M., & Kern, D. (2002). Enzyme dynamics during catalysis. *Science*, *295*(5559), 1520–1523.

Fraser, J. S., Clarkson, M. W., Degnan, S. C., Erion, R., Kern, D., & Alber, T. (2009). Hidden alternative structures of proline isomerase essential for catalysis. *Nature*, *462*(7273), 669–673.

Fraser, J. S., van den Bedem, H., Samelson, A. J., Lang, P. T., Holton, J. M., Echols, N., et al. (2011). Accessing protein conformational ensembles using room-temperature X-ray crystallography. *Proceeding of the National Academy of Sciences of the United States of America*, *108*(39), 16247–16252.

Frauenfelder, H., & Leeson, D. T. (1998). The energy landscape in non-biological and biological molecules. *Nature Structural Biology*, *5*(9), 757–759.

Goodey, N. M., & Benkovic, S. J. (2008). Allosteric regulation and catalysis emerge via a common route. *Nature Chemical Biology*, *4*(8), 474–482.

Haliloglu, T., Bahar, I., & Erman, B. (1997). Gaussian dynamics of folded proteins. *Physical Review Letters*, *79*(16), 3090–3093.

Hathorn, B. C., Sumpter, B. G., Noid, D. W., Tuzun, R. E., & Yang, C. (2002). Vibrational normal modes of polymer nanoparticle dimers using the time-averaged normal coordinate analysis method. *Journal of Physical Chemistry A*, *106*(40), 9174–9180.

Henzler-Wildman, K., & Kern, D. (2007). Dynamic personalities of proteins. *Nature*, *450*(7172), 964–972.

Henzler-Wildman, K. A., Lei, M., Thai, V., Kerns, S. J., Karplus, M., & Kern, D. (2007). A hierarchy of timescales in protein dynamics is linked to enzyme catalysis. *Nature*, *450*(7171), 913–916.

Ishima, R. (2014). CPMG relaxation dispersion. *Methods in Molecular Biology*, *1084*, 29–49.

Kay, L. E. (2015). New views of functionally dynamic proteins by solution NMR spectroscopy. *Journal of Molecular Biology*, *428*, 323–331.

Kleckner, I. R., & Foster, M. P. (2011). An introduction to NMR-based approaches for measuring protein dynamics. *Biochimica et Biophysica Acta*, *1814*(8), 942–968.

Knowles, J. R. (1991). Enzyme catalysis: Not different, just better. *Nature*, *350*(6314), 121–124.

Leach, A. R. (2001). *Molecular modeling: Principles and applications* (2nd ed.). Essex, UK: Pearson Education Limited/Prentice Hall.

Lisi, G. P., & Loria, J. P. (2016). Solution NMR spectroscopy for the study of enzyme allostery. *Chemical Reviews*, *116*(11), 6323–6369.

Palmer, A. G., 3rd. (2015). Enzyme dynamics from NMR spectroscopy. *Accounts Chemical Research*, *48*(2), 457–465.

Palmer, A. G., 3rd, Kroenke, C. D., & Loria, J. P. (2001). Nuclear magnetic resonance methods for quantifying microsecond-to-millisecond motions in biological macromolecules. *Methods in Enzymology, 339*, 204–238.

Perahia, D., Levy, R. M., & Karplus, M. (1990). Motions of an alpha-helical polypeptide—Comparison of molecular and harmonic dynamics. *Biopolymers, 29*(4–5), 645–677.

Qui, C., & Bahar, I. (Eds.), (2005). *Normal mode analysis: Theory and applications to biological and chemical systems*. Boca Raton, FL: Chapman and Hall/CRC.

Ramanathan, A., & Agarwal, P. K. (2009). Computational identification of slow conformational fluctuations in proteins. *Journal of Physical Chemistry B, 113*(52), 16669–16680.

Ramanathan, A., Agarwal, P. K., Kurnikova, M., & Langmead, C. J. (2010). An online approach for mining collective behaviors from molecular dynamics simulations. *Journal of Computational Biology, 17*(3), 309–324.

Ramanathan, A., Savol, A. J., Agarwal, P. K., & Chennubhotla, C. S. (2012). Event detection and sub-state discovery from biomolecular simulations using higher-order statistics: Application to enzyme adenylate kinase. *Proteins, 80*(11), 2536–2551.

Ramanathan, A., Savol, A., Burger, V., Chennubhotla, C. S., & Agarwal, P. K. (2014). Protein conformational populations and functionally relevant substates. *Accounts of Chemical Research, 47*(1), 149–156.

Ramanathan, A., Savol, A. J., Langmead, C. J., Agarwal, P. K., & Chennubhotla, C. S. (2011). Discovering conformational sub-states relevant to protein function. *PLoS One, 6*(1), e15827.

Savol, A. J., Burger, V. M., Agarwal, P. K., Ramanathan, A., & Chennubhotla, C. S. (2011). QAARM: Quasi-anharmonic autoregressive model reveals molecular recognition pathways in ubiquitin. *Bioinformatics, 27*(13), I52–I60.

Tuzun, R. E., Noid, D. W., Sumpter, B. G., & Yang, C. (2002). Normal coordinate analysis for polymer systems: Capabilities and new opportunities. *Macromolecular Theory and Simulations, 11*(7), 711–728.

Warshel, A., Sharma, P. K., Kato, M., Xiang, Y., Liu, H. B., & Olsson, M. H. M. (2006). Electrostatic basis for enzyme catalysis. *Chemical Reviews, 106*(8), 3210–3235.

CHAPTER THIRTEEN

Computation of Rate Constants for Diffusion of Small Ligands to and from Buried Protein Active Sites

P.-H. Wang*, D. De Sancho†,‡, R.B. Best§, J. Blumberger¶,1
*RIKEN Theoretical Molecular Science Laboratory, Wako-shi, Saitama, Japan
†CIC NanoGUNE, Donostia-San Sebastián, Spain
‡Ikerbasque, Basque Foundation for Science, Bilbao, Spain
§Laboratory of Chemical Physics, National Institute of Diabetes and Digestive and Kidney Diseases, National Institutes of Health, Bethesda, MD, United States
¶University College London, London, United Kingdom
[1]Corresponding author: e-mail address: j.blumberger@ucl.ac.uk

Contents

Abstract

The diffusion of ligands to actives sites of proteins is essential to enzyme catalysis and many cellular signaling processes. In this contribution we review our recently developed methodology for calculation of rate constants for diffusion and binding of small molecules to buried protein active sites. The diffusive dynamics of the ligand obtained from molecular dynamics simulation is coarse grained and described by a Markov state model. Diffusion and binding rate constants are then obtained either from the reactive flux formalism or by fitting the time-dependent population of the Markov state model to a phenomenological rate law. The method is illustrated by applications to diffusion of substrate and inhibitors in [NiFe] hydrogenase, CO-dehydrogenase, and myoglobin. We

Methods in Enzymology, Volume 578
ISSN 0076-6879
http://dx.doi.org/10.1016/bs.mie.2016.05.039

also discuss a recently developed sensitivity analysis that allows one to identify hot spots in proteins, where mutations are expected to have the strongest effects on ligand diffusion rates.

1. INTRODUCTION

The transport of ligands to and from protein active sites is a ubiquitous process in biology, biochemistry, and biophysics. It plays a crucial role in the initial and final steps of enzyme catalysis that result in the binding of substrate molecules to catalytic sites and the unbinding of products. Similarly, inhibition of enzymatic function is induced by the diffusion and binding of inhibitors to the active site. This can have desirable effects like the blocking of enzyme function in pharmacological drug targets, or unwelcome effects like the irreversible damage of hydrogen-producing enzymes (hydrogenases) by molecular oxygen. Intraprotein ligand transport is also important for cellular signal transduction and allosteric control of biochemical processes.

In this contribution we focus on the diffusion of small ligands comprised of only a few atoms, such as the gas molecules O_2, CO, or CO_2, to buried protein active sites, although the approach should also be applicable to small drug molecules. Much of our current understanding of this process is inferred from protein X-ray crystallography. In particular, Xe-binding X-ray diffraction measurements could be used to map out the cavities inside the protein that small ligands with a diameter similar to Xe may occupy on their way to the active site (Doukov, Blasiak, Seravalli, Ragsdale, & Drennan, 2008; Johnson et al., 2007; Montet et al., 1997; Murray, Maghlaoui, Kargul, Sugiura, & Barber, 2008). Additionally, picosecond time-resolved X-ray crystallography has resulted in interesting insights into the kinetics of ligand unbinding and rebinding for small proteins like myoglobin (Barends et al., 2015; Hummer, Schotte, & Anfinrud, 2004). Despite these advances, the information that experimental techniques provide is in general insufficient to obtain a full molecular-scale understanding of ligand transport, since they probe only the populated sites within the protein and the change in their population with time. What are the diffusion paths that ligands take to access buried protein active sites? What is the timescale for diffusion into the active site pocket and how long does it take until the ligand is chemically bound once in the active site pocket? What are the key residues that determine the diffusion and binding kinetics? How can we predict mutations that are most likely to interfere with these processes, make them faster or slower?

Several molecular dynamics (MD) simulation techniques have been developed in recent years to investigate these questions. Equilibrium MD simulations (Baron et al., 2009; D'Abramo, Di Nola, & Amadei, 2009; Ruscio et al., 2008; Teixeira, Baptista, & Soares, 2006) as well as enhanced sampling methods such as locally enhanced sampling (Cohen, Kim, King, Seibert, & Schulten, 2005; Cohen, Kim, Posewitz, et al., 2005; Elber & Karplus, 1990), implicit ligand sampling (Cohen, Arkhipov, Braun, & Schulten, 2006; Cohen & Schulten, 2007), metadynamics (Ceccarelli, Anedda, Casu, & Ruggerone, 2008; Nishihara, Hayashi, & Kato, 2008), or the single sweep/temperature accelerated MD method (Maragliano, Cottone, Ciccotti, & Vanden-Eijnden, 2010) were developed to probe the diffusion process of small ligands in proteins. Energetic properties including activation barriers (Ceccarelli et al., 2008; Nishihara et al., 2008) and global free energy surfaces (Cohen et al., 2006; Cohen & Schulten, 2007; Maragliano et al., 2010) were reported. In addition, diffusion rates were calculated using MD simulations (D'Abramo et al., 2009; Ruscio et al., 2008), transition network analysis (Mishra & Meuwly, 2010), or a combination of MD simulations and Markov state modeling (MSM) (De Sancho, Kubas, Wang, Blumberger, & Best, 2015; Wang, Best, & Blumberger, 2011a, 2011b; Wang & Blumberger, 2012; Wang, Bruschi, De Gioia, & Blumberger, 2013).

Here we review in detail our combined MD + MSM approach (De Sancho et al., 2015; Wang et al., 2011a, 2011b; Wang & Blumberger, 2012; Wang et al., 2013), which uses recent advances in the derivation of MSMs from equilibrium MD simulations (Chodera & Noé, 2014). The method was motivated by our initial analysis of MD trajectories for diffusion of H_2 molecules to the buried active site of NiFe hydrogenase (Wang et al., 2011b). We observed that H_2 moves within the protein toward the active site via diffusive jumps between empty protein cavities. The diffusive hopping of ligands in the protein interior was then coarse grained onto a time series of jumps between discrete sites (Markov states) corresponding to the cavities within the protein. The computed transition rates between these states were used to obtain the overall rate for diffusion from solution to the active site and for the reverse process either via the reactive flux formalism or via a fit of the time-dependent state populations to a phenomenological rate law. Applications of our methodology to a number of different gas processing enzymes were also successful, reproducing experimental rates where available, and giving a molecular scale insight into the diffusion process in wild-type and mutant enzymes.

In Section 2.1, we briefly review classical master equations and Markovian dynamics to give some theoretical background to the computational methodology developed. Section 2.2 is the centerpiece of our contribution. We explain in detail every step of our computational scheme, from MD simulation to the definition of Markov states, the construction of the rate matrix for transitions between the Markov states and the calculation of diffusion rates from the transition rate matrix. Some of the technical challenges are explained and possible ways to address them are suggested. In this section we also explain our recently developed sensitivity analysis, which allows us to make predictions for protein residues that, when mutated, are expected to have the greatest effect on ligand diffusion. Application of our methodology to diffusion of small ligands like H_2, O_2, CO, and CO_2 in hydrogenases, dehydrogenases/acetyl co-enzyme A synthase (CODH/ACS), and myoglobin is discussed in Section 3. The method is summarized in Section 4, where we also give a short outlook on possible extensions of our methodology.

2. METHODOLOGY

2.1 Classical Master Equations and Markovian Dynamics

We review very briefly classical master equations (Hummer, 2005; Zwanzig, 1983), the Markovian approximation, and some useful properties of the transition rate matrix, which will be referred to in later sections. In the master equation formalism configuration space is divided into nonoverlapping cells (index i), and the time evolution of the probability p_i for finding the system in cell i is obtained using the generalized Langevin equation. The result is the familiar expression

$$\dot{p}_i(t) = -\int_0^t dt' \sum_j K_{ij}(t-t')p_j(t'), \tag{1}$$

where $K_{ij}(t - t')$ are the elements of the transition memory kernel. If the dynamics becomes Markovian (ie, memoryless) after some period of time, Eq. (1) reduces to

$$\dot{p}_i(t) = \sum_j k_{ij}p_j(t), \tag{2}$$

where the k_{ij} are the rate constants for transitions $i \leftarrow j$. They obey the following three conditions:

$$k_{ij} \geq 0 \quad \forall i \neq j, \tag{3}$$

$$k_{ii} \leq 0 \quad \forall i, \tag{4}$$

$$\sum_i k_{ij} = 0. \tag{5}$$

The equilibrium (or stationary) distribution $p_{\mathrm{eq},i}$ is then given by

$$\frac{\partial}{\partial t} p_{\mathrm{eq},i} = \sum_j k_{ij} p_{\mathrm{eq},j} = 0, \tag{6}$$

and the system satisfies the condition of *detailed balance*,

$$k_{ji} p_{\mathrm{eq},i} = k_{ij} p_{\mathrm{eq},j}. \tag{7}$$

As a result, the matrix formed by the rate constants k_{ij}, denoted $\mathbf{K}$, is related to a symmetric matrix $\widetilde{\mathbf{K}}$ by a similarity transformation

$$\widetilde{\mathbf{K}} = \mathbf{D}^{-1/2} \mathbf{K} \mathbf{D}^{1/2} = \mathbf{D}^{1/2} \mathbf{K}^{\mathrm{T}} \mathbf{D}^{-1/2} = \widetilde{\mathbf{K}}^{\mathrm{T}}, \tag{8}$$

where $\mathbf{D} = \mathrm{diag}(p_{\mathrm{eq}})$ is the diagonal matrix of equilibrium probabilities. $\widetilde{\mathbf{K}}$ is therefore orthogonally diagonalizable

$$\widetilde{\mathbf{K}} = \widetilde{\mathbf{U}} \widetilde{\Lambda} \widetilde{\mathbf{U}}^{\mathrm{T}}, \tag{9}$$

where $\widetilde{\mathbf{U}} \widetilde{\mathbf{U}}^{\mathrm{T}} = \mathbf{I}$ and $\widetilde{\Lambda} = \mathrm{diag}(\widetilde{\lambda}_1, \widetilde{\lambda}_2, \ldots, \widetilde{\lambda}_M)$ is the diagonal matrix of eigenvalues, sorted such that $\widetilde{\lambda}_1 \geq \widetilde{\lambda}_2 \geq \ldots \geq \widetilde{\lambda}_M$. $\mathbf{K}$ shares the same eigenvalues as $\widetilde{\mathbf{K}}$:

$$\mathbf{K} = \mathbf{D}^{1/2} \widetilde{\mathbf{K}} \mathbf{D}^{-1/2} = (\mathbf{D}^{1/2} \widetilde{\mathbf{U}}) \widetilde{\Lambda} (\mathbf{D}^{1/2} \widetilde{\mathbf{U}})^{-1} = \mathbf{U} \Lambda \mathbf{U}^{-1}, \tag{10}$$

but the eigenvectors differ by a factor of $\mathbf{D}^{1/2}$. All the eigenvalues are real. λ_1 is zero and all others are negative. We further note that Eq. (2) can be solved formally in terms of a matrix exponential,

$$p_i(t) = \sum_j (e^{t\mathbf{K}})_{ij} p_j(0). \tag{11}$$

The propagators can also be identified in this relation,

$$p(i,t|j,0) = (e^{t\mathbf{K}})_{ij}, \tag{12}$$

where $p(i,t|j,0)$ is the conditional probability that a trajectory starting in cell j is in cell i at a later time t.

2.2 Diffusion and Binding Rate Constants

2.2.1 Diffusion–Reaction Model

We consider the binding of a ligand (L) to an enzyme active site (E) and the reverse process, the dissociation of the ligand-bound enzyme (B),

$$\mathrm{E} + \mathrm{L} \underset{k_{\mathrm{out}}}{\overset{k_{\mathrm{in}}[\mathrm{L}]}{\rightleftharpoons}} \mathrm{B}, \tag{13}$$

where [L] is the concentration of ligand in the solvent, and k_{in}, k_{out} the corresponding phenomenological rate constants. Reaction Eq. (13) is assumed to be a diffusion–reaction process. It is divided into a transport process describing diffusion of L from the solvent through the protein to the active site pocket with rate constants k_{+1} and k_{-1}, respectively,

$$\mathrm{E} + \mathrm{L} \underset{k_{-1}}{\overset{k_{+1}[\mathrm{L}]}{\rightleftharpoons}} \mathrm{G}, \tag{14}$$

and a chemical binding step with rate constants k_{+2} and k_{-2}, respectively,

$$\mathrm{G} \underset{k_{-2}}{\overset{k_{+2}}{\rightleftharpoons}} \mathrm{B}. \tag{15}$$

In the geminate state G the ligand diffuses in the active site pocket but is still chemically unchanged, and in state B the ligand is chemically bound to the active site.

In the following we describe the computational methodology that we have developed to compute the rate constants k_{+1} and k_{-1}, as well as k_{in} and k_{out} by combining data from MD simulation with MSM. The steps of our computational scheme are summarized in Fig. 1.

2.2.2 MD Simulation

In a first step a long equilibrium MD simulation is carried out for a system containing the solvated protein and ligand molecules initially inserted in the solvent, outside the protein, at random positions. The purpose of the simulation is to identify the cavities inside the protein that can be readily occupied by ligand molecules. To ensure good sampling, the ligand concentration is typically chosen to be in the order of 100 mM, which is

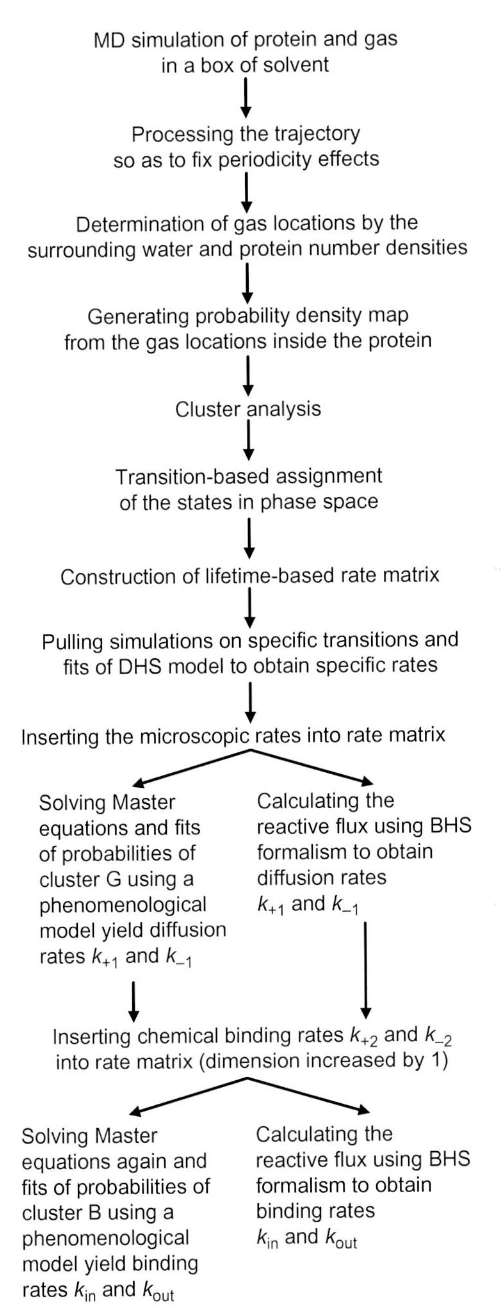

Fig. 1 Flowchart for calculation of ligand diffusion and binding rate constants (see Section 2.2).

about 100 times larger than under experimental conditions at 1 bar gas pressure. As we will see later, a ligand concentration of about 100 m*M* is still sufficiently low so that concentration effects on the rate constants can be safely ignored. Small ligands like H_2, CO, and CO_2 penetrate the protein relatively fast, within a few 100 ps and the fraction of ligand molecules inside the protein is usually converged after about 10 ns. Importantly, after about 100 ns simulation time we typically observe one or a few events where a small ligand molecule enters the active site.

2.2.3 Definition of Location of Ligands

We developed an approach for distinguishing ligand locations between two regions: inside the protein and outside the protein. To accomplish this, the trajectory had to be processed as follows. All the frames of the trajectory were translated so as to center the protein in the box. They were least-square-fitted relative to the protein structure of the first frame. We then defined a spherical region inside the protein which was centered at the active site. This region has a radius of certain distance in order to enclose most protein atoms in this sphere. In the case of [NiFe]-hydrogenase, this (optimized) radius is set at 25 Å. (Wang et al., 2011a, 2011b). It guarantees that ligands inside the protein are correctly identified. After the trajectory was processed, a grid was defined producing small cubes of edge lengths 0.5 Å. Each atom in the system is assigned to a cube. If a cube contains the center of mass of a ligand molecule, then the number density of protein and solvent atoms in the 15^3 nearest cubes (± 7 grid spacing) is calculated. If the protein number density around the given ligand molecule is zero, it would be defined as outside the protein. Other ligand molecules would be defined as inside the protein. For our application to myoglobin a similar procedure was used, but in this case without using a grid for the assignment (De Sancho et al., 2015). Simply, the gas molecule was set to be outside the protein when it was further than 6 Å from any protein heavy atom.

2.2.4 Cluster Analysis

In a next step a cluster analysis is used to help us identify the cavities in the protein that are frequently visited. For this purpose we have used the gromos algorithm (Daura et al., 1999), where for each ligand pose sampled, the number of neighbors is counted within a given cutoff radius. The pose with the largest number of neighbors is selected and the position of its center of mass identified with the position of cluster 1. This pose and its neighbors are then eliminated from the ensemble of poses (Daura et al., 1999). Repeating this procedure for the remaining poses in the ensemble one generates an ensemble of N clusters, ranked according to their population, ie, number of neighbors within the cutoff distance. This way a large number of clusters are produced, typically hundreds. In practice about 90–95% of the most populated clusters are used for kinetic modeling. At this stage, two special clusters are defined corresponding to the initial and final states for diffusion: the solvent cluster, S, corresponding to the initial state, in which the ligand occupies the solvent, defined as regions with zero protein number density as mentioned in the previous section, and the "geminate" cluster, G, corresponding to the final state, in which the ligand occupies the binding pocket. In state G the ligand is not yet chemically bound to the protein but diffuses within the active site pocket, as explained in Section 2.2.1.

2.2.5 Construction of Rate Matrix

Having defined the clusters or states of our model for diffusion, we are now in the position to calculate the transition rates from MD simulation. We have used two methods for this purpose, one based on estimating the lifetime of ligands in a cavity (Buchete & Hummer, 2008), and the other based on the lag-time dependent transition probability matrix. In both approaches we assume that all transitions have been observed in the long equilibrium MD simulation in step 1 of our computational scheme (the statistics of certain slow transitions can be improved using enhanced sampling methods; see Section 2.2.8). Moreover, we assume that the dynamics is ergodic and Markovian, ie, obeys detailed balance (Eq. 7).

In the lifetime-based method we define N_{ji} to be the number of transitions $j \leftarrow i$ observed during the long MD run. For a long trajectory, detailed balance and microscopic time reversibility ensure that $N_{ij} = N_{ji}$. To maximize the counts of events, the matrix of transitions is symmetrized, $N_{ij}^{sym} = N_{ij} + N_{ji}$. The rate for transition from state i to state j is then equal to the inverse lifetime of the ligand in state i, T_i, times the fraction of transitions out of i into j, $N_{ji}^{sym} / \sum_{l=1}^{N} N_{li}^{sym}$,

$$k_{ji} = \frac{N_{ji}^{sym}}{T_i \sum_l N_{li}^{sym}}, \quad \forall i \neq j, \tag{16}$$

$$k_{ii} = -T_i^{-1} = -\sum_{j \neq i} k_{ji}, \quad \forall i. \tag{17}$$

The lifetime T_i is the average time the ligand spends in state i when it occupies i. It is obtained by averaging over the times during which the ligand is assigned to state i. The proper assignment of a given trajectory of poses to states still remains challenging (Buchete & Hummer, 2008) because an improper assignment can lead to an unphysically large number of recrossings and non-Markovian dynamics. Further below we discuss an assignment that addresses this issue.

In the lag time-based approach we first calculate the transition probability matrix $\mathbf{T}$ for different values of the lag time Δt, with elements t_{ji} defined as

$$t_{ji}(\Delta t) = \frac{N_{ji}}{\sum_j N_{ji}}. \tag{18}$$

A lag time is then chosen so that the slowest relaxation times of $\mathbf{T}(\Delta t)$ are reasonably converged. Then the rate matrix is computed using a maximum likelihood propagator based method (Buchete & Hummer, 2008). If short lag times can be chosen, the rate matrix can be approximated using a Taylor expansion (De Sancho, Mittal, & Best, 2013). In this case the rate matrix elements are

$$k_{ji} = t_{ji}(\Delta t)/\Delta t, \quad \forall i \neq j, \tag{19}$$

$$k_{ii} = -\sum_{j \neq i} k_{ji}, \quad \forall i. \tag{20}$$

The last expression is obtained by enforcing mass conservation.

2.2.6 Assignment of States

Rates for transitions between clusters or states are calculated according to Eqs. (16) and (17) or Eqs. (19) and (20). A major difficulty we encountered with these approaches are the apparent "recrossings" caused by the imperfectly drawn boundaries of the stable states (Buchete & Hummer, 2008; Sarich, Noé, & Schütte, 2010). Drawing on ideas from transition-path sampling (Best & Hummer, 2005; Bolhuis, Chandler, Dellago, & Geissler, 2002) and milestoning (Faradjian & Elber, 2004; Vanden-Eijnden, Venturoli, Ciccotti, & Elber, 2008), we define compact "core" regions for each cluster. A trajectory crossing out of one of these regions is only deemed to have made a transition to another cluster when it enters the "core" region of the new cluster, hence the term "transition-based assignment" (TBA (Buchete & Hummer, 2008)). This way we avoid apparent transitions when the ligand crosses approximately drawn state boundaries, but in fact remains in the original metastable state. The "core" region for each cluster i was defined using a spherical radius, r_{TBA}, so that a given fraction of the population was within that sphere.

2.2.7 Adjustment for Ligand Concentration

While transitions between clusters within the protein are unimolecular, those from the solvent to protein clusters are bimolecular. Therefore, the microscopic rates from solution to protein clusters depend on ligand concentration. Here we assert that the concentration of ligands in the solvent remains constant. Consequently, all solvent to protein cluster transitions become unimolecular (ie, pseudo-first order in the protein concentration), and therefore the kinetics of all transitions remains unimolecular. However, since the transition rates are calculated from the average residence time of a *single* ligand, the rate matrix **K** constructed from the simulation data is for a ligand concentration $1/V_{H_2O}^{sim}$, where $V_{H_2O}^{sim}$ is the volume of the simulation box outside the protein. In order to obtain the rate matrix for a reference (eg, experimental) ligand concentration $[L]^\circ$, we scale all pseudo-first-order solvent-to-protein rate constants by a factor $V_{H_2O}^{sim}/V_{H_2O}{}^\circ$, where $V_{H_2O}{}^\circ$ is the volume per molecule of ligand at a ligand concentration $[L]^\circ$. For typical protein simulations (10,000 water molecules) and experimental ligand concentrations ($[L]^\circ = 1\ mM$), the scaling factor is $V_{H_2O}^{sim}/V_{H_2O}{}^\circ = 0.2$. This illustrates that adjustment for ligand concentration is important and should not be neglected.

2.2.8 Enhanced Sampling of Transitions

Some transitions may not be well sampled in the long equilibrium MD simulations. Active sites deeply buried in the protein interior may only be visited once or at most a few times within a few 100 ns of simulation time. Clearly, in this case the statistics is insufficient for estimation of transition rates. Standard rare-event sampling techniques may be employed to improve the accuracy of the rate estimate. We have investigated two techniques, one based on the calculation of the mean force for a preselected transition path bridging initial and final state cluster, and one based on nonequilibrium pulling simulation. Of course, other approaches like umbrella sampling, transition path sampling, or metadynamics could also be used. We found the pulling simulation to be superior over the mean force approach. A disadvantage of the latter method is that the mean force needs to be calculated for a very large number of points in 3D Cartesian space in order to be able to fit a reasonably smooth free energy surface to the data points. Moreover, it is unclear a priori whether standard transition state theory gives a good approximation to the kinetics of all of the transitions. Dynamical effects such as barrier recrossings may occur and would necessitate the calculation of transmission coefficients for each transition. This is both tedious and computationally highly demanding, if not unfeasible.

The explicit construction of a free energy surface is completely avoided in the nonequilibrium pulling simulation technique. In this approach the ligand is "pulled" by an external force from the initial to the final state and the time for the transition is recorded. The direction of the pulling force is defined by a suitable combination of atomic positions of residues lining the initial and final state cluster. Several simulations with equal magnitude of external pulling force are carried out for different initial conditions of the ligand (position within the initial state cluster and velocity) and the transition time averaged to what is referred to as the mean first passage time, τ. A series of such calculations are carried out for decreasing magnitude of external pulling force acting on the ligand. The mean first passage times obtained for different external forces are converted to rates, $k(F) = 1/\tau(F)$, and fit to the Dudko–Hummer–Szabo (DHS) model (Dudko, Hummer, & Szabo, 2006) for force-dependent kinetics,

$$k(F) = k_0\left(1 - \frac{\nu\beta F q^{\ddagger}}{\Delta G^{\ddagger}}\right)^{(1/\nu)-1} e^{\Delta G^{\ddagger}[1-(1-(\frac{\nu\beta F q^{\ddagger}}{\Delta G^{\ddagger}}))^{1/\nu}]}, \tag{21}$$

where $k_0 = k(0)$ is the rate at zero force, $\Delta G^{\ddagger}$ is the height of the barrier, and $q^{\ddagger}$ is the distance to the transition state. The parameter β is $(k_B T)^{-1}$, where k_B is the Boltzmann constant and T is the temperature. In the applications that we will discuss in Section 3, the parameter ν was set equal to 2/3.

The DHS model is a one-dimensional model for force-dependent kinetics derived from Kramers theory (Kramers, 1940). Two of its main advantages are completeness and robustness (Best, Paci, Hummer, & Dudko, 2008). The Kramers-based description is theoretically better founded than, eg, the phenomenological Bell model (Bell, 1978; Evans & Ritchie, 1997), which can be obtained from the DHS model by setting $\nu = 1$. Importantly, the DHS model accounts for the dependence of the transition state location on force and thus explains the curvature in the force dependence, as frequently observed in MD simulations. A disadvantage of the pulling simulation approach is that the rate constant of interest, k_0, needs to be obtained by *extrapolation* of the MFPTs at finite force to the MFTP at zero force using Eq. (21). This can lead to uncertainties in k_0, in particular for very slow events where high external forces must be employed to enforce the transition within a reasonable window of simulation time.

2.2.9 Diffusion Rate Constants from Phenomenological Kinetic Model

Once the rate matrix is constructed by combining rates from equilibrium and nonequilibrium pulling simulations, one can propagate the initial populations in time according to Eq. (11). One is typically interested in the diffusion of the ligand from the solvent to the active site cluster "G". Hence we set the population of the solvent equal to 1 and the populations of all other clusters equal to 0 at $t=0$ and monitor the time-dependent population of cluster G. For all protein–ligand systems investigated so far, the increase in p_G is of sigmoidal shape. In this case the population of $p_G(t)$ follows monoexponential kinetics and can be effectively described by reaction equation (14) with [L] being constant. The time-dependent population of [G] in Eq. (14) is given by

$$p_G(t) = \frac{k_{+1}[\mathrm{L}]}{k_{+1}[\mathrm{L}] + k_{-1}}[1 - \exp(-(k_{+1}[\mathrm{L}] + k_{-1})t)]. \tag{22}$$

Hence, in the final step of our computational scheme, the populations $p_G(t)$ obtained via propagation of the matrix exponential Eq. (11) are fit to equation (22). The fit parameters are then identified with the diffusion rate constants k_{+1} and k_{-1}, respectively. Applications of this kinetic model to a diverse set of protein–ligand systems have typically resulted in excellent fits with correlation coefficients in excess of 0.99 (see Section 3).

2.2.10 Diffusion Rate Constants from Reactive Flux

Alternatively, the diffusion rate constants can be obtained from the Berezhkovskii–Hummer–Szabo (BHS) formalism for calculation of committors and reactive fluxes (Berezhkovskii, Hummer, & Szabo, 2009). Here the clusters are divided in three regions, one containing only the solvent cluster "S," one containing only the "G" cluster, and the region I between S and G, containing all other clusters. The committor of each cluster x, $\phi_G(x)$, is defined as the probability of reaching the active site cluster G before reaching the solvent cluster S when the ligand is initially in cluster x. Obviously, $\phi_G(G)=1$ and $\phi_G(S)=0$. For all other clusters the committor can be calculated from the rate matrix **K** by solving the linear equation

$$\sum_{i\in I}\phi_G(i)k_{ij} + \sum_{i\in G}k_{ij} = 0, \quad j\in I. \tag{23}$$

The total flux through an isocommittor surface $\sum$ can be calculated from

$$J_{S\to G} = \sum_{i\in G^*, j\in S^*} k_{ij} p_{eq}(j)[\phi_G(i) - \phi_G(j)], \tag{24}$$

where G^* and S^* denote the sets of clusters in the I region located on the G- and S-sides of $\sum$ that are connected by a single transition, respectively, ie, G^* and S^* lie adjacent to $\sum$. Although it is proven that J is independent of the choice of $\sum$ (Berezhkovskii et al., 2009), we can specify a $\sum$ lying between $\phi_G(j)$ and $\phi_G(i)$. In this way, all contributions to J have the same sign, and the local flux flowing in the S→G direction from cluster i to j that has a greater ϕ_G can be expressed as

$$J_{i\to j} = k_{ji} p_{eq}(i)[\phi_G(j) - \phi_G(i)], \tag{25}$$

which is a measure for the importance of each transition based on the magnitude of its contribution to the total flux. By repeating this procedure using different $\sum$, one can evaluate the flux percentage of each diffusion pathway.

Importantly, the diffusion rate k_{+1} is directly proportional to the flux normalized to unit solvent population,

$$k_{+1} = \frac{J_{S\to G}}{f_S}, \tag{26}$$

where f_S is the fractional population of the solvent state in equilibrium. Similarly, the diffusion rate from the active site to the solvent is

$$k_{-1} = \frac{J_{S\leftarrow G}}{f_G}. \tag{27}$$

2.2.11 Binding and Dissociation Rate Constants

In the geminate state G the ligand diffuses in the active site pocket but is not yet chemically bound to the active site. It is straightforward to include the chemical binding step, reaction equation (15), in the Markov state model by defining an additional state B and increasing the dimension of the rate matrix **K** from N to $N+1$. The rate constants for chemical binding and detachment, k_{+2} and k_{-2}, respectively, may be obtained with quantum chemical methods using transition state theory to convert the calculated free energy barriers in reaction rates.

Finally, the effective rate constants for ligand binding and dissociation, k_{in} and k_{out}, respectively, may be obtained similarly as k_{+1} and k_{-1} by propagating the exponential of the rate matrix (now with dimension $N+1$) and fitting the resultant populations $p_B(t)$ to the phenomenological kinetic expression for Eq. (13),

$$p_B(t) = \frac{k_{in}[L]}{k_{in}[L] + k_{out}}[1 - \exp(-(k_{in}[L] + k_{out})t)]. \tag{28}$$

Alternatively, k_{in} and k_{out} may be obtained from the reactive flux formalism, as discussed for k_{+1} and k_{-1} in Section 2.2.10.

2.3 Prediction of Mutations Affecting Diffusion Rate Constant

Based on the BHS formalism (see Section 2.2.10) for the calculation of fluxes, we have developed a simple method for determining the sites of a protein that are sensitive to perturbations (De Sancho et al., 2015). We define a sensitivity parameter, α_s, for a perturbation on a microstate s as a partial derivative of the diffusion rate k_{+1} with respect to a change in the free energy g_s of the microstate,

$$\alpha_s = \frac{\partial}{\partial g_s} \ln(k_{+1}). \tag{29}$$

The dependence of the transition rate constants k_{is} on changes in the stability of a microstate is approximated using linear free energy relationships. For a mutation causing a change δg_s to the free energy of microstate s, we assume that the transition state for the $s \rightleftharpoons i$ transition is halfway between microstates s and i. Hence the perturbed forward ($s \rightarrow i$) and backward ($s \leftarrow i$) rate constants are expressed as

$$k_{is}(\delta g_s) = k_{is} e^{\beta \delta g_s/2} \text{ and } k_{si}(\delta g_s) = k_{si} e^{-\beta \delta g_s/2}, \tag{30}$$

respectively. According to this description, a negative δg_s corresponds to a stabilizing mutation and would result in a slowdown of the $s \rightarrow i$ transition, whereas a positive δg_s, ie, a destabilizing mutation, would lead to a speedup of the $s \rightarrow i$ transition. For the calculation of the sensitivity parameter α_s one inserts Eq. (30) in Eq. (24) and Eq. (26) in Eq. (29). Due to the linear free energy relation equation (30), the derivative of k_{+1} with respect to g_s can be obtained in analytic form (see De Sancho et al., 2015 for an explicit expression). This allows for a very fast calculation of the sensitivity parameter for each cavity/microstate.

The estimate of the change in the diffusion rate upon perturbation of one of the microstates can be used in a quantitative way, provided a reasonable model for the effects of protein mutations on the stability of the microstates can be found. The simplest such model consists in assuming that the change in the free energy of a state due to mutation, Δg_s, is proportional to the change in cavity volume induced by the mutation, ie, purely entropic (see De Sancho et al., 2015 for details). The predicted change in the diffusion rate constant can then be obtained from the finite difference version of Eq. (29),

$$\Delta \ln k_{+1} = \Delta g_s \times \alpha_s. \tag{31}$$

3. APPLICATIONS

3.1 Ligand Diffusion in Hydrogenase, CODH/ACS, and Myoglobin

We applied the computational approach described in Section 2.2 to two gas processing metalloenzymes, [NiFe]-hydrogenase (Montet et al., 1997) and carbon monoxide dehydrogenase/acetyl-coenzyme A synthase (CODH/ACS) (Doukov et al., 2008). In both enzymes the catalytic site is buried deep inside the protein, a few 10 Å away from the protein surface. For hydrogenases we carried out simulations for the substrate H_2 (Wang et al., 2011b) and for inhibitors CO (Wang et al., 2011a; Wang & Blumberger, 2012) and O_2 (Wang et al., 2011b), whereas for CODH/ACS we simulated the diffusion and chemical binding of CO_2 (Wang et al., 2013). Calculations were also carried out for a smaller protein, myoglobin (De Sancho et al., 2015), for which a wealth of experimental kinetic data are available (Olson, Soman, & Phillips, 2007).

3.1.1 Microstates and Ligand Diffusion Paths from MD

All MD simulations were carried out for aqueous solutions of the proteins with ligand molecules initially placed at random positions in the solvent outside the protein. We found that small ligands like H_2, O_2, CO, and CO_2 penetrate the proteins relatively fast, within a few 100 ps and the fraction of ligand molecules inside the protein was usually converged after about 10 ns. For each protein we observed one or a few events where a ligand molecule enters and leaves the active site pocket after about 50–100 ns simulation time. The states obtained after clustering the ligand positions are shown in Fig. 2 for each protein. For [NiFe]-hydrogenase and myoglobin the simulations recovered all the protein cavities that were detected earlier in Xe-binding X-ray crystallography experiments, and many additional

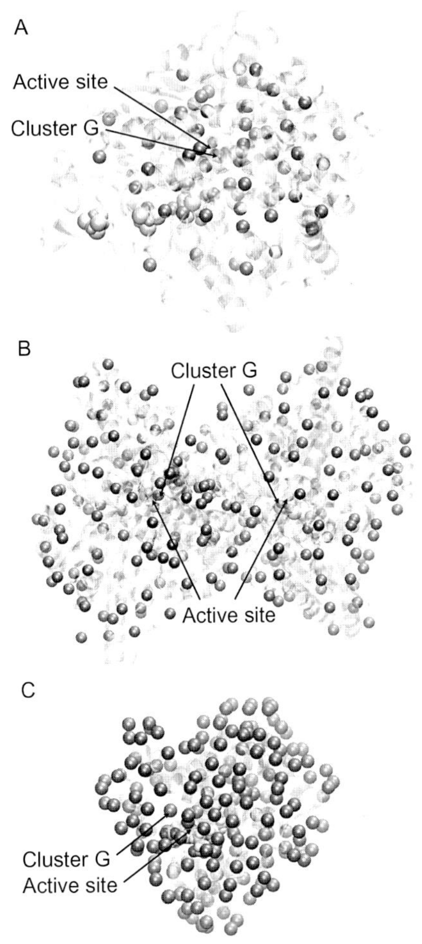

Fig. 2 Location of microstates (clusters) and active sites in (A) [NiFe]-hydrogenase, (B) CODH, and (C) myoglobin. Protein secondary structures are shown in *white ribbon* representation, *while clusters* are indicated as *dark yellow spheres. Color codes*: Fe, *pink*; Ni, *blue*; S, *yellow*; O, *red*; C, *cyan. Blue spheres* in (C) denote N atoms. *Panel (A): Adapted with permission from Wang, P., Best, R. B., & Blumberger, J. (2011b). Multiscale simulation reveals multiple pathways for H_2 and O_2 transport in a [NiFe]-hydrogenase.* Journal of the American Chemical Society, 133, *3548–3556. Copyright 2011 American Chemical Society. Panel (B): Adapted with permission from Wang, P., Bruschi, M., De Gioia, L., & Blumberger, J. (2013). Uncovering a dynamically formed substrate access tunnel in carbon monoxide dehydrogenase/acetyl-CoA synthase.* Journal of the American Chemical Society, 135*(25), 9493–9502. Copyright 2013 American Chemical Society. Panel (C): Adapted with permission from De Sancho, D., Kubas, A., Wang, P., Blumberger, J., & Best, R. B. (2015). Identification of mutational hot spots for substrate diffusion: Application to myoglobin.* Journal of Chemical Theory and Computation, 11*(4), 1919–1927. Copyright 2015 American Chemical Society.* (See the color plate.)

cavities that were not detected experimentally. For CODH/ACS the simulations revealed a dynamically formed access channel for CO_2 that was not visible in Xe-binding experiments. The picture that emerged from these simulations is that the ligands quickly diffuse toward one or a few cavities in close proximity to the active site (denoted "central cavity" in the following) from where a slow final transition into the active site pocket occurs. Very small ligands like H_2 can take several routes toward the central cavity as shown for hydrogenase, whereas larger ligands like CO and CO_2 are more likely to diffuse along a single preferred path ("channel") toward the central cavity.

3.1.2 Final Transition: Pulling Simulation

The final transition from the central cavity into the active site was found to be the bottleneck for diffusion in each protein investigated. The side chains of amino acid residues lining the transition path partially block ligand access to the active site. As a consequence, these final transitions could not be sufficiently well sampled in equilibrium MD simulations. We used non-equilibrium pulling simulations to obtain estimates for the rates of these transitions. The MFPTs obtained for different pulling forces are shown in Fig. 3. The data fit the DHS model equation (21) for force-dependent kinetics very well in each case with correlation coefficients typically in excess of 0.98. Nevertheless, relatively large forces of at least 50 kJ mol^{-1} $Å^{-1}$ were necessary to afford a sufficiently large number of pulling trajectories (a few hundreds) to converge the MFPT. The extrapolation of the data to zero force using the DHS model is the step that carries the highest uncertainty in our computational scheme. Depending on the system investigated, the error in the transition rate may be up to about an order of magnitude, with the uncertainty increasing with decreasing transition rate.

3.1.3 Diffusion Rate Constants

The time-dependent populations of the active site cluster G, obtained by propagation of the matrix exponential equation (11), are shown in Fig. 4. For each protein and ligand investigated, we obtained a sigmoidal curve that fits well the time-dependent population equation (22) of the effective unimolecular reaction equation (14), with correlation coefficients in excess of 0.99. The diffusion rate constants obtained from the fit, k_{+1} and k_{-1}, respectively, are summarized in Table 1. k_{+1} is typically in the order of 10^4–10^5 s^{-1} mM^{-1} and k_{-1} is about 10^7 s^{-1}. For myoglobin the diffusion rates were also calculated from the reactive flux formalism (Eqs. 26 and 27).

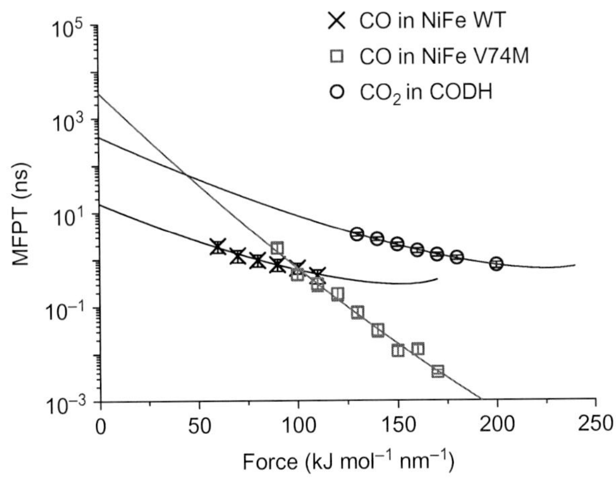

Fig. 3 Mean first passage times (MFPT) vs pulling force from constant-force pulling simulations for the final transition into the active site. See Fig. 2 for the location of the clusters within the protein. Statistical error bars for each pulling simulation are indicated but may be smaller than the symbol size. Fits to Eq. (21) are shown in *solid lines. Adapted with permission from Wang, P., & Blumberger, J. (2012). Mechanistic insight into the blocking of CO diffusion in [NiFe]-hydrogenase mutants through multiscale simulation.* Proceedings of the National Academy of Sciences of the United States of America, 109, *6399–6404. Data for CO_2 in CODH are taken from Wang, P., Bruschi, M., De Gioia, L., & Blumberger, J. (2013). Uncovering a dynamically formed substrate access tunnel in carbon monoxide dehydrogenase/acetyl-CoA synthase.* Journal of the American Chemical Society, 135*(25), 9493–9502.*

The agreement with the values obtained from the fit is reasonably good. For [NiFe]-hydrogenase, CO diffusion rates were computed for the wild-type enzyme and for a number of mutant enzymes in which the size of amino acid side chains close to the active site was varied. Here the computed CO diffusion rates could be directly compared to the experimental CO binding rate because it was shown experimentally that the binding process is limited by diffusion (Liebgott et al., 2010). The large changes in the diffusion rate for the mutant enzymes, of up to four orders of magnitude, could be very well reproduced (Wang & Blumberger, 2012) (see Table 1).

3.1.4 Dependence on Ligand Concentration

A point that deserves further attention is the effect of ligand concentration on the diffusion rates. The ligand concentration we have chosen for equilibrium MD simulations was typically a few 100 m*M* so as to ensure good sampling of the protein cavities. This is about 100 times larger than at experimental conditions at 1 bar gas pressure. The effect of the concentration on

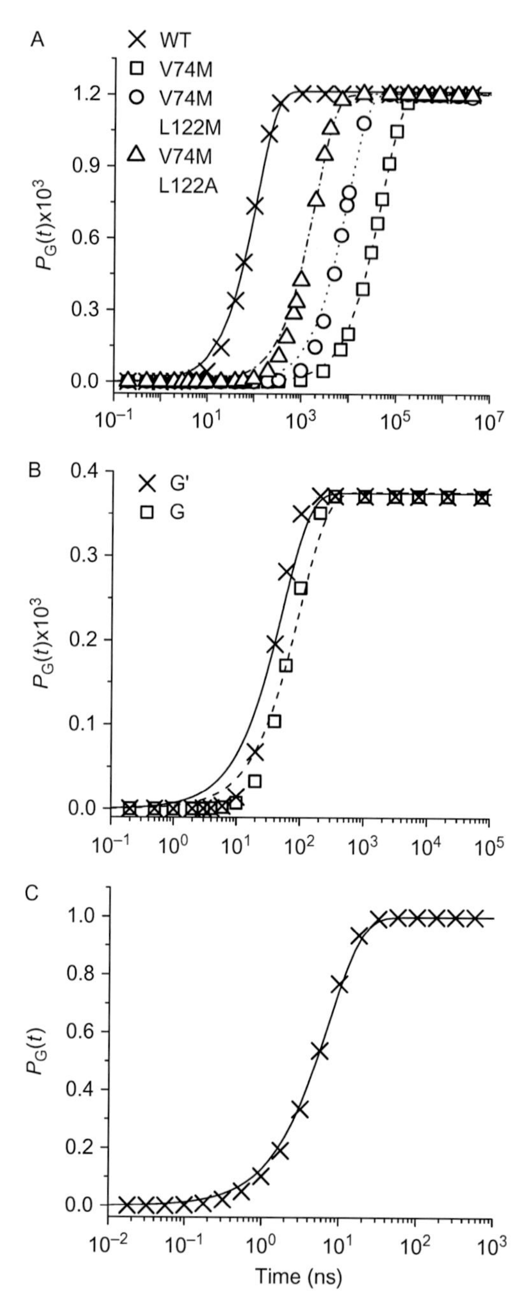

Fig. 4 Diffusion kinetics of (A) CO in the wild type (WT), V74M, V74M L122M, and V74M L122A [NiFe]-hydrogenases, (B) CO_2 in CODH, and (C) CO in myoglobin. Data points are obtained by solving the master equation (Eq. 11) for the population of the active site cluster G, p_G. At $t = 0$, the probability for the solvent cluster was set equal to one, and the

the diffusion rates was tested in case of myoglobin. First we checked whether the rate of making transitions from the solvent S to any of the protein microstates i in the network, $k_P = \sum_{i \neq S} k_{iS} p_S$, was dependent of the ligand concentration. We found that it can be safely ignored up to concentrations of about 500 mM (see Fig. 5A). At higher concentrations the dynamics of the transitions becomes concentration dependent, which resulted in a modest

Table 1 Summary of Computed (Comp.) and Experimental (Exp.) Diffusion Rates

Enzyme	Ligand	k_{+1} (Comp.)[a,b]	k_{-1} (Comp.)[a,c]	k_{in} (Exp.)[b]
NiFe WT	H_2	99,000	1900	
NiFe WT	O_2	17,000	110	
NiFe WT	CO	11,000	88	10,000–20,000[d]
				63,000[e]
NiFe MA	CO	600	5.0	1490[e]
NiFe MM	CO	130	1.10	220[e]
NiFe V74M	CO	24	0.20	19[e]
CODH WT	CO_2	4800	150	
Mb WT	CO	306,700	136	12,000[f]
		646,000[g]	150[g]	

[a]Obtained from the fit of the data shown in Fig. 4 to Eq. (22), unless stated otherwise.
[b]In units s^{-1} mM(ligand)$^{-1}$.
[c]In units 10^5 s^{-1}.
[d]Leroux et al. (2008), k_{+1}(exp.) $\approx$ k_{in}(exp.), 303K. Values given in units s^{-1} atm. (CO)$^{-1}$, but the difference with respect to s^{-1}mM(CO)$^{-1}$ is negligibly small.
[e]Liebgott et al. (2010), k_{+1}(exp.) $\approx$ k_{in}(exp.), 313K.
[f]Carver et al. (1990).
[g]Obtained from the reactive flux, Eqs. (26) and (27).

probabilities for all other clusters were set equal to zero. Fits to the phenomenological first-order equation (Eq. 22) are shown in *lines. Panel (A): Adapted with permission from Wang, P., & Blumberger, J. (2012). Mechanistic insight into the blocking of CO diffusion in [NiFe]-hydrogenase mutants through multiscale simulation.* Proceedings of the National Academy of Sciences of the United States of America, 109*, 6399–6404. Panel (B): Adapted with permission from Wang, P., Bruschi, M., De Gioia, L., & Blumberger, J. (2013). Uncovering a dynamically formed substrate access tunnel in carbon monoxide dehydrogenase/acetyl-CoA synthase.* Journal of the American Chemical Society, 135*(25), 9493–9502. Copyright 2013 American Chemical Society. Panel (C): Adapted with permission from De Sancho, D., Kubas, A., Wang, P., Blumberger, J., & Best, R. B. (2015). Identification of mutational hot spots for substrate diffusion: Application to myoglobin.* Journal of Chemical Theory and Computation, 11*(4), 1919–1927. Copyright 2015 American Chemical Society.*

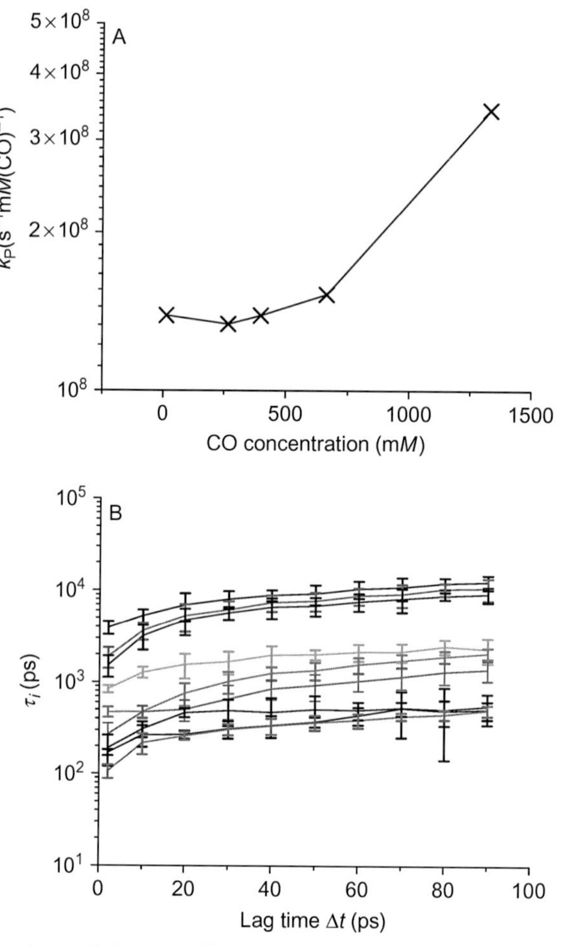

Fig. 5 (A) Comparison of the CO diffusion rate from solvent to protein microstates within myoglobin, k_P, for different CO concentrations. The rate is virtually unchanged up to CO concentrations of about 500 m*M*. (B) Dependence of the slowest relaxation times with the lag time (Δt) for the construction of the MSM. At lag times $\Delta t \geq$ 10 ps the slowest modes are converged within the relatively large error of the calculation. *Both panels are adapted with permission from De Sancho, D., Kubas, A., Wang, P., Blumberger, J., & Best, R. B. (2015). Identification of mutational hot spots for substrate diffusion: Application to myoglobin.* Journal of Chemical Theory and Computation, 11*(4), 1919–1927. Copyright 2015 American Chemical Society.*

increase in k_P. Additionally we tested whether the slowest modes of the system had a strong concentration dependence, and again, we found the result to be independent of the number of ligand molecules within the simulation box (De Sancho et al., 2015).

3.1.5 MSM Validation

We also tested the dependence of transition rates on the lag-time Δt in Eq. (18) and found that the relaxation times τ_i obtained from the rate matrix are relatively insensitive to the lag time (see Fig. 5B), supporting the assumption of Markovian dynamics. More stringent tests for the validity of the MSM model were also carried out and we refer here to the supporting information of De Sancho et al. (2015) for details. Taken together, the validity of the MSM, the insensitivity of diffusion rates in a wide range of ligand concentrations, and the fact that we could reproduce very well experimental mutations effects on ligand diffusion are a strong confirmation of the validity of our methodology.

3.2 Sensitivity Analysis for Myoglobin

In most recent work, we applied the sensitivity analysis discussed in Section 2.3 to myoglobin, for which a plethora of experimental kinetic data on mutant proteins are available (Olson et al., 2007). The diffusion paths and rate constants obtained for wild-type myoglobin were in good agreement with other simulations (Ruscio et al., 2008) and experiment (see Elber, 2010). The sensitivity parameter α_s, Eq. (29), was calculated for each cavity/microstate, requiring only the transition rate matrix for the wild-type enzyme. Besides, the change in the free energy of a microstate upon mutation, Δg_s in Eq. (31), was estimated based on a very simple model that accounted only for the change in entropy due to the change in cavity volume.

The predicted and experimental changes in the diffusion rate upon mutation are shown in Fig. 6. The overall trend is well reproduced, though the agreement is semiquantitative. For validation of the sensitivity model we explicitly simulated one of the most sensitive myoglobin mutants (L29W). The change in the diffusion rate obtained was in excellent agreement with the value predicted by the sensitivity model (compare filled circle and triangle for L29W in Fig. 6) and in good agreement with the experimental value. There are, however, two outliers involving mutations to tryptophan residues, V68W and H64W. We think that the deviation is mainly due to the simple model used for estimation of Δg_s. Taken together, these results suggest that our sensitivity analysis can become a useful screening method for the identification of hot spots for mutations. The accuracy may be further increased by improving estimations for Δg_s.

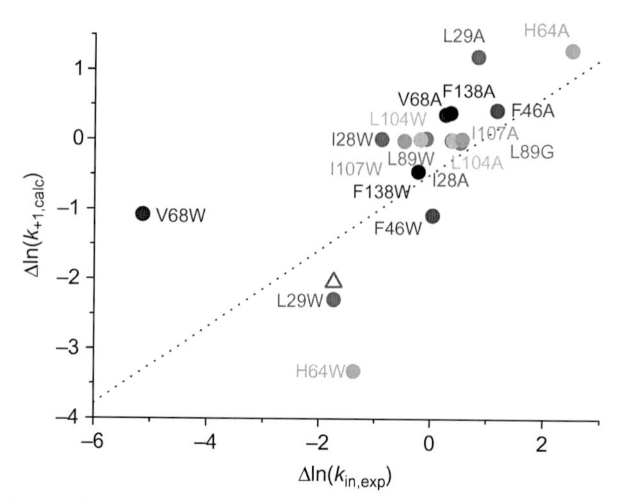

Fig. 6 Correlation of experimental and calculated changes in the logarithm of the CO diffusion rate constant in myoglobin. The triangle corresponds to the result from the explicit simulation of the L29W mutant. Adapted with permission from De Sancho, D., Kubas, A., Wang, P., Blumberger, J., & Best, R. B. (2015). Identification of mutational hot spots for substrate diffusion: Application to myoglobin. *Journal of Chemical Theory and Computation, 11*(4), 1919–1927. Copyright 2015 American Chemical Society.

4. CONCLUDING REMARKS

We have reviewed the methodology we have developed for the calculation of rate constants for the diffusion and binding of small ligands to buried protein active sites. At first MD simulations are used to map out the cavities within the protein. Then the ligand distribution within the protein is coarse grained to discrete Markov states and the transition rates between these states estimated from MD data. Once the full rate matrix is available, the rate constants for diffusion and binding to the active site can be obtained either from a fit of the time-dependent population of the active site to a phenomenological rate law or from the reactive flux formalism.

Applications to a diverse but limited set of ligand/protein systems suggest that the present methodology gives an order of magnitude estimate for diffusion rates. Given the complexity of the problem, the agreement is good and promising for future diffusion studies in proteins. There are a number of issues that hamper a more accurate rate calculation and they may be addressed in future work (i) the estimation of slow transition rates via extrapolation of the MFPTs to zero force, (ii) the limited accuracy of ligand/protein interactions in typical force fields, and (iii) the side-chain dynamics described by protein force fields.

For the calculation of the binding rates, one needs to take into account also the errors made in DFT calculations of the rate of chemical attachment/ detachment. This error may be as large as a few kcal mol^{-1}, in particular for metalloenzymes whose active sites typically have a complicated electronic structure. Experiments typically probe the total binding rate constant (diffusion + chemical binding), which makes it difficult to disentangle the error in the absolute binding rate due to MD and DFT calculations. Therefore, comparison to experimental mutation effects is probably the best way to validate the MD simulations.

Finally, one could ask whether the methodology developed could be extended to the binding kinetics of larger ligands, possibly drug-like molecules. Since the methodology is general and no particular assumption is made regarding the nature of the ligand and the protein, the answer is yes, in principle. Note that most proteins that bind larger molecules have their binding site closer to or at the protein surface for the obvious reason that larger molecules will not be able to significantly penetrate the protein. Hence, a Markov state model for the binding kinetics of larger molecules may not include many cavity states within the protein as was the case for the di- and triatomic ligands investigated here, but other states related to different binding sites at the protein surface and possibly different protein conformations. The challenges are to sample both the intramolecular degrees of freedom of the ligand, which was not necessary for the small, rigid molecules investigated here, and the relevant protein conformations, possibly on multiple timescales.

ACKNOWLEDGMENTS

D.D.S. acknowledges a Research Fellowship from Ikerbasque. R.B.B. was supported by the Intramural Research Program of the National Institute of Diabetes and Digestive and Kidney Diseases of the National Institutes of Health. Some of the work reviewed here was carried out, while D.D.S. was supported by EPSRC Grant EP/J016764/1 and J.B. by EPSRC Grant EP/J015571/1.

REFERENCES

Barends, T. R. M., Foucar, L., Ardevol, A., Nass, K., Aquila, A., Botha, S., ... Schlichting, I. (2015). Direct observation of ultrafast collective motions in CO myoglobin upon ligand dissociation. *Science*, *350*, 445–450.

Baron, R., Riley, C., Chenprakhon, P., Thotsaporn, K., Winter, R., Alfieri, A., ... Fraaije, M. (2009). Multiple pathways guide oxygen diffusion into flavoenzyme active sites. *Proceedings of the National Academy of Sciences of the United States of America, 106*(26), 10603–10608.

Bell, G. (1978). Models for the specific adhesion of cells to cells. *Science, 200*(4342), 618–627.

Berezhkovskii, A., Hummer, G., & Szabo, A. (2009). Reactive flux and folding pathways in network models of coarse-grained protein dynamics. *Journal of Chemical Physics, 130*, 205102.

Best, R. B., & Hummer, G. (2005). Reaction coordinates and rates from transition paths. *Proceedings of the National Academy of Sciences of the United States of America, 102*(19), 6732–6737.

Best, R. B., Paci, E., Hummer, G., & Dudko, O. K. (2008). Pulling direction as a reaction coordinate for the mechanical unfolding of single molecules. *Journal of Physical Chemistry B, 112*(19), 5968–5976.

Bolhuis, P. G., Chandler, D., Dellago, C., & Geissler, P. L. (2002). Transition path sampling: Throwing ropes over rough mountain passes, in the dark. *Annual Review of Physical Chemistry, 53*, 291–318.

Buchete, N., & Hummer, G. (2008). Coarse master equations for peptide folding dynamics. *Journal of Physical Chemistry B, 112*(19), 6057–6069.

Carver, T. E., Rohlfs, R. J., Olson, J. S., Gibson, Q. H., Blackmore, R. S., Springer, B. A., & Sligar, S. G. (1990). Analysis of the kinetic barriers for ligand binding to sperm whale myoglobin using site-directed mutagenesis and laser photolysis techniques. *Journal of Biological Chemistry, 265*(32), 20007–20020.

Ceccarelli, M., Anedda, R., Casu, M., & Ruggerone, P. (2008). CO escape from myoglobin with metadynamics simulations. *Proteins: Structure, Function, and Bioinformatics, 71*(3), 1231–1236.

Chodera, J. D., & Noé, F. (2014). Markov state models of biomolecular conformational dynamics. *Current Opinion in Structural Biology, 25*, 135–144. (Theory and simulation/macromolecular machines).

Cohen, J., Arkhipov, A., Braun, R., & Schulten, K. (2006). Imaging the migration pathways for O_2, CO, NO, and Xe inside myoglobin. *Biophysical Journal, 91*(5), 1844–1857.

Cohen, J., Kim, K., King, P., Seibert, M., & Schulten, K. (2005). Finding gas diffusion pathways in proteins: application to O_2 and H_2 transport in CpI [FeFe]-hydrogenase and the role of packing defects. *Structure, 13*(9), 1321–1329.

Cohen, J., Kim, K., Posewitz, M., Ghirardi, M., Schulten, K., Seibert, M., & King, P. (2005). Molecular dynamics and experimental investigation of H_2 and O_2 diffusion in [Fe]-hydrogenase. *Biochemical Society Transactions, 33*, 80–82.

Cohen, J., & Schulten, K. (2007). O_2 migration pathways are not conserved across proteins of a similar fold. *Biophysical Journal, 93*(10), 3591–3600.

D′Abramo, M., Di Nola, A., & Amadei, A. (2009). Kinetics of carbon monoxide migration and binding in solvated myoglobin as revealed by molecular dynamics simulations and quantum mechanical calculations. *Journal of Physical Chemistry B, 113*(51), 16346–16353.

Daura, X., Gademann, K., Jaun, B., Seebach, D., van Gunsteren, W., & Mark, A. (1999). Peptide folding: When simulation meets experiment. *Angewandte Chemie International Edition, 38*(1–2), 236–240.

De Sancho, D., Kubas, A., Wang, P., Blumberger, J., & Best, R. B. (2015). Identification of mutational hot spots for substrate diffusion: Application to myoglobin. *Journal of Chemical Theory and Computation, 11*(4), 1919–1927.

De Sancho, D., Mittal, J., & Best, R. B. (2013). Folding kinetics and unfolded state dynamics of the GB1 hairpin from molecular simulation. *Journal of Chemical Theory and Computation, 9*(3), 1743–1753.

Doukov, T. I., Blasiak, L. C., Seravalli, J., Ragsdale, S. W., & Drennan, C. L. (2008). Xenon in and at the end of the tunnel of bifunctional carbon monoxide dehydrogenase/acetyl-CoA synthase. *Biochemistry, 47*(11), 3474–3483.

Dudko, O. K., Hummer, G., & Szabo, A. (2006). Intrinsic rates and activation free energies from single-molecule pulling experiments. *Physical Review Letters*, *96*(10), 108101.

Elber, R. (2010). Ligand diffusion in globins: Simulations versus experiment. *Current Opinion in Structural Biology*, *20*(2), 162–167.

Elber, R., & Karplus, M. (1990). Enhanced sampling in molecular dynamics: Use of the time-dependent Hartree approximation for a simulation of carbon monoxide diffusion through myoglobin. *Journal of the American Chemical Society*, *112*(25), 9161–9175.

Evans, E., & Ritchie, K. (1997). Dynamic strength of molecular adhesion bonds. *Biophysical Journal*, *72*(4), 1541–1555.

Faradjian, A., & Elber, R. (2004). Computing time scales from reaction coordinates by milestoning. *Journal of Chemical Physics*, *120*, 10880–10889.

Hummer, G. (2005). Position-dependent diffusion coefficients and free energies from Bayesian analysis of equilibrium and replica molecular dynamics simulations. *New Journal of Physics*, 7, 34.

Hummer, G., Schotte, F., & Anfinrud, P. A. (2004). Unveiling functional protein motions with picosecond X-ray crystallography and molecular dynamics simulations. *Proceedings of the National Academy of Sciences of the United States of America*, *101*(43), 15330–15334.

Johnson, B. J., Cohen, J., Welford, R. W., Pearson, A. R., Schulten, K., Klinman, J. P., & Wilmot, C. M. (2007). Exploring molecular oxygen pathways in Hansenula polymorpha copper-containing amine oxidase. *Journal of Biological Chemistry*, *282*, 17767–17776.

Kramers, H. (1940). Brownian motion in a field of force and the diffusion model of chemical reactions. *Physica*, 7, 284–303.

Leroux, F., Dementin, S., Burlat, B., Cournac, L., Volbeda, A., Champ, S., … Léger, C. (2008). Experimental approaches to kinetics of gas diffusion in hydrogenase. *Proceedings of the National Academy of Sciences of the United States of America*, *105*(32), 11188–11193.

Liebgott, P., Leroux, F., Burlat, B., Dementin, S., Baffert, C., Lautier, T., … Meynial-Salles, I. (2010). Relating diffusion along the substrate tunnel and oxygen sensitivity in hydrogenase. *Nature Chemical Biology*, *6*(1), 63–70.

Maragliano, L., Cottone, G., Ciccotti, G., & Vanden-Eijnden, E. (2010). Mapping the network of pathways of CO diffusion in myoglobin. *Journal of the American Chemical Society*, *132*, 1010–1017.

Mishra, S., & Meuwly, M. (2010). Quantitative analysis of ligand migration from transition networks. *Biophysical Journal*, *99*(12), 3969–3978.

Montet, Y., Amara, P., Volbeda, A., Vernede, X., Claude, E., Hatchikian, M., … Fontecilla-Camps, J. (1997). Gas access to the active site of Ni-Fe hydrogenases probed by X-ray crystallography and molecular dynamics. *Nature Structural Biology*, *4*(7), 523–526.

Murray, J., Maghlaoui, K., Kargul, J., Sugiura, M., & Barber, J. (2008). Analysis of xenon binding to photosystem II by X-ray crystallography. *Photosynthesis Research*, *98*, 523–527.

Nishihara, Y., Hayashi, S., & Kato, S. (2008). A search for ligand diffusion pathway in myoglobin using a metadynamics simulation. *Chemical Physics Letters*, *464*(4–6), 220–225.

Olson, J. S., Soman, J., & Phillips, G. N., Jr. (2007). Ligand pathways in myoglobin: A review of Trp cavity mutations. *Life*, *59*, 552.

Ruscio, J., Kumar, D., Shukla, M., Prisant, M., Murali, T., & Onufriev, A. (2008). Atomic level computational identification of ligand migration pathways between solvent and binding site in myoglobin. *Proceedings of the National Academy of Sciences of the United States of America*, *105*(27), 9204–9209.

Sarich, M., Noé, F., & Schütte, C. (2010). On the approximation error of markov state models. *SIAM Multiscale Modeling and Simulation*, *8*, 1154–1177.

Teixeira, V., Baptista, A., & Soares, C. (2006). Pathways of H_2 toward the active site of [NiFe]-hydrogenase. *Biophysical Journal*, *91*(6), 2035–2045.

Vanden-Eijnden, E., Venturoli, M., Ciccotti, G., & Elber, R. (2008). On the assumptions underlying milestoning. *Journal of Chemical Physics*, *129*, 174102.

Wang, P., Best, R. B., & Blumberger, J. (2011a). A microscopic model for gas diffusion dynamics in a [NiFe]-hydrogenase. *Physical Chemistry Chemical Physics*, *13*, 7708–7719.
Wang, P., Best, R. B., & Blumberger, J. (2011b). Multiscale simulation reveals multiple pathways for H_2 and O_2 transport in a [NiFe]-hydrogenase. *Journal of the American Chemical Society*, *133*, 3548–3556.
Wang, P., & Blumberger, J. (2012). Mechanistic insight into the blocking of CO diffusion in [NiFe]-hydrogenase mutants through multiscale simulation. *Proceedings of the National Academy of Sciences of the United States of America*, *109*, 6399–6404.
Wang, P., Bruschi, M., De Gioia, L., & Blumberger, J. (2013). Uncovering a dynamically formed substrate access tunnel in carbon monoxide dehydrogenase/acetyl-CoA synthase. *Journal of the American Chemical Society*, *135*(25), 9493–9502.
Zwanzig, R. (1983). From classical dynamics to continuous time random walks. *Journal of Statistical Physics*, *30*(2), 255–262.

CHAPTER FOURTEEN

Calculation of Enzyme Fluctuograms from All-Atom Molecular Dynamics Simulation

T.H. Click*, N. Raj*, J.-W. Chu*,†,‡,1

*Institute of Bioinformatics and Systems Biology, National Chiao Tung University, Hsinchu, Taiwan, ROC
†Department of Biological Science and Technology, National Chiao Tung University, Hsinchu, Taiwan, ROC
‡Institute of Molecular Medicine and Bioengineering, National Chiao Tung University, Hsinchu, Taiwan, ROC
[1]Corresponding author: e-mail address: jwchu@nctu.edu.tw

Contents

Abstract

In this work, a computational framework is presented to compute the time evolution of force constants for a coarse grained (CG) elastic network model along an all-atom molecular dynamics trajectory of a protein system. Obtained via matching distance fluctuations, these force constants represent strengths of mechanical coupling between CG beads. Variation of coupling strengths with time is hence termed the fluctuogram of protein dynamics. In addition to the schematic procedure and implementation considerations, several ways of combining force constants and data analysis are presented to illustrate the potential application of protein fluctuograms. The unique angle provided by the fluctuogram expands the scope of atomistic simulations and is expected to impact upon fundamental understanding of protein dynamics as well as protein engineering technologies.

Methods in Enzymology, Volume 578
ISSN 0076-6879
http://dx.doi.org/10.1016/bs.mie.2016.05.024

1. INTRODUCTION

Proteins display conformational changes in almost all activities in the cell, including sensing (Lee et al., 2008) and signal transduction (Changeux & Edelstein, 2005; Mouillet-Richard et al., 2000) as well as energy conversion (Kaila, Wikström, & Hummer, 2014) and force generation (Piazzesi et al., 2002; Schnauß, Glaser, Schuldt, Golde, & Händler, 2015). Dynamics is thus an essential component in enzymology. Structural transitions at the nanoscale also exhibit intriguing phenomena due to interplay of stochasticity and complex molecular interactions. For example, conformational changes can be triggered and modulated by molecular signals at protein sites distal to the response region though allosteric coupling (Hilser & Thompson, 2007; Lin et al., 2011; Silvestre-Ryan, Lin, & Chu, 2011). However, the physics underlying its mechanistic cause has remained elusive. Establishing fundamental understanding for unique features of protein dynamics is not only of scientific interest but also critical in advancing technological areas such as drug development and biomolecular engineering.

Solely relying on experimental methods to resolve protein conformational changes, though, is very difficult. Under significant thermal forces, signals of dynamics are inevitably noisy. Ensemble experiments (Borgia et al., 2015; Yu et al., 2012) that measure kinetics may reach adequate signal-to-noise ratios, but information of dynamics is doomed to be washed away. Since transition processes of complex systems often involve sequential changes of several order parameters, elucidation of mechanism via averaged dynamics (kinetics) tends to have the issue of underdetermination. Although single-molecule experiments (Jahn, Buchner, Hugel, & Rief, 2016; McCann, Zheng, Chiantia, & Bowen, 2011; Stockmar, Kobitski, & Nienhaus, 2016) can be employed to escape the constraints of ensemble averages, accessible degrees of freedom for following dynamics are limited to mostly one or a couple at best. Although invaluable insight on protein dynamics can be gained from single-molecule data, their ability to quantitatively construct a comprehensive picture of conformational changes requires further development.

Therefore, molecular simulations offer a fruitful route to supplement details that experimental methods cannot observe. Applied with all-atom force fields of biomolecules, molecular dynamics (MD) simulations can potentially be used to resolve conformational changes at the atomistic level (Abraham et al., 2015; Brooks et al., 2009; Case et al., 2005; Phillips et al., 2005).

However, the system size and timescale involved in structure transitions of biomolecules are mostly beyond the reach of direct all-atom MD simulations with explicit solvent molecules. Enhanced sampling approaches and free energy simulation methods are usually necessary to overcome the timescale challenge (Carvalho et al., 2014; Simonson, Archontis, & Karplus, 2002; Sugita & Okamoto, 1999; Tirion, 1996). For example, identification of an appropriate order parameter via reaction path optimization followed by thermodynamic integration is a powerful approach to study protein conformational changes with all-atom MD (Brokaw, Haas & Chu, 2009; Cho, Gross, & Chu, 2011; Haas & Chu, 2009; Lin, Beckham, Himmel, Crowley, & Chu, 2013). Although the approach of coarse graining (CG) a complicated model into a lower resolution one may extend the timescale of MD simulation to some extent (Cho & Chu, 2009; Izvekov & Voth, 2005; Izvekov, Parrinello, Burnham, & Voth, 2004), immense amount of effort is required for testing and validation to ensure that the CG model is a useful tool to address the questions of interest.

For instance, in formulating intermolecular interactions as free energies expanded by CG degrees of freedom (Gross, Bell, & Chu, 2013; Lu, Dama, & Voth, 2013; Shi, Liu, & Voth, 2008), the results will be specific to how atoms are grouped into CG sites and the functional forms of interaction potentials (Cho & Chu, 2009). The process of coarse graining and performing a CG simulation hence projects out a particular free energy landscape of the underlying system. Selection of a CG scheme to best address the system being studied and the question being asked is highly nontrivial and does not enjoy a universal principle. Indeed, the process of designing and improving coarse graining schemes is a useful route to accumulate specific knowledge such that reduced system sizes can be exploited to gain understanding via more effective sampling.

The notion of employing coarse graining as a platform to dissect the free energy landscape of complex molecular systems can be particularly powerful if model parameters can be computed from the results of all-atom MD simulation (Das, Lu, Andersen, & Voth, 2012; Hills, Lu, & Voth, 2010; Saunders & Voth, 2012; Zhang & Voth, 2010). With this multiscale coarse graining, interaction parameters of the CG model serve as reporters of the effective free energy surface based on information of atomistic simulation. Here, we discuss in detail a method of this kind—the fluctuogram for modeling protein dynamics.

The fluctuogram calculation is based on an elastic network model (ENM) (Tirion, 1996) as CG scale representation. Typically, the Cα positions of a protein are used to form a network of beads, which are connected via harmonic bonds if a pair of beads are within a cutoff distance, say 10 Å. With a simple energetic function, ENM stresses the importance of native protein structures in governing the patterns of dynamical movements. By applying normal mode analysis (NMA), low-frequency eigenvectors of the Hessian matrix of the ENM Hamiltonian can be visualized to explore potential directions of conformational changes related to function. This structure-based approach leads to the common practice of employing the same value of force constant for all harmonic bonds in an ENM (Atilgan et al., 2001; Haliloglu, Bahar, & Erman, 1997; Kondrashov, Van Wynsberghe, Bannen, Cui, & Phillips, 2007), but some work has been done using heterogeneous harmonic bonds (Chu & Voth, 2006; Li, Zhang, & Xia, 2015; López-Blanco & Chacon, 2015; Lyman, Pfaendtner, & Voth, 2008; Silvestre-Ryan et al., 2011; Xia, Tong, & Lu, 2013).

To connect to the atomistic scale via multiscale coarse graining, one way is to focus on the variances of bond fluctuations in a CG ENM computed via NMA. From the atomistic scale, variances of bond fluctuations can also be calculated from the trajectories of all-atom MD because coordinates of CG beads can be calculated from all-atom configurations. Matching atomistic and CG scales can then be achieved by adjusting force constants in the ENM to reproduce the bond fluctuations observed in all-atom MD. This scheme is termed fluctuation matching (Chu & Voth, 2006; Lyman et al., 2008; Silvestre-Ryan et al., 2011). The calculated force constants represent effective strengths of mechanical coupling between CG beads in the network, which are resulted from specific molecular interactions composed of the protein sequence, solvent environment, and thermodynamic state.

An issue in this matching of atomistic and CG scales is that the energy function of ENM is harmonic with respect to a reference structure whereas that of molecular mechanical force fields is not. During MD, a protein structure is constantly varying on an anharmonic energy surface. Fluctuation matching with ENM thus approximates the structure variation in an all-atom MD trajectory as the Boltzmann distribution of a harmonic potential with the averaged structure being the reference. For nanosecond timescale during which protein dynamics usually does not cause a large conformational change, this assumption is less problematic. For a longer trajectory, a practical way to overcome this issue is breaking the long trajectory into segments, and fluctuation matching is applied to compute the averaged structure and ENM force constants of each window. This sequential analysis of an all-atom MD trajectory via fluctuation matching and ENM is termed as the fluctuogram of protein dynamics (Silvestre-Ryan et al., 2011).

In this work, we analyze the dependence of protein fluctuogram on different schemes of mapping a snapshot in an atomistic MD trajectory onto a CG configuration. To facilitate the implementation that each bond in the ENM has a distinct force constant, a general computational framework is established. The potential ways of employing force constants to infer protein function will also be discussed.

2. CALCULATION OF PROTEIN FLUCTUOGRAMS

The procedure of computing the fluctuogram from an all-atom MD trajectory is presented in the following with discussion of methodological details.

1. Set the mapping from a snapshot in an all-atom trajectory into the corresponding coarse grained configuration. Fig. 1 illustrates two examples of calculating CG coordinates from an atomistic structure of amino acids.

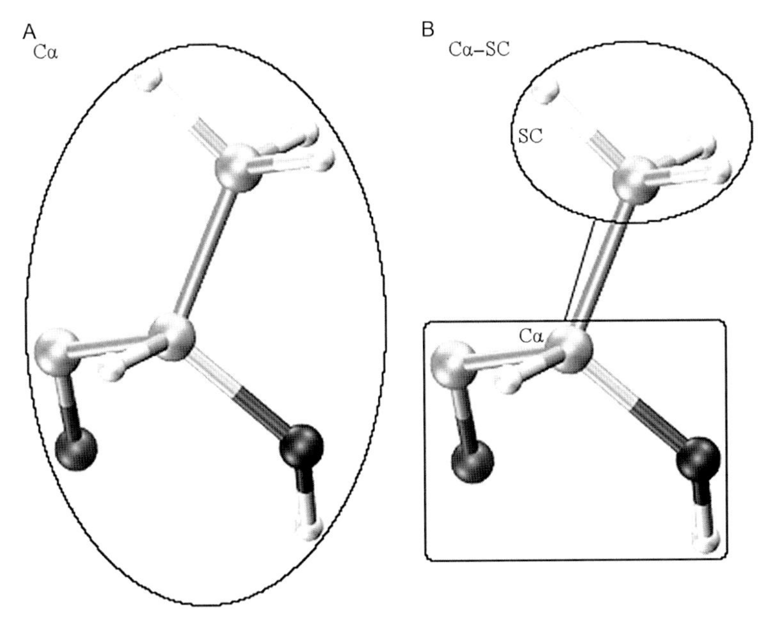

Fig. 1 Coarse-grain model of an alanine. (A) The Cα model utilizes the Cα coordinates of each residue and the total mass of the residue. (B) The Cα–SC model separates the model into the backbone (Cα coordinates and total mass of square region) and the sidechain (sidechain center of mass and total mass of circular region). (See the color plate.)

2. Split an all-atom MD trajectory into segments of Δt in a nanosecond timescale range. An overlap between segments can be introduced for continuity of the fluctuogram.
3. Transform each all-atom MD segment into snapshots of CG coordinates via the mapping scheme of 1.
4. Set a cutoff distance for including a harmonic bond between CG beads based on a reference structure. This cutoff and the reference structure are only for setting an initial ENM, for which a single value of force constant may be used for all bonds. The cutoff is usually set to be longer than necessary, eg, 10 Å, since unnecessary bonds will have a zero value of force constant in the end of a fluctuation matching calculation.
5. Compute the averaged bond lengths between CG beads in a selected trajectory segment that was transformed from all-atom MD. These averaged bond lengths are employed as equilibrium bond lengths in the ENM.

$$b_0 = r_{ij}^{\text{avg}} \tag{1}$$

6. Compute the variances of bond lengths of the CG trajectory segment in step 5. These variances of bond lengths are employed as targeted values for fitting force constants. After energy minimization, perform NMA of the CG model.
7. Compute the variances of bond-length fluctuations of the ENM via the normal modes. Temperature is set to the same value as that used in the all-atom MD simulation.
8. Update the force constant of each harmonic bond in the ENM according to the following equation. The principle is that the force constant needs to be increased/reduced if the bond fluctuation is over-/under-estimated at the CG scale. The minimum value a force constant can assume is zero. A value of 0.012 kcal/mol for α was found to give stable and effective iterations of fluctuation matching.

$$k_{ij}^{m+1} = k_{ij}^{m} - \alpha \left(\frac{1}{\left\langle \delta r_{ij}^2 \right\rangle^{\text{CG}}} - \frac{1}{\left\langle \delta r_{ij}^2 \right\rangle^{\text{AA}}} \right) \tag{2}$$

9. Repeat steps 7–9 until convergence of all force constants. The root-of-mean-square-difference of force constants in comparison with the previous step typically decreases below 0.005 kcal/mol/Å^2 within 200 steps of iteration.

10. Go to step 5 for another segment of trajectory and stop when the fluctuation-matched ENM of every segment is calculated.

3. IMPLEMENTATION CONSIDERATIONS

The fluctuogram calculations require linking the CG ENM to the atomistic scale and treating each harmonic bond with a different force constant and equilibrium length (Chu & Voth, 2006; Lyman et al., 2008; Silvestre-Ryan et al., 2011). We integrate the series of distinct tasks listed earlier via a Python 2.7 code that utilizes several freely available libraries, including MDAnalysis 0.10 (Michaud-Agrawal, Denning, Woolf, & Beckstein, 2011), Numpy 1.10 (van der Walt, Colbert, & Varoquaux, 2011), Scipy 0.16.0 (Jones, Oliphant, & Peterson, 2001), and pandas 0.16.0 (McKinney, 2010). The structure of scripts were designed such that implementing different ways of mapping an all-atom configuration onto a CG structure can be accomplished with a high level of automation. To facilitate analysis of fluctuograms, data structure is developed to document features of each harmonic bond. For example, the types of CG beads (sidechain and backbone) that a bond connects and the sequence separation of CG beads that a bond connects are both accessible attributes. As a result, the profiles of coupling strengths for different harmonic bonds of different types can be determined quickly. The composition of bonds that gives a specific range of coupling strengths can be easily obtained as well. To determine the mechanical coupling associated with a CG bead, we employ the convention that half of the force constant of a bond gives to each of the beads it connects. The force constants associated with CG beads may then be summed to determine the coupling strength of a protein residue, a domain, or the entire protein molecule. The coupling strength of a protein–ligand complex, for example, can be determined by summing all force constants that connect the CG beads of the two molecular entities. Therefore, the strengths of mechanical coupling for different aspects of the biomolecule system during an all-atom MD simulation can be determined with the fluctuogram calculation. The CHARMM software (Brooks et al., 2009) is employed for calculations of NMA, getting statistics of bond fluctuations of the CG ENM, and computing bond averaged bond lengths and variances from transformed all-atom trajectories. The fluctuation matching for different trajectory segments can be conducted in parallel. All in-house developed scripts may be obtained upon request.

3.1 Analysis of a Fluctuogram

As an example for illustrating the calculation and usage of protein fluctuogram, we consider rat anionic trypsin (RAT) (Pasternak, Ringe, & Hedstrom, 1999), a protease with well-characterized information on structure and function. The structure of pdb-id 3tgi includes RAT and the bound bovine pancreatic trypsin inhibitor (BPTI). All-atom MD of 300 ns was performed for both the BPTI-bound and apo forms using the CHARMM36 force field (Huang & MacKerell, 2013) and the Gromacs 5.1 (Abraham et al., 2015) software. The system was simulated with a protonation state corresponding to pH 7 in an orthorhombic dodecahedron water box with NaCl ions added at a 0.15 *M* ionic strength. After heating up to 300 K and 2 ns equilibration at 300 K and 1 atm, the production run was continued for 300 ns with a frame saved every picosecond.

To conduct fluctuogram calculations, the system is coarse grained into two resolutions. First, the position of a Cα atom sets the coordinate of a CG bead for each amino acid residue. Second, in addition to the Cα site representing the backbone, the center of mass of sidechain atoms is set as a CG bead representing the sidechain site. In the following, the first resolution is referred to as the Cα model and the second as the Cα–SC model.

Following the procedure of fluctuogram calculation listed earlier, the all-atom trajectory was transformed into both Cα and Cα–SC ENM models with a 4-ns window size yielding 149 CG snapshots. A 10 Å cutoff was selected to determine inclusion of bond interactions in the initial ENM. A Ca^{2+} ion was coordinated with Glu70, Glu77, and Glu80 on the sidechain carboxyl oxygens and the backbone carboxyl oxygens of Asn72 and Val75, which was included as a single site in the CG models. Details of the number of sites used for fluctuation matching are listed in Table 1. In the following, several ways of analyzing the fluctuograms of the apo protein (apo-RAT) and BPTI-bound form (BPTI–RAT) are presented for illustration of usage.

For BPTI–RAT, distributions of all intraresidue and interresidue coupling strengths are shown in Fig. 2 for both Cα and Cα–SC models.

Table 1 Numbers of CG Beads for Different Schemes of All-Atom-to-CG Mapping

	Cα	Cα–SC
apo-RAT	224	424
RAT–BPTI complex	280	531

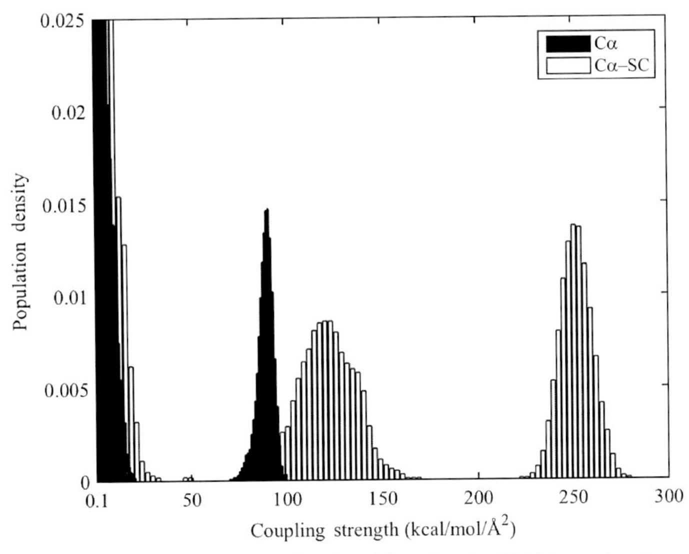

Fig. 2 Distribution of coupling strengths of residues for the BPTI-bound serine protease.

Interresidue coupling strength is the sum of all force constants for the bonds connecting two amino acid residues. Intraresidue coupling strength is the force constant of the bond connecting the Cα and SC beads for the Cα–SC model. Therefore, the Cα model does not contain intraresidue coupling strength by construction. For the Cα–SC model, intraresidue coupling represents the peak group of largest strengths in Fig. 2 in the range of 220–270 kcal/mol/Å^2. The peak group of the second largest coupling strengths for the Cα–SC model is in the range of 85–160 kcal/mol/Å^2, which comes from the couplings between the first nearest neighbor residues, ie, between I and $I+1$. For the Cα model, the first nearest neighbor couplings are more narrowly distributed in the 75–100 kcal/mol/Å^2 range. Most of the other interresidue coupling strengths are lower than 60 and 25 kcal/mol/Å^2 in the Cα–SC and Cα model, respectively. As shown in Fig. 2, the additional sidechain bead of the Cα–SC model gives rise to more widely distributed coupling strengths compared to the single-site model. A similar trend is seen in the distributions of coupling strengths between CG sites, ie, before combining force constants according to residue numbers (data not shown).

To focus on intermolecular couplings in a protein matrix, we sum all harmonic interactions for residue pairs separated by three or more amino acids. This total coupling strength of each protein system is shown in

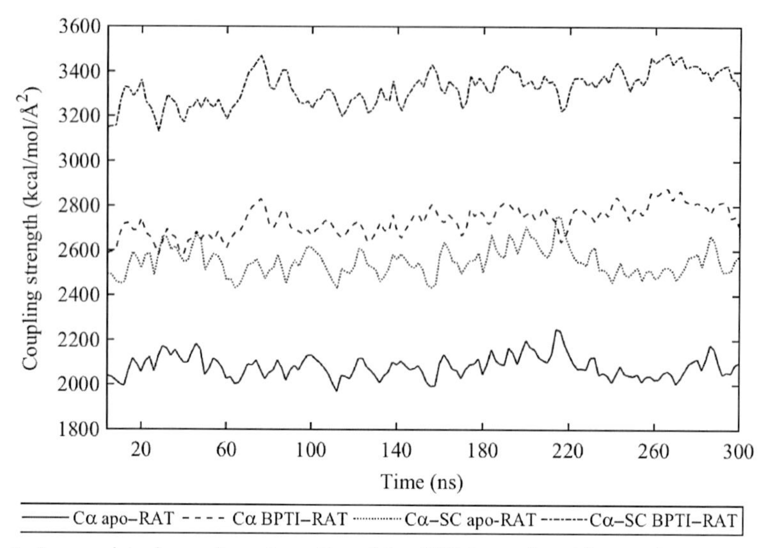

Fig. 3 Accumulated coupling strengths of the BPTI-bound and free serine protease in both the Cα and Cα–SC models.

Fig. 3. For both apo-RAT and BPTI–RAT, the Cα–SC model has a higher value for its higher resolution with an additional sidechain bead. Comparing to the apo-RAT system, BPTI–RAT has a bound substrate, and the total coupling strength is hence larger. In all cases, the total coupling strength assumes a rather constant value despite significant fluctuations in the protien structure. As illustrated later, conformational changes do cause variation in coupling strengths within the protein matrix, but the total coupling strength appears to be a fairly conserved quantity throughout the all-atom MD trajectories. This result implies that changes in coupling strengths of different parts in the protein system tend to compensate one another.

Fig. 4 plots the total coupling strength between the protein and ligand pair through all-atom MD trajectories. Both Cα–SC and Cα models exihibit a similar pattern of protein–ligand coupling as they come from the same reference atomistic trajectory. The Cα–SC model with an additional sidechain bead has a higher coupling strength as expected. Fig. 4 shows that the calculated fluctuograms can be employed to reveal variation of the coupling strength between RAT and BPTI over the MD trajectory.

By decomposing a force constant to the residues that it connects and summing the values of harmonic bonds that are associated with an amino acid, coupling strengths of individual residues can be determined from the fluctuogram. Focusing on the Cα–SC model, Fig. 5 shows the evolution

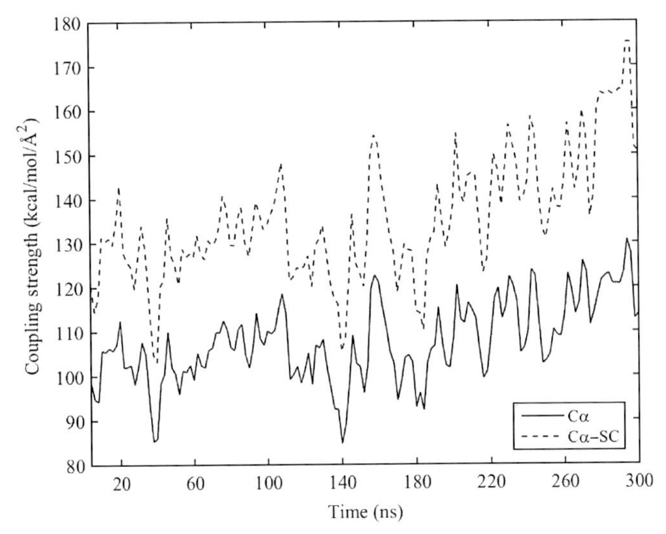

Fig. 4 Comparison of total coupling strengths between serine protease (RAT) and the inhibitor (BPTI) for the Cα and Cα–SC models.

of mechanical coupling strengths of residues during all-atom MD simulations. It is clear that protein residues exhibit different levels of changes in their coupling strengths during dynamics. Dynamical variation of mechanical coupling thus becomes accessible quantities with the fluctuogram calculation. For apo-RAT and BPTI–RAT, it can be seen in Fig. 5 that variation in mechanical coupling occurs throughout the protein matrix, not limited to residues at the binding pocket. Therefore, a potential application of computing fluctuograms is analysis of protein allostery.

4. DISCUSSION

In this chapter, we present the framework of fluctuogram calculations to parameterize a heterogeneous ENM over consecutive segments of snapshots along a MD trajectory. In addition to the averaged CG structures in each time window, fluctuation-matched force constants are obtained. Strengths of mechanical couplings in the protein system hence become observables of all-atom MD simulation. Analysis of biomolecular simulations is primarily based on structural properties; the method presented here offers an orthogonal angle on mechanical couplings to deduce molecular mechanisms. A modeler can devise specific schemes of all-atom-to-CG mapping to conduct fluctuogram calculations, providing a general and

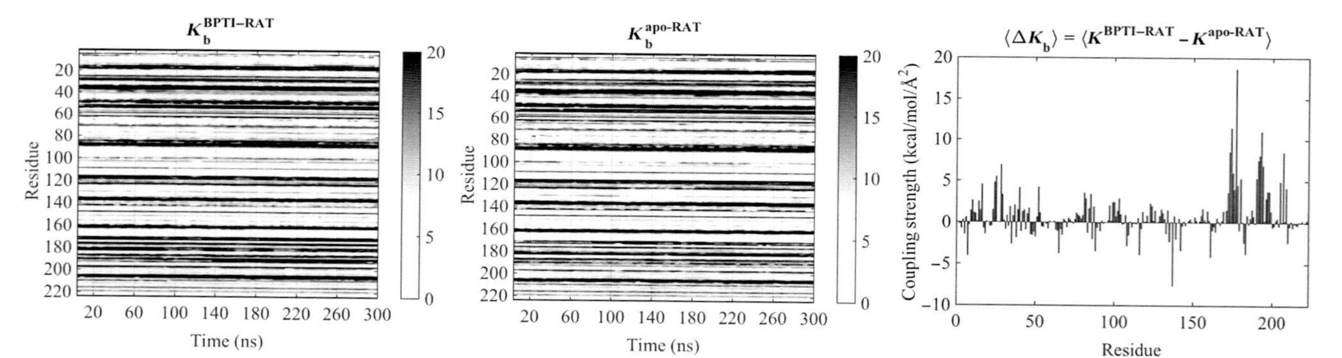

Fig. 5 Total coupling strength per residue over time for both the RAT–BPTI and apo-RAT (*left* and *middle panel*), and the average coupling strength difference between RAT–BPTI and apo-RAT (*right panel*).

flexible platform of analysis. Several ways of combining force constants and packaging data are presented to illustrate potential ways of applying the fluctuogram. For example, higher values of force constants that a residue assumes indicate stronger levels of couplings to other residues in the protein matrix during dynamics. Other properties of mechanical coupling strengths that may be of interest include degrees of variation over a trajectory and changes in responding to ligand binding. Comparison of such properties of the fluctuogram indicates significant correlation with the patterns of sequence correlation (Lin et al., 2011; Silvestre-Ryan et al., 2011), illustrating the ability to capture functionally important protein residues. As discussed earlier, the coupling strengths can also be employed to provide a quantitative picture on protein and ligand interactions. Fluctuogram is thus a general computational framework to gain fundamental understanding of protein dynamics and may also be employed to identify important residues in the structure network in comparison with experiments.

ACKNOWLEDGMENTS

This work was supported by the National Chiao Tung University, Taiwan, Republic of China. The funding from the Ministry of Science and Technology of Taiwan, Republic of China via Grant numbers 102-281-B-009-005-, 102-2113-M-009-022-MY2, and 103-2628-M-009-003-MY3 is acknowledged. We also thank MDAnalysis developers Oliver Beckstein and Richard Gowers for their advice in code development.

REFERENCES

Abraham, M. J., Murtola, T., Schulz, R., Páll, S., Smith, J. C., Hess, B., et al. (2015). GROMACS: High performance molecular simulations through multi-level parallelism from laptops to supercomputers. *SoftwareX*, *1–2*, 19–25. http://doi.org/10.1016/j.softx.2015.06.001.

Atilgan, A. R., Durell, S. R., Jernigan, R. L., Demirel, M. C., Keskin, O., & Bahar, I. (2001). Anisotropy of fluctuation dynamics of proteins with an elastic network model. *Biophysical Journal*, *80*(1), 505–515. http://doi.org/10.1016/S0006-3495(01)76033-X.

Borgia, A., Kemplen, K. R., Borgia, M. B., Soranno, A., Shammas, S., Wunderlich, B., et al. (2015). Transient misfolding dominates multidomain protein folding. *Nature Communications*, *6*, 8861. http://doi.org/10.1038/ncomms9861.

Brokaw, J. B., Haas, K. R., & Chu, J.-W. (2009). Reaction path optimization with holonomic constraints and kinetic energy potentials. *Journal of Chemical Theory and Computation*, *5*(8), 2050–2061. http://doi.org/10.1021/ct9001398.

Brooks, B. R., Brooks, C. L., MacKerell, A. D., Nilsson, L., Petrella, R. J., Roux, B., et al. (2009). CHARMM: The biomolecular simulation program. *Journal of Computational Chemistry*, *30*(10), 1545–1614. http://doi.org/10.1002/jcc.21287.

Carvalho, A. T. P., Barrozo, A., Doron, D., Kilshtain, A. V., Major, D. T., & Kamerlin, S. C. L. (2014). Challenges in computational studies of enzyme structure, function and dynamics. *Journal of Molecular Graphics and Modelling*, *54*, 62–79. http://doi.org/10.1016/j.jmgm.2014.09.003.

Case, D. A., Cheatham, T. E., Darden, T., Gohlke, H., Luo, R., Merz, K. M., et al. (2005). The Amber biomolecular simulation programs. *Journal of Computational Chemistry, 26*(16), 1668–1688. http://doi.org/10.1002/jcc.20290.

Changeux, J.-P., & Edelstein, S. J. (2005). Allosteric mechanisms of signal transduction. *Science, 308*(5727), 1424–1428. http://doi.org/10.1126/science.1108595.

Cho, H. M., & Chu, J.-W. (2009). Inversion of radial distribution functions to pair forces by solving the Yvon-Born-Green equation iteratively. *The Journal of Chemical Physics, 131*(13), 134107. http://doi.org/10.1063/1.3238547.

Cho, H. M., Gross, A. S., & Chu, J.-W. (2011). Dissecting force interactions in cellulose deconstruction reveals the required solvent versatility for overcoming biomass recalcitrance. *Journal of the American Chemical Society, 133*(35), 14033–14041. http://doi.org/10.1021/ja2046155.

Chu, J.-W., & Voth, G. A. (2006). Coarse-grained modeling of the actin filament derived from atomistic-scale simulations. *Biophysical Journal, 90*(5), 1572–1582. http://doi.org/10.1529/biophysj.105.073924.

Das, A., Lu, L., Andersen, H. C., & Voth, G. A. (2012). The multiscale coarse-graining method. X. Improved algorithms for constructing coarse-grained potentials for molecular systems. *The Journal of Chemical Physics, 136*, 194115. http://doi.org/10.1063/1.4705420.

Gross, A. S., Bell, A. T., & Chu, J.-W. (2013). Preferential interactions between lithium chloride and glucan chains in N, N-dimethylacetamide drive cellulose dissolution. *Journal of Physical Chemistry B, 117*(12), 3280–3286. http://doi.org/10.1021/jp311770u.

Haas, K., & Chu, J.-W. (2009). Decomposition of energy and free energy changes by following the flow of work along reaction path. *The Journal of Chemical Physics, 131*(14), 144105. http://doi.org/10.1063/1.3243080.

Haliloglu, T., Bahar, I., & Erman, B. (1997). Gaussian dynamics of folded proteins. *Physical Review Letters, 79*(16), 3090–3093. http://doi.org/10.1103/PhysRevLett.79.3090.

Hills, R. D., Jr., Lu, L., & Voth, G. A. (2010). Multiscale coarse-graining of the protein energy landscape. *PLoS Computational Biology, 6*(6), e1000827. http://doi.org/10.1371/journal.pcbi.1000827.

Hilser, V. J., & Thompson, E. B. (2007). Intrinsic disorder as a mechanism to optimize allosteric coupling in proteins. *Proceedings of the National Academy of Sciences of the United States of America, 104*(20), 8311–8315. http://doi.org/10.1073/pnas.0700329104.

Huang, J., & MacKerell, A. D. (2013). CHARMM36 all-atom additive protein force field: Validation based on comparison to NMR data. *Journal of Computational Chemistry, 34*(25), 2135–2145. http://doi.org/10.1002/jcc.23354.

Izvekov, S., Parrinello, M., Burnham, C. J., & Voth, G. A. (2004). Effective force fields for condensed phase systems from ab initio molecular dynamics simulation: A new method for force-matching. *The Journal of Chemical Physics, 120*(23), 10896. http://doi.org/10.1063/1.1739396.

Izvekov, S., & Voth, G. A. (2005). A multiscale coarse-graining method for biomolecular systems. *Journal of Physical Chemistry B, 109*(7), 2469–2473. http://doi.org/10.1021/jp044629q.

Jahn, M., Buchner, J., Hugel, T., & Rief, M. (2016). Folding and assembly of the large molecular machine Hsp90 studied in single-molecule experiments. *Proceedings of the National Academy of Sciences of the United States of America, 113*(5), 1232–1237. http://doi.org/10.1073/pnas.1518827113.

Jones, E., Oliphant, T. E., & Peterson, P. (2001). *SciPy: Open source scientific tools for Python [Computer software]*. Retrieved from, https://www.scipy.org.

Kaila, V. R. I., Wikström, M., & Hummer, G. (2014). Electrostatics, hydration, and proton transfer dynamics in the membrane domain of respiratory complex I. *Proceedings of the*

National Academy of Sciences of the United States of America, 111(19), 6988–6993. http://doi.org/10.1073/pnas.1319156111.

Kondrashov, D. A., Van Wynsberghe, A. W., Bannen, R. M., Cui, Q., & Phillips, G. N. (2007). Protein structural variation in computational models and crystallographic data. *Structure (London, England: 1993), 15*(2), 169–177. http://doi.org/10.1016/j.str.2006.12.006.

Lee, J., Natarajan, M., Nashine, V. C., Socolich, M., Vo, T., Russ, W. P., et al. (2008). Surface sites for engineering allosteric control in proteins. *Science, 322*(5900), 438–442. http://doi.org/10.1126/science.1159052.

Li, M., Zhang, J. Z. H., & Xia, F. (2015). Heterogeneous elastic network model improves description of slow motions of proteins in solution. *Chemical Physics Letters, 618*, 102–107. http://doi.org/10.1016/j.cplett.2014.11.006.

Lin, Y., Beckham, G. T., Himmel, M. E., Crowley, M. F., & Chu, J.-W. (2013). Endoglucanase peripheral loops facilitate complexation of glucan chains on cellulose via adaptive coupling to the emergent substrate structures. *Journal of Physical Chemistry B, 117*(37), 10750–10758. http://doi.org/10.1021/jp405897q.

Lin, Y., Silvestre-Ryan, J., Himmel, M. E., Crowley, M. F., Beckham, G. T., & Chu, J.-W. (2011). Protein allostery at the solid-liquid interface: Endoglucanase attachment to cellulose affects glucan clenching in the binding cleft. *Journal of the American Chemical Society, 133*(41), 16617–16624. http://doi.org/10.1021/ja206692g.

López-Blanco, J. R., & Chacon, P. (2015). New generation of elastic network models. *Current Opinion in Structural Biology, 37*, 46–53. http://doi.org/10.1016/j.sbi.2015.11.013.

Lu, L., Dama, J. F., & Voth, G. A. (2013). Fitting coarse-grained distribution functions through an iterative force-matching method. *The Journal of Chemical Physics, 139*(12), 121906. http://doi.org/10.1063/1.4811667.

Lyman, E., Pfaendtner, J., & Voth, G. A. (2008). Systematic multiscale parameterization of heterogeneous elastic network models of proteins. *Biophysical Journal, 95*(9), 4183–4192. http://doi.org/10.1529/biophysj.108.139733.

McCann, J. J., Zheng, L., Chiantia, S., & Bowen, M. E. (2011). Domain orientation in the N-Terminal PDZ tandem from PSD-95 is maintained in the full-length protein. *Structure (London, England: 1993), 19*(6), 810–820. http://doi.org/10.1016/j.str.2011.02.017.

McKinney, W. (2010). Data structures for statistical computing in Python. In S. van der Walt & K. J. Millman (Eds.), *Presented at the Proceedings of the 9th Python in Science Conference* (pp. 51–56).

Michaud-Agrawal, N., Denning, E. J., Woolf, T. B., & Beckstein, O. (2011). MDAnalysis: A toolkit for the analysis of molecular dynamics simulations. *Journal of Computational Chemistry, 32*, 2319–2327. http://doi.org/10.1002/jcc.21787.

Mouillet-Richard, S., Ermonval, M., Chebassier, C., Laplanche, J., Lehmann, S., Launay, J., et al. (2000). Signal transduction through prion protein. *Science, 289*(5486), 1925–1928.

Pasternak, A., Ringe, D., & Hedstrom, L. (1999). Comparison of anionic and cationic trypsinogens: The anionic activation domain is more flexible in solution and differs in its mode of BPTI binding in the crystal structure. *Protein Science, 8*(1), 253–258. http://doi.org/10.1110/ps.8.1.253.

Phillips, J. C., Braun, R., Wang, W., Gumbart, J., Tajkhorshid, E., Villa, E., et al. (2005). Scalable molecular dynamics with NAMD. *Journal of Computational Chemistry, 26*(16), 1781–1802. http://doi.org/10.1002/jcc.20289.

Piazzesi, G., Reconditi, M., Linari, M., Lucii, L., Sun, Y. B., Narayanan, T., et al. (2002). Mechanism of force generation by myosin heads in skeletal muscle. *Nature, 415*(6872), 659–662. http://doi.org/10.1038/415659a.

Saunders, M. G., & Voth, G. A. (2012). Coarse-graining of multiprotein assemblies. *Current Opinion in Structural Biology, 22*(2), 144–150. http://doi.org/10.1016/j.sbi.2012.01.003.

Schnauß, J., Glaser, M., Schuldt, C., Golde, T., & Händler, T. (2015). Motor-free force generation in biological systems. *diffusion-fundamentalsorg*, *23*(5), 1–15.
Shi, Q., Liu, P., & Voth, G. A. (2008). Coarse-graining in interaction space: An analytical approximation for the effective short-ranged electrostatics. *Journal of Physical Chemistry B*, *112*(50), 16230–16237. http://doi.org/10.1021/jp807205q.
Silvestre-Ryan, J., Lin, Y., & Chu, J.-W. (2011). "Fluctuograms" reveal the intermittent intra-protein communication in subtilisin Carlsberg and correlate mechanical coupling with co-evolution. *PLoS Computational Biology*, 7(3), e1002023. http://doi.org/10.1371/journal.pcbi.1002023.
Simonson, T., Archontis, G., & Karplus, M. (2002). Free energy simulations come of age: Protein-ligand recognition. *Accounts of Chemical Research*, *35*(6), 430–437.
Stockmar, F., Kobitski, A. Y., & Nienhaus, G. U. (2016). Fast folding dynamics of an intermediate state in RNase H measured by single-molecule FRET. *Journal of Physical Chemistry B*, *120*(4), 641–649. http://doi.org/10.1021/acs.jpcb.5b09336.
Sugita, Y., & Okamoto, Y. (1999). Replica-exchange molecular dynamics method for protein folding. *Chemical Physics Letters*, *314*(1–2), 141–151.
Tirion, M. M. (1996). Large amplitude elastic motions in proteins from a single-parameter, atomic analysis. *Physical Review Letters*, 77(9), 1905.
van der Walt, S., Colbert, S. C., & Varoquaux, G. (2011). The NumPy array: A structure for efficient numerical computation. *Computing in Science & Engineering*, *13*(2), 22–30.
Xia, F., Tong, D., & Lu, L. (2013). Robust heterogeneous anisotropic elastic network model precisely reproduces the experimental B-factors of biomolecules. *Journal of Chemical Theory and Computation*, *9*(8), 3704–3714. http://doi.org/10.1021/ct4002575.
Yu, H., Gupta, A. N., Liu, X., Neupane, K., Brigley, A. M., Sosova, I., et al. (2012). Energy landscape analysis of native folding of the prion protein yields the diffusion constant, transition path time, and rates. *Proceedings of the National Academy of Sciences of the United States of America*, *109*(36), 14452–14457. http://doi.org/10.1073/pnas.1206190109.
Zhang, Z., & Voth, G. A. (2010). Coarse-grained representations of large biomolecular complexes from low-resolution structural data. *Journal of Chemical Theory and Computation*, *6*(9), 2990–3002. http://doi.org/10.1021/ct100374a.

CHAPTER FIFTEEN

Constructing Kinetic Network Models to Elucidate Mechanisms of Functional Conformational Changes of Enzymes and Their Recognition with Ligands

L. Zhang*, H. Jiang*, F.K. Sheong*, F. Pardo-Avila*[,1], P.P.-H. Cheung*, X. Huang*[,†,2]

*The Hong Kong University of Science and Technology, Kowloon, Hong Kong
[†]State Key Laboratory of Molecular Neuroscience, Center for System Biology and Human Health, School of Science and Institute for Advance Study, The Hong Kong University of Science and Technology, Kowloon, Hong Kong
[2]Corresponding author: e-mail address: xuhuihuang@ust.hk

Contents

[1] Current address: Department of Structural Biology, Stanford University School of Medicine, Stanford, CA 94305-5126, United States

Methods in Enzymology, Volume 578
ISSN 0076-6879
http://dx.doi.org/10.1016/bs.mie.2016.05.026

Abstract

Enzymes are biological macromolecules that catalyze complex reactions in life. In order to perform their functions effectively and efficiently, enzymes undergo conformational changes between different functional states. Therefore, elucidating the dynamics between these states is essential to understand the molecular mechanisms of enzymes. Although experimental methods such as X-ray crystallography and cryoelectron microscopy can produce high-resolution structures, the detailed conformational dynamics of many enzymes still remain obscure. While molecular dynamics (MD) simulations are able to complement the experiments by providing structure-based dynamics at atomic resolution, it is usually difficult for them to reach the biologically relevant timescales (hundreds of microseconds or longer). Kinetic network models (KNMs), in particular Markov state models (MSMs), hold great promise to overcome this challenge because they can bridge the timescale gap between MD simulations and experimental observations. In this chapter, we review the procedure of constructing KNMs to elucidate the molecular mechanisms of enzymes. First, we will give a general introduction of MSMs, including the methods to construct and validate MSMs. Second, we will present the applications of KNMs to study two important enzymes: the human Argonaute protein and the RNA polymerase II. We conclude by discussing the future perspectives regarding the potential of KNMs to investigate the dynamics of enzymes' functional conformational changes.

1. INTRODUCTION

Enzymes are biological catalysts that accelerate chemical reactions within cells (Benkovic & Hammes-Schiffer, 2003). Enzymes often cycle through several metastable conformational states when they perform their catalytic activities under physiological conditions (Hammes, Benkovic, & Hammes-Schiffer, 2011). To understand how enzymes perform their functions, great efforts have been made to investigate these metastable states. In the past decades, high-resolution structures provided by X-ray crystallography (Kendrew et al., 1958) and cryoelectron microscopy (Bock et al., 2013; Cheng, 2015; Doerr, 2015; Frank et al., 1995; Li et al., 2015) have improved our understanding of enzyme functions. However, these experimental techniques can only provide static snapshots, and the information on real-time conformational change is missing. To monitor the dynamics of enzymes, alternative spectroscopic experimental methods have been developed, including temperature jump relaxation spectroscopy (Callender & Dyer, 2002), single-molecule fluorescence spectroscopy (Ha et al., 1999; Xia, Li, & Fang, 2013), and nuclear magnetic resonance spectroscopy (including paramagnetic relaxation enhancement (Clore, 2008; Clore, Tang, &

Iwahara, 2007) and residual dipolar coupling (Camilloni & Vendruscolo, 2015; Lipsitz & Tjandra, 2004)). Despite advances in recent experimental methods, it is still difficult to capture the details of the conformational changes of enzymes at the atomic level.

Molecular dynamics (MD) simulation is a computational technique used to simulate the physical movement of atoms and molecules by numerically solving the Newton's equations of motion (Karplus & McCammon, 2002; Levitt, 1983). MD simulation is capable of providing the structures and dynamics of enzymes with higher spatial and temporal resolution than most of the experimental methods. In MD simulations, the interactions between particles are determined by the potential energy with parameters defined in molecular mechanics force fields. Over the past decades, major effort has been spent on developing and improving the accuracy of potential functions in the force fields in order to better describe the dynamics of biological molecules (Brooks et al., 1983, 2009; Cornell et al., 1995; Jorgensen & Tiradorives, 1988; Lindorff-Larsen et al., 2010; van Gunsteren & Berendsen, 1987).

Although MD simulation has shown its great capacity to elucidate the atomic details of structure-based functions of enzymes, it remains a challenge to directly apply it to investigate the essential conformational changes of biological macromolecules, which usually occur at the timescale of hundreds of microseconds or even longer. Therefore, there exists a timescale gap between conventional MD simulations and experimental observations (Fig. 1).

In order to overcome this gap, kinetic network model (KNM), such as Markov state model (MSM), is one popular method to model long-timescale dynamics of biomolecules. Recently, this method has been successfully applied to study various biological processes, for example, investigation of conformational changes of proteins (Da et al., 2015; Da, Pardo-Avila,

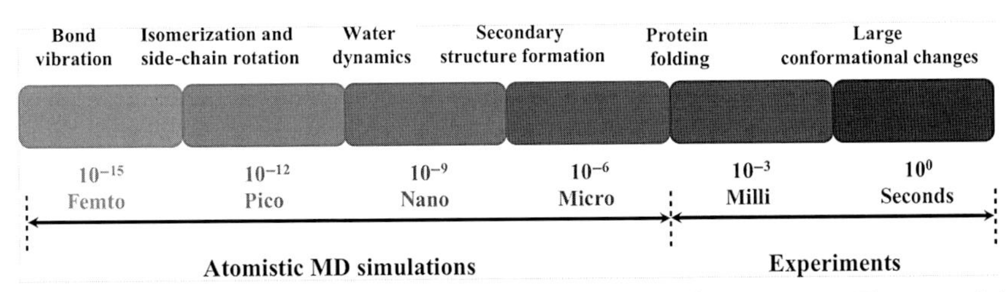

Fig. 1 Timescale gap between atomistic MD simulations and experiments. The essential conformational changes of protein observable in experiments happen at the timescale of hundreds of microseconds or even longer. However, the timescale reachable by atomistic MD simulations is usually limited to microseconds.

Wang, & Huang, 2013; Da, Wang, & Huang, 2012; Malmstrom, Kornev, Taylor, & Amaro, 2015; Silva et al., 2014), discovery of the cryptic allosteric sites for a folded protein (Bowman & Geissler, 2012), identification of the key structural intermediates for kinase activation (Shukla, Meng, Roux, & Pande, 2014), illumination of protein–ligand binding mechanism (Gu, Silva, Meng, Yue, & Huang, 2014; Jiang et al., 2015; Kohlhoff et al., 2014; Silva, Bowman, Sosa-Peinado, & Huang, 2011), elucidation of the protein-folding dynamics (Bowman, Voelz, & Pande, 2011; Ensign, Kasson, & Pande, 2007; Huang et al., 2010; Noe, Schutte, Vanden-Eijnden, Reich, & Weikl, 2009; Qiao, Bowman, & Huang, 2013; Voelz, Bowman, Beauchamp, & Pande, 2010), and modeling the time-resolved spectroscopy of protein conformational changes (Zhuang, Cui, Silva, & Huang, 2011).

In this chapter, we provide a general review of how we have applied KNMs to characterize the molecular mechanisms underlying enzyme dynamics. We will first introduce the basic background of MSMs and illustrate how to construct and validate MSMs. Next, we will demonstrate how MSMs can extract long-timescale kinetic information from short MD simulations using our previous works on two enzymes as case studies: the microRNA (miRNA) loading mechanism of human Argonaute protein (Ago) (Jiang et al., 2015) and the translocation dynamics of RNA Polymerase II (Pol II) (Silva et al., 2014). In the third case study, we adopted KNMs to combine the individual steps of the nucleotide addition cycle (NAC) in order to describe the overall dynamics of transcription elongation for Pol II (Yu, Da, & Huang, 2015). Finally, we discuss the future perspectives on applying KNMs to study the functional conformational changes of enzymes.

2. METHODOLOGY TO ELUCIDATE LONG-TIMESCALE DYNAMICS OF ENZYMES

We adopt the following workflow to investigate the dynamics of an enzyme (Fig. 2). The initial structural models for different functional states are built based on crystal structures and molecular modeling. Next, because of the timescale gap as depicted in Fig. 1, the direct transition between the initial and final states is difficult to observe in conventional MD simulations. To circumvent this, preliminary pathways between states are generated by applying external energies/forces to the system. Afterward, representative conformations along the pathways are selected to seed unbiased MD simulations. Subsequently, conformations from MD simulations are clustered into a number of microstates. The steps of MD simulations and clustering

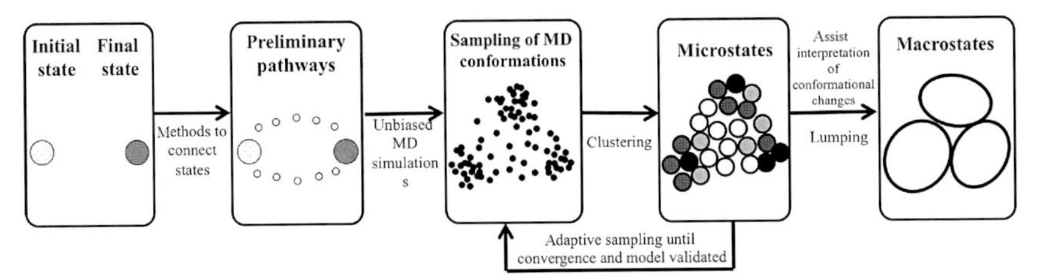

Fig. 2 Workflow to construct MSMs based on MD simulations. First, structural models for initial and final states are constructed based on experimental structures or structural modeling. Second, external forces/energies are applied to connect the initial and final states by generating preliminary pathways. Third, unbiased MD simulations are performed on each of the representative conformations selected from the preliminary pathways. Fourth, MD conformations are clustered into a number of microstates. This clustering step and the previous step of unbiased MD simulations are repeated until the conformational space is converged and the MSM is validated. The final step is to lump the microstates to a few macrostates (metastable states) to assist the interpretation of conformational changes.

are recursively performed until we can achieve a converged conformational space and a validated MSM. The final step is to kinetically lump the microstates into several metastable states to facilitate the interpretation of molecular mechanisms. In the case where only the initial state is available, instead of constructing preliminary pathways, we initiate a first round of MD simulations directly from this structure. Then, we perform adaptive sampling with subsequent rounds of MD simulations to fully sample the conformational space of the system and construct MSMs. An example of this case is illustrated in Section 3.1.

2.1 Generating Preliminary Pathways to Connect the Functional States and Selecting Representative Conformations to Seed MD Simulations

Establishment of the preliminary pathways connecting the initial and final states is an essential step to investigate conformational changes using MSMs. Several methods can serve for this purpose. For example, the modified Climber algorithm (Silva et al., 2014; Weiss & Levitt, 2009) adds a restraining energy to the system while keeping its high flexibility, which enables the system to move around high-energy barriers. The applied external energy is linearly dependent on the distance deviation of the current conformation from the target configuration. This method has been successfully used to study the dynamic translocation of Pol II during transcription elongation (Silva et al., 2014; see Section 3.2 for details). Steered MD simulation (Isralewitz, Gao, & Schulten, 2001) is useful to generate initial pathways for investigating small molecule/ion diffusion. In practice, an external force is applied to pull the small molecule/ion toward different directions to consider all possible diffusion pathways. This method has been adopted to study the pyrophosphate ion (PPi) release in RNA polymerases (Da et al., 2015, 2013, 2012). The program Caver (Chovancova et al., 2012) can also be used to study diffusion processes of molecule/ion, as it can detect the possible pathways that the small molecule/ion (simulated as a sphere) can navigate without any steric clash. This method has been used to discover the nucleoside triphosphate (NTP) entry channels for Pol II (Zhang, Silva, Pardo-Avila, Wang, & Huang, 2015). The spheres constituting the channels have been used as reference to place NTP molecules along the pathways and determine the feasibility of the channels as NTP entry pathways. As external energies/forces are introduced to drive the system to move from the initial to the final state in most of these methods, the produced pathways may not faithfully represent the minimum free energy pathway connecting the initial and final state. Nevertheless, they may serve as a set of preliminary pathways for further sampling using unbiased MD simulations. There also exist more sophisticated algorithms to obtain most probable pathways without introducing any bias. For example, the string method (Maragliano, Fischer, Vanden-Eijnden, & Ciccotti, 2006; Pan, Sezer, & Roux, 2008) can identify the most probable transition pathways for complex systems by refining the paths in a multidimensional space. The pathways are described by a set of collective variables and are being updated according to their average dynamic drift. Recently, the string method has been successfully employed to identify the activation pathway of the Src kinase (Shukla et al., 2014).

After obtaining the conformations along the preliminary pathways, it is necessary to select the representative conformations to seed conventional MD simulations. A common approach for selecting the conformations is to use geometric clustering algorithms such as k-centers (Hochbaum & Shmoys, 1985; Zhao, Sheong, Sun, Sander, & Huang, 2013) and k-medoids (Kaufman & Rousseeuw, 1987) algorithms to group all the conformations into a number of clusters. For each cluster, we randomly select a certain number of conformations as the representative conformations. This number may vary according to the complexity of the process.

2.2 Constructing MSMs: "Splitting-and-Lumping" Protocol

After several rounds of MD simulations to fully sample the conformational space, MSMs are constructed to predict the long-timescale transition between states (Fig. 2). The basic idea behind MSMs is to partition the

MD conformations into a number of states (process named as "clustering"), provided that the Markovian property of the states is ensured. Under this circumstance, the long-timescale dynamics can be predicted by propagating the transition probability matrix T (Eq. 1):

$$P(n\tau) = T(\tau)^n P(0) \tag{1}$$

where τ is the lag time, $P(0)$ is the initial populations of states, and $P(n\tau)$ is the state populations at time $n\tau$. The transition probability matrix $T(\tau)$ at lag time τ is computed by symmetrizing and column normalizing the transition count matrix, which is built by counting the number of transitions between each pair of states from the MD trajectories. In this subsection, we will illustrate the commonly used "splitting-and-lumping" protocol to construct MSMs based on MD simulations.

In order to combine the information from a large number of MD trajectories initiated from different regions of the conformational space, we need the "clustering" process to obtain statistical quantities describing the system (Bowman, Beauchamp, Boxer, & Pande, 2009). In practice, we can simply apply clustering algorithms such as k-centers (Hochbaum & Shmoys, 1985; Zhao et al., 2013) or k-medoids (Kaufman & Rousseeuw, 1987) to partition all the sampled conformations into microstates according to their geometric similarity (Fig. 2). The intuitive way to decompose MD conformations is to use root-mean-square deviation (of heavy atoms or Cα atoms) between pairs of conformations as the distance metric. However, this measurement might be overwhelmed by noises caused by the structural fluctuations during MD simulations. In this regard, dimension reduction techniques are helpful to identify the essential degrees of freedom. One can project MD conformations based on directions with maximum variance determined by principal component analysis (Amadei, Linssen, & Berendsen, 1993). Alternatively, new algorithms (Perez-Hernandez, Paul, Giorgino, De Fabritiis, & Noe, 2013; Schwantes & Pande, 2013) such as time-lagged independent component analysis have been developed to determine a set of key components that can best approximate the slowest dynamics of the system (Perez-Hernandez et al., 2013). With this algorithm, pairwise distances between conformations can be measured in the reduced dimensional space composed of these key components. An alternative method for clustering is to characterize the cluster centers by density peaks (Rodriguez & Laio, 2014). Density peaks are the regions that have higher density than the neighboring environment, and also far from other dense regions. This algorithm is capable to recognize the non-convex clusters, which cannot be identified by k-center algorithm.

However, it is computationally expensive, thus difficult to be applied to study macromolecules.

Conformations within each microstate should be geometrically similar, so that we can assume they share kinetic similarity. In this regard, we usually have a large number of "microstates" (Bowman, Huang, & Pande, 2009) to ensure that conformations within each microstate have only very short memories (Bowman, Beauchamp, et al., 2009). Microstate-MSM, which satisfies the Markovian property, is usually used to calculate the thermodynamic and kinetic quantities.

Although quantities calculated from the microstate-MSM can be directly compared with experimental observables (Bowman, Beauchamp, et al., 2009), a large number of microstates often hinder the physical interpretation of molecular mechanisms (Pande, Beauchamp, & Bowman, 2010). To facilitate the visualization and interpretation of the underlying molecular mechanism, microstates are kinetically lumped into a few macrostates based on the transition probability matrix built on the microstate level (Fig. 2). Several algorithms can accomplish this lumping task, including Perron-cluster cluster analysis (PCCA) and its improved version (PCCA+) (Dangkulwanich et al., 2014; Deuflhard, Huisinga, Fischer, & Schutte, 2000; Noe & Fischer, 2008), spectral clustering (Ng, Jordan, & Weiss, 2002; Sheong, Silva, Meng, Zhao, & Huang, 2015), Bayesian agglomerative clustering engine (BACE) (Bowman, 2012), the most probable paths algorithm (Jain & Stock, 2012), superlevel-set hierarchical clustering algorithm (Yao et al., 2009), etc. It is worth noting that the Hierarchical Nystrom Extension Graph (HNEG) method developed by us (Yao et al., 2013) can better handle the sampling noise that are prominent in enzymatic systems by lumping only the statistically significant basins and then incorporating the poorly sampled regions back to the basins. Benchmarking results based on a Bayesian model comparison (Bowman, Meng, & Huang, 2013) indicate that HNEG, together with BACE, routinely performs better than other alternative methods.

In addition to the aforementioned discrete-state models, the variational approach has been developed to obtain the key ingredients of the slow kinetics (Nuske, Keller, Perez-Hernandez, Mey, & Noe, 2014).

2.3 Alternative Method to Construct MSMs: Automatic State Partitioning for Multibody Systems

While the "splitting-and-lumping" protocol works well for many protein systems, it might not be adequate in effectively decomposing the conformational states when dealing with cases such as ligand binding to enzymes, which involve multibody or multitimescale problems (Fig. 3A). The enzyme–ligand interaction affects the motions of both the enzyme and the ligand. When the ligand is far away from the enzyme, the ligand can diffuse relatively fast; when the ligand is close to the enzyme, the motion of the ligand is hindered (Silva et al., 2011). This will cause a dilemma as k-centers algorithm in the traditional protocol tends to generate states of similar sizes, in the sense that a large number of clusters may cause the diffusive region to be oversplit, resulting in a large number of states only containing a few conformations with poor statistical significance, whereas a small number of clusters may not be able to properly separate the bound conformations from the encounter complex in the dense region and the MSM will become invalid if conformations within a state have internal barriers (Fig. 3A).

To address this issue, we have proposed an adaptive algorithm, known as the automatic state partitioning for multibody (APM) systems (Sheong et al., 2015). The main idea behind the algorithm is to include an escape probability in clustering to describe the timescale of system dynamics, determined by the population of the system escaping from a specific state after a predefined lag time τ. If the state has a low escape probability, it is likely to be in a region of slow dynamics and further splitting is performed.

The protocol of APM can be summarized as:

- Recursively splitting the conformational space until all microstates have escape probability $> 1 - \frac{1}{e}$ at the lag time τ.
- Lumping the microstates into macrostates.
- Repeat the above procedures for a prespecified number of iterations, and the recursive splitting procedure is performed for each macrostate independently.

As illustrated in Fig. 3B, this protocol will result in clusters of varying sizes according to the local kinetics. The outer diffusive region has much larger states than the central region, ensuring the statistical significance of the diffusive region and the high resolution in the bound region at the same time. It should also be noted that, even though this algorithm is originally designed for multibody systems, it is also helpful in investigating single-body system with multitimescale dynamics, such as protein folding, where the dynamics of the unfolded conformations is much faster than that of the folded conformations. This kind of example is illustrated in Section 3.1, where APM is used to elucidate the dynamics of human Ago.

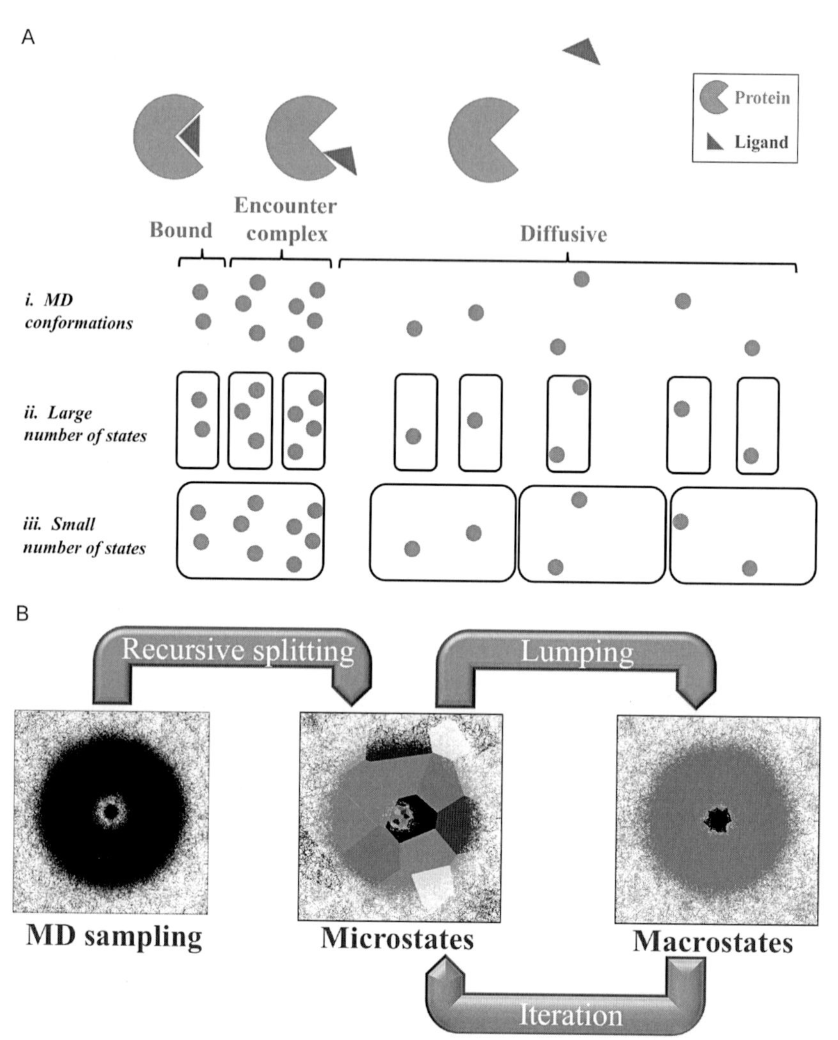

Fig. 3 Illustration of the automatic state partitioning for multibody (APM) algorithm to address the limitation of k-centers algorithm in two-body system. (A) Limitation of k-center algorithm used to investigate the protein–ligand binding mechanisms. There are three types of interaction between protein and ligand: bounded form when the ligand is correctly oriented with the protein binding site, encounter complex form when the ligand is not correctly oriented with the protein binding site, and diffusive form when the ligand is far from the protein. In the bounded and encounter complex forms, the dynamics of the protein and ligand is much slower than that in the diffusive form. This difference in dynamics will usually result in a large deviation in sampling densities as shown in (i) (each MD conformation is represented by a *blue sphere*). Under such condition, the k-centers algorithm is not appropriate: as shown in (ii), partitioning the conformational space with a large number of clusters using the k-centers algorithm will generate many states with poor statistics in the diffusive region; as shown in (iii),

2.4 Validation of MSMs

In order to comply with the Markovian property of the model, one needs to choose a proper lag time τ. The ith implied timescale τ_i at lag time τ is defined as (Da, Sheong, Silva, & Huang, 2014; Swope, Pitera, & Suits, 2004):

$$\tau_i(\tau) = -\frac{\tau}{\ln \lambda_{i+1}} \quad (2)$$

where λ_{i+1} is the $(i+1)$th eigenvalue (sorted from largest to smallest) of the transition probability matrix. Because the largest eigenvalue of transition probability matrix is always 1, the implied timescale corresponding to λ_1 is undefined. The model is Markovian if the implied timescale of the slowest modes becomes constant.

To further validate MSMs, one can apply the Chapman–Kolmogorov test and check whether the propagation of states' population counted from the MD trajectory matches with that calculated from the MSM (Chodera, Singhal, Pande, Dill, & Swope, 2007; Prinz et al., 2011). The statistical error of the MD trajectories can be estimated by simply applying bootstrapping algorithm to the original MD dataset or by applying a Bayesian approach (Noe, 2008). The model is valid if the model prediction lies within the error bars of the population counted from bootstrapped MD simulations. Residence probability test is an alternative method to validate MSMs. It is a simpler derivative from the Chapman–Kolmogorov test and examines the probability of the system to stay in a certain state based on the diagonal components of the transition matrix (Noe et al., 2009; Prinz et al., 2011).

2.5 Quantities Computed from Validated MSMs

Useful thermodynamic and kinetic quantities can be calculated from validated MSMs. We can determine the equilibrium population of states based

---

low clustering resolution due to the small number of states will lead to the mixing of the bounded and the encounter complex forms. (B) Illustration of APM algorithm. We recursively split the MD conformations into microstates such that the escape probability is $> 1 - \frac{1}{e}$. This state decomposition method will generate microstates with different sizes. Microstates are then lumped into a few macrostates, and the recursive splitting is repeated for a predetermined number of times for each macrostate. *Panel B adapted with permission from Sheong, F. K., Silva, D.-A., Meng, L., Zhao, Y., & Huang, X. (2015). Automatic state partitioning for multibody systems (APM): An efficient algorithm for constructing Markov state models to elucidate conformational dynamics of multibody systems.* Journal of Chemical Theory and Computation, 11*(1), 17–27. doi:10.1021/Ct5007168. Copyright (2015) American Chemical Society.* (See the color plate.)

on the first eigenvector of the transition probability matrix (Noe & Fischer, 2008). Besides, we can estimate the transition timescales between states based on the Markovian property of the model. Mean-first-passage-time (MFPT) is one of the most commonly used measurements to describe interstate transitions. Furthermore, we can identify the major pathways connecting the initial and final states with the help of transition path theory (E & Vanden-Eijnden, 2010). Additionally, we can obtain the long-time behavior of the system by synthesizing trajectories much longer than the original MD trajectory via Markov-chain-Monte-Carlo (Silva et al., 2014). Moreover, we can acquire ensemble averages of particular observables and make comparison with experimental data.

3. CASE STUDIES

3.1 The Mechanism of Molecular Recognition Between Human Argonaute-2 and miRNA

miRNAs are short noncoding RNAs that are critical for the regulation of gene expression and the defense against viruses (Bartel, 2004; Obbard, Gordon, Buck, & Jiggins, 2009; Wilson & Doudna, 2013). Since they are estimated to regulate over 50% of human genes (Bartel, 2009; Schirle, Sheu-Gruttadauria, & MacRae, 2014), miRNAs play essential roles in both normal physiological functions (Chivukula et al., 2014; Wang, Heegaard, & Orum, 2012) and progression of disease such as cancer (Jansson & Lund, 2012). In order to achieve its function, a miRNA has to be loaded into an Ago, forming the RNA-induced silencing complex that recognizes and inhibits the target messenger RNA (mRNA) with high sequence specificity (Joshua-Tor & Hannon, 2011; Kawamata & Tomari, 2010; Meister, 2013; Sasaki & Tomari, 2012). Among all the four human Agos, hAgo2 is the most extensively studied because of its RNaseH-like enzymatic function and the selective binding of miRNAs (Deerberg, Willkomm, & Restle, 2013; Elkayam et al., 2012; Frank, Sonenberg, & Nagar, 2010; MacRae, 2012; Schirle et al., 2014). Therefore, elucidating the mechanism of the molecular recognition between hAgo2 and miRNA is of great importance for both the in-depth understanding of miRNA functions and the development of new therapeutics for human diseases.

Despite the intensive experimental studies, the detailed dynamics of hAgo2–miRNA recognition still remain elusive. By combining MD simulations with MSMs and protein–RNA docking, we identified an open metastable state of apo hAgo2 that can accommodate miRNA. Furthermore, we revealed that miRNA loading into hAgo2 adopts a two-step mechanism: selective binding followed by structural rearrangement (Jiang et al., 2015).

We followed the workflow as illustrated in Section 2 to study the conformational space of apo hAgo2. The structural model was built based on the crystal structure (PDB ID: 4F3T) (Elkayam et al., 2012). Initiated from this conformation, we performed three rounds of MD simulations (~8 μs in total) and obtained over 394,000 conformations. We then used the APM algorithm as mentioned in Section 2.3 (Sheong et al., 2015) to construct MSMs based on the MD conformations. The flexible protein moieties (PIWI loops, L1L2 linking regions, and the PAZ domain) were used to define the metric for the geometric clustering. We obtained a 480-microstate-MSM with 20 ns as the lag time. The microstates were subsequently lumped into seven macrostates to facilitate the visualization.

The seven macrostates can be classified into three groups: an open state (19.0%), a partially open state (55.8%), and several closed states (Fig. 4A). Conformations in the open state of hAgo2 have largely exposed miRNA binding grooves, as indicated by their large center-of-mass (c.o.m.) distances between PIWI loops and the PAZ domain (Fig. 4A). The open state is further distinguished from the partially open state by the major PIWI loop angle, a variable that describes the conformation of this loop (Fig. 4B). It is also worth noting that despite the significant structural differences, the MFPTs derived from the validated 480-microstate-MSM suggest that transitions between open, partially open, and closed states take only tens of microseconds. Such fast dynamics ensures that the open conformations can be easily accessed by the diffusing miRNA during hAgo2–miRNA recognition.

Assessment of the largely open hAgo2 conformations using protein–RNA docking simulations suggests that only the open state can accommodate miRNA (de Vries et al., 2007; Dominguez, Boelens, & Bonvin, 2003). In Fig. 4B, the best-scoring pose from each docking simulation was plotted against the successful docking pose marked in red (a successful pose has both 5′ and 3′ miRNA termini accurately anchored in the binding groove of hAgo2). Conformations that can produce successful docking pose all belong to the open state, suggesting that only the open state is compatible for miRNA loading (Fig. 4B). Interestingly, short MD simulations started from the successfully docked structures display significant structural rearrangement that produces conformations with high structural similarity to the crystal structure (Fig. 4C).

Based on the above results, we proposed a two-step model of hAgo2–miRNA recognition: selective binding followed by structural rearrangement

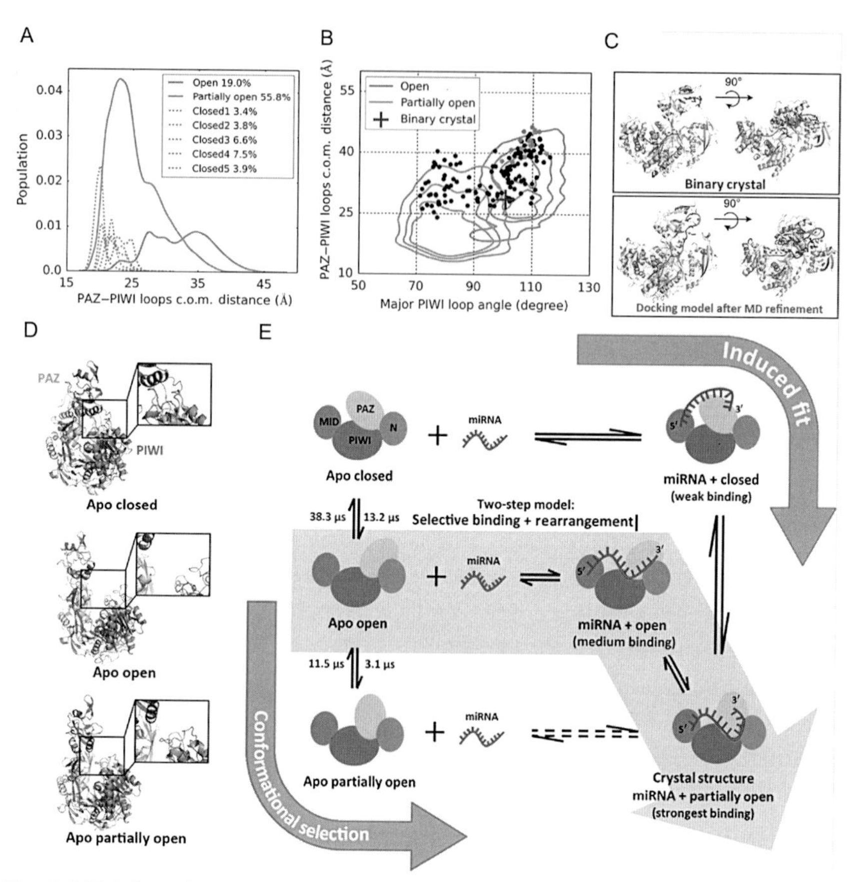

Fig. 4 MSMs based on MD simulations coupled with protein–RNA docking elucidate a two-step mechanism of hAgo2–miRNA recognition. (A) Distribution of distances between the center-of-mass (c.o.m.) of the PAZ domain and PIWI loops for each macrostate. The equilibrium population of each macrostate is presented. A large distance implies an open conformation and small distance implies a closed conformation. (B) Projections of hAgo2–miRNA successful (*red dots*) and unsuccessful (*black dots*) docking models built from the selected structures of open microstates. (C) Structural comparison between the binary crystal structure (*upper panel*) and a representative model after refinement by MD simulations (*lower panel*) with hAgo2 colored in *gray* and miRNA in *red*. (D) Representative structures of closed, open, and partially open states of apo hAgo2 from the MSM. Enlarged view of the interdomain region between PAZ (*pink*) and PIWI (*green*) of each structure is presented in the *inset panels*. (E) The proposed two-step model (highlighted by the *cyan arrow*) of miRNA loading into hAgo2: selective binding followed by structural rearrangement. The two-step model is compared with the induced fit mechanism (highlighted by the *upper right gray arrow*) and the conformational selection mechanism (highlighted by the *lower left gray arrow*). The average transition times between states are denoted next to the *arrows*. *Panel A adapted from fig. 1A of, panel B adapted from fig. 2 of, panel C adapted from fig. 3C*

(Fig. 4D and E). As suggested by our MSMs, apo hAgo2 undergoes fast transitions among open, partially open, and closed conformations (Fig. 4D and E). The selective binding of miRNA to open hAgo2 is a plausible first step, as the protein–RNA docking suggests that only the open conformations can accommodate miRNA. Upon the initial binding, the hAgo2–miRNA complex performs structural rearrangement to adopt the stable crystal conformation (Fig. 4E). Our approach of combining MSMs with MD simulations and large-scale molecular docking holds great potential to study the dynamics of protein–protein, protein–DNA/RNA, and protein–ligand recognitions.

3.2 Dynamics of Translocation for RNA Polymerase II

Pol II is the biological enzyme that catalyzes transcription in yeast (Cheung & Cramer, 2012; Kornberg, 2007). The information encoded in the DNA is read by Pol II to synthesize mRNA. During transcription elongation, Pol II undergoes the NAC to add one NTP to the nascent RNA strand (Brueckner, Ortiz, & Cramer, 2009; Kornberg, 2007) (Fig. 5A). In general, NAC involves conformational changes between six different functional states of Pol II. The cycle starts from the posttranslocation state (state I) (Bernecky, Herzog, Baumeister, Plitzko, & Cramer, 2016; Westover, Bushnell, & Kornberg, 2004b). NTP diffuses into the Pol II and binds to a preinsertion site (state II, "E-site"; Batada, Westover, Bushnell, Levitt, & Kornberg, 2004; Kettenberger, Armache, & Cramer, 2004; Landick, 2005; Westover, Bushnell, & Kornberg, 2004a). In the next step, an important protein motif named trigger loop (TL) changes its conformation to close up the active site and reorientate NTP (state III) (Wang, Bushnell, Westover, Kaplan, & Kornberg, 2006). Afterward, chemical catalysis occurs and NTP is incorporated into the RNA strand, followed by the release of the by-product PPi (state IV). The TL then changes its conformation from the closed to the open state, leaving Pol II in the pretranslocation state (state V) (Gnatt, Cramer, Fu, Bushnell, & Kornberg, 2001). To free the active site for the next round of NTP incorporation, nucleotides move forward by one register position (process named as "translocation") to return to

---

of, panel D adapted from fig. 1C of, and panel E adapted from fig. 4 of Jiang, H., Sheong, F. K., Zhu, L., Gao, X., Bernauer, J., & Huang, X. (2015). Markov state models reveal a two-step mechanism of miRNA loading into the human argonaute protein: Selective binding followed by structural re-arrangement. PLoS Computational Biology, 11*(7), e1004404. doi:10.1371/journal.pcbi.1004404.* (See the color plate.)

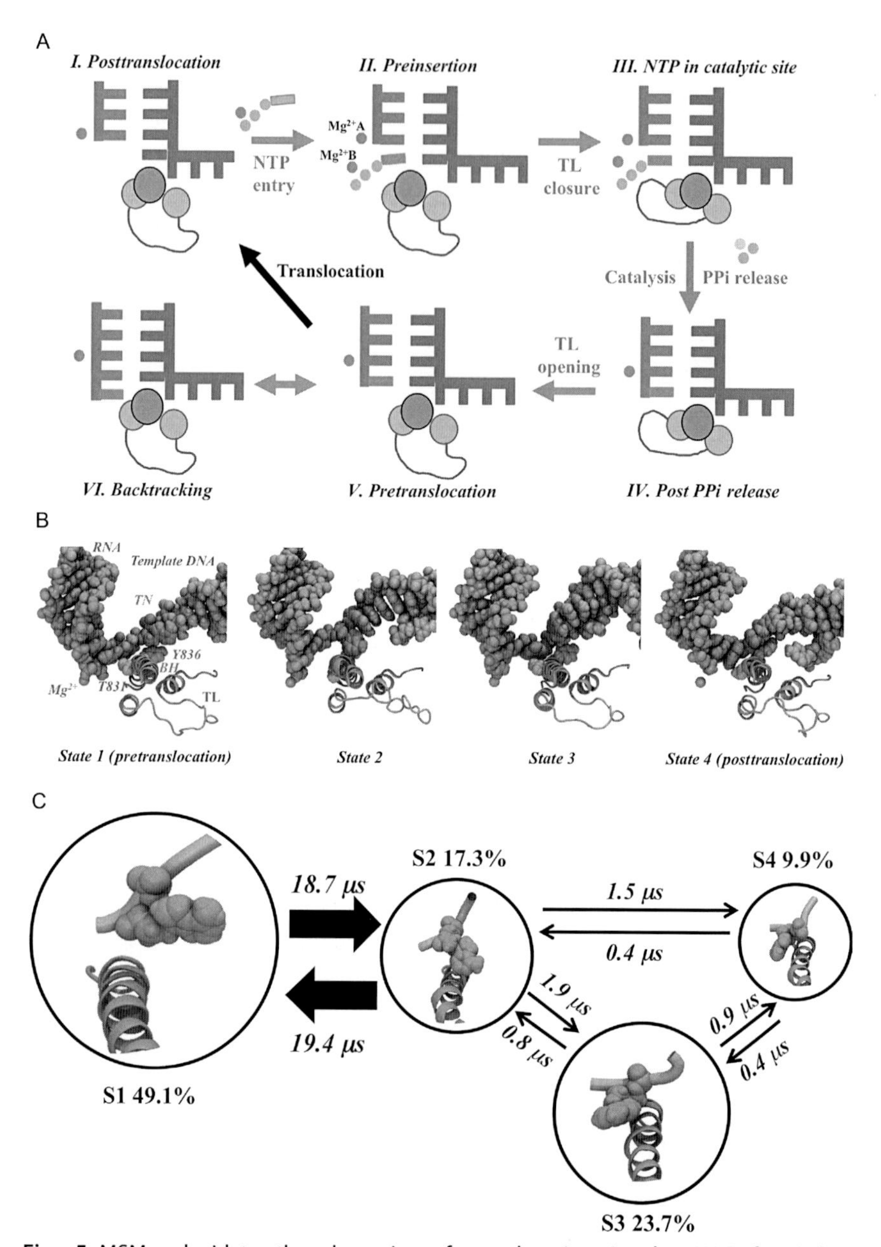

Fig. 5 MSMs elucidate the dynamics of translocation in the NAC for Pol II. (A) A schematic representation of the NAC with six states. The template DNA and nascent RNA strands are shown in *cyan* (*dark gray* in the print version) and *red* (*gray* in the print version), respectively. The NTP molecule is shown with the base and sugar ring in *yellow* (*gray* in the print version) *rectangle* and the triphosphate tail in *orange* (*light gray* in the print version) *spheres*. Two critical protein motifs BH and TL are shown

the posttranslocation state (state I) (Cheung & Cramer, 2012; Kornberg, 2007). To maintain transcription fidelity, Pol II has a backtracking mechanism when the mismatched NTP is incorporated. During backtracking, the nucleotides move backward (the opposite direction to the translocation) by leaving the mismatched terminal 3′-RNA exposed for cleavage (state VI) (Cheung & Cramer, 2011; Sydow et al., 2009; Wang et al., 2009). Then, Pol II goes back to the posttranslocation state (state I) after correcting the misincorporation.

Although X-ray crystallographic studies have provided the structural basis of elongation (Cheung & Cramer, 2011; Gnatt et al., 2001; Kettenberger et al., 2004; Sydow et al., 2009; Wang et al., 2009, 2006; Westover et al., 2004a, 2004b), the molecular details of the dynamic transitions between states are missing. Elucidating the functional conformational changes at atomic resolution can greatly facilitate our understanding of the underlying mechanism of Pol II transcription elongation. In one of our previous studies, we have applied MSMs to investigate the dynamics of translocation (between state V and state I) (Fig. 5A) (Silva et al., 2014). We identified four metastable states, including the two states similar to the crystal structures and another two intermediate states that were not found previously (Fig. 5B). Populations of the four states were reported and the rate-limiting step was determined by calculating the MFPTs between states (Fig. 5C). In general, our results support a Brownian-ratchet model, in which translocation is purely driven by the thermal energy of the system.

in *green* (*gray* in the print version) and *purple* (*light gray* in the print version), respectively. Mg ions are shown with *magenta* (*gray* in the print version) *spheres.* (B) Atomic representation of the four metastable states elucidated by the MSM. State 1 and state 4 correspond to pre- and posttranslocation states, respectively. States 2 and 3 are two intermediate states. The transition nucleotide is shown in *orange* (*gray* in the print version). Two BH residues are essential in translocation: Y836 (*yellow* (*gray* in the print version)) and T831 (*light gray*). The nontemplate DNA strand is shown in *green* (*light gray* in the print version). The color codes for other structural components are the same as in (A). (C) Populations and transition timescale for the four-state model. The position of the transition nucleotide is shown for each of the states, with the BH as a reference. The size of the bubble containing the conformation of each state is proportional to the population. The MFPT between states (S1–S4) is shown next to the *arrows.* The rate-limiting step is between S1 and S2, denoted by the *thickest arrows. Panel B adapted from fig. 1B–E of and in Panel C the figures for the conformations within the bubble were adapted from fig. 5 of Silva, D.-A., Weiss, D. R., Pardo-Avila, F., Da, L. T., Levitt, M., Wang, D., & Huang, X. (2014). Millisecond dynamics of RNA polymerase II translocation at atomic resolution.* Proceedings of the National Academy of Sciences of the United States of America, 111*(21), 7665–7670. doi:10.1073/Pnas.1315751111.*

The simulation system contains ~426,000 atoms. As translocation requires the movement of all nucleotides, it is impossible to observe translocation events directly from straightforward MD simulations of nanoseconds timescale. To address this problem, we followed the workflow as introduced in Section 2 (Fig. 2). In particular, structural models of Pol II in pre- and posttranslocation states were built based on available crystal structures (Cramer, Bushnell, & Kornberg, 2001; Gnatt et al., 2001; Wang et al., 2006; Westover et al., 2004b). Two independent pathways (one for forward translocation from pre- to posttranslocation and the other for the reverse process) were generated using modified Climber algorithm (Silva et al., 2014; Weiss & Levitt, 2009). Representative conformations from both pathways were selected for the unbiased MD simulations. The bias resulted from the external energy exerted into the preliminary pathways was removed after a few rounds of MD simulations (fig. S9 of Silva et al., 2014). Conformations from accumulated simulation time >1.5 μs were used to build MSMs. In particular, MD conformations were divided into 976 microstates and further lumped into 4 metastable states using PCCA algorithm with lag time of 5 ns.

MSM analysis elucidates four metastable states. Besides the pre- (S1) and posttranslocation (S4) states, two intermediates (S2 and S3) were identified (Fig. 5B). Consistent with the experimental observations (Abbondanzieri, Greenleaf, Shaevitz, Landick, & Block, 2005; Hein, Palangat, & Landick, 2011), Pol II can oscillate between pre- and posttranslocation states. Pretranslocation (S1) state is the most populated (49.1%). Posttranslocation state (S4, 9.9%) shares a wide energy basin with the two intermediates (17.3% and 23.7% for S2 and S3, respectively). MFPT calculations show that the transition between S1 and S2 is the rate-limiting step, and S2 is the essential joint state, meaning that all the major pathways connecting S1 and S4 must go through S2 (Fig. 5C).

Postanalysis based on the four-state model constructed from our workflow demonstrates an asynchronous translocation pattern of nucleotides: RNA:DNA hybrid does not translocate simultaneously with the transition nucleotide (TN) (fig. 2 of Silva et al., 2014). In particular, the RNA and DNA in the hybrid translocate simultaneously and complete their transition in one single step. That is, during the transition from pretranslocation (S1) to the intermediate state S2, the hybrid reaches its final position as in the final state (S4). On the contrary, the TN adopts three-step mechanism to accomplish the translocation. That is, from S1 to S2, TN moves from the

$i+2$ site to the halfway position above the bridge helix (BH) and stacks with the BH amino acid Y836; from S2 to S3, TN crosses the BH while still a few Angstroms away from the canonical $i+1$ site; the TN translocation is completed with TN relocating to the active state from S3 to S4. It is noted that the conformation of intermediate state S2 with TN stacking with Y836 is similar to the experimental structures captured by trapping the polymerase by either α-amanitin or DNA damage (Brueckner & Cramer, 2008; Walmacq et al., 2012, 2015; Wang, Zhu, Huang, & Lippard, 2010; Wang et al., 2015).

Structural analysis suggests that BH motion can promote translocation by interacting with the TN. During the MD simulations, the central region of the BH demonstrates higher flexibility compared to the two BH tips and the dynamics of the residues located in the BH center correlates with the movement of the nucleotides (fig. 3A–C of Silva et al., 2014). In particular, TN forms stacking interaction with BH Y836 in the intermediate state S2. The mutant MD simulations reveal that this stacking is key to facilitate translocation. On one hand, mutating Y836 to phenylalanine (Y836F) by maintaining the stacking shows similarity to the wild type; on the other hand, Y836V mutant by disrupting the stacking interaction can hinder translocation (fig. 4C of Silva et al., 2014). In addition, we found that the BH motion correlates with the dynamics of TL (fig. 3D of Silva et al., 2014), suggesting that TL motion may affect translocation by influencing the thermal fluctuation of the BH. This observation is also consistent with the previous experimental studies (Brueckner & Cramer, 2008; Kaplan, Jin, Zhang, & Belyanin, 2012; Wang et al., 2006; Weixlbaumer, Leon, Landick, & Darst, 2013).

In summary, our previous work provides a solid example to illustrate how to apply the workflow (Section 2 and Fig. 2) to elucidate the atomic details of conformational changes of enzymes. Following the same procedures, MSMs have also been used to study the PPi release in Pol II and other RNA polymerases (Da et al., 2015, 2013, 2012).

3.3 Dynamics of the Whole NAC for RNA Polymerase II

The workflow we have described so far focuses on elucidating the conformational changes between states one step at a time. However, in addition to investigating only single steps during the NAC of Pol II (Section 3.2), we have also combined all the individual steps and applied KNMs to interpret the dynamics of the whole NAC (Yu et al., 2015) (Fig. 6). By fitting to

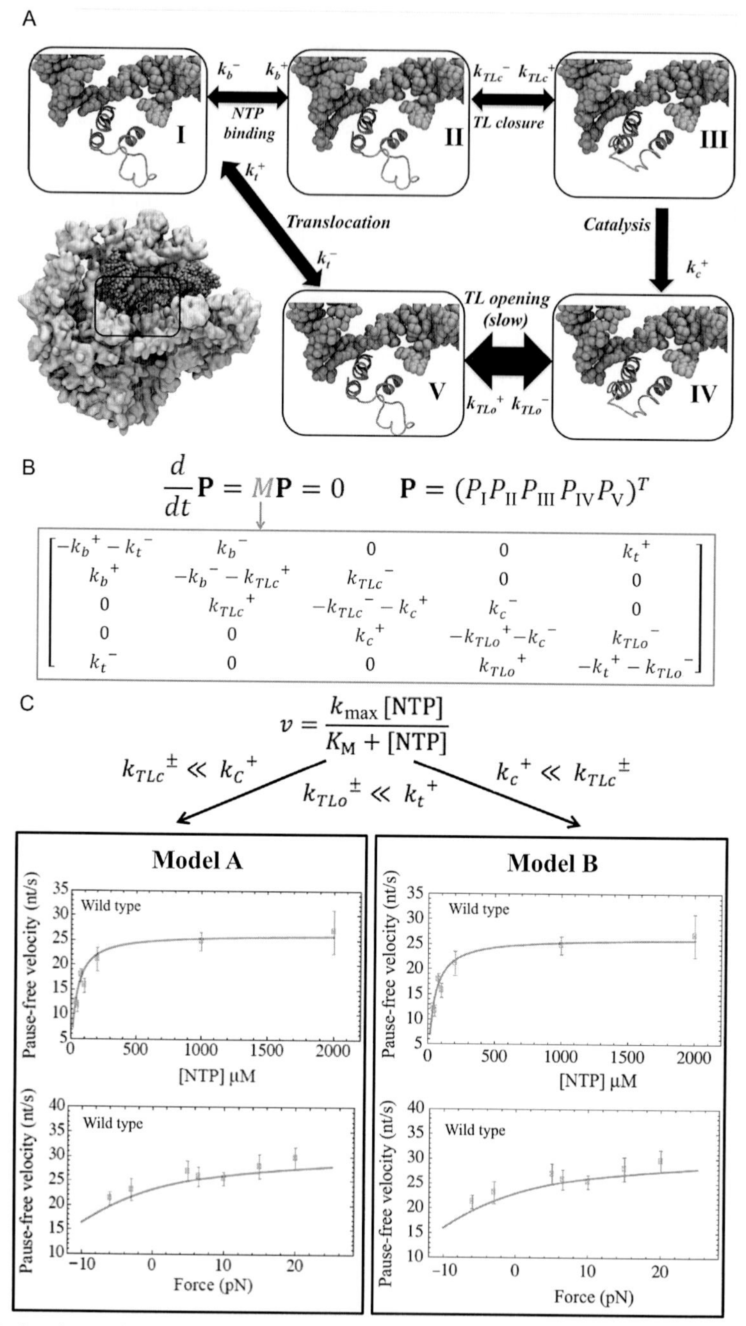

Fig. 6 See legend on opposite page.

single-molecule experimental data (Dangkulwanich et al., 2014), we found that the one slow step corresponds to TL opening (between state IV and state V). The other slow step still remains to be determined. On one hand, it can be assigned to the TL closing upon NTP binding (model A); on the other hand, it can be interpreted as a precatalytic structural rearrangement after TL closing (model B).

In particular, we included five states in the KNM and defined the transition rates between states with the catalysis step assumed to be irreversible (Fig. 6A). Following the concept of net rate constant (Cleland, 1975), the time Pol II takes to complete one NAC can be derived by solving the master equation (Dangkulwanich et al., 2014) and expressed by using the predefined transition rates (Fig. 6B). Both model A and model B were built with the assumption that translocation is fast, while the TL opening (preceding translocation) is much slower ($k_{TLo}^{\pm} \ll k_t^{+}$) (Fig. 6C). In model A, we made another assumption that TL closing (and its reversal) after NTP binding happens much slower than the catalytic transition ($k_{TLc}^{\pm} \ll k_C^{+}$). Under this condition, the elongation rate fits well the experimental data, suggesting a high NTP affinity or a low NTP dissociation constant. Contrary to model A, the catalysis step is assumed to be the rate-limiting step ($k_c^{+} \ll k_{TLc}^{\pm}$) in model B. Fitting to the experimental data based on this assumption suggests a relatively low dissociation constant of the NTP. Comparison between model A and model B indicates that accurate measurements of the NTP binding affinity or intrinsic NTP dissociation constant can help to resolve a more accurate model for transcription elongation.

Fig. 6 KNMs elucidate slow steps during the transcription elongation for Pol II. (A) The five-state KNM to investigate the dynamics of the whole NAC. The *left bottom corner* shows a structural model of the Pol II with the active site circled. The atomic representation of the five states in the active site is similar to those used in Fig. 5B. The transition rates between states are denoted next to the *arrows*. One slow step corresponds to the TL opening (between state IV and V). (B) The master equation to study the whole NAC using the five-state model, with the rate matrix shown inside the *red* (*light gray* in the print version) *frame*. The rate matrix is expressed with the rate parameters defined in (A). (C) The assumption used in deriving the elongation rate. Both model A and model B show consistency with the single-molecule experimental data. The *blue* (*light gray* in the print version) data points are from single-molecule experiments. The *red* (*gray* in the print version) *curves* are obtained by the KNM fitting to the experimental data. *Panel C adapted from figs. S2 and 3 of reference Yu, J., Da, L. T., & Huang, X. (2015). Constructing kinetic models to elucidate structural dynamics of a complete RNA polymerase II elongation cycle.* Physical Biology, 12*(1). doi:10.1088/1478-3975/12/1/016004.*

This work not only provides insights into the dynamics of the whole NAC but also serves as a critical link between the individual steps of the NAC and the overall transcription elongation, of which the quantities (such as kinetic rates) can be measured by experiments. This work also opens up the opportunity to investigate the effect of the local structural and dynamical perturbations (such as DNA damages and epigenetics modifications) on the overall elongation dynamics. Furthermore, introduction of the sequence dependency into the KNM can help to pinpoint the error-prone sequence motifs of the transcriptome (Gout, Thomas, Smith, Okamoto, & Lynch, 2013; Imashimizu, Oshima, Lubkowska, & Kashlev, 2013).

4. REMARKS AND FUTURE PERSPECTIVES

In this chapter, we have reviewed the fundamental concept of MSMs and the methodology to construct and validate MSMs. The most significant advantage of MSMs is to predict long-timescale dynamics of biological macromolecules (including enzymes) based on a large number of short MD simulations. We used three case studies to illustrate the application of KNMs to study the dynamics of two enzymes: (i) investigation of the recognition mechanism between hAgo2 and miRNA; (ii) elucidation of the translocation dynamics for Pol II; (iii) construction of a KNM by combining the individual steps in the NAC to study the overall dynamics of Pol II transcription elongation. These studies are only a few examples that demonstrate the power of KNMs, including MSMs, in solving the problems that were previously inaccessible. We anticipate that KNMs hold great potential to elucidate molecular mechanisms of conformational changes of more complex biological systems with the advent of computing power.

ACKNOWLEDGMENTS

This work was supported by the Hong Kong Research Grant Council (Grant numbers 16302214, 609813, HKUST C6009-15G, AoE/M-09/12, M-HKUST601/13, and T13-607/12R to X.H.) and the National Science Foundation of China (Grant number 21273188 to X.H.).

REFERENCES

Abbondanzieri, E. A., Greenleaf, W. J., Shaevitz, J. W., Landick, R., & Block, S. M. (2005). Direct observation of base-pair stepping by RNA polymerase. *Nature*, *438*(7067), 460–465. http://dx.doi.org/10.1038/nature04268.

Amadei, A., Linssen, A. B. M., & Berendsen, H. J. C. (1993). Essential dynamics of proteins. *Proteins*, *17*(4), 412–425. http://dx.doi.org/10.1002/Prot.340170408.

Bartel, D. P. (2004). MicroRNAs: Genomics, biogenesis, mechanism, and function. *Cell, 116*(2), 281–297. http://dx.doi.org/10.1016/S0092-8674(04)00045-5.

Bartel, D. P. (2009). MicroRNAs: Target recognition and regulatory functions. *Cell, 136*(2), 215–233. http://dx.doi.org/10.1016/j.cell.2009.01.002.

Batada, N. N., Westover, K. D., Bushnell, D. A., Levitt, M., & Kornberg, R. D. (2004). Diffusion of nucleoside triphosphates and role of the entry site to the RNA polymerase II active center. *Proceedings of the National Academy of Sciences of the United States of America, 101*(50), 17361–17364. http://dx.doi.org/10.1073/pnas.0408168101.

Benkovic, S. J., & Hammes-Schiffer, S. (2003). A perspective on enzyme catalysis. *Science, 301*(5637), 1196–1202. http://dx.doi.org/10.1126/Science.1085515.

Bernecky, C., Herzog, F., Baumeister, W., Plitzko, J. M., & Cramer, P. (2016). Structure of transcribing mammalian RNA polymerase II. *Nature, 529*, 551–554. http://dx.doi.org/10.1038/nature16482. Advance online publication.

Bock, L. V., Blau, C., Schroder, G. F., Davydov, I. I., Fischer, N., Stark, H., … Grubmuler, H. (2013). Energy barriers and driving forces in tRNA translocation through the ribosome. *Nature Structural & Molecular Biology, 20*(12), 1390–1396. http://dx.doi.org/10.1038/nsmb.2690.

Bowman, G. R. (2012). Improved coarse-graining of Markov state models via explicit consideration of statistical uncertainty. *Journal of Chemical Physics, 137*(13), 134111. http://dx.doi.org/10.1063/1.4755751.

Bowman, G. R., Beauchamp, K. A., Boxer, G., & Pande, V. S. (2009). Progress and challenges in the automated construction of Markov state models for full protein systems. *Journal of Chemical Physics, 131*(12), 124101. http://dx.doi.org/10.1063/1.3216567.

Bowman, G. R., & Geissler, P. L. (2012). Equilibrium fluctuations of a single folded protein reveal a multitude of potential cryptic allosteric sites. *Proceedings of the National Academy of Sciences of the United States of America, 109*(29), 11681–11686. http://dx.doi.org/10.1073/pnas.1209309109.

Bowman, G. R., Huang, X., & Pande, V. S. (2009). Using generalized ensemble simulations and Markov state models to identify conformational states. *Methods, 49*(2), 197–201. http://dx.doi.org/10.1016/J.Ymeth.2009.04.013.

Bowman, G. R., Meng, L., & Huang, X. (2013). Quantitative comparison of alternative methods for coarse-graining biological networks. *Journal of Chemical Physics, 139*(12), 121905. http://dx.doi.org/10.1063/1.4812768.

Bowman, G. R., Voelz, V. A., & Pande, V. S. (2011). Atomistic folding simulations of the five-helix bundle protein lambda(6-85). *Journal of the American Chemical Society, 133*(4), 664–667. http://dx.doi.org/10.1021/ja106936n.

Brooks, B. R., Brooks, C. L., Mackerell, A. D., Nilsson, L., Petrella, R. J., Roux, B., … Karplus, M. (2009). CHARMM: The biomolecular simulation program. *Journal of Computational Chemistry, 30*(10), 1545–1614. http://dx.doi.org/10.1002/jcc.21287.

Brooks, B. R., Bruccoleri, R. E., Olafson, B. D., States, D. J., Swaminathan, S., & Karplus, M. (1983). CHARMM: A program for macromolecular energy, minimization, and dynamics calculations. *Journal of Computational Chemistry, 4*(2), 187–217. http://dx.doi.org/10.1002/Jcc.540040211.

Brueckner, F., & Cramer, P. (2008). Structural basis of transcription inhibition by alpha-amanitin and implications for RNA polymerase II translocation. *Nature Structural & Molecular Biology, 15*(8), 811–818. http://dx.doi.org/10.1038/Nsmb.1458.

Brueckner, F., Ortiz, J., & Cramer, P. (2009). A movie of the RNA polymerase nucleotide addition cycle. *Current Opinion in Structural Biology, 19*(3), 294–299. http://dx.doi.org/10.1016/J.Sbi.2009.04.005.

Callender, R., & Dyer, R. B. (2002). Probing protein dynamics using temperature jump relaxation spectroscopy. *Current Opinion in Structural Biology, 12*(5), 628–633. http://dx.doi.org/10.1016/S0959-440x(02)00370-6.

Camilloni, C., & Vendruscolo, M. (2015). A tensor-free method for the structural and dynamical refinement of proteins using residual dipolar couplings. *Journal of Physical Chemistry B, 119*(3), 653–661. http://dx.doi.org/10.1021/jp5021824.

Cheng, Y. F. (2015). Single-particle cryo-EM at crystallographic resolution. *Cell, 161*(3), 450–457. http://dx.doi.org/10.1016/j.cell.2015.03.049.

Cheung, A. C. M., & Cramer, P. (2011). Structural basis of RNA polymerase II backtracking, arrest and reactivation. *Nature, 471*(7337), 249–253. http://dx.doi.org/10.1038/nature09785.

Cheung, A. C. M., & Cramer, P. (2012). A movie of RNA polymerase II transcription. *Cell, 149*(7), 1431–1437. http://dx.doi.org/10.1016/J.Cell.2012.06.006.

Chivukula, R. R., Shi, G. L., Acharya, A., Mills, E. W., Zeitels, L. R., Anandam, J. L., … Mendell, J. T. (2014). An essential mesenchymal function for miR-143/145 in intestinal epithelial regeneration. *Cell, 157*(5), 1104–1116. http://dx.doi.org/10.1016/j.cell.2014.03.055.

Chodera, J. D., Singhal, N., Pande, V. S., Dill, K. A., & Swope, W. C. (2007). Automatic discovery of metastable states for the construction of Markov models of macromolecular conformational dynamics. *Journal of Chemical Physics, 126*(15), 155101. http://dx.doi.org/10.1063/1.2714538.

Chovancova, E., Pavelka, A., Benes, P., Strnad, O., Brezovsky, J., Kozlikova, B., … Damborsky, J. (2012). CAVER 3.0: A tool for the analysis of transport pathways in dynamic protein structures. *PLoS Computational Biology, 8*(10), e1002708. http://dx.doi.org/10.1371/Journal.Pcbi.1002708.

Cleland, W. W. (1975). Partition analysis and concept of net rate constants as tools in enzyme-kinetics. *Biochemistry, 14*(14), 3220–3224. http://dx.doi.org/10.1021/Bi00685a029.

Clore, G. M. (2008). Visualizing lowly-populated regions of the free energy landscape of macromolecular complexes by paramagnetic relaxation enhancement. *Molecular Biosystems, 4*(11), 1058–1069. http://dx.doi.org/10.1039/b810232e.

Clore, G. M., Tang, C., & Iwahara, J. (2007). Elucidating transient macromolecular interactions using paramagnetic relaxation enhancement. *Current Opinion in Structural Biology, 17*(5), 603–616. http://dx.doi.org/10.1016/j.sbi.2007.08.013.

Cornell, W. D., Cieplak, P., Bayly, C. I., Gould, I. R., Merz, K. M., Ferguson, D. M., … Kollman, P. A. (1995). A second generation force field for the simulation of proteins, nucleic acids, and organic molecules (vol 117, pg 5179, 1995). *Journal of the American Chemical Society, 118*(9), 2309.

Cramer, P., Bushnell, D. A., & Kornberg, R. D. (2001). Structural basis of transcription: RNA polymerase II at 2.8 angstrom resolution. *Science (New York, N.Y.), 292*(5523), 1863–1876. http://dx.doi.org/10.1126/science.1059493.

Da, L. T., E, C. E., Duan, B., Zhang, C., Zhou, X., & Yu, J. (2015). A Jump-from-cavity pyrophosphate ion release assisted by a key lysine residue in T7 RNA polymerase transcription elongation. *PLoS Computational Biology, 11*(11), e1004624. http://dx.doi.org/10.1371/journal.pcbi.1004624.

Da, L. T., Pardo-Avila, F., Wang, D., & Huang, X. (2013). A two-state model for the dynamics of the pyrophosphate ion release in bacterial RNA polymerase. *PLoS Computational Biology, 9*(4), e1003020. http://dx.doi.org/10.1371/Journal.Pcbi.1003020.

Da, L. T., Sheong, F. K., Silva, D.-A., & Huang, X. (2014). Application of Markov state models to simulate long timescale dynamics of biological macromolecules. *Protein Conformational Dynamics, 805*, 29–66. http://dx.doi.org/10.1007/978-3-319-02970-2_2.

Da, L. T., Wang, D., & Huang, X. (2012). Dynamics of pyrophosphate ion release and its coupled trigger loop motion from closed to open state in RNA polymerase II. *Journal of the American Chemical Society, 134*(4), 2399–2406. http://dx.doi.org/10.1021/ja210656k.Dynamics.

Dangkulwanich, M., Ishibashi, T., Liu, S. X., Kireeva, M. L., Lubkowska, L., Kahlev, M., & Bustamante, C. J. (2014). Complete dissection of transcription elongation reveals slow translocation of RNA polymerase II in a linear ratchet mechanism. *Biophysical Journal, 106*(2), 485A–486A.

de Vries, S. J., van Dijk, A. D., Krzeminski, M., van Dijk, M., Thureau, A., Hsu, V., … Bonvin, A. M. (2007). HADDOCK versus HADDOCK: New features and performance of HADDOCK2.0 on the CAPRI targets. *Proteins: Structure, Function, and Bioinformatics, 69*(4), 726–733. http://dx.doi.org/10.1002/prot.21723.

Deerberg, A., Willkomm, S., & Restle, T. (2013). Minimal mechanistic model of siRNA-dependent target RNA slicing by recombinant human Argonaute 2 protein. *Proceedings of the National Academy of Sciences of the United States of America, 110*(44), 17850–17855. http://dx.doi.org/10.1073/pnas.1217838110.

Deuflhard, P., Huisinga, W., Fischer, A., & Schutte, C. (2000). Identification of almost invariant aggregates in reversible nearly uncoupled Markov chains. *Linear Algebra and Its Applications, 315*(1–3), 39–59. http://dx.doi.org/10.1016/S0024-3795(00)00095-1.

Doerr, A. (2015). Structural biology: Cryo-EM goes high-resolution. *Nature Methods, 12*(7), 598–599. http://dx.doi.org/10.1038/nmeth.3469.

Dominguez, C., Boelens, R., & Bonvin, A. M. (2003). HADDOCK: A protein-protein docking approach based on biochemical or biophysical information. *Journal of the American Chemical Society, 125*(7), 1731–1737. http://dx.doi.org/10.1021/ja026939x.

E, W., & Vanden-Eijnden, E. (2010). Transition-path theory and path-finding algorithms for the study of rare events. *Annual Review of Physical Chemistry, 61*, 391–420. http://dx.doi.org/10.1146/annurev.physchem.040808.090412.

Elkayam, E., Kuhn, C. D., Tocilj, A., Haase, A. D., Greene, E. M., Hannon, G. J., & Joshua-Tor, L. (2012). The structure of human argonaute-2 in complex with miR-20a. *Cell, 150*(1), 100–110. http://dx.doi.org/10.1016/j.cell.2012.05.017.

Ensign, D. L., Kasson, P. M., & Pande, V. S. (2007). Heterogeneity even at the speed limit of folding: Large-scale molecular dynamics study of a fast-folding variant of the villin headpiece. *Journal of Molecular Biology, 374*(3), 806–816. http://dx.doi.org/10.1016/j.jmb.2007.09.069.

Frank, F., Sonenberg, N., & Nagar, B. (2010). Structural basis for 5′-nucleotide base-specific recognition of guide RNA by human AGO2. *Nature, 465*(7299), 818–822. http://dx.doi.org/10.1038/nature09039.

Frank, J., Zhu, J., Penczek, P., Li, Y. H., Srivastava, S., Verschoor, A., … Agrawal, R. K. (1995). A model of protein-synthesis based on cryoelectron microscopy of the *E. coli* ribosome. *Nature, 376*(6539), 441–444. http://dx.doi.org/10.1038/376441a0.

Gnatt, A. L., Cramer, P., Fu, J. H., Bushnell, D. A., & Kornberg, R. D. (2001). Structural basis of transcription: An RNA polymerase II elongation complex at 3.3 angstrom resolution. *Science, 292*(5523), 1876–1882. http://dx.doi.org/10.1126/Science.1059495.

Gout, J. F., Thomas, W. K., Smith, Z., Okamoto, K., & Lynch, M. (2013). Large-scale detection of in vivo transcription errors. *Proceedings of the National Academy of Sciences of the United States of America, 110*(46), 18584–18589. http://dx.doi.org/10.1073/pnas.1309843110.

Gu, S., Silva, D.-A., Meng, L. M., Yue, A., & Huang, X. H. (2014). Quantitatively characterizing the ligand binding mechanisms of choline binding protein using Markov state model analysis. *PLoS Computational Biology, 10*(8), e1003767. http://dx.doi.org/10.1371/Journal.Pcbi.1003767.

Ha, T. J., Ting, A. Y., Liang, J., Caldwell, W. B., Deniz, A. A., Chemla, D. S., … Weiss, S. (1999). Single-molecule fluorescence spectroscopy of enzyme conformational dynamics and cleavage mechanism. *Proceedings of the National Academy of Sciences of the United States of America, 96*(3), 893–898. http://dx.doi.org/10.1073/Pnas.96.3.893.

Hammes, G. G., Benkovic, S. J., & Hammes-Schiffer, S. (2011). Flexibility, diversity, and cooperativity: Pillars of enzyme catalysis. *Biochemistry*, *50*(48), 10422–10430. http://dx.doi.org/10.1021/bi201486f.

Hein, P. P., Palangat, M., & Landick, R. (2011). RNA transcript 3′-proximal sequence affects translocation bias of RNA polymerase. *Biochemistry*, *50*(32), 7002–7014. http://dx.doi.org/10.1021/b1200437q.

Hochbaum, D. S., & Shmoys, D. B. (1985). A best possible heuristic for the k-center problem. *Mathematics of Operations Research*, *10*(2), 180–184. http://dx.doi.org/10.1287/Moor.10.2.180.

Huang, X., Yao, Y., Bowman, G. R., Sun, J., Guibas, L. J., Carlsson, G., & Pande, V. S. (2010). Constructing multi-resolution Markov state models (MSMs) to elucidate RNA hairpin folding mechanisms. *Pacific Symposium on Biocomputing*, *15*, 228–239.

Imashimizu, M., Oshima, T., Lubkowska, L., & Kashlev, M. (2013). Direct assessment of transcription fidelity by high-resolution RNA sequencing. *Nucleic Acids Research*, *41*(19), 9090–9104. http://dx.doi.org/10.1093/nar/gkt698.

Isralewitz, B., Gao, M., & Schulten, K. (2001). Steered molecular dynamics and mechanical functions of proteins. *Current Opinion in Structural Biology*, *11*(2), 224–230. http://dx.doi.org/10.1016/S0959-440x(00)00194-9.

Jain, A., & Stock, G. (2012). Identifying metastable states of folding proteins. *Journal of Chemical Theory and Computation*, *8*(10), 3810–3819. http://dx.doi.org/10.1021/ct300077q.

Jansson, M. D., & Lund, A. H. (2012). MicroRNA and cancer. *Molecular Oncology*, *6*(6), 590–610. http://dx.doi.org/10.1016/j.molonc.2012.09.006.

Jiang, H., Sheong, F. K., Zhu, L., Gao, X., Bernauer, J., & Huang, X. (2015). Markov state models reveal a two-step mechanism of miRNA loading into the human argonaute protein: Selective binding followed by structural re-arrangement. *PLoS Computational Biology*, *11*(7), e1004404. http://dx.doi.org/10.1371/journal.pcbi.1004404.

Jorgensen, W. L., & Tiradorives, J. (1988). The OPLS potential functions for proteins: Energy minimizations for crystals of cyclic-peptides and crambin. *Journal of the American Chemical Society*, *110*(6), 1657–1666.

Joshua-Tor, L., & Hannon, G. J. (2011). Ancestral roles of small RNAs: An ago-centric perspective. *Cold Spring Harbor Perspectives in Biology*, *3*(10), a003772. http://dx.doi.org/10.1101/cshperspect.a003772.

Kaplan, C. D., Jin, H., Zhang, I. L., & Belyanin, A. (2012). Dissection of Pol II trigger loop function and Pol II activity-dependent control of start site selection in vivo. *PLoS Genetics*, *8*(4), e1002627. http://dx.doi.org/10.1371/journal.pgen.1002627.

Karplus, M., & McCammon, J. A. (2002). Molecular dynamics simulations of biomolecules. *Nature Structural Biology*, *9*(9), 646–652. http://dx.doi.org/10.1038/Nsb0902-646.

Kaufman, L., & Rousseeuw, P. J. (1987). Clustering by means of medoids. In Y. Dodge (Ed.), *Statistical data analysis based on the L1–norm and related methods* (pp. 405–416). Amsterdam, the Netherlands: North-Holland.

Kawamata, T., & Tomari, Y. (2010). Making RISC. *Trends in Biochemical Sciences*, *35*(7), 368–376. http://dx.doi.org/10.1016/j.tibs.2010.03.009.

Kendrew, J. C., Bodo, G., Dintzis, H. M., Parrish, R. G., Wyckoff, H., & Phillips, D. C. (1958). A three-dimensional model of the myoglobin molecule obtained by X-ray analysis. *Nature*, *181*(4610), 662–666.

Kettenberger, H., Armache, K. J., & Cramer, P. (2004). Complete RNA polymerase II elongation complex structure and its interactions with NTP and TFIIS. *Molecular Cell*, *16*, 955–965.

Kohlhoff, K. J., Shukla, D., Lawrenz, M., Bowman, G. R., Konerding, D. E., Belov, D., … Pande, V. S. (2014). Cloud-based simulations on Google Exacycle reveal ligand modulation of GPCR activation pathways. *Nature Chemistry*, *6*(1), 15–21. http://dx.doi.org/10.1038/NCHEM.1821.

Kornberg, R. D. (2007). The molecular basis of eukaryotic transcription. *Proceedings of the National Academy of Sciences of the United States of America*, *104*(32), 12955–12961. http://dx.doi.org/10.1073/Pnas.0704138104.

Landick, R. (2005). NTP-entry routes in multi-subunit RNA polymerases. *Trends in Biochemical Sciences*, *30*(12), 651–654. http://dx.doi.org/10.1016/J.Tibs.2005.10.001.

Levitt, M. (1983). Protein folding by restrained energy minimization and molecular-dynamics. *Journal of Molecular Biology*, *170*(3), 723–764. http://dx.doi.org/10.1016/S0022-2836(83)80129-6.

Li, N. N., Zhai, Y. L., Zhang, Y. X., Li, W. Q., Yang, M. J., Lei, J. L., ... Gao, N. (2015). Structure of the eukaryotic MCM complex at 3.8 angstrom. *Nature*, *524*(7564), 186–191. http://dx.doi.org/10.1038/nature14685.

Lindorff-Larsen, K., Piana, S., Palmo, K., Maragakis, P., Klepeis, J. L., Dror, R. O., & Shaw, D. E. (2010). Improved side-chain torsion potentials for the Amber ff99SB protein force field. *Proteins*, *78*(8), 1950–1958. http://dx.doi.org/10.1002/Prot.22711.

Lipsitz, R. S., & Tjandra, N. (2004). Residual dipolar couplings in NMR structure analysis. *Annual Review of Biophysics and Biomolecular Structure*, *33*, 387–413. http://dx.doi.org/10.1146/Annurev.Biophys.33.110502.140306.

MacRae, I. (2012). Structural and mechanism of human argonaute-2. *Nucleic Acid Therapeutics*, *22*(6), A5.

Malmstrom, R. D., Kornev, A. P., Taylor, S. S., & Amaro, R. E. (2015). Allostery through the computational microscope: cAMP activation of a canonical signalling domain. *Nature Communications*. *6*. Article no. 7588. http://dx.doi.org/10.1038/Ncomms8588.

Maragliano, L., Fischer, A., Vanden-Eijnden, E., & Ciccotti, G. (2006). String method in collective variables: Minimum free energy paths and isocommittor surfaces. *Journal of Chemical Physics*, *125*(2), 024106. http://dx.doi.org/10.1063/1.2212942.

Meister, G. (2013). Argonaute proteins: Functional insights and emerging roles. *Nature Reviews Genetics*, *14*(7), 447–459. http://dx.doi.org/10.1038/nrg3462.

Ng, A. Y., Jordan, M. I., & Weiss, Y. (2002). On spectral clustering: Analysis and an algorithm. *Advances in Neural Information Processing Systems*, *14*, 849–856.

Noe, F. (2008). Probability distributions of molecular observables computed from Markov models. *Journal of Chemical Physics*, *128*(24), 244103. http://dx.doi.org/10.1063/1.2916718.

Noe, F., & Fischer, S. (2008). Transition networks for modeling the kinetics of conformational change in macromolecules. *Current Opinion in Structural Biology*, *18*(2), 154–162. http://dx.doi.org/10.1016/J.Sbi.2008.01.008.

Noe, F., Schutte, C., Vanden-Eijnden, E., Reich, L., & Weikl, T. R. (2009). Constructing the equilibrium ensemble of folding pathways from short off-equilibrium simulations. *Proceedings of the National Academy of Sciences of the United States of America*, *106*(45), 19011–19016. http://dx.doi.org/10.1073/Pnas.0905466106.

Nuske, F., Keller, B. G., Perez-Hernandez, G., Mey, A. S. J. S., & Noe, F. (2014). Variational approach to molecular kinetics. *Journal of Chemical Theory and Computation*, *10*(4), 1739–1752. http://dx.doi.org/10.1021/ct4009156.

Obbard, D. J., Gordon, K. H. J., Buck, A. H., & Jiggins, F. M. (2009). The evolution of RNAi as a defence against viruses and transposable elements. *Philosophical Transactions of the Royal Society of London. Series B, Biological Sciences*, *364*(1513), 99–115. http://dx.doi.org/10.1098/rstb.2008.0168.

Pan, A. C., Sezer, D., & Roux, B. (2008). Finding transition pathways using the string method with swarms of trajectories. *Journal of Physical Chemistry B*, *112*(11), 3432–3440. http://dx.doi.org/10.1021/jp0777059.

Pande, V. S., Beauchamp, K., & Bowman, G. R. (2010). Everything you wanted to know about Markov state models but were afraid to ask. *Methods*, *52*(1), 99–105. http://dx.doi.org/10.1016/j.ymeth.2010.06.002.

Perez-Hernandez, G., Paul, F., Giorgino, T., De Fabritiis, G., & Noe, F. (2013). Identification of slow molecular order parameters for Markov model construction. *Journal of Chemical Physics, 139*(1), 015102. http://dx.doi.org/10.1063/1.4811489.

Prinz, J. H., Wu, H., Sarich, M., Keller, B., Senne, M., Held, M., ... Noe, F. (2011). Markov models of molecular kinetics: Generation and validation. *Journal of Chemical Physics, 134*(17), 174105. http://dx.doi.org/10.1063/1.3565032.

Qiao, Q., Bowman, G. R., & Huang, X. H. (2013). Dynamics of an intrinsically disordered protein reveal metastable conformations that potentially seed aggregation. *Journal of the American Chemical Society, 135*(43), 16092–16101. http://dx.doi.org/10.1021/Ja403147m.

Rodriguez, A., & Laio, A. (2014). Clustering by fast search and find of density peaks. *Science, 344*(6191), 1492–1496. http://dx.doi.org/10.1126/science.1242072.

Sasaki, H. M., & Tomari, Y. (2012). The true core of RNA silencing revealed. *Nature Structural & Molecular Biology, 19*(7), 657–660. http://dx.doi.org/10.1038/Nsmb.2302.

Schirle, N. T., Sheu-Gruttadauria, J., & MacRae, I. J. (2014). Structural basis for microRNA targeting. *Science, 346*(6209), 608–613. http://dx.doi.org/10.1126/science.1258040.

Schwantes, C. R., & Pande, V. S. (2013). Improvements in Markov state model construction reveal many non-native interactions in the folding of NTL9. *Journal of Chemical Theory and Computation, 9*(4), 2000–2009. http://dx.doi.org/10.1021/ct300878a.

Sheong, F. K., Silva, D.-A., Meng, L., Zhao, Y., & Huang, X. (2015). Automatic state partitioning for multibody systems (APM): An efficient algorithm for constructing Markov state models to elucidate conformational dynamics of multibody systems. *Journal of Chemical Theory and Computation, 11*(1), 17–27. http://dx.doi.org/10.1021/Ct5007168.

Shukla, D., Meng, Y. L., Roux, B., & Pande, V. S. (2014). Activation pathway of Src kinase reveals intermediate states as targets for drug design. *Nature Communications. 5*. Article no. 3397. http://dx.doi.org/10.1038/Ncomms4397.

Silva, D.-A., Bowman, G. R., Sosa-Peinado, A., & Huang, X. (2011). A role for both conformational selection and induced fit in ligand binding by the LAO protein. *PLoS Computational Biology*, 7(5), e1002054. http://dx.doi.org/10.1371/Journal.Pcbi.1002054.

Silva, D.-A., Weiss, D. R., Pardo-Avila, F., Da, L. T., Levitt, M., Wang, D., & Huang, X. (2014). Millisecond dynamics of RNA polymerase II translocation at atomic resolution. *Proceedings of the National Academy of Sciences of the United States of America, 111*(21), 7665–7670. http://dx.doi.org/10.1073/Pnas.1315751111.

Swope, W. C., Pitera, J. W., & Suits, F. (2004). Describing protein folding kinetics by molecular dynamics simulations. 1. Theory. *Journal of Physical Chemistry B, 108*(21), 6571–6581. http://dx.doi.org/10.1021/Jp037421y.

Sydow, J. F., Brueckner, F., Cheung, A. C. M., Damsma, G. E., Dengl, S., Lehmann, E., ... Cramer, P. (2009). Structural basis of transcription: Mismatch-specific fidelity mechanisms and paused RNA polymerase II with frayed RNA. *Molecular Cell, 34*(6), 710–721. http://dx.doi.org/10.1016/j.molcel.2009.06.002.

van Gunsteren, W. F., & Berendsen, H. J. C. (1987). *Groningen molecular simulation (GROMOS) library manual, BIOMOS b.v.* Groningen, The Netherlands: The University of Groningen.

Voelz, V. A., Bowman, G. R., Beauchamp, K., & Pande, V. S. (2010). Molecular simulation of ab initio protein folding for a millisecond folder NTL9(1-39). *Journal of the American Chemical Society, 132*(5), 1526–1528. http://dx.doi.org/10.1021/ja9090353.

Walmacq, C., Cheung, A. C. M., Kireeva, M. L., Lubkowska, L., Ye, C. C., Gotte, D., ... Kashlev, M. (2012). Mechanism of translesion transcription by RNA polymerase II and its role in cellular resistance to DNA damage. *Molecular Cell, 46*(1), 18–29. http://dx.doi.org/10.1016/j.molcel.2012.02.006.

Walmacq, C., Wang, L., Chong, J., Scibelli, K., Lubkowska, L., Gnatt, A., ... Kashlev, M. (2015). Mechanism of RNA polymerase II bypass of oxidative cyclopurine DNA lesions. *Proceedings of the National Academy of Sciences of the United States of America, 112*(5), E410–E419. http://dx.doi.org/10.1073/pnas.1415186112.

Wang, D., Bushnell, D. A., Huang, X., Westover, K. D., Levitt, M., & Kornberg, R. D. (2009). Structural basis of transcription: Backtracked RNA polymerase II at 3.4 angstrom resolution. *Science (New York, N.Y.)*, *324*(5931), 1203–1206. http://dx.doi.org/10.1126/science.1168729.

Wang, D., Bushnell, D. A., Westover, K. D., Kaplan, C. D., & Kornberg, R. D. (2006). Structural basis of transcription: Role of the trigger loop in substrate specificity and catalysis. *Cell*, *127*(5), 941–954. http://dx.doi.org/10.1016/j.cell.2006.11.023.

Wang, X. W., Heegaard, N. H. H., & Orum, H. (2012). MicroRNAs in liver disease. *Gastroenterology*, *142*(7), 1431–1443. http://dx.doi.org/10.1053/j.gastro.2012.04.007.

Wang, L. F., Zhou, Y., Xu, L., Xiao, R., Lu, X. Y., Chen, L., … Wang, D. (2015). Molecular basis for 5-carboxycytosine recognition by RNA polymerase II elongation complex. *Nature*, *523*(7562), 621–625. http://dx.doi.org/10.1038/nature14482.

Wang, D., Zhu, G. Y., Huang, X., & Lippard, S. J. (2010). X-ray structure and mechanism of RNA polymerase II stalled at an antineoplastic monofunctional platinum-DNA adduct. *Proceedings of the National Academy of Sciences of the United States of America*, *107*(21), 9584–9589. http://dx.doi.org/10.1073/pnas.1002565107.

Weiss, D. R., & Levitt, M. (2009). Can morphing methods predict intermediate structures? *Journal of Molecular Biology*, *385*(2), 665–674. http://dx.doi.org/10.1016/J.Jmb.2008.10.064.

Weixlbaumer, A., Leon, K., Landick, R., & Darst, S. A. (2013). Structural basis of transcriptional pausing in bacteria. *Cell*, *152*(3), 431–441. http://dx.doi.org/10.1016/J.Cell.2012.12.020.

Westover, K. D., Bushnell, D. A., & Kornberg, R. D. (2004a). Structural basis of transcription: Nucleotide selection by rotation in the RNA polymerase II active center. *Cell*, *119*(4), 481–489. http://dx.doi.org/10.1016/j.cell.2004.10.016.

Westover, K. D., Bushnell, D. A., & Kornberg, R. D. (2004b). Structural basis of transcription: Separation of RNA from DNA by RNA polymerase II. *Science*, *303*(5660), 1014–1016. http://dx.doi.org/10.1126/Science.1090839.

Wilson, R. C., & Doudna, J. A. (2013). Molecular mechanisms of RNA interference. *Annual Review of Biophysics*, *42*, 217–239. http://dx.doi.org/10.1146/annurev-biophys-083012-130404.

Xia, T., Li, N., & Fang, X. H. (2013). Single-molecule fluorescence imaging in living cells. *Annual Review of Physical Chemistry*, *64*, 459–480. http://dx.doi.org/10.1146/annurev-physchem-040412-110127.

Yao, Y., Cui, R. Z., Bowman, G. R., Silva, D.-A., Sun, J., & Huang, X. (2013). Hierarchical Nystrom methods for constructing Markov state models for conformational dynamics. *Journal of Chemical Physics*, *138*(17), 174106. http://dx.doi.org/10.1063/1.4802007.

Yao, Y., Sun, J., Huang, X., Bowman, G. R., Singh, G., Lesnick, M., … Carlsson, G. (2009). Topological methods for exploring low-density states in biomolecular folding pathways. *Journal of Chemical Physics*, *130*(14), 144115. http://dx.doi.org/10.1063/1.3103496.

Yu, J., Da, L. T., & Huang, X. (2015). Constructing kinetic models to elucidate structural dynamics of a complete RNA polymerase II elongation cycle. *Physical Biology*, *12*(1), 016004. http://dx.doi.org/10.1088/1478-3975/12/1/016004.

Zhang, L., Silva, D.-A., Pardo-Avila, F., Wang, D., & Huang, X. (2015). Structural model of RNA polymerase II elongation complex with complete transcription bubble reveals NTP entry routes. *PLoS Computational Biology*, *11*(7), e1004354. http://dx.doi.org/10.1371/journal.pcbi.1004354.

Zhao, Y., Sheong, F. K., Sun, J., Sander, P., & Huang, X. (2013). A fast parallel clustering algorithm for molecular simulation trajectories. *Journal of Computational Chemistry*, *34*(2), 95–104. http://dx.doi.org/10.1002/jcc.23110.

Zhuang, W., Cui, R. Z., Silva, D.-A., & Huang, X. (2011). Simulating the T-jump-triggered unfolding dynamics of trpzip2 peptide and its time-resolved IR and two-dimensional IR signals using the Markov state model approach. *Journal of Physical Chemistry B*, *115*(18), 5415–5424. http://dx.doi.org/10.1021/jp109592b.

CHAPTER SIXTEEN

Microscopic Characterization of Membrane Transporter Function by In Silico Modeling and Simulation

J.V. Vermaas*,†,‡, N. Trebesch*,†,‡, C.G. Mayne†,‡, S. Thangapandian†,‡, M. Shekhar*,†,‡, P. Mahinthichaichan†,‡, J.L. Baylon*,†,‡, T. Jiang*,†,‡, Y. Wang*,†,‡, M.P. Muller*,†,‡,§, E. Shinn*,†,‡, Z. Zhao*,†,‡, P.-C. Wen†,‡, E. Tajkhorshid*,†,‡,§,1

*Center for Biophysics and Quantitative Biology, University of Illinois at Urbana-Champaign, Urbana, IL, United States
†University of Illinois at Urbana-Champaign, Urbana, IL, United States
‡Beckman Institute for Advanced Science and Technology, University of Illinois at Urbana-Champaign, Urbana, IL, United States
§College of Medicine, University of Illinois at Urbana-Champaign, Urbana, IL, United States
[1]Corresponding author: e-mail address: emad@life.illinois.edu

Contents

Methods in Enzymology, Volume 578
ISSN 0076-6879
http://dx.doi.org/10.1016/bs.mie.2016.05.042

Abstract

Membrane transporters mediate one of the most fundamental processes in biology. They are the main gatekeepers controlling active traffic of materials in a highly selective and regulated manner between different cellular compartments demarcated by biological membranes. At the heart of the mechanism of membrane transporters lie protein conformational changes of diverse forms and magnitudes, which closely mediate critical aspects of the transport process, most importantly the coordinated motions of remotely located gating elements and their tight coupling to chemical processes such as binding, unbinding and translocation of transported substrate and cotransported ions, ATP binding and hydrolysis, and other molecular events fueling uphill transport of the cargo. An increasing number of functional studies have established the active participation of lipids and other components of biological membranes in the function of transporters and other membrane proteins, often acting as major signaling and regulating elements. Understanding the mechanistic details of these molecular processes require methods that offer high spatial and temporal resolutions. Computational modeling and simulations technologies empowered by advanced sampling and free energy calculations have reached a sufficiently mature state to become an indispensable component of mechanistic studies of membrane transporters in their natural environment of the membrane. In this article, we provide an overview of a number of major computational protocols and techniques commonly used in membrane transporter modeling and simulation studies. The article also includes practical hints on effective use of these methods, critical perspectives on their strengths and weak points, and examples of their successful applications to membrane transporters, selected from the research performed in our own laboratory.

1. NANOSCALE EFFECTS GOVERNING MEMBRANE TRANSPORTER FUNCTION

Lipid bilayers are an impermeable barrier, actively compartmentalizing life into cells and organelles that are clearly distinct at the nanoscale (Medalia, 2002; Schneider et al., 2010). Membrane transporters are incredibly specific in their capacity as the conduit of molecular transit across the membrane, serving as the Maxwellian "demon" that selectively permits specific substrates to cross while barring the path for others (Thomson, 1874). Unlike Maxwell's fictional demon, membrane transporters function by well-established thermodynamic principles, exploiting cellular sources of chemical energy such as ATP or preestablished ion gradients to drive conformational or enzymatic changes that facilitate the movement of substrate. Transporter-driven processes are found throughout biology and are used to drive not only exchange of nutrients, ions, and metabolites across the membrane, but also more complex processes such as ATP synthesis (Weber, Wilke-Mounts, & Senior, 2003), cellular signaling (Blakely & Edwards, 2012), and the excretion of cellular toxins (DeGorter, Xia, Yang, & Kim, 2012).

The structure of the membrane transport proteins themselves are as diverse as their function, with over 10,000 different transporters classified into 49 superfamilies (Saier, Reddy, Tamang, & Vastermark, 2014). Despite their large diversity in function, there are, however, also similar constraints that evolution has placed upon membrane transporters. Transport is an active process that requires chemical energy to be fueled, either a preexisting electrochemical gradient as in passive carriers and secondary transporters or ATP hydrolysis in primary transporters. Thus unregulated transport can effectively act like a short circuit in biology, insidiously draining the capacity for the cell to do work, ultimately leading to disease (Hediger, Clémençon, Burrier, & Bruford, 2013) or death (Ajao et al., 2015; Feng et al., 2015) if not ameliorated. The centrality of transporters to biological function has made them an attractive topic of study to a broad field of researchers, including computational scientists using techniques detailed in this chapter after a brief overview of the fundamental question of membrane transporter function.

1.1 Alternating Access Mechanism in Membrane Transporters

Membrane transporters mediate the translocation of specific substrate from one side of the membrane to the other. Although seemingly straightforward, this task cannot be accomplished by simply creating an open pathway at the membrane to allow free passage of the substrate. Under physiological conditions, many membrane transporter substrates move against their concentration gradient across the membrane. Uncontrolled flow of the substance through a wide open pathway, therefore, would be detrimental and even deadly to the cell.

To fulfill this important requirement, virtually all membrane transporters utilize the "alternating access" mechanism (Jardetzky, 1966) to carry out their function. According to this mechanism, as the transport protein

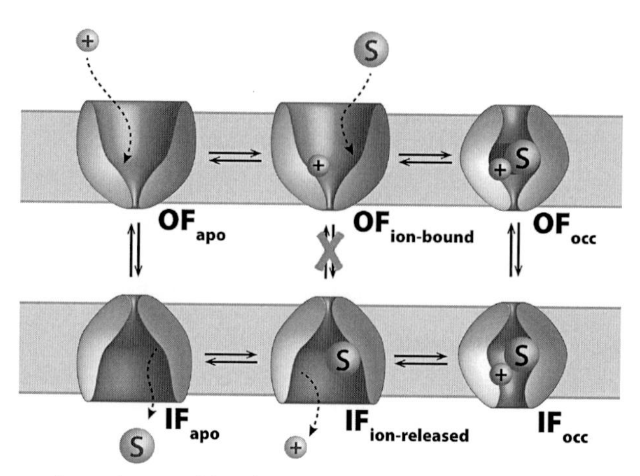

Fig. 1 An exemplary scheme of the alternating access mechanism adopted by a cation-coupled symporter, where ions and substrate move in tandem. Here, the coupling mechanism only permits conformational transitions when the binding site is either completely vacant or bound with both chemical species, forbidding transitions of partially bound states (red X). Different types of transporters have different coupling mechanisms; however, all share this feature of certain forbidden transitions to regulate substrate transit and to prevent draining membrane potential. (See the color plate.)

undergoes conformational changes to move the substrate from one side to the other, coordinated closing and opening motions of specific gating elements within the protein ensure that the bound substrate is only accessible from one side of the membrane at any given time (Fig. 1).

At least two major conformational states are visited by a membrane transporter during its functional cycle: the "inward-facing" (IF) and the "outward-facing" (OF) states. In a microscopic view, however, a series of events including the coordinated substrate binding/release, the engagement/discharge of the energy source, and the interconversion between these two major conformational states, need to take place in a particular order during each transport cycle. Key functional aspects, such as transport directionality, mechanism, and efficiency, rely on free energy landscapes controlling the interconversion of these states and how they are affected by chemical events within the protein.

The requirement for global conformational changes renders membrane transporter structures highly flexible and dynamic, and often difficult to capture experimentally. Nevertheless, detailed, atomistic descriptions of the structures and dynamics of these molecular machines are highly relevant and pivotal to studies of the activities/pathologies of membrane transporters. Thus, simulating these movements with molecular dynamics and characterizing their thermodynamic properties using free energy calculations can be of great value to mechanistic studies of membrane transporters (J. Li, Wen, Moradi, & Tajkhorshid, 2015).

1.2 Augmenting Mechanistic Studies of Membrane Transporters Using Simulation

In addition to traditional experimental approaches, simulation with classical molecular dynamics (MD) (Hug, 2013) is a compelling addition to the scientific arsenal, going beyond the resolution and interpretation limitations of conventional experiment to provide models of these systems at arbitrary resolution. Crucially, MD permits atomic events (such as a side chain rotation or the formation of a salt bridge) to be observed on a single molecule level, allowing the effect of specific mutations or binding events to be captured for biological assemblies (Perilla et al., 2015).

The resolution offered by MD is particularly important because transporters are inherently dynamic, which can cause difficulties for experimental observation. Substrate arrives on one side of the bilayer, finds a transporter, and follows it as the transporter undergoes a conformational change and unbinds on the other side of the bilayer. Thus, while static methods such as X-ray crystallography (Barends et al., 2014) or cryo-EM (Jackson, McCoy, Terwilliger, Read, & Wiedenheft, 2015; R. Y.-R. Wang et al., 2015) are indispensable to determining a 3D structure, they are missing some of the rich details that can only be obtained by exploiting the unparalleled simultaneous spatial and temporal resolution of MD simulation (Dror, Dirks, Grossman, Xu, & Shaw, 2012; E. H. Lee, Hsin, Sotomayor, Comellas, & Schulten, 2009). Instead, ensemble spectroscopy techniques such as NMR (Murray, Das, & Cross, 2013; Oxenoid & Chou, 2013) or EPR (McHaourab, Steed, & Kazmier, 2011) as well as single molecule techniques such as FRET (Vitrac, MacLean, Jayaraman, Bogdanov, & Dowhan, 2015) have had the greatest success in probing transporter conformational change. However, due to resolution restrictions on these techniques, computational modeling and MD also have a role to play.

The relationship between simulation and experiment can best be thought of as a symbiotic one (Karplus & McCammon, 2002; Papaleo, 2015). Simulation is impossible without the high-resolution experimentally derived structures. Likewise, experiment is driven forward by understanding of the interactions present in the system at the nanoscale. While later sections of the chapter will emphasize computational techniques used to probe a transporter system of interest, fruitful collaboration with experiment only improves the final scientific result.

2. MODELING MEMBRANE TRANSPORTERS IN THEIR NATIVE ENVIRONMENT

Having established the role of molecular simulation in investigating the mechanism of transporter function, actually carrying out the simulations involves a number of decisions that need to be made and procedures that need to be carried out prior to simulation. The aim of this section is to serve as a guide through the process of setting up membrane transporter simulations, complete with recommendations for state of the art tools and techniques that are expected in the larger field of biomolecular simulation, and a special emphasis on where membrane systems would differ from their soluble counterparts. The set of steps for complete system construction are expanded below and are broadly applicable to any membrane-embedded protein.

1. Decide on a level of simulation detail.
2. Obtain a starting conformation for the membrane protein of interest.
   (a) Choose a 3D structure.
   (b) Fill in missing elements of the chosen structure.
   (c) Determine protonation and termination.
3. Assemble a complete simulation system.
   (a) Choose a membrane composition.
   (b) Build the membrane.
   (c) Embed the model structure into the generated membrane.
4. Equilibrate the system under appropriate simulation conditions.

2.1 Choosing a Level of Detail: Atomistic or Coarse Grained?

The first decision prior to any simulation is to choose the level of detail for the simulation. Since classical MD simulations propagate Newton's equations of motion forward in time for each particle by using an empirical force field (Hug, 2013; Karplus & McCammon, 2002), stable integration of the position demands that the timestep between force evaluations be approximately 10$\times$ shorter than the timescale of the fastest degree of freedom (Hess, 2008). In the original formulation of biomolecular MD, each particle in the simulation system represents a specific atom, and the fastest degrees of freedom are the bond vibrations to hydrogen. The relevant vibrational modes are observed at $\sim$3000 cm^{-1} in IR spectra (de Vries & Hobza, 2007), implying $\sim 10^{14}$ vibrations of the bond per second, limiting unconstrained atomic MD timesteps to 1 fs. If the length of bonds to hydrogen atoms are fixed, the next fastest modes are heavy atom vibrations, which limit the timesteps to 2–2.5 fs each.

Many biological processes happen on the ns–ms timescale, so for atomic simulation, approximately $10^6 - 10^{12}$ timesteps need to be taken. At a reasonable simulation rate of ~10 ms of wallclock time per simulation timestep, a rate dictated by current hardware communication performance, a μs of simulated time would take approximately 4 months for a fully unconstrained atomic system, and a ms would be unrealistic except on specialized hardware (Shaw et al., 2014). To achieve longer to timescales economically, coarse grained alternatives to atomic simulation have been developed, which fundamentally depart from an atomic representation to make larger particles that represent multiple atoms (Marrink & Tieleman, 2013), multiple residues (Saunders & Voth, 2012, 2013), simply omit hydrogen (S. Lee et al., 2014), or a combination of these approaches (Han & Schulten, 2012). As a consequence of the increased particle mass, coarse-grained particle vibration is retarded, allowing for substantially longer timesteps of 20–40 fs in these representations, lengthening the timescale accessible to the simulation.

Both atomic and coarse-grained representations are frequently used for production simulations and have different strengths. Coarse-grained simulations are frequently favored when the timescale required for the process is large and the interaction specificity is not central to the result, such as simulations of lipid membrane mixing (Ingólfsson et al., 2014) or sampling the organization of larger lipids around membrane supercomplexes (C.-K. Lee, Pao, & Smit, 2015). Atomic scale simulations are indispensable if specific interactions are essential to the function, such as salt bridges breaking to precipitate a large conformational change in a transporter (Moradi, Enkavi, & Tajkhorshid, 2015) or a single hydrogen bond tuning the redox state of a protein cofactor (Vermaas, Taguchi, Dikanov, Wraight, & Tajkhorshid, 2015). Again emphasizing that both approaches are accepted practice, it should be noted that there is always a danger of oversimplifying the system and missing essential features in a coarse-grained simulation. If more detailed interactions are required, reverse coarse graining approaches have been developed to convert simulation systems back to an atomic representation (Stansfeld & Sansom, 2011; Wassenaar, Pluhackova, Böckmann, Marrink, & Tieleman, 2014). The examples in the sections to follow will all focus on atomic simulation.

2.2 Initial Structural Model Construction and Refinement

For any level of biomolecular simulation, a correct starting structure is essential to the validity of the results. Structural resources such as the protein databank (PDB) (H. Berman, Henrick, & Nakamura, 2003; H. M. Berman et al., 2000) provide vital starting points for a 3D structure. However, choosing a single structure for the protein of interest can be quite difficult simply due to the number of options available. Generally speaking, there are no hard and fast rules about which structure is "best", although low resolution structures are best avoided if at all possible. Additionally, the structure should make sense in terms of membrane topology. For some membrane transporters, crystal contacts between neighbors can cause significant artifacts, as discussed by Y.-J. Chen et al. (2007) and Wisedchaisri, Park, Iadanza, Zheng, and Gonen (2014) when comparing their crystal structures against other structures. Once a starting point is obtained, there are a number of additional technical elements that should be checked prior to starting simulation, including the completeness of the structure and protonation.

PDB structures from crystallographic data are often missing pieces of the native protein, which were too floppy or dynamic to be well resolved in the crystal lattice. Many tools exist to fill in these gaps, and broadly speaking come in two different flavors, those that use population-based statistics to generate candidate models, and those that directly use additional experimental inputs to refine the structure. Given a protein structure with gaps, tools such as MODELLER (Webb & Sali, 2014), Rosetta (Chaudhury, Lyskov, & Gray, 2010; Das & Baker, 2008), or particularly SuperLooper (Hildebrand et al., 2009), which has been optimized for membrane protein loops, can fill in those gaps and complete the protein structure. This obviates the need to restrain the geometry of what would otherwise be loose ends, although they should be monitored during simulation. Additionally, PDB structures may contain cis-peptide bonds or chirality errors (Croll, 2015; Schreiner, Trabuco, Freddolino, & Schulten, 2011; Touw, Joosten, & Vriend, 2015), which can be detected using tools included in webservers such as MolProbity (V. B. Chen et al., 2010) or WHAT IF (Rodriguez, Chinea, Lopez, Pons, & Vriend, 1998), and corrected through model rebuilding and/or re-refinement of the crystal structure. These stereochemical errors, nevertheless, can be easier addressed during system construction using the Chirality/Cispeptide (Schreiner et al., 2011) plugins of VMD (Humphrey, Dalke, & Schulten, 1996).

Refinement methods depend on experimental observables to apply additional constraints to the protein structure. CryoEM or X-ray derived electron densities can be used as a biasing potential for the protein structure using molecular dynamics flexible fitting (MDFF) (McGreevy et al., 2014; Trabuco, Villa, Schreiner, Harrison, & Schulten, 2009), where the classical MD force field is supplemented by forces acting on the atoms to bring them to electron dense regions of space. EPR/DEER data can similarly be incorporated into simulation structures using restrained ensemble simulations applying a series of distance restraints (Islam, Stein, Mchaourab, & Roux, 2013). In principle, with the advent of complex collective variables available in easy to use packages (Fiorin, Klein, & Hénin, 2013; Tribello, Bonomi, Branduardi, Camilloni, & Bussi, 2014), NMR observables such as NOEs (nuclear Overhauser effect) and RDCs (residual dipolar couplings) can also be included in structure determination protocols (Camilloni & Vendruscolo, 2015; Fu et al., 2014).

Another often overlooked feature of the structure preparation process is the determination of the protonation states for each and every ionizable residue. Now the N- and C-termini tend to be fairly clear decisions, as they are nearly always protonated and deprotonated, respectively, according to their pK_a and general environment. However, if the protein is incomplete, such as if only one domain is being simulated, it may be appropriate to use a neutral terminating patch instead. For ionizable residues such as histidine or aspartate, their pK_a can shift dramatically depending on their protein environment. If the ionization state is unknown, the pK_a of each residue can be estimated using tools such as PROPKA (Olsson, Søndergaard, Rostkowski, & Jensen, 2011; Søndergaard, Olsson, Rostkowski, & Jensen, 2011), or H++ (Gordon et al., 2005), allowing the protonation state to be assigned. PROPKA is particularly suitable to this task, as when combined with the PDB2PQR originally designed for Poisson–Boltzmann electrostatics calculations (Dolinsky et al., 2007), it generates an output that assigns protonation locations, including the always ambiguous histidine, which has two neutral forms. Neither PROPKA or H++ takes into account which residues interact with the membrane, and instead assumes the proteins are soluble, so caution should be exercised with their output on the protein periphery and inspected carefully for transporter systems.

2.3 Membrane Composition and Construction Considerations

Prior to assembling a membrane-embedded transporter simulation system, it is important to consider what lipid composition is most appropriate for the transporter in question. Due to the desire to improve performance by limiting the size of the simulation, there may be fewer than 100 lipids per leaflet for modeled membranes surrounding transport proteins. As a result, single composition membranes form the zero-order approximation of a cell membrane in vivo. Phosphatidylcholine (PC) and phosphatidylethanolamine (PE) lipids are commonly used because they represent the largest components of steady state mammalian and bacterial cell membranes, respectively (Dowhan, 1997; van Meer, Voelker, & Feigenson, 2008).

A more complete approach, however, would be to include other membrane actors, such as sterols or signaling lipids that have been experimentally shown to influence transporter function (Hamilton et al., 2014; Hong & Amara, 2010). Sterols are generally abundant, and can be included in mixed systems without difficulty at their experimental concentration (Jo, Lim, Klauda, & Im, 2009; E. Wu et al., 2014), keeping in mind that lipid mixing processes are slow, and require extensive simulation for full equilibration (Ingólfsson et al., 2014). Rare signaling lipids represent significantly below 1% of membrane phospholipid in vivo (van Meer et al., 2008), and should be strongly enriched in typical simulations so that the embedded protein may have a chance to interact with the lipid. This enrichment depends on the assumption that the transporters locally enhance the concentration of certain lipids from the heterogeneous lipid distribution in live cells and is likely true for proteins with specific lipid binding sites.

The location of the transporter must also be considered; mitochondrion, for example, have high levels of cardiolipin, a bacterial lipid not found elsewhere in mammalian cells (Paradies, Paradies, Benedictis, Ruggiero, & Petrosillo, 2014). Similarly, yeast and bacterial cells have a membrane composition that can differ substantially from mammalian cells, and the membrane composition must be changed accordingly. The context in which the transporter operates should be taken into account as well. If, for example, a transporter functions on activated platelets or cells undergoing apoptosis, it is likely to interact more frequently with charged lipids flipped from the inner membrane (Fadok et al., 1992). Before determining a lipid composition for the system, the following steps should be taken:

1. Review the available literature to determine if the transporter function is influenced by specific lipids. If a specific lipid plays a major role in the activity of the transporter, it may be advisable for a larger-than-physiological concentration of the moiety be included to ensure the critical interaction is sampled. Multiple initial membrane configurations should be generated, as the protein–lipid interactions will be biased towards the initial conformation.
2. Review the membrane context for implications as to the lipid environment surrounding the transporter, including the species from which the transporter was isolated and the localization of the transporter within the

cell. If a particular lipid has a significant presence, it may be advisable to include it to reproduce the conditions of the specific functional environment of interest.

3. Review experimental studies on the transporter to inform the design of the simulation. Since it is known that the choice of lipid changes the behavior of transporters in vitro, direct comparisons between simulation and experiment are best made at similar membrane compositions.

One final thing to consider is that once the lipid composition for the simulation has been determined, a lipid patch of appropriate size for the system must be generated. Many tools exist for this step, including both webservers and stand-alone programs (Bovigny, Tamò, Lemmin, Maïno, & Dal Peraro, 2015; Jo, Kim, Iyer, & Im, 2008; Wassenaar, Ingólfsson, Böckmann, Tieleman, & Marrink, 2015). The CHARMM-GUI membrane builder is an exemplary tool for this step, as it includes most physiological mammalian and bacterial phospholipids and sterols (E. Wu et al., 2014). CHARMM-GUI also permits the membrane geometry to be selected and contains an interface to alternative membrane representations that accelerates lipid diffusion (Qi et al., 2015). This Highly Mobile Membrane Mimetic (HMMM) representation (Ohkubo, Pogorelov, Arcario, Christensen, & Tajkhorshid, 2012) can be used to accelerate sampling of the lipid environment around proteins, capturing headgroup-specific interactions (Baylon et al., 2016; Vermaas, Baylon, et al., 2015).

2.4 Membrane-Embedded Transporter System Assembly

Aside from membrane considerations, the protein itself needs to be oriented and placed correctly relative to this membrane. Transporter proteins typically have a very distinct belt-like region in contact with the membrane. This region may not be obvious merely through visual inspection alone. Commonly, a web server such as the PPM (Positioning of Proteins in Membrane) (A. L. Lomize, Pogozheva, Lomize, & Mosberg, 2006, 2007; A. L. Lomize, Pogozheva, & Mosberg, 2011; M. A. Lomize, Lomize, Pogozheva, & Mosberg, 2006), is used to automatically identify the hydrophobic belt region of a transmembrane protein. This is done by minimizing the transfer energies of the membrane protein from water to an artificial lipid bilayer (A. L. Lomize et al., 2011). This is a crucial step because the function of the membrane protein is greatly affected by the lipid–protein interactions. Misplacement of the orientation of the protein and mismatch of the protein hydrophobic belt and membrane bilayer may lead to long equilibration requirements or misleading conclusions.

Following the orientation of the membrane protein, it must be inserted into the bilayer. For simple cylindrical transporters, the protein can simply be superimposed onto the membrane. Lipids that overlap with the protein can be programmatically removed via structure building tools such as VMD (Humphrey et al., 1996) or CHARMM (Brooks et al., 2009) (Fig. 2A). Usually this will leave large gaps at the interface between the protein and the lipids. This artifact can be equilibrated away, as the lipids will naturally pack against the transporter during simulation. With the new CHARMM36 lipid force field parameters (Klauda et al., 2010) and constant pressure condition applied to the lateral (membrane plane) and perpendicular directions, the area per lipids will match the experimental values once equilibrated, ie, $\sim$70 Å^2 lipid (Leftin, Molugu, Job, Beyer, & Brown, 2014). However, this protocol may change the protein structure, which is frequently ameliorated by restraining the protein backbone via a harmonic potential to its initial position. The restrained equilibration relaxes the lipid tails prior to production simulation. This step should continue until the unit cell dimension of the membrane settle to near a fixed value, when the restraints can be released. Alternatively, a repulsive force centered around the protein and pointing outwards can be applied to the lipids to remove clashes (Faraldo-Gómez, Smith, & Sansom, 2002; Shen, Bassolino, & Stouch, 1997; Tieleman & Berendsen, 1998).

For noncylindrical transporters, many other methods exist (Javanainen, 2014; Jefferys et al., 2015; Kandt, Ash, & Tieleman, 2007; Stansfeld et al., 2015; Wolf et al., 2010; E. Wu et al., 2014). In all these methods, the lipids that have severe overlaps with the membrane protein are removed first. They differ in the strategy for removing minor clashes and wrapping lipids around the protein. In method from Kandt et al. (2007), the X and Y coordinates of the preequilibrated membrane are first scaled up to remove clashes and then gradually scaled towards the target lipid density to accommodate lipids around the membrane protein. In the method of Wolf et al. (2010), the X and Y coordinates (along the membrane plane) of the transporter protein are first scaled down and then scaled back to the original to allow the lipids to tightly fit to the hydrophobic belt of the transporter using molecular dynamics (Fig. 2B). In Javanainen's method (Javanainen, 2014), the lateral pressure along the membrane plane is used to wrap the lipids around the restrained target membrane protein (Fig. 2D). In the MemProtMD method developed by Stansfeld et al. (2015), the key idea to change the representation of the target system to a coarse-grained version and take advantage of the fast dynamics in new resolution-reduced system to

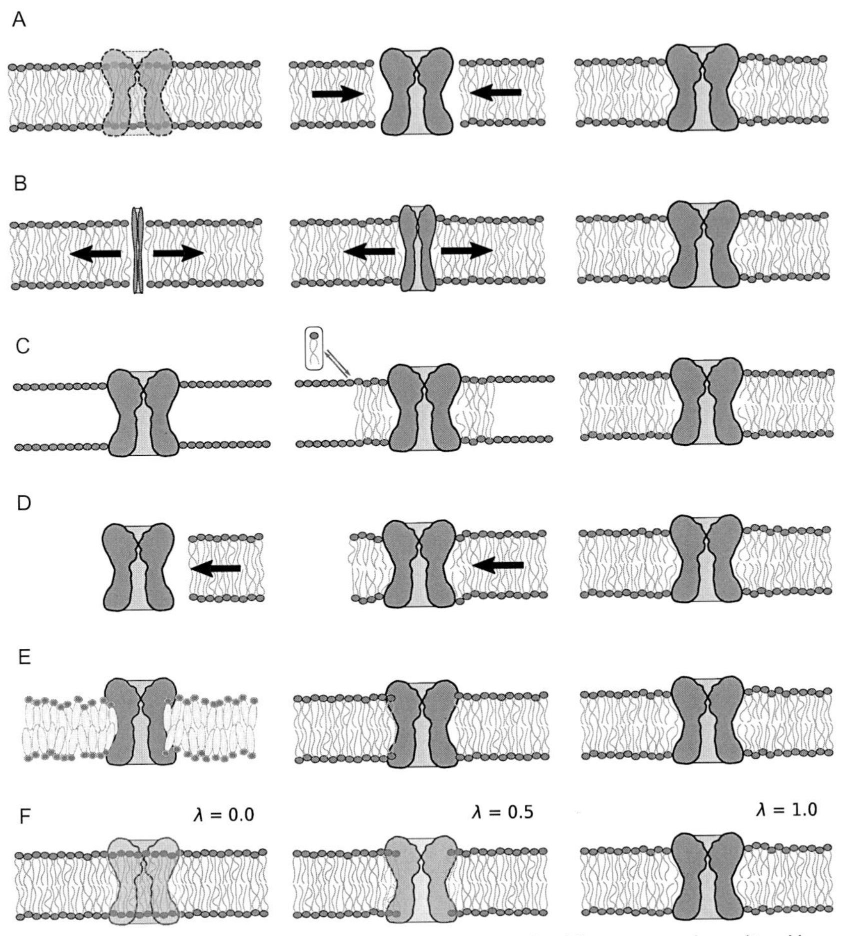

Fig. 2 Schematic diagrams of the many membrane embedding protocols outlined here. The protein is colored blue (gray in the print version) and the phospholipid bilayer is shown with a purple (gray in the print version) head group and yellow (light gray in the print version) lipid tails. (A) The naive lipid deletion method carried out within VMD (Humphrey et al., 1996) or CHARMM (Brooks et al., 2009), where overlapping lipids are removed from the bilayer and equilibration eliminates the gap as denoted by the arrows. (B) The method of Wolf, Hoefling, Aponte-Santamaría, Grubmüller, and Groenhof (2010), where the protein, initially a thin rod, expands to make space for itself within the membrane. Outward displacement of lipids is illustrated by the arrows. (C) Pseudo atoms replaced by lipid conformers taken from a lipid library that results in a nearly equilibrated membrane–protein complex (Jo, Kim, & Im, 2007). (D) The membrane is introduced to the protein by a pressure applied to the membrane, resulting in an embedded complex (Javanainen, 2014). Flow of the membrane is indicated by the direction of the arrow. (E) Stansfeld et al. (2015) use coarse-grained simulation to reach equilibrium. Reversion of coarse grain lipids to atomistic allows further equilibration. (F) Alchemical techniques (Jefferys, Sands, Shi, Sansom, & Fowler, 2015) gradually introduce protein–lipid interactions by using a soft-core Van der Waals potential as indicated by the λ parameter.

accelerate the lipid diffusion process (Fig. 2E). This coarse-grained system is subsequently converted back to the all-atom system with some equilibration. In the Alchembed method developed by Jefferys et al. (2015), interactions between the protein and the lipids are replaced by alchemical soft-core potentials with gradually increasing intensity to remove clashes between lipids and the protein including aromatic ring piercing (Fig. 2F). Last but not least, a module named the "Membrane Builder" from the web application CHARMM-GUI (Jo et al., 2007, 2008; E. Wu et al., 2014) can also be used for preparing the membrane–protein system. The algorithm used by CHARMM-GUI "Membrane Builder" first determines the lipid head group position on the two membrane surfaces via a simulation with pseudo atoms that surround the embedded protein (Jo et al., 2007) (Fig. 2C). The pseudo atoms are then replaced by full lipids selected from a conformer library with 2000 distinct conformations taken from a MD trajectory, generating a membrane embedded membrane protein.

All the methods mentioned above have been demonstrated to be well suited for building simulation systems with membrane proteins embedded in pure or mixed biological membranes. However, each of them have a different learning curve and CHARMM-GUI stands out for its simple interface and rich customization features for different lipid types (E. Wu et al., 2014).

2.5 Simulation Condition Considerations

Membrane protein simulation conditions differ from their soluble counterparts in a number of aspects. The membrane is an anisotropic medium, with distinct stress moduli along the membrane normal and membrane parallel directions. Barostats in simulation should take this asymmetry into account. MD engines include an option for a barostat that can change the shape of the dimensions of the periodic cell independently between the membrane normal and membrane parallel axes, based on the seminal work of Martyna, Tobias, and Klein (1994). The two membrane parallel axes should grow and shrink in a constant ratio with one another, lest the membrane deform anisotropically and allow the embedded protein to contact itself across the periodic boundary.

While the thermostat used for simulation of membrane transporters follows the same general advice as soluble proteins, using a Nosé–Hoover (Evans & Holian, 1985) or Langevin thermostat (Brünger, Brooks III, & Karplus, 1984; J. C. Phillips et al., 2005), the target temperature of membrane protein systems needs to be carefully chosen. Membranes undergo phase transitions near physiological temperature, which have been hypothesized by Gray, Díaz-Vázquez, and Veatch (2015) to be a way by which the cell can reorganize its membranes in response to different growth conditions. Thus unlike soluble proteins, which are frequently simulated near room temperature (300 K) to mimic experimental conditions, membrane protein simulations are typically a bit warmer (310 K) to stay above the liquid-gel transition temperature for the lipids being simulated and to mimic natural disordered lipids (Coppock & Kindt, 2010).

Membrane protein simulation conditions differ from solution simulations in one final respect: rather than only needing a force field to describe the interactions within the protein and between the protein and water, membrane protein simulations also require accurate parameters for protein–membrane interactions. Amber (Dickson, Rosso, Betz, Walker, & Gould, 2012), Gromos (Reif, Winger, & Oostenbrink, 2013), OPLS (Siu, Pluhackova, & Böckmann, 2012), and CHARMM (Klauda et al., 2010) lipid force fields have all been validated for these mixed interactions at the atomic level, and all continue to be improved and updated as more experimental data becomes available. Currently, either the combination of the CHARMM36 lipid (Klauda et al., 2010) and protein (Best et al., 2012) force fields or the SLIPID (Jämbeck & Lyubartsev, 2013) and AMBER99SB-ILDN (Aliev et al., 2014) force fields would be recommended for new simulation having been consistently found to outperform their competitors in membrane properties (Paloncýová, DeVane, Murch, Berka, & Otyepka, 2014) and correctly capture membrane–protein interaction (Sun, Forsman, & Woodward, 2015).

3. MODELING SUBSTRATE BINDING AND UNBINDING PROCESSES IN MEMBRANE TRANSPORTERS

The activity of membrane transporters is tightly coupled to their substrates and may additionally be modulated by other small molecules present in the system under study. While experimental quantities such as a dissociation constant (K_d) are often readily available for many compounds and transporter combinations, those are missing the details of the interaction, and thereby have difficulty making predictions on how the dynamics change once a substrate is bound. Since the binding fundamentally lowers the barrier to conformational transition, determining the interactions that take place and quantifying their effect through free energy calculation is a frequent

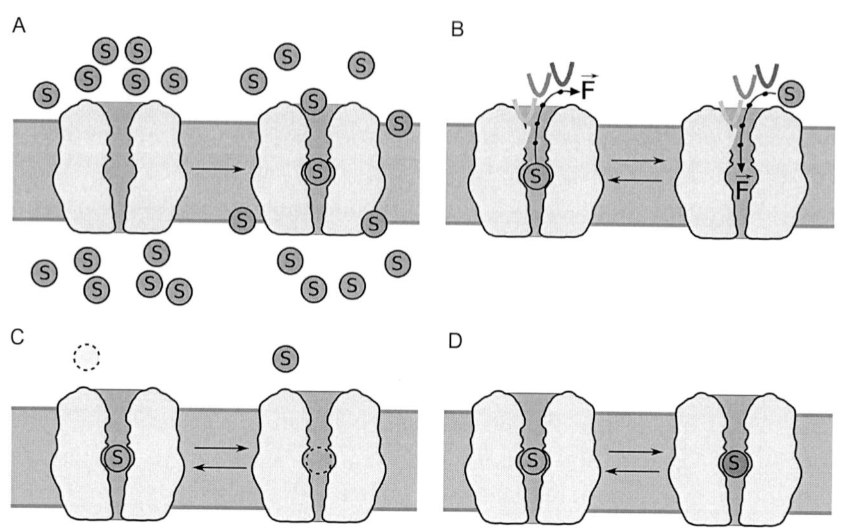

Fig. 3 Schematic of widely employed techniques for characterizing substrate association and dissociation, including their energetics. (A) Flooding simulations are where a high concentration of substrate (green (gray in the print version) circle) is allowed to interact with the transporter. (B) Steered molecular dynamics (SMD) and umbrella sampling (US) use biasing forces applied along a reaction coordinate (curved arrows) to steer a binding or unbinding process. (C and D) Alchemical methods such as free energy perturbation (FEP) or thermodynamic integration (TI) permit absolute binding (C) and relative binding (D) free energies to be computed without knowing the binding or unbinding pathway.

addition to traditional equilibrium simulation of the unbound state. Furthermore, quantifying the energy change is frequently used to compare with experiment and connect back in silico observables with those determined in vitro and is a crucial sanity check on the validity of the results. An additional complication to adding substrates or drugs to the system is that they may be missing parameters that describe their interaction with other elements of the system. Techniques commonly used to add these features to transporter simulation are discussed in this section (Fig. 3) and should be used in conjunction with equilibrium MD of the *apo* state to elucidate how substrates trigger conformational change.

3.1 Determining Substrate Force Field Parameters

Classical MD simulations are founded on molecular mechanics (MM) force fields, a mathematical description of interatomic (comprising intra- and intermolecular) interactions (Guvench & MacKerell, 2008; Mackerell,

2004; Monticelli & Tieleman, 2013; Ponder & Case, 2003). MM potential energy functions are comprised of terms to describe each individual topological and intermolecular element, in which each term is tuned for a specific interaction using parameters. For the transporters themselves and their environment, decades of research has led to the availability of robust parameter sets for commonly studied biopolymers (eg, proteins (Best et al., 2012; Cornell et al., 1995; Kamisnki, Friesner, Tirado-Rives, & Jorgensen, 2001; D.-W. Li & Brüschweiler, 2010; Lindorff-Larsen et al., 2010; MacKerell, Jr. et al., 1998; MacKerell, Jr., Feig, & Brooks, III, 2004; Oostenbrink, Villa, Mark, & Van Gunsteren, 2004), DNA/RNA (Aduri et al., 2007; Denning, Priyakumar, Nilsson, & Mackerell, Jr., 2011; Hart et al., 2012; MacKerell Jr. & Banavali, 2000; Pérez et al., 2007; Soares et al., 2005), carbohydrates (Damm, Frontera, Tirado-Rives, & Jorgensen, 1997; Glennon & Merz, Jr., 1997; Guvench, Hatcher, Venable, Pastor, & Mackerell, Jr., 2009; Guvench et al., 2011; Kirschner, Yongye, Tschampel, & González-Outeiriño, 2008; Kony, Damm, Stoll, & Van Gunsteren, 2002; Raman, Guvench, & Mackerell, Jr., 2010)) and other biological structures for which dynamic studies are required (eg, lipids; Dickson et al., 2014; Feller, Yin, Pastor, & MacKerell Jr., 1997; Klauda et al., 2010; Maciejewski, Pasenkiewicz-Gierula, Cramariuc, Vattulainen, & Ròg, 2014; Skjevik, Made, Walker, & Teigen, 2012). The substrates of membrane transporters are generally classified as small molecules which, in contrast to biopolymers, frequently contain unique functional groups not shared with the biopolymers they interact with, and demand parameters not necessarily covered by existing parameter sets. Despite lacking an inherent repeated structure, transporter substrates often share common substructure or functional groups (eg, aromatic rings, alcohols, amines, amides, alkyl chains, and olefins). Efforts towards developing parameter sets covering these common elements are the basis for generalized extensions to popular biopolymer force fields, such as the CHARMM General Force Field (CGenFF) (Vanommeslaeghe et al., 2010) and the Generalized AMBER Force Field (GAFF) (J. Wang, Wolf, Caldwell, Kollman, & Case, 2004).

While such parameter sets provide coverage for common substructures, it is unreasonable to expect that they can provide coverage for the exponential combinations required to describe small molecule parameter space (estimated at 10^{18}–10^{200} molecules!) (Drew, Baiman, Khwaounjoo, Yu, & Reynisson, 2011). The focus of obtaining parameters is thus shifted from parameterizing entire molecules, to focusing on linkages between substructures that lack parameters. Although this does not represent a complete solution to obtaining parameters, it significantly reduces the complexity and practical challenges associated with parametrization. There are several tools available that facilitate filling in the remaining missing parameters for small molecules and take one of two approaches: assigning parameters by analogy to molecules for which parameters are already known, or by optimizing parameters against target data, typically high level quantum mechanical calculations. Traditionally, comparisons to determine molecular analogy has been performed by hand. While this allows for a high degree of control, it requires that the researcher has an in-depth knowledge of chemical similarity and what types of molecules are already present in the parameter set. Recently, several programs have been developed that automate the process of determining parameters by analogy, such as the CGenFF Program (Vanommeslaeghe & MacKerell, Jr., 2012; Vanommeslaeghe, Raman, & MacKerell, Jr., 2012) and MATCH (Yesselman, Price, Knight, & Brooks, III, 2012) for the CGenFF force field, AnteChamber (J. Wang, Wang, Kollman, & Case, 2006) and R.E.D. (Vanquelef et al., 2011) for the AMBER force field, and the Automated Topology Builder (ATB) (Malde et al., 2011) and PRODRG (Schüettelkopf & van Aalten, 2004) for the GROMOS force field.

In addition to targeting different force fields, each of these resources uses its own methodology to make parameter assignments. It is critical to understand how each tool arrives at the a parameter set and to make sure that this method is consistent with the target force field for the desired application, as this has been shown to strongly impact the quality of subsequent simulations (Lemkul, Allen, & Bevan, 2010). Each resource also relies upon a different database of parameter knowledge, which can impact the results. The CGenFF program, for instance only crosschecks against the main CGenFF parameter set, excluding many chemotypes for which parameters are well known but are distributed as add-on specialty parameters, such as parameters for lipids, sterols, carbohydrates, fluoroalkanes, and amines, amongst many others that are contained in separate "stream files" and are not part of the CGenFF Program similarity search.

Assigning parameters by analogy is often expedient and in some cases sufficient; however, there are many cases which require computing the parameters directly, such as refining parameters for a specific system or obtaining parameters for novel chemotypes where no suitable analogy exists. Computing parameters directly is a significantly more involved process that requires multiple calculations to fit parameters to target data than for the analogy method (Fig. 4). The availability of toolsets for performing these tasks are

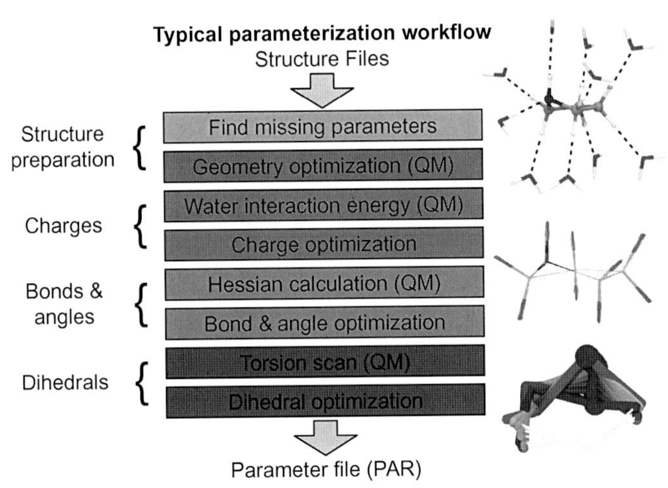

Fig. 4 General parameterization workflow for developing CHARMM-compatible parameters. These steps are implemented in the Force Field Toolkit (ffTK) (Mayne et al., 2013) and are grouped by color (different gray shades in the print version) to highlight how quantum mechanical target data feeds into each step of the parameter optimization process. Molecules showing the water interactions, bond stretching, and torsional scanning are presented on the right beside the workflow.

less widespread, with the most options available for the CHARMM family of force fields (ie, CHARMM, CGenFF). The General Automated Atomic Model Parametrization (GAAMP) webserver (L. Huang & Roux, 2013) is the simplest to use, requiring only modest input and employing a highly automated, albeit blackbox, approach to optimizing parameters. The Force Field Toolkit (ffTK) (Mayne, Saam, Schulten, Tajkhorshid, & Gumbart, 2013), distributed as a plugin within the popular VMD molecular modeling package (Humphrey et al., 1996), is designed specifically with usability in mind by automating tedious and error-prone tasks, providing reasonable defaults for novice users, and featuring a set of analytical tools to assess the details of the optimization calculations and allow for extensive customization by expert users. Expert users may be interested in the stand-alone program lsfitpar (Vanommeslaeghe, Yang, & MacKerell, Jr., 2015), which provides users with the same routines employed by the developers of CGenFF in a command-line environment, or the Force Balance program (L.-P. Wang, Chen, & Voorhis, 2013), which is written in Python and furnishes an infrastructure that accommodates a diversity of target data, such as experimental data, and optimization techniques, such as MD-simulated properties, to fit parameters. In addition to these tools, the Paramfit program

(Betz & Walker, 2015) supports parametrization of for the AMBER force field, and the ForceFit program (Waldher, Kuta, Chen, Henson, & Clark, 2010) provides general routines that are not specific to any particular force field.

Following the steps to obtain parameters, it is critical to assess the quality of the parameters. Depending on the molecule in question, the availability of experimental data, assessments towards the parameter quality range greatly in terms of complexity and rigor. The CGenFF Program provides a "penalty score" for each parameter which describes how close the term matches by analogy to the existing term (Vanommeslaeghe et al., 2012). Using this penalty score, the authors of the tool suggest that the parameter is fair, requires basic validation, or extensive validation/optimization. When computing parameters directly, ffTK provides a significant amount of data and analytical tools to assess the degree of fit between the MM-computed properties and the driving target data (Mayne et al., 2013). The most rigorous assessment of parameter performance, however, is the ability to reproduce condensed phase properties from MD simulations. The simplest calculations are to compute the density and enthalpy of vaporization (Vanommeslaeghe et al., 2010). These properties are only relevant to molecules that are liquids under simulation conditions and are largely defined by the Lennard–Jones term of the force field; therefore, they are not ideal for assessing charge assignments or bonded parameters. Currently, the most rigorous approach is to compute the free energy of solvation, where experimental values are frequently available (Mayne et al., 2013; Mobley & Guthrie, 2014; Mobley, Liu, Cerutti, Swope, & Rice, 2011; Zhang, Tuguldur, & van der Spoel, 2015). Experimental partition coefficients for octanol/water solvent systems are also frequently available for lipophillic compounds (CRC, 2015; Sangster, 1989) and can be approximated from free energy calculations using alchemical simulations (Garrido, Economou, Queimada, Jorge, & Macedo, 2012; Vermaas, Taguchi, et al., 2015).

3.2 Ensemble Docking to Identify Putative Substrate Binding Sites

In cases where the substrate binding site of the transporter is unknown, small molecule docking can be used to search for possible binding conformations of a ligand in the binding site of a protein (Dolghih, Bryant, Renslo, & Jacobson, 2011; Mihasan, 2012; Strynadka et al., 1996). The identified favorable protein–ligand conformations through docking provide a clear picture of the key molecular interactions that facilitate binding. Not all docking programs use the same scoring functions and docking protocols, so users should evaluate the reliability of available software for the ligands and proteins of interest to compare their results (Cross et al., 2009; Plewczynski, Łaźniewski, Augustyniak, & Ginalski, 2011). Following are a few ways to obtain reliable docking results, (i) a consensus approach (Clark, Strizhev, Leonard, Blake, & Matthew, 2002), where a few docking programs are used to predict the protein/ligand complex and a predominant pose is selected, (ii) redocking a crystallographically solved similar protein/ligand complex of interest to validate the programs in predicting the X-ray identified complex (Thangapandian, John, Lee, Arulalapperumal, & Lee, 2012), and (iii) using a benchmark dataset (Hevener et al., 2009), which includes known binders and nonbinders of the protein of interest, to evaluate the predictive ability of the docking programs.

Molecular docking was first carried out on static or nearly static structures, where only a few protein side chains were allowed to change their orientation (Meng, Zhang, Mezei, & Cui, 2011; Morris & Lim-Wilby, 2008; Thangapandian, John, Sakkiah, & Lee, 2010). However, this static approach misses binding sites that might become accessible only due to the motion of a protein or only exist after membrane interaction changes protein conformation, a feature that is common among membrane-associated proteins (Denisov, Grinkova, Baylon, Tajkhorshid, & Sligar, 2015; Lin, Perryman, Schames, & McCammon, 2002; Rinne, Mobarec, Mahaut-Smith, Kolb, & Bünemann, 2015; Rodríguez, Ranganathan, & Carlsson, 2015). To take these conformational changes into account, a so called "ensemble docking" technique is used. Ensemble docking is performed against a series of protein conformations taken from a MD trajectory or experimentally solved conformations of a protein (Amaro, Baron, & McCammon, 2008; Campbell, Lamb, & Joseph-McCarthy, 2014; Ellingson, Miao, Baudry, & Smith, 2015; S.-Y. Huang & Zou, 2007; Korb et al., 2012; Lin et al., 2002; Tian et al., 2014). The breathing motions from MD may find even more poses and states, rather than relying on a single or a few experimental snapshots to be representatives of the ensemble of conformations encountered in vivo.

As an example, ensemble docking of a drug substrate on to P-glycoprotein (P-gp), an ATP-binding cassette (ABC) transporter that exports toxins and drugs out of the cell (Aller et al., 2009; J. Li, Jaimes, & Aller, 2014), is demonstrated. Here, an ensemble of 1000 protein conformations obtained from a long MD trajectory (Fig. 5A) is used. Then the docking region is defined, keeping in mind that it should comprise the complete translocation

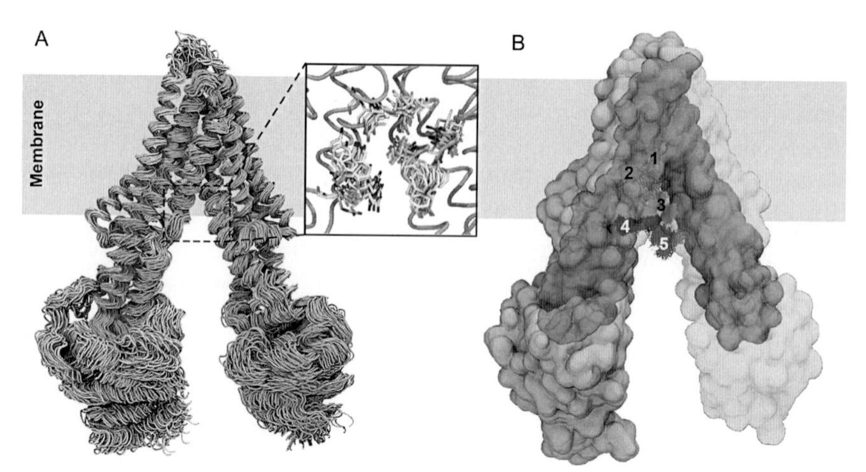

Fig. 5 Ensemble docking of a drug substrate into the translocation lumen of P-gp. (A) MD simulation-generated ensemble of conformations used in molecular docking. Two pseudosymmetric halves of P-gp are colored in cyan and orange, respectively. Only 100 out of 1000 conformations used in ensemble docking are shown. A set of translocation path residues is shown to display the diverse side chain conformations represented by the MD-generated ensemble (inset). (B) Top 5 RMSD-based clusters of docked poses of the drug molecule are shown in different colors with P-gp in surface representation. One half of the protein is shown transparent to clearly show the clusters of the docked drug molecule. (See the color plate.)

compartment and not just the final binding site, as transport proteins usually have multiple binding sites (Dolghih et al., 2011; Martin et al., 2000; Safa, 2004; Sharom, 2014). A drug substrate was docked in to the defined docking region using AutoDock (Morris et al., 2009), generating 10 docked poses for each protein conformation. For this example, ensemble docking has resulted a total of 10,000 docked poses of the drug molecule (1000 conformations × 10 poses) covering the entire translocation compartment. A clustering method was then used to cluster all the docked poses, which in turn provided a set of possible binding sites (Fig. 5B). Repeating the same protocol for both end states of the transport cycle would provide additional insight as to the complete path of a substrate during the large-scale transition of the transporter.

3.3 Substrate Binding and Unbinding from Unbiased Simulation

The docking methods presented in the previous section provide starting points for substrates bound to their binding site. However, only through

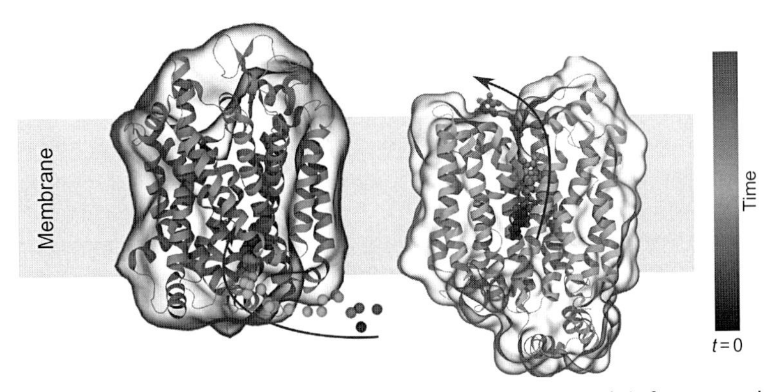

Fig. 6 Binding of an ion (left) or unbinding of substrate (right) from membrane transporters captured in unbiased equilibrium simulations. The color of the ion and substrate changes with time to show movement in the direction of the arrows. (See the color plate.)

simulation is it possible to explore the detailed dynamics of substrate binding or unbinding (Fig. 6). Through extended equilibrium MD simulation, important dynamical elements of substrate binding have been observed, including the specific interactions that bind the substrate to the transporter, and gating residues that prevent premature translocation (Andersson et al., 2012; Enkavi & Tajkhorshid, 2010; Z. Huang & Tajkhorshid, 2008; X. Jiang, Villafuerte, Andersson, White, & Kaback, 2014; Simmons et al., 2014; Y. Wang, Ohkubo, & Tajkhorshid, 2008; Watanabe et al., 2010; Yin, Jensen, Tajkhorshid, & Schulten, 2006; Zhao et al., 2011; Zomot & Bahar, 2012). Conceptually, these simulations are simple equilibrium simulations, although an unprepared investigator can find analyzing their results surprisingly counterintuitive.

For instance, despite calculating forces based on a potential energy function, the exact value of that function for interaction strength between substrate and protein has no meaning in isolation, as it neglects the contribution of water and other species to the sum of the forces acting upon the substrate. Instead, dedicated free energy methods are required to compute experimentally relevant quantities and are detailed in subsequent sections. However, these types of analyses can be useful in identifying specific interactions, such a hydrogen bonds, that restrain substrate dynamics and are liable to play a larger role in governing the substrate binding and unbinding behavior, as exemplified by numerous applications (Cheng & Bahar, 2015; Kantcheva et al., 2013; Koldsø, Autzen, Grouleff, & Schiøtt, 2013). In cases where the substrate bound form is being studied, it is recommended to release any equilibrating restraints placed upon the substrate binding site first, such that the residues surrounding the binding site can adopt their final favored conformation (J. Li & Tajkhorshid, 2012; Tavoulari et al., 2016). If this is insufficient, additional harmonic distance restraints can be applied to enforce a hypothesized binding geometry, though it may not prove to be stable during unrestrained simulation.

Studies of binding or unbinding processes can focus on specific binding residues, such as those that form an intricate H-bonding network between the substrate and the binding site that needs to be disrupted for the substrate to be released from the binding site (Watanabe et al., 2010; Zomot & Bahar, 2012). However in some instances, other protein side chains can block binding or unbinding based on their rotameric state, acting as "gate" controlling the flow of substrate (Watanabe et al., 2010; Zomot & Bahar, 2010). The coupling between the substrate rearrangement and gate opening is frequently accomplished by water solvating the binding site and lubricating the unbinding event (Cheng & Bahar, 2014; Choe, Rosenberg, Abramson, Wright, & Grabe, 2010). Analyzing the trajectories for these features is frequently first done by eye, using visualization tools such as VMD (Humphrey et al., 1996), Chimera (Pettersen et al., 2004), or PyMOL (Schrödinger, LLC, 2015), and then measuring quantities of interest via trajectory analysis tools such as VMD (Humphrey et al., 1996), MDTraj (McGibbon et al., 2015), or MDAnalysis (Michaud-Agrawal, Denning, Woolf, & Beckstein, 2011).

Unlike substrate unbinding, binding events are difficult to capture using unbiased MD simulation, as the entropy decreases substantially upon binding relative to the unbound state. Additionally, substrate binding requires the formation of interactions between the substrate and the binding site which is generally accompanied by local conformation reorganization of the amino acids in the binding site. Nevertheless, binding of small ions (Zomot, Gur, & Bahar, 2015) and gaseous molecules (CO and O_2) has been observed previously in MD simulations (Baron, McCammon, & Mattevi, 2009; Ruscio et al., 2008). Furthermore, strong interactions, such as those between charged substrates and their binding site, have also been captured spontaneously in MD simulations (Dehez, Pebay-Peyroula, & Chipot, 2008; Enkavi & Tajkhorshid, 2010; Y. Wang et al., 2008).

For uncharged substrates, the simplest simulation approach to apply to the transporter system to identify substrate binding sites and pathways would be to increase the concentration of the substrate, which may increase the probability of binding at the simulation timescale. In this "flooding" approach, a high concentration of substrate is placed in the simulation system and allowed to diffuse in an unbiased manner (Brannigan, LeBard, Hénin, Eckenhoff, & Klein, 2010; Murali, Wallner, Trudell, Bertaccini, & Lindahl, 2011). Copies of substrate can be generated by substituting water molecules or using softwares such as PACKMOL (Martínez, Andrade, Birgin, & Martínez, 2009). Binding sites and transport pathways of substrate can be identified by visualizing simulated trajectories and by clustering analysis (Brannigan et al., 2010; Buch, Giorgino, & De Fabritiis, 2011; LeBard, Hénin, Eckenhoff, Klein, & Brannigan, 2012; Ruscio et al., 2008; P. H. Wang, Best, & Blumberger, 2011). As an example of such analysis, the entire trajectory of a simulation can be clustered into a three-dimensional occupancy map using tools integrated into VMD (Humphrey et al., 1996), which indicates regions within the protein where substrate is frequently sampled, and can be applied to a diverse set of transmembrane proteins (Arcario, Mayne, & Tajkhorshid, 2014; Mahinthichaichan, Gennis, & Tajkhorshid, 2016).

Implicit ligand sampling (ILS) is an alternative and complementary approach to these flooding simulations to characterize substrate binding and transport pathways in proteins (Cohen, Arkhipov, Braun, & Schulten, 2006; Saam, Ivanov, Walther, Holzhutter, & Kuhn, 2007; Saam et al., 2010; Y. Wang, Cohen, Boron, Schulten, & Tajkhorshid, 2007; Y. Wang & Tajkhorshid, 2010). Rather than explicitly simulating copies of the substrate, ILS is a postprocessing method where substrates are probed to see if they would fit and bind favorably in different regions of a dense grid during a substrate-free simulation. ILS is most suitable for small hydrophobic gases, such as O_2, NO and CO_2, since the approach implicitly assumes that there are no strong interactions between the substrate and protein that might perturb the overall protein structure dynamics (Cohen et al., 2006). In this manner, ILS can be viewed as "systematic docking", in that snapshots of the protein taken from a trajectory of MD simulation in the absence of a targeting molecules may be used to quantitatively identify high affinity sites for substrate binding over the entire structure (Fig. 7). ILS has been successfully employed to study gas transport in membrane proteins, which include aquaporins (Y. Wang et al., 2007; Y. Wang & Tajkhorshid, 2010) and bioenergetic proteins (photosystem II and cytochrome c oxidase) (Mahinthichaichan et al., 2016; Vassiliev, Zaraiskaya, & Bruce, 2013).

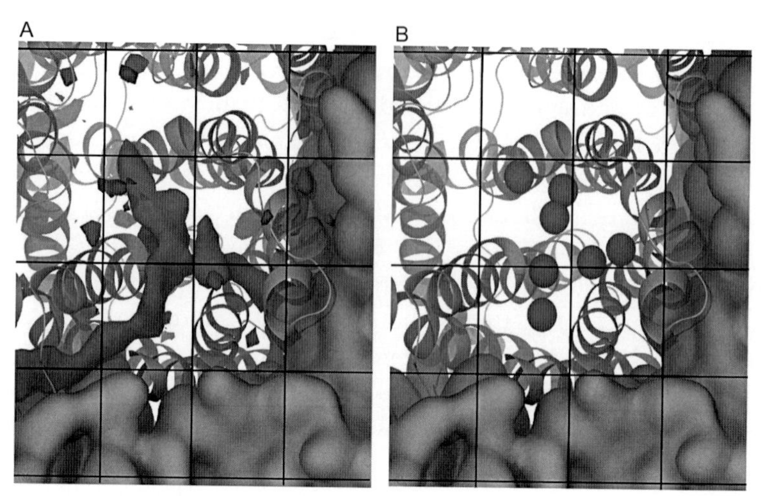

Fig. 7 Probing substrate binding sites and pathways using implicit ligand sampling (ILS) technique. (A) Substrate binding sites predicted by ILS are shown in red isosurfaces. Blue helices represent the protein. The brown surface represents lipid bilayer. Black lines represent the grid on which the substrate is placed systematically during ILS analysis. (B) Experimental observed substrate binding sites taken from crystal structures. Substrates are shown in balls. (See the color plate.)

3.4 Substrate Binding Pathway and Mechanism from Biased Simulation

Binding and particularly unbinding are relatively rare events in the lifetime of a transporter. Over a transport cycle of approximately 10 ms, a small substrate such as a sugar will travel $\sim$ 1 μm if left to freely diffuse, and so transporters bind their substrates tightly to confine the substrate during the cycle. Thus the barrier to unbinding is sufficiently high such that equilibrium MD may not capture and unbinding event. Similarly, the high entropy of a single substrate in solution can render binding too slow of a process to capture with equilibrium MD. By applying forces in addition those of the force field, the membrane binding and unbinding process can be examined in detail.

The simplest of these approaches is steered molecular dynamics (SMD; Fig. 3B), where a force is applied to induce a change within the simulation, in this case to force the binding or unbinding of the substrate. SMD comes in two major flavors, a constant force mode that was originally implemented, and a constant velocity mode analogous to that adopted in atomic force microscopy (AFM) which records the force required over a chosen pathway.

In order to be consistent with an AFM setup, the pulling reaction coordinate should be in a constant direction, although the freedom of working in silico permits other reaction coordinates to be explored as well that might better reflect the unbinding or binding reaction coordinate, such as coordination number or the distance to a binding site (Fiorin et al., 2013; Tribello et al., 2014). In either context, the choice of a proper force constant determines the accuracy of the SMD simulation. The force constant must be high enough so that the free energy barrier for the substrate transport in the membrane transporter is overcome, but ideally not so high that the measurement is far from equilibrium (Isralewitz, Baudry, Gullingsrud, Kosztin, & Schulten, 2001). A common rule of thumb is to make sure that the thermal fluctuations as a result of pulling lies in the order of 0.5 Å (Isralewitz et al., 2001).

For SMD, it is critical to think statistically. A single SMD pull has limited significance in elucidating the binding and unbinding pathway. The nonequilibrium work from repeated pulls places an upper bound on the free energy difference of the process at hand (Jarzynski, 1997), thereby providing a method by which different candidate pathways can be evaluated. For as far as is possible, these pulls should start from different starting configurations, such as by taking different timepoints from a trajectory as the starting point for individual pulls.

SMD does not overcome the fundamental problem of equilibrium simulation, however, in that high energy states are still only rarely sampled. To sample these rarely visited states of a binding process, an additional external potential can be added to constrain the reaction coordinate to force sampling of a small region of reaction coordinate space (Kästner, 2011), typically taken along a SMD trajectory to seed the initial positions (Fig. 3B). Since the added potential is known, the underlying free energy profile near this highly sampled region can be deduced through self-consistently solving for the unweighted free energies given the weighted population distribution in each local environment. By repeating this procedure for many adjacent regions of reaction coordinate space, the total free energy profile can be estimated in a self-consistent manner (Kumar, Bouzida, Swendsen, Kollman, & Rosenberg, 1992) via a number of different packages (Chodera, Swope, Pitera, Seok, & Dill, 2007; Grossfield, 2013; Hub, de Groot, & van der Spoel, 2010). Since the individual applied potential looks like an umbrella, and each "umbrella" spans a region of the reaction coordinate, this sampling method is called umbrella sampling (Torrie & Valleau, 1977).

Setting up umbrella sampling calculations to study substrate binding to transporters requires carefully balancing the force constants used to add the external, usually harmonic, potential. If the force constant is too weak, the underlying potential energy surface can overpower the biasing potential, causing specific regions of the free energy landscape to be undersampled that remain too high in energy to be thermally accessible. If the force constant is too strong, each umbrella will span only a small amount of the reaction coordinate. Since the sampled regions must overlap for a complete profile, strong force constants can dramatically increase the computational cost by forcing additional umbrellas to be placed in undersampled gaps. Umbrella placement and strength can be approached algorithmically (Sabri Dashti & Roitberg, 2013), although as a rule of thumb, the spacing between adjacent umbrellas should be such that the potential bias is ~3 kT at the center of the adjacent umbrellas.

A more modern development in the field has been the proliferation of replica exchange umbrella sampling (REUS) studies, based on studies by Sugita, Kitao, and Okamoto (2000), where the biasing potentials are exchanged over the course of the simulation set. The goal of this approach is to overcome a shortcoming of conventional umbrella sampling studies, where the simulations may be exploring conformational space orthogonal to the measured reaction coordinate, thereby changing the free energy profile in a hidden way (Neale, Rodinger, & Pomès, 2008). Replica exchange methods eliminate the effect of these hidden degrees of freedom, and can dramatically accelerate free energy profile convergence (W. Jiang, Luo, Maragliano, & Roux, 2012; Kokubo, Tanaka, & Okamoto, 2013).

Less structured exploration methods have also been employed to study substrate dynamics in membrane transporters, namely metadynamics (Laio & Parrinello, 2002) and the adaptive biasing force (ABF) method (Darve & Pohorille, 2001; Darve, Wilson, & Pohorille, 2002). The basic approach in these biased sampling techniques is to smooth the energy landscape of the system along a predefined reaction coordinate by the addition of biases during the simulation that let the system overcome energetic barriers, with the goal of uniformly sampling the entire reaction coordinate (Darve, Rodríguez-Gómez, & Pohorille, 2008). One advantage of these methods over umbrella sampling is that they can be significantly cheaper computationally, requiring only a single simulation to produce a result. However, the robustness of both methods improves substantially with multiple copies, whose results can be agglomerated into a single free energy profile (Minoukadeh, Chipot, & Leliévre, 2010; Raiteri, Laio, Gervasio, Micheletti, & Parrinello, 2006).

3.5 Alchemical Perturbation Applied to Substrate Binding

There are times when rather than an unbinding or binding process, the relative or absolute binding free energies are the critical observable comparable to experiment (Chipot & Pearlman, 2002; Chipot & Pohorille, 2007). For example, many transporters use ionic gradients to fuel their transition, and so the relative affinity of two similar ions (eg, Na^+ or Li^+) may be the quantity of interest (Thompson et al., 2009). Alchemical method such as free energy perturbation (FEP) (Zwanzig, 1954) and thermodynamic integration (TI) (Kirkwood, 1936) exploit the fact that free energy is a state function (ie, it is path independent) by computing a free energy difference from in silico alchemical transmutation to complete a thermodynamic cycle (Pohorille, Jarzynski, & Chipot, 2010). By carefully choosing the thermodynamic cycle, complicated absolute (Fig. 3C) or relative (Fig. 3D) binding free energies can be computed far more efficiently than via other methods (W. L. Jorgensen & Thomas, 2008; Tembre & Cammon, 1984).

From the perspective of membrane transporter binding calculations, there is very little difference between FEP and TI methods, and the two methods become equivalent in the limit of infinitesimal step sizes (Christ, Mark, & van Gunsteren, 2010; W. L. Jorgensen & Thomas, 2008). They differ in their formalism of how the free energy difference is computed, but generally speaking both operate by slowly decoupling one set of atoms from the calculation while coupling an originally decoupled set of atoms. The progress of this alchemical process is characterized by the scaling parameter λ, which varies between 0 and 1 to represent the initial and final states (Beveridge & DiCapua, 1989). This setup makes the free energy change a continuous function of λ between the initial and final states (Frenkel & Smit, 2002). The intermediate values of λ must be chosen for both methods and are frequently not equally spaced between the end points, as the largest free energy changes occur when a particle is being grown into or out of existence, where the nonbonded interactions may introduce large values as the weakly-coupled atoms overlaps with other atoms in the system (Goette & Grubmüller, 2009). These "end-point catastrophes" (Lu, Kofke, & Woolf, 2004) are generally handled by altering the nonbonded interactions when λ is near extrema; however, it is still a good idea to sample more in these regions (W. L. Jorgensen & Thomas, 2008).

The principle differences between FEP and TI comes in analysis. FEP takes discrete steps, and so one basic estimate of the error comes from taking the steps forward (from $\lambda = 0$ to $\lambda = 1$) and backward (from $\lambda = 1$ to $\lambda = 0$) and evaluating the hysteresis (W. Jorgensen & Ravimohan, 1985).

This approach is implemented with the ParseFEP tool in VMD to analyze NAMD simulations (Liu, Dehez, Cai, & Chipot, 2012), and other MD packages have other tools such as `g_bar` in GROMACS (Pronk et al., 2013) to evaluate the output of alchemical simulations.

4. EMERGING TECHNIQUES TO SIMULATE LARGE-SCALE STRUCTURAL TRANSITIONS IN MEMBRANE TRANSPORTERS

Large-scale structural transitions are perhaps the defining functional features of membrane transporters, enabling them to regulate the accessibility of their binding site(s) to one side of the membrane at a time and allowing a chemical energy source to be used to drive the uphill motion of substrates. Unfortunately, while conventional MD simulation is well suited to work out the details of the interactions that might drive these processes, the millisecond or (often) longer time scales required for large-scale conformational changes rule out unbiased equilibrium MD simulation as a technique for studying these changes. We present here a recently developed procedure (Moradi et al., 2015; Moradi & Tajkhorshid, 2013, 2014) involving an array of advanced non-equilibrium MD techniques that can be used to study these conformational changes. In essence, this procedure simplifies to the following set of steps:

1. Obtain atomistic models of the states.
2. Choose collective variables and biasing protocols to drive the system from one state to another.
3. Relax the transition pathway with a refinement technique.
4. Sample along the pathway to obtain a free energy profile.
5. Analyze the profile quality and return to Step 2 if need be.

4.1 Defining Target End States

Before any kind of transition can be investigated, the end states of that transition must be defined. For a few transporters (eg, a glutamate transporter homolog called Glt_{Ph} (Reyes, Ginter, & Boudker, 2009; Yernool, Boudker, Jin, & Gouaux, 2004) and a bacterial ABC exporter called MsbA (Ward, Reyes, Yu, Roth, & Chang, 2007)), X-ray crystal structures have captured the transporter in multiple states in their transport cycles. In these select few cases, two crystal structures can be used as the end states, and the conformational transition between the structures can be investigated. However, for most transporters, X-ray crystal structures are only available for a single state, so the structure of a second state must be modeled before the large-scale structural changes can be investigated. There are currently two main approaches available to model unknown states in transporters: homology modeling and repeat-swap modeling.

Homology modeling is a general tool in which an unknown structure for a protein is generated using a known structure of a homologous protein as a template (Martí-Renom et al., 2000). Homology modeling can be performed by many programs, and MODELLER (Webb & Sali, 2014) and SWISS-MODEL (Biasini et al., 2014) are popular examples. While the specific techniques used by different software packages differ, the basic procedure they use is the same (Martí-Renom et al., 2000). First, a sequence alignment is used to identify corresponding portions of the two proteins. The local conformations of these portions in the protein with unknown structure are then made to match the conformations in the template, which causes the global conformation of the entire protein to approximately match that of the template. Finally, the resulting model is refined to increase its quality (eg, by eliminating steric clashes between residues).

Recently, it has been found that many transporters are composed of pairs of pseudosymmetric structural elements called inverted repeats (Crisman, Qu, Kanner, & Forrest, 2009; Forrest et al., 2008; Vergara-Jaque, Fenollar-Ferrer, Kaufmann, & Forrest, 2015). When inverted with respect to the plane of the membrane, the topology of one repeat is (nearly) identical to the topology of the other (Fig. 8A). While their topologies are (nearly) identical, the conformations of the repeats are quite different, and it has been shown that swapping the conformations of the repeats will swap the overall conformation of the transporter from an IF to OF state or *vice versa* (Fig. 8B)

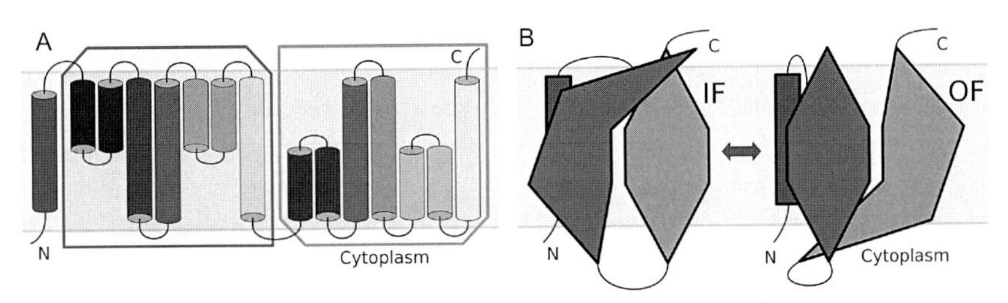

Fig. 8 Schematic of the inverted repeats of a transporter. (A) The topologies of the inverted repeats, boxed in blue (gray in the print version) and orange (light gray in the print version), are identical when inverted with respect to the plane of the membrane. The color (different gray shades in the print version) lightness of the helices can be used to identify corresponding helices in the two repeats. (B) In any given state, the conformations of the repeats differ. When the IF conformations of the repeats are swapped, they form the OF state and *vice versa*.

(Crisman et al., 2009; Forrest et al., 2008; Vergara-Jaque et al., 2015). The sequence similarity between the individual repeats is usually fairly low, but it is high enough that homology modeling techniques can be used to swap the conformations of the repeats (Crisman et al., 2009; Forrest et al., 2008; Vergara-Jaque et al., 2015). Once the conformations of the repeats have been swapped, the repeats can be fit back together, and the resulting global conformation of the model can be empirically refined to generate a model of a missing state in a transport cycle (Crisman et al., 2009; Forrest et al., 2008; Vergara-Jaque et al., 2015).

Before investigating the transition between a known and a modeled state, rigorous investigation of the quality of the modeled state is essential. If the modeled state does not correspond sufficiently well to a physiological state visited by the transporter, the subsequent steps involved in this procedure will yield results that are of limited value at an extremely high computational cost. Standard modeling techniques attempt to maximize the internal quality of the model but cannot take into account the transporter's environment. The stability of any model in the context of a membrane should be verified through equilibrium simulations (Moradi et al., 2015). If the quality of the model is not sufficient, it is common for the transporter to revert from the modeled state back to the known crystallographic state within a few dozen nanoseconds of simulation.

4.2 Generating a Biasing Protocol to Induce Transporter Transitions

One common way of inducing a structural transition is through a targeted MD (TMD) simulation, which makes use of the root-mean-square displacement (RMSD) of a selection of atoms in the transporter (Fiorin et al., 2013; Schlitter, Engels, Krüger, Jacoby, & Wollmer, 1993). RMSD is a common measure of differences in protein conformations (the lower the RMSD, the more similar the conformations) and is calculated using the equation $\sqrt{\frac{1}{N}\Sigma_j \| \mathbf{x}_j - \mathbf{x}_j' \|^2}$, where N represents the number of atoms in the selection, $\mathbf{x}_j$ represents the 3D coordinates of atom j in one conformation, and $\mathbf{x}_j'$ represents the 3D coordinates of the atom in a reference conformation. When an RMSD calculation is performed for large-scale conformational differences, only the protein's heavy or C_α backbone atoms are selected, as side chain rotations and the movement of lighter atoms are small-scale conformational changes that may obscure the progress of the global transition. In a TMD simulation, the starting conformation is driven to the final (target) conformation by inducing a linear reduction of the RMSD, measured with reference to the target conformation, from its initial value to zero (Fiorin et al., 2013; Schlitter et al., 1993). This is done by applying forces to the selected atoms, and the forces are calculated using the gradient of a harmonic energy potential of the form $U = \frac{k}{2}(\zeta - \zeta')^2$ (Fiorin et al., 2013; Schlitter et al., 1993). Here, k is a user-specified force constant, ζ represents the RMSD of the transporter's conformation and ζ' represents a target RMSD value that changes throughout the simulation.

The TMD approach of investigating transporter transitions is popular due to its simplicity, but its use commonly results in transitions that approximately correspond to linear interpolations between the end states. Unfortunately, such transitions are often inconsistent with the alternating-access mechanism (ie, substrate is accessible from both sides of the membrane at the same time), which makes them nonphysiological and therefore of limited scientific value (Moradi & Tajkhorshid, 2013, 2014). To avoid this problem, it is better to induce changes in ways specific to the transporter under investigation. By investigating the large-scale differences in conformation between the two end states and any experimental clues about the structural nature of the transporter's mechanism, one can qualitatively predict the kinds of motion that must be induced in the transporter to transition it from one state to the next (Moradi et al., 2015; Moradi & Tajkhorshid, 2013, 2014). To induce these motions in a simulation, forces derived from harmonic potentials involving so-called "collective variables" (ie, reaction coordinates) can be applied to the atoms in the transporter in the same way as is done with RMSD in TMD simulations (Fiorin et al., 2013). Generally, a collective variable is a quantity that is calculated from the positions of the atoms and is used to represent a specific conformational element in the transporter (Moradi et al., 2015; Moradi & Tajkhorshid, 2013, 2014). Besides RMSD, examples of collective variables include distances between centers of mass of two atom selections within the transporter along a user-defined axis, which can be useful for translating the selections with respect to one another, and the orientation quaternion (Coutsias, Seok, & Dill, 2004; Fiorin et al., 2013; Horn, 1987), which quantifies the differences in orientation between two states (Fig. 9A).

By modeling each of the structural differences between the end states with a separate collective variable, one can represent these structural differences quantitatively and induce the predicted motions associated with the transporter's mechanism using a biasing protocol. Nonequilibrium driven

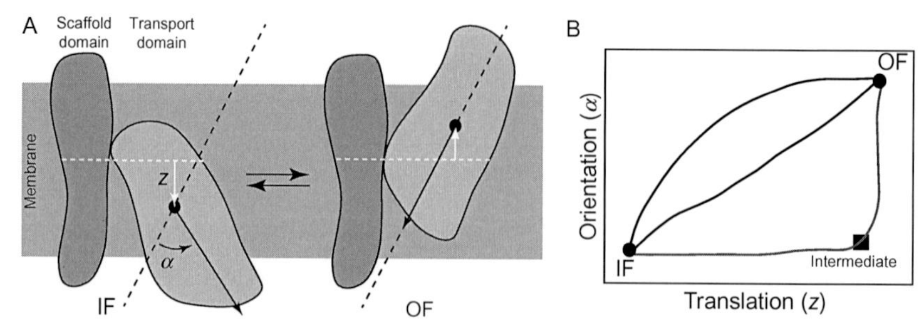

Fig. 9 Two important considerations in the design of a biasing protocol. (A) The choice of collective variables. The differences between an exemplary IF and OF state are decomposed into two collective variables. One, z, represents the position of the center of mass of the transport domain relative to the center of mass of the scaffold domain along the membrane normal axis. The other, α, represents the change in orientation of the transport domain. (B) The choice of the transition pathway. Three possible transition pathways are shown in the collective variable space. If a crystal structure or other experimental evidence identified an intermediate state, two of the transition pathways could immediately be eliminated from consideration.

MD simulations are defined by the biasing protocols they use, which include the set of collective variables utilized to induce transitions as well as the parameters associated with the applied harmonic potentials. Transporters' large-scale structural transitions are complex, so it is necessary to test a variety of biasing protocols to refine initial mechanistic predictions and to find good candidates for transition pathways through the collective variable space (Fig. 9B). Three criteria are used to judge the quality of a biasing protocol and the associated transition pathway (Moradi et al., 2015; Moradi & Tajkhorshid, 2013, 2014). First, it is necessary to check whether the desired conformational changes were actually induced by the biasing protocol by ensuring that the collective variables reach their final target values during the simulation. Second, the functional relevance of the pathway should be verified. For example, the pathway must be consistent with the alternating-access mechanism, and unphysical deformations of the membrane (which tend to arise as a result of inducing large-scale conformational changes in unrealistically short time scales) should be minimized. Finally, the nonequilibrium work required to induce the transition (defined as $W(t) = \int_0^t \frac{\partial U(t')}{\partial t'} dt'$) should be minimized. A pathway that reproducibly requires less work corresponds to a pathway with a lower free energy barrier between the end states. The lower the free energy barrier, the easier and more likely it is for the transporter to use the associated pathway in reality.

Several refinements to the initial biasing protocol can and should be made when investigating a transition pathway. For example, the initial choice of collective variables may not adequately represent the conformational elements of the transporter, in which case different collective variables will need to be tested. The ideal force constants k associated with the harmonic potentials must also be determined empirically for each collective variable (Moradi & Tajkhorshid, 2013, 2014). Force constants should be high enough that the desired conformational changes are induced but low enough that the transporter has some freedom to avoid high energy barriers (eg, steric clashes) and keep the work low. With the choice of collective variables and force constants established, changes in the collective variables should be induced in different orders and at different rates, and transitions that explore a variety of areas in the collective variable space should be tested (Moradi & Tajkhorshid, 2013, 2014). The relative quality of different pathways is never known *a priori*, so it is necessary to be as comprehensive as possible in this exploratory phase. It must be stated that the extreme complexity involved in this problem necessarily prohibits one from conducting a systematic search to find an ideal biasing protocol and transition pathway. There are an infinite number of possible biasing protocols that use different combinations of collective variables and induce changes in them in unique ways. Intuition and knowledge of the transporter are the best guides for empirically designing a finite number of biasing protocols that have the best chance of capturing an acceptable transition pathway (Moradi & Tajkhorshid, 2013, 2014).

Due to the number of different biasing protocols that should be tested (eg, ~200 in previous transporter studies (Moradi & Tajkhorshid, 2013, 2014)), it is advisable that preliminary simulations be made quite short (~ 5 ns). Progressively fewer candidate transition pathways can be extended to longer and longer simulation time scales as the best pathways are identified at shorter time scales. Also, since these simulations are stochastic, the best candidate biasing protocols should be repeated several times to ensure their best qualities (especially low work) are reproducible (Moradi & Tajkhorshid, 2013, 2014). Before proceeding to the following steps involved in characterizing a transition pathway, it is advisable to reduce the work to less than ~100 kcal·mol^{-1}. This is desirable because reducing the work required by the transition in this step greatly reduces the computational cost of subsequent steps. Once a satisfactory biasing protocol and transition pathway have been found, the pathway can be refined using a more advanced computational technique.

4.3 Refining the Transition Pathway

Refinement of the transition pathway found in the previous step can be accomplished by applying a path refinement technique like the string method with swarms of trajectories (SMwST). SMwST is a method that relaxes the initial pathway to the nearest minimum free energy pathway through the following steps (Gan, Yang, & Roux, 2009; Pan, Sezer, & Roux, 2008). First, a set of images (ie, conformations) are obtained at evenly spaced intervals along the initial pathway. Distances along the pathway are calculated using a metric (eg, weighted Euclidean distance) based on the collective variables used in the driven transition. Multiple independent copies of the images (swarm of trajectories) are then harmonically restrained near their initial collective variable values before being released into short equilibrium simulations. The average drift observed during the equilibrium simulations is used to calculate a new image center for each swarm of trajectories, and the positions of the image centers are refined such that they remain evenly spaced along the new relaxed pathway. Iterations of this restrain-release cycle are performed until the pathway has converged and no longer changes (Fig. 10). A convenient method of evaluating the convergence of SMwST is through the use of the Fréchet distance, which is able to quantify the similarity between two curves (W. Jiang et al., 2014; Seyler, Kumar, Thorpe, & Beckstein, 2015).

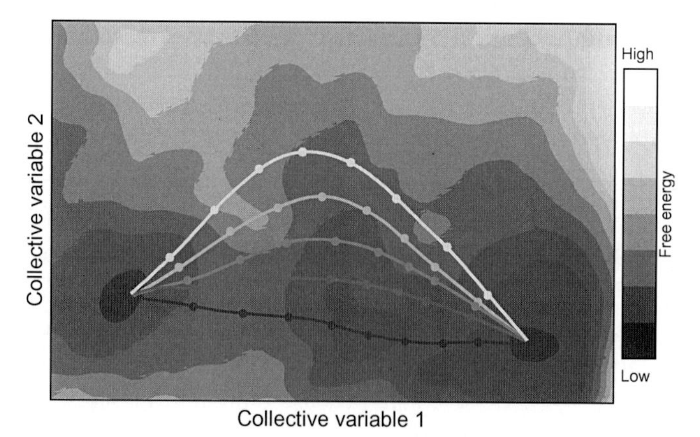

Fig. 10 Schematic representation of SMwST. A series of paths, iteratively refined by SMwST, connecting two minima in a schematic 2D free energy surface are shown. The color gradient represents different iterations, while the red (dark gray in the print version) line represents the converged lowest free energy path.

In a recent study on the large-scale conformational changes in a transporter called GlpT (glycerol-3-phosphate:phosphate antiporter), 50 images along the initial pathway were used, and 20 copies per image were simulated (Moradi et al., 2015). In each iteration, the copies were restrained for 5 ps and released for 5 ps, and ~20 iterations were required for convergence (Moradi et al., 2015). Such a small number of iterations was required due to the high quality (low free energy) of the initial pathway. Because SMwST is so computationally expensive, careful refinement of the initial transition pathway by evaluating a variety of different biasing protocols should be performed before using SMwST. SMwST is also quite sensitive to the collective variables used, so great care should be taken to ensure that the collective variables chosen in the previous step adequately capture the transporter's motion (Moradi & Tajkhorshid, 2014). Finally, it is worth noting that a convenient feature of SMwST is that it can be implemented either serially or in parallel (Moradi et al., 2015). That is, each transporter copy associated with each window can either be simulated one after the other on a small computing cluster or at the same time on a massive supercomputer.

4.4 Obtaining a Free Energy Profile

As described previously, US is a technique that is used to obtain free energy profiles and other thermodynamic information through enhanced conformational sampling along a reaction coordinate. By using the refined transition pathway found previously as the reaction coordinate, US can be applied not only to binding processes but also to the structural changes involved in transport cycles. To obtain an accurate free energy profile along such a transition pathway, it is essential to use REUS since there are many orthogonal degrees of freedom in the transporter that are not directly sampled. When REUS is applied to large-scale conformational changes in transporters, it is more commonly called bias-exchange umbrella sampling (BEUS) (Moradi et al., 2015; Moradi & Tajkhorshid, 2013, 2014). Just as for a conventional US calculation, the number of umbrellas and force constants applied during a BEUS simulation are important parameters, and empirical or systematic efforts to optimize them must be made. Additionally, all umbrellas should have roughly equal exchange rates with their neighbors in a BEUS simulation to ensure convergence of the free energy profile (Moradi et al., 2015; Moradi & Tajkhorshid, 2013, 2014). Once the BEUS simulation has been run, the free energy profile along the pathway can be reconstructed using the weighted histogram analysis method (WHAM)

(Kumar et al., 1992), a method which is considered standard, or other techniques (Shirts & Chodera, 2008; H. Wu, Mey, Rosta, & Noé, 2014). If multiple transition pathways within a transport cycle are investigated, it is possible to structurally elucidate and thermodynamically characterize the entire transport cycle for a transporter. Using this approach, an engineered cycle for GlpT, composed of three natural transition pathways from the transport cycle and one that is considered forbidden, has recently been characterized. (Moradi et al., 2015). In Fig. 11, a schematic of the engineered cycle is shown alongside the free energy profile obtained for the entire cycle. Importantly, the location and height of major barriers in this profile are quantitatively available, and it is clear that the structural transition induced along the natural pathways is significantly more energetically favorable than the one induced along the forbidden pathway.

Before the free energy profile resulting from a BEUS simulation can be considered accurate, the BEUS data must first be analyzed for poor sampling, which would make the free energy profile unreliable. If poor sampling is observed along any of the collective variables used, a better biasing protocol must be employed (Moradi & Tajkhorshid, 2014). Evaluation of sampling is fairly straightforward, as one simply needs to create a histogram, using a collective variable as the independent variable for the plot, from the

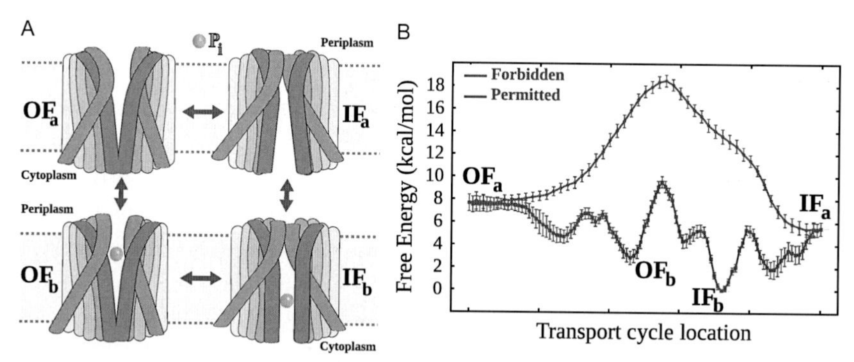

Fig. 11 Thermodynamic characterization of the transport mechanism of GlpT. (A) Schematic representation of the engineered cycle with gating helices highlighted in pink (gray in the print version) and blue (dark gray in the print version). The substrate is inorganic phosphate (P_i), and the cycle consists of transitions between four states: the apo OF (OF_a) and IF (IF_a) states and the bound OF (OF_b), and IF (IF_b) states. (B) Free energy profile along the entire cycle with the important states denoted explicitly. *Adapted from Moradi, M., Enkavi, G., & Tajkhorshid, E. (2015). Atomic-level characterization of transport cycle thermodynamics in the glycerol-3-phosphate:phosphate transporter.* Nature Communications, 6, *8393.*

conformations visited in the BEUS simulation and verify that it is uniform over the range of collective variable values between the end states (Moradi & Tajkhorshid, 2014). Whether or not poor sampling along one of the collective variables is observed, it is also useful to identify the important degrees of freedom present in the BEUS simulation. One way this can be done is by using principal component analysis (PCA) (Amadei, Linnsen, & Berendsen, 1993), which is a method used to determine the orthogonal combinations of degrees of freedom that show the greatest variation within a data set, on the conformations generated in the BEUS simulation. If n collective variables were used in a biasing protocol, the first n principal components (PCs) should correspond roughly to these collective variables (Moradi & Tajkhorshid, 2014). If more than n significant PCs are found, the result indicates that there are combinations of degrees of freedom that must be accounted for to ensure reliable sampling (Moradi & Tajkhorshid, 2014). To account for these degrees of freedom, collective variables must be added to a revised biasing protocol, the steps required to obtain a new free energy profile must be completed, and PCA must be performed once more.

In addition to BEUS simulations along a refined pathway, which are inherently one dimensional, it is also possible to perform multidimensional BEUS simulations to obtain informative free energy landscapes. While more general configurations are possible, these landscapes are most commonly generated by performing two dimensional (2D) BEUS simulations on a rectangular grid (Moradi & Tajkhorshid, 2014). Due to the computational cost associated with such simulations, only approximate free energy landscapes can be generated (Moradi & Tajkhorshid, 2014). Before such a simulation can be performed, the collective variables that represent the transporter's important structural elements must be well established, but an extensive exploration of the collective variable space does not need to have been performed. A 2D BEUS simulation can either be used to guide pathway exploration, or it can be used after exploration is finished to generate a more complete understanding of the thermodynamics that govern the transporter's transition. To set up a 2D BEUS simulation, conformations corresponding to the points in the BEUS grid must be generated, and the grid points must be spaced following the same considerations as for 1D BEUS simulations. If large areas of the collective variable space have been explored during pathway exploration, these grid conformations can be generated by driving nearby conformations from previous simulations to the grid points. If such data is not available, conformations representing the grid points can also be generated systematically using driven simulations.

After verifying the quality of the sampling and reconstructing the free energy landscape, one can then identify the minimum free energy path connecting any two points in the landscape using a variety of path-finding algorithms (Moradi et al., 2015), and this path can be further refined using SMwST.

5. THE DIRECTION OF FUTURE MEMBRANE PROTEIN STUDIES

The membrane protein field is progressing rapidly, currently experiencing exponential growth in the number of available 3D structures of distinct membrane proteins (White, 2009). With the growing success of cryo-EM in determining membrane protein structures to high resolution (Vinothkumar, 2015), this trend is expected to continue for the foreseeable future. Similarly, the methods of the modeling community will not remain static. While much of the groundwork for current simulation techniques was established in the 1990s, the continued advancement of computing power granted by Moore's Law, replica exchange and other enhanced sampling techniques have become much more common and have become far more routine to a degree not foreseen even a few years ago.

The wealth of available structural information is also pushing simulation in new directions. For instance, crowding on the membrane surface is generally currently neglected, despite suggestions that interplay between different membrane actors is critical to membrane protein function in general (Guigas & Weiss, 2015; R. Phillips, Ursell, Wiggins, & Sens, 2009) and to understand cellular membrane topology (Stachowiak et al., 2012). This will require a substantial increase in the simulation scope in the years to come, as the community moves from studying individual transporters to collective action by the membrane-embedded community of proteins.

ACKNOWLEDGMENTS

This work was supported in part by the National Institutes of Health (Grants R01-GM086749, R01-GM101048, U54-GM087519, and P41-GM104601 to E.T.). We also acknowledge computing resources provided by Blue Waters, INCITE, XSEDE (grant TG-MCA06N060 to E.T.), and PSC Anton, which were instrumental for some of the simulations results presented and discussed in this article. J.V.V. acknowledges support from the Sandia National Laboratories Campus Executive Program, which is funded by the Laboratory Directed Research and Development (LDRD) Program. Sandia is a multiprogram laboratory managed and operated by Sandia Corporation, a wholly owned subsidiary of Lockheed Martin Corporation, for the US Department of Energy's National Nuclear Security Administration under Contract No. DE-AC04-94AL85000. N.T. acknowledges support by the National Science Foundation Graduate Research Fellowship Program under Grant No. 1144245.

REFERENCES

Aduri, R., Psciuk, B. T., Saro, P., Taniga, H., Schlegel, H. B., & SantaLucia, J. (2007). AMBER force field parameters for the naturally occurring modified nucleosides in RNA. *Journal of Chemical Theory and Computation*, *3*, 1464–1475.

Ajao, C., Andersson, M. A., Teplova, V. V., Nagy, S., Gahmberg, C. G., Andersson, L. C., … Salkinoja-Salonen, M. (2015). Mitochondrial toxicity of triclosan on mammalian cells. *Toxicology Reports*, *2*, 624–637.

Aliev, A. E., Kulke, M., Khaneja, H. S., Chudasama, V., Sheppard, T. D., & Lanigan, R. M. (2014). Motional timescale predictions by molecular dynamics simulations: Case study using proline and hydroxyproline sidechain dynamics. *Proteins*, *82*(2), 195–215.

Aller, S. G., Yu, J., Ward, A., Weng, Y., Chittaboina, S., Zhuo, R., … Chang, G. (2009). Structure of P-glycoprotein reveals a molecular basis for poly-specific drug binding. *Science*, *323*(5922), 1718–1722.

Amadei, A., Linnsen, A. B. M., & Berendsen, H. J. C. (1993). Essential dynamics of proteins. *PROTEINS: Structure, Function, and Genetics*, *17*, 412–425.

Amaro, R. E., Baron, R., & McCammon, J. A. (2008). An improved relaxed complex scheme for receptor flexibility in computer-aided drug design. *Journal of Computer-Aided Molecular Design*, *22*(9), 693–705.

Andersson, M., Bondar, A. N., Freites, J. A., Tobias, D. J., Kaback, H. R., & White, S. H. (2012). Proton-coupled dynamics in lactose permease. *Structure*, *20*(11), 1893–1904.

Arcario, M. J., Mayne, C. G., & Tajkhorshid, E. (2014). Atomistic models of general anesthetics for use in *in silico* biological studies. *Journal of Physical Chemistry B*, *118*, 12075–12086.

Barends, T. R. M., Foucar, L., Botha, S., Doak, R. B., Shoeman, R. L., Nass, K., … Schlichting, I. (2014). De novo protein crystal structure determination from X-ray free-electron laser data. *Nature*, *505*(7482), 244–247.

Baron, R., McCammon, J. A., & Mattevi, A. (2009). The oxygen-binding vs. oxygen-consuming paradigm in biocatalysis: Structural biology and biomolecular simulation. *Current Opinion in Structural Biology*, *19*, 672–679.

Baylon, J. L., Vermaas, J. V., Muller, M. P., Arcario, M. J., Pogorelov, T. V., & Tajkhorshid, E. (2016). Accelerating membrane dynamics to study protein–lipid interactions at an atomic level. *Biochimica et Biophysica Acta – Biomembranes*, *1858*(7), 1573–1583.

Berman, H., Henrick, K., & Nakamura, H. (2003). Announcing the worldwide Protein Data Bank. *Nature Structural Biology*, *10*(12), 980.

Berman, H. M., Westbrook, J., Feng, Z., Gilliland, G., Bhat, T. N., Weissig, H., … Bourne, P. E. (2000). The protein data bank. *Nucleic Acids Research*, *28*, 235–242.

Best, R. B., Zhu, X., Shim, J., Lopes, P. E. M., Mittal, J., Feig, M., & MacKerell, A. D. (2012). Optimization of the additive charmm all-atom protein force field targeting improved sampling of the backbone ϕ, ψ and side-chain χ_1 and χ_2 dihedral angles. *Journal of Chemical Theory and Computation*, *8*(9), 3257–3273.

Betz, R. M., & Walker, R. C. (2015). Paramfit: Automate optimizatino of force field parameters for molecular dyanmics simulations. *Journal of Computational Chemistry*, *36*, 79–87.

Beveridge, D. L., & DiCapua, F. M. (1989). Free energy via molecular simulation: Applications to chemical and biological systems. *Annual Review of Biophysics and Biophysical Chemistry*, *18*, 431–492.

Biasini, M., Bienert, S., Waterhouse, A., Arnold, K., Studer, G., Schmidt, T., … Schwede, T. (2014). SWISS-MODEL: Modeling protein tertiary and quaternary structure using evolutionary information. *Nucleic Acids Research*, *42*(W1), W252–W258.

Blakely, R. D., & Edwards, R. H. (2012). Vesicular and plasma membrane transporters for neurotransmitters. *Cold Spring Harbor Perspectives in Biology*, *4*(2).

Bovigny, C., Tamò, G., Lemmin, T., Maïno, N., & Dal Peraro, M. (2015). LipidBuilder: A framework to build realistic models for biological membranes. *Journal of Chemical Information and Modeling, 55*(12), 2491–2499.

Brannigan, G., LeBard, D. N., Hénin, J., Eckenhoff, R. G., & Klein, M. L. (2010). Multiple binding sites for the general anesthetic isoflurane identified in the nicotinic acetylcholine receptor transmembrane domain. *Proceedings of the National Academy of Sciences, USA, 107*, 14122–14127.

Brooks, B. R., Brooks, C. L., Mackerell, A. D., Nilsson, L., Petrella, R. J., Roux, B., … Karplus, M. (2009). Charmm: The biomolecular simulation program. *Journal of Computational Chemistry, 30*, 1545–1614.

Brünger, A. T., & Brooks III, C. L..Karplus, M. (1984). Stochastic boundary conditions for molecular dynamics simulations of ST2 water. *Chemical Physics Letters, 105*(5), 495–498.

Buch, I., Giorgino, T., & De Fabritiis, G. (2011). Complete reconstruction of an enzyme-inhibitor binding process by molecular dynamics simulations. *Proceedings of the National Academy of Sciences, USA, 108*(25), 10184–10189.

Camilloni, C., & Vendruscolo, M. (2015). Using pseudocontact shifts and residual dipolar couplings as exact NMR restraints for the determination of protein structural ensembles. *Biochemistry, 54*(51), 7470–7476.

Campbell, A. J., Lamb, M. L., & Joseph-McCarthy, D. (2014). Ensemble-based docking using biased molecular dynamics. *Journal of Chemical Information and Modeling, 54*(7), 2127–2138.

Chaudhury, S., Lyskov, S., & Gray, J. J. (2010). PyRosetta: A script-based interface for implementing molecular modeling algorithms using rosetta. *Bioinformatics, 26*(5), 689–691.

Chen, V. B., Bryan Arendall, III, W., Headd, J. J., Keedy, D. A., Immormino, R. M., Kapral, G. J., … Richardson, D. C. (2010). MolProbity: All-atom structure validation for macromolecular crystallography. *Acta Crystallographica D, 66*, 12–21.

Chen, Y.-J., Pornillos, O., Lieu, S., Ma, C., Chen, A. P., & Chang, G. (2007). X-ray structure of EmrE supports dual topology model. *Proceedings of the National Academy of Sciences, USA, 104*, 18999–19004.

Cheng, M. H., & Bahar, I. (2014). Complete mapping of substrate translocation highlights the role of LeuT N-terminal segment in regulating transport cycle. *PLoS Comput Biol, 10*(10), e1003879.

Cheng, M. H., & Bahar, I. (2015). Molecular mechanism of dopamine transport by human dopamine transporter. *Structure, 23*(11), 2171–2181.

Chipot, C., & Pearlman, D. A. (2002). Free energy calculations. The long and winding gilded road. *Molecular Simulation, 28*, 1–12.

Chipot, C., & Pohorille, A. (2007). *Free energy calculations: Theory and applications in chemistry and biology*. Berlin: Springer.

Chodera, J. D., Swope, W. C., Pitera, J. W., Seok, C., & Dill, K. A. (2007). Use of the weighted histogram analysis method for the analysis of simulated and parallel tempering simulations. *Journal of Chemical Theory and Computation, 3*(1), 26–41.

Choe, S., Rosenberg, J. M., Abramson, J., Wright, E. M., & Grabe, M. (2010). Water permeation through the sodium-dependent galactose cotransporter vSGLT. *Biophysical Journal, 99*(7), L56–L58.

Christ, C. D., Mark, A. E., & van Gunsteren, W. F. (2010). Basic ingredients of free energy calculations: A review. *Journal of Computational Chemistry, 31*(8), 1569–1582.

Clark, R. D., Strizhev, A., Leonard, J. M., Blake, J. F., & Matthew, J. B. (2002). Consensus scoring for ligand/protein interactions. *Journal of Molecular Graphics and Modelling, 20*(4), 281–295.

Cohen, J., Arkhipov, A., Braun, R., & Schulten, K. (2006). Imaging the migration pathways for O_2, CO, NO, and Xe inside myoglobin. *Biophysical Journal, 91*, 1844–1857.

Coppock, P. S., & Kindt, J. T. (2010). Determination of phase transition temperatures for atomistic models of lipids from temperature-dependent stripe domain growth kinetics. *Journal of Physical Chemistry B, 114*(35), 11468–11473.

Cornell, W. D., Cieplak, P., Bayly, C. I., Gould, I. R., Merz, K. M., Jr., Ferguson, D. M., ... Kollman, P. A. (1995). A second generation force field for the simulation of proteins, nucleic acids, and organic molecules. *Journal of the American Chemical Society, 117*, 5179–5197.

Coutsias, E. A., Seok, C., & Dill, K. A. (2004). Using quaternions to calculate RMSD. *Journal of Chemical Physics, 25*(15), 1849–1857.

Crisman, T. J., Qu, S., Kanner, B. I., & Forrest, L. R. (2009). Inward-facing conformation of glutamate transporters as revealed by their inverted-topology structural repeats. *Proceedings of the National Academy of Sciences, USA, 106*, 20752–20757.

Croll, T. I. (2015). The rate of *cis–trans* conformation errors is increasing in low-resolution crystal structures. *Acta Crystallographica D, 71*(3), 706–709.

Cross, J. B., Thompson, D. C., Rai, B. K., Baber, J. C., Fan, K. Y., Hu, Y., & Humblet, C. (2009). Comparison of several molecular docking programs: Pose prediction and virtual screening accuracy. *Journal of Chemical Information and Modeling, 49*(6), 1455–1474.

Damm, W., Frontera, A., Tirado-Rives, J., & Jorgensen, W. L. (1997). OPLS all-atom force field for carbohydrates. *Journal of Computational Chemistry, 18*, 1955–1970.

Darve, E., & Pohorille, A. (2001). Calculating free energies using average force. *Journal of Chemical Physics, 115*, 9169–9183.

Darve, E., Rodríguez-Gómez, D., & Pohorille, A. (2008). Adaptive biasing force method for scalar and vector free energy calculations. *Journal of Chemical Physics, 128*, 144120.

Darve, E., Wilson, D., & Pohorille, A. (2002). Calculating free energies using a scaled-force molecular dynamics algorithm. *Molecular Simulation, 28*, 113–144.

Das, R., & Baker, D. (2008). Macromolecular modeling with rosetta. *Annual Review of Biochemistry, 77*, 363–382.

DeGorter, M., Xia, C., Yang, J., & Kim, R. (2012). Drug transporters in drug efficacy and toxicity. *Annual Review of Pharmacology and Toxicology, 52*(1), 249–273.

Dehez, F., Pebay-Peyroula, E., & Chipot, C. (2008). Binding of ADP in the mitochondrial ADP/ATP carrier is driven by an electrostatic funnel. *Journal of the American Chemical Society, 130*, 12725–12733.

Denisov, I. G., Grinkova, Y. V., Baylon, J. L., Tajkhorshid, E., & Sligar, S. G. (2015). Mechanism of drug-drug interactions mediated by human cytochrome P450 CYP3A4 monomer. *Biochemistry, 54*(13), 2227–2239.

Denning, E. J., Priyakumar, U. D., Nilsson, L., & Mackerell, A. D., Jr. (2011). Impact of 2'-hydroxyl smapling on the conformational properties of RNA: Update of the CHARMM all-atom additive force field for RNA. *Journal of Computational Chemistry, 32*, 1929–1943.

de Vries, M. S., & Hobza, P. (2007). Gas-phase spectroscopy of biomolecular building blocks. *Annual Review of Physical Chemistry, 58*(1), 585–612.

Dickson, C. J., Madej, B. D., Skjevik, Å. A., Betz, R. M., Teigen, K., Gould, I. R., & Walker, R. C. (2014). Lipid14: The amber lipid force field. *Journal of Chemical Theory and Computation, 10*(2), 865–879.

Dickson, C. J., Rosso, L., Betz, R. M., Walker, R. C., & Gould, I. R. (2012). GAFFlipid: A general amber force field for the accurate molecular dynamics simulation of phospholipid. *Soft Matter, 8*(37), 9617.

Dolghih, E., Bryant, C., Renslo, A. R., & Jacobson, M. P. (2011). Predicting binding to P-glycoprotein by flexible receptor docking. *PLoS Computational Biology*, 7(6), e1002083.

Dolinsky, T. J., Czodrowski, P., Li, H., Nielsen, J. E., Jensen, J. H., Klebe, G., & Baker, N. A. (2007). PDB2PQR: Expanding and upgrading automated preparation of biomolecular structures for molecular simulations. *Nucleic Acids Research, 35* (Web Server), W522–W525.

Dowhan, W. (1997). Molecular basis for membrane phospholipid diversity: Why are there so many lipids? *Annual Review of Biochemistry, 66*(1), 199–232.

Drew, K. L. M., Baiman, H., Khwaounjoo, P., Yu, B., & Reynisson, J. (2011). Size estimation of chemical space: How big is it? *Journal of Pharmacy and Pharmacology, 64*, 490–495.

Dror, R. O., Dirks, R. M., Grossman, J., Xu, H., & Shaw, D. E. (2012). Biomolecular simulation: A computational microscope for molecular biology. *Annual Review of Biophysics, 41*(1), 429–452.

Ellingson, S. R., Miao, Y., Baudry, J., & Smith, J. C. (2015). Multi-conformer ensemble docking to difficult protein targets. *Journal of Physical Chemistry B, 119*(3), 1026–1034.

Enkavi, G., & Tajkhorshid, E. (2010). Simulation of spontaneous substrate binding revealing the binding pathway and mechanism and initial conformational response of GlpT. *Biochemistry, 49*, 1105–1114.

Evans, D. J., & Holian, B. L. (1985). The nose–hoover thermostat. *Journal of Chemical Physics, 83*(8), 4069.

Fadok, V. A., Voelker, D. R., Campbell, P. A., Cohen, J. J., Bratton, D. L., & Henson, P. M. (1992). Exposure of phosphatidylserine on the surface of apoptotic lymphocytes triggers specific recognition and removal by macrophages. *Journal of Immunology, 148*, 2207–2216.

Faraldo-Gómez, J. D., Smith, G. R., & Sansom, M. S. (2002). Setting up and optimization of membrane protein simulations. *European Biophysics Journal, 31*, 217–227.

Feller, S. E., Yin, D., Pastor, R. W., & MacKerell, A. D., Jr. (1997). Molecular dynamics simulation of unsaturated lipids at low hydration: Parametrization and comparison with diffraction studies. *Biophysical Journal, 73*, 2269–2279.

Feng, X., Zhu, W., Schurig-Briccio, L. A., Lindert, S., Shoen, C., Hitchings, R., … Oldfield, E. (2015). Antiinfectives targeting enzymes and the proton motive force. *Proceedings of the National Academy of Sciences, USA, 112*(51), E7073–E7082.

Fiorin, G., Klein, M. L., & Hénin, J. (2013). Using collective variables to drive molecular dynamics simulations. *Molecular Physics, 111*(22–23), 3345–3362.

Forrest, L. R., Zhang, Y.-W., Jacobs, M. T., Gesmonde, J., Xie, L., Honig, B. H., & Rudnick, G. (2008). Mechanism for alternating access in neurotransmitter transporters. *Proceedings of the National Academy of Sciences, USA, 105*(30), 10338–10343.

Frenkel, D., & Smit, B. (2002). *Understanding molecular simulation from algorithms to applications*. California: Academic Press.

Fu, B., Sahakyan, A. B., Camilloni, C., Tartaglia, G. G., Paci, E., Caflisch, A., … Cavalli, A. (2014). ALMOST: An all atom molecular simulation toolkit for protein structure determination. *Journal of Computational Chemistry, 35*(14).

Gan, W., Yang, S., & Roux, B. (2009). Atomistic view of Src kinase conformational activation using string method with swarms-of-trajectories. *Biophysical Journal, 97*, L8–L10.

Garrido, N. M., Economou, I. G., Queimada, A. J., Jorge, M., & Macedo, E. A. (2012). Prediction of the n-hexane/water and 1-octanol/water partition coefficients for environmentally relevant compounds using molecular simulation. *AIChE Journal, 58*(6), 1929–1938.

Glennon, T. M., & Merz, K. M., Jr. (1997). A carbohydrate force field for AMBER and its application to the study of saccharide to surface adsorption. *Journal of Molecular Structure, 395–396*, 157–171.

Goette, M., & Grubmüller, H. (2009). Accuracy and convergence of free energy differences calculated from nonequilibrium switching processes. *Journal of Computational Chemistry, 30*, 447–456.

Gordon, J. C., Myers, J. B., Folta, T., Shoja, V., Heath, L. S., & Onufriev, A. (2005). H++: A server for estimating pK_as and adding missing hydrogens to macromolecules. *Nucleic Acids Research, 33*, W368–W371.

Gray, E. M., Díaz-Vázquez, G., & Veatch, S. L. (2015). Growth conditions and cell cycle phase modulate phase transition temperatures in RBL-2H3 derived plasma membrane vesicles. *PLoS One*, *10*(9), e0137741.

Grossfield, A. (2013). *WHAM: The weighted histogram analysis method, version 2.0.9*. http://membrane.urmc.rochester.edu/content/wham.

Guigas, G., & Weiss, M. (2015). Effects of protein crowding on membrane systems. *Biochimica et Biophysica Acta—Biomembranes*.

Guvench, O., Hatcher, E., Venable, R. M., Pastor, R. W., & Mackerell, A. D., Jr. (2009). CHARMM additive all-atom force field for glycosidic linkages between hexopyranoses. *Journal of Chemical Theory and Computation*, *5*, 2353–2370.

Guvench, O., & MacKerell, A. D. (2008). Comparison of protein force fields for molecular dynamics simulations. *Methods Mol. Biol.*, *443*, 63–88.

Guvench, O., Mallajosyula, S. S., Raman, E. P., Hatcher, E., Vanommeslaeghe, K., Foster, T. J., … MacKerell, A. D. (2011). CHARMM additive all-atom force field for carbohydrate derivatives and its utility in polysaccharide and carbohydrate protein modeling. *Journal of Chemical Theory and Computation*, 7(10), 3162–3180.

Hamilton, P. J., Belovich, A. N., Khelashvili, G., Saunders, C., Erreger, K., Javitch, J. A., … Galli, A. (2014). PIP_2 regulates psychostimulant behaviors through its interaction with a membrane protein. *Nature Chemical Biology*, *10*(7), 582–589.

Han, W., & Schulten, K. (2012). Further optimization of a hybrid united-atom and coarse-grained force field for folding simulations: Improved backbone hydration and interactions between charged side chains. *Journal of Chemical Theory and Computation*, *8*, 4413–4424.

Hart, K., Foloppe, N., Baker, C. M., Denning, E. J., Nilsson, L., & Mackerell, A. D., Jr. (2012). Optimization of the CHARMM additive force field for DNA: Improved treatment of the BI/BII conformational equilibrium. *Journal of Chemical Theory and Computation*, *8*, 348–362.

Haynes, W., Lide, D., & Bruno, T. (2015). *CRC handbook of chemistry and physics*. Boca Raton, Florida: CRC Press.

Hediger, M. A., Clémençon, B., Burrier, R. E., & Bruford, E. A. (2013). The ABCs of membrane transporters in health and disease (SLC series): Introduction. *Molecular Aspects of Medicine*, *34*(2-3), 95–107.

Hess, B. (2008). P-lincs: A parallel linear constraint solver for molecular simulation. *Journal of Chemical Theory and Computation*, *4*(1), 116–122.

Hevener, K. E., Zhao, W., Ball, D. M., Babaoglu, K., Qi, J., White, S. W., & Lee, R. E. (2009). Validation of molecular docking programs for virtual screening against dihydropteroate synthase. *Journal of Chemical Information and Modeling*, *49*(2), 444–460.

Hildebrand, P. W., Goede, A., Bauer, R. A., Gruening, B., Ismer, J., Michalsky, E., & Preissner, R. (2009). SuperLooper—A prediction server for the modeling of loops in globular and membrane proteins. *Nucleic Acids Research*, *37*(Web Server), W571–W574.

Hong, W. C., & Amara, S. G. (2010). Membrane cholesterol modulates the outward facing conformation of the dopamine transporter and alters cocaine binding. *Journal of Biological Chemistry*, *285*(42), 32616–32626.

Horn, B. K. P.. (1987). Closed-form solution of absolute orientation using unit quaternions. *Journal of the Optical Society of America A*, *4*(4), 629–642.

Huang, L., & Roux, B. (2013). Automated force field parameterization for nonpolarizable and polarizable atomic models based on ab initio target data. *Journal of Chemical Theory and Computation*, *9*, 3543–3556.

Huang, S.-Y., & Zou, X. (2007). Ensemble docking of multiple protein structures: Considering protein structural variations in molecular docking. *PROTEINS: Structure, Function, and Bioinformatics*, *66*(2), 399–421.

Huang, Z., & Tajkhorshid, E. (2008). Dynamics of the extracellular gate and ion-substrate coupling in the glutamate transporter. *Biophysical Journal, 95*, 2292–2300.

Hub, J. S., de Groot, B. L., & van der Spoel, D. (2010). g_wham-a free weighted histogram analysis implementation including robust error and autocorrelation estimates. *Journal of Chemical Theory and Computation, 6*, 3713–3720.

Hug, S. (2013). Classical molecular dynamics in a nutshell. In L. Monticelli & E. Salonen (Eds.), *Biomolecular simulations: Methods and protocols* (pp. 127–152). Totowa, NJ: Humana Press.

Humphrey, W., Dalke, A., & Schulten, K. (1996). VMD – Visual Molecular Dynamics. *Journal of Molecular Graphics, 14*(1), 33–38.

Ingólfsson, H. I., Melo, M. N., van Eerden, F. J., Arnarez, C., López, C. A., Wassenaar, T. A., ... Marrink, S. J. (2014). Lipid n of the plasma membrane. *Journal of the American Chemical Society, 136*(14), 14554–14559.

Islam, S. M., Stein, R. A., Mchaourab, H. S., & Roux, B. (2013). Structural refinement from restrained-ensemble simulations based on EPR/DEER data: Application to T4 lysozyme. *Journal of Physical Chemistry B, 117*(17), 4740–4754.

Isralewitz, B., Baudry, J., Gullingsrud, J., Kosztin, D., & Schulten, K. (2001). Steered molecular dynamics investigations of protein function. *Journal of Molecular Graphics and Modeling, 19*, 13–25.

Jackson, R. N., McCoy, A. J., Terwilliger, T. C., Read, R. J., & Wiedenheft, B. (2015). X-ray structure determination using low-resolution electron microscopy maps for molecular replacement. *Nature Protocols, 10*(9), 1275–1284.

Jämbeck, J. P. M., & Lyubartsev, A. P. (2013). Another piece of the membrane puzzle: Extending Slipids further. *Journal of Chemical Theory and Computation, 9*(1), 774–784.

Jardetzky, O. (1966). Simple allosteric model for membrane pumps. *Nature, 211*(5052), 969–970.

Jarzynski, C. (1997). Nonequilibrium equality for free energy differences. *Physical Review Letters, 78*, 2690–2693.

Javanainen, M. (2014). Universal method for embedding proteins into complex lipid bilayers for molecular dynamics simulations. *Journal of Chemical Theory and Computation, 10*, 2577–2587.

Jefferys, E., Sands, Z. A., Shi, J., Sansom, M. S., & Fowler, P. W. (2015). Alchembed: A computational method for incorporating multiple proteins into complex lipid geometries. *Journal of Chemical Theory and Computation, 11*, 2743–2754.

Jiang, W., Luo, Y., Maragliano, L., & Roux, B. (2012). Calculation of free energy landscape in multi-dimensions with hamiltonian-exchange umbrella sampling on petascale supercomputer. *Journal of Chemical Theory and Computation, 8*(11), 4672–4680.

Jiang, W., Phillips, J., Huang, L., Fajer, M., Meng, Y., Gumbart, J., ... Roux, B. (2014). Generalized scalable multiple copy algorithms for molecular dynamics simulations in NAMD. *Computer Physics Communications, 185*, 908–916.

Jiang, X., Villafuerte, M. K. R., Andersson, M., White, S. H., & Kaback, H. R. (2014). Galactoside-binding site in lacy. *Biochemistry, 53*(9), 1536–1543.

Jo, S., Kim, T., & Im, W. (2007). Automated builder and database of protein/membrane complexes for molecular dynamics simulations. *PLoS One, 2*, e880.

Jo, S., Kim, T., Iyer, V. G., & Im, W. (2008). CHARMM-GUI: A web-based graphical user interface for CHARMM. *Journal of Computational Chemistry, 29*, 1859–1865.

Jo, S., Lim, J. B., Klauda, J. B., & Im, W. (2009). CHARMM-GUI membrane builder for mixed bilayers and its application to yeast membranes. *Biophysical Journal, 97*, 50–58.

Jorgensen, W., & Ravimohan, C. (1985). Monte Carlo simulation of differences in free energies of hydration. *Journal of Chemical Physics, 83*, 3050.

Jorgensen, W. L., & Thomas, L. L. (2008). Perspective on free-energy perturbation calculations for chemical equilibria. *Journal of Chemical Theory and Computation, 4*(6), 869–876.

Kamisnki, G. A., Friesner, R. A., Tirado-Rives, J., & Jorgensen, W. L. (2001). Evaluation and reparameterization of the OPLS-AA force field for protiens via comparison with accurate quantum chemical calculations on peptides. *Journal of Physical Chemistry B, 105*(28), 6476–6487.

Kandt, C., Ash, W. L., & Tieleman, D. P. (2007). Setting up and running molecular dynamics simulations of membrane proteins. *Methods, 41*(4), 475–488.

Kantcheva, A. K., Quick, M., Shi, L., Winther, A.-M. L., Stolzenberg, S., Weinstein, H., ... Nissen, P. (2013). Chloride binding site of neurotransmitter sodium symporters. *Proceedings of the National Academy of Sciences, USA, 110*(21), 8489–8494.

Karplus, M., & McCammon, J. A. (2002). Molecular dynamics simulations of biomolecules. *Nature Structural Biology, 265*, 646–652.

Kästner, J. (2011). Umbrella sampling. *Wiley Interdisciplinary Reviews: Computational Molecular Science, 1*(6), 932–942.

Kirkwood, J. G. (1936). Statistical mechanics of fluid mixtures. *Chemical Reviews, 19*, 275.

Kirschner, K. N., Yongye, A. B., Tschampel, S. M., & González-Outeiriño, J. (2008). GLYCAM06: A generalizable biomolecular force field. carbohydrates. *Journal of Computational Chemistry, 29*, 622–655.

Klauda, J. B., Venable, R. M., Freites, J. A., O'Connor, J. W., Tobias, D. J., Mondragon-Ramirez, C., ... Pastor, R. W. (2010). Update of the CHARMM all-atom additive force field for lipids: Validation on six lipid types. *Journal of Physical Chemistry B, 114*(23), 7830–7843.

Kokubo, H., Tanaka, T., & Okamoto, Y. (2013). Two-dimensional replica-exchange method for predicting protein-ligand binding structures. *Journal of Computational Chemistry, 34*(30), 2601–2614.

Koldsø, H., Autzen, H. E., Grouleff, J., & Schiøtt, B. (2013). Ligand induced conformational changes of the human serotonin transporter revealed by molecular dynamics simulations. *PLoS One, 8*(6), e63635.

Kony, D., Damm, W., Stoll, S., & Van Gunsteren, W. F. (2002). An improved OPLS force field for carbohydrates. *Journal of Computational Chemistry, 23*, 1416–1429.

Korb, O., Olsson, T. S. G., Bowden, S. J., Hall, R. J., Verdonk, M. L., Liebeschuetz, J. W., & Cole, J. C. (2012). Potential and limitations of ensemble docking. *Journal of Chemical Information and Modeling, 52*(5), 1262–1274.

Kumar, S., Bouzida, D., Swendsen, R. H., Kollman, P. A., & Rosenberg, J. M. (1992). The weighted histogram analysis method for free-energy calculations on biomolecules. I. The method. *Journal of Computational Chemistry, 13*, 1011–1021.

Laio, A., & Parrinello, M. (2002). Escaping free energy minima. *Proceedings of the National Academy of Sciences, USA, 99*(20), 12562–12566.

LeBard, D. N., Hénin, J., Eckenhoff, R. G., Klein, M. L., & Brannigan, G. (2012). General anesthetics predicted to block the GLIC pore with micromolar affinity. *PLoS Computational Biology, 8*, e1002532.

Lee, C.-K., Pao, C.-W., & Smit, B. (2015). PSII-LHCII supercomplex organizations in photosynthetic membrane by coarse-grained simulation. *Journal of Physical Chemistry B, 119*, 3999–4008.

Lee, E. H., Hsin, J., Sotomayor, M., Comellas, G., & Schulten, K. (2009). Discovery through the computational microscope. *Structure, 17*, 1295–1306.

Lee, S., Tran, A., Allsopp, M., Lim, J. B., Hénin, J., & Klauda, J. B. (2014). CHARMM36 united atom chain model for lipids and surfactants. *Journal of Physical Chemistry B, 118*(2), 547–556.

Leftin, A., Molugu, T. R., Job, C., Beyer, K., & Brown, M. F. (2014). Area per lipid and cholesterol interactions in membranes from separated local-field ^{1}3C NMR spectroscopy. *Biophysical Journal, 107*, 2274–2286.

Lemkul, J. A., Allen, W. J., & Bevan, D. R. (2010). Practical considerations for building GROMOS-compatible small molecule topologies. *Journal of Chemical Information and Modeling, 50*, 2221–2235.

Li, D.-W., & Brüschweiler, R. (2010). NMR-based protein potentials. *Angewandte Chemie – International Edition in English, 49*, 6778–6780.

Li, J., Jaimes, K. F., & Aller, S. G. (2014). Refined structures of mouse p-glycoprotein. *Protein Science, 23*(1), 34–36.

Li, J., & Tajkhorshid, E. (2012). A gate-free pathway for substrate release from the inward-facing state of the Na^{+}-galactose transporter. *Biochimica et Biophysica Acta – Biomembranes, 1818*, 263–271.

Li, J., Wen, P.-C., Moradi, M., & Tajkhorshid, E. (2015). Computational characterization of structural dynamics underlying function in active membrane transporters. *Current Opinion in Structural Biology, 31*, 96–105.

Lin, J.-H., Perryman, A. L., Schames, J. R., & McCammon, J. A. (2002). Computational drug design accommodating receptor flexibility: The relaxed complex scheme. *Journal of the American Chemical Society, 124*(20), 5632–5633.

Lindorff-Larsen, K., Piana, S., Palmo, K., Maragakis, P., Klepeis, J. L., Dror, R. O., & Shaw, D. E. (2010). Improved side-chain torsion potentials for amber ff99SB protein force field. *Proteins, 78*, 1950–1958.

Liu, P., Dehez, F., Cai, W., & Chipot, C. (2012). A toolkit for the analysis of free-energy perturbation calculations. *Journal of Chemical Theory and Computation, 8*(8), 2606–2616.

Lomize, A. L., Pogozheva, I. D., Lomize, M. A., & Mosberg, H. I. (2006). Positioning of proteins in membranes: A computational approach. *Protein Science, 15*, 1318–1333.

Lomize, A. L., Pogozheva, I. D., Lomize, M. A., & Mosberg, H. I. (2007). The role of hydrophobic interactions in positioning of peripheral proteins in membranes. *BMC Structural Biology*, 7, 44.

Lomize, A. L., Pogozheva, I. D., & Mosberg, H. I. (2011). Anisotropic solvent model of the lipid bilayer. 2. Energetics of insertion of small molecules, peptides, and proteins in membranes. *Journal of Chemical Information and Modeling, 51*, 930–946.

Lomize, M. A., Lomize, A. L., Pogozheva, L. D., & Mosberg, H. I. (2006). OPM: Orientations of proteins in membranes database. *Bioinformatics, 22*, 623–625.

Lu, N., Kofke, D. A., & Woolf, T. B. (2004). Improving the efficiency and reliability of free energy perturbation calculations using overlap sampling methods. *Journal of Computational Chemistry, 25*(1), 28–40.

Maciejewski, A., Pasenkiewicz-Gierula, M., Cramariuc, O., Vattulainen, I., & Ròg, T. (2014). Refined OPLS all-atom force field for saturated phosphatidylcholine bilayers at full hydration. *Journal of Physical Chemistry B, 118*, 4571–4581.

Mackerell, A. D. (2004). Empirical force fields for biological macromolecules: Overview and issues. *Journal of Computational Chemistry, 25*, 1584–1604.

MacKerell, A. D., Jr., & Banavali, N. K. (2000). All-atom empirical force field for nucleic acids: II. Application to molecular dynamics simulations of DNA and RNA in solution. *Journal of Computational Chemistry, 21*(2), 105–120.

MacKerell, A. D., Jr., Bashford, D., Bellott, M., Dunbrack, Jr., R. L..Evanseck, J. D., Field, M. J., … Karplus, M. (1998). All-atom empirical potential for molecular modeling and dynamics studies of proteins. *Journal of Physical Chemistry B, 102*, 3586–3616.

MacKerell, Jr., A. D..Feig, M., & Brooks, III, C. L. (2004). Extending the treatment of backbone energetics in protein force fields: Limitations of gas-phase quantum mechanics in

reproducing protein conformational distributions in molecular dynamics simulations. *Journal of Computational Chemistry*, *25*(11), 1400–1415.
Mahinthichaichan, P., Gennis, R., & Tajkhorshid, E. (2016). All the O_2 consumed by *Thermusthermophilus* cytochrome ba_3 is delivered to the active site through a long, open hydrophobic tunnel with entrances within the lipid bilayer. *Biochemistry*, *55*(8), 1265–1278.
Malde, A. K., Zuo, L., Breeze, M., Stroet, M., Poger, D., Nair, P. C., ... Mark, A. E. (2011). An automated force field topology builder (ATB) and repository: Version 1.0. *Journal of Chemical Theory and Computation*, 7, 4026–4037.
Marrink, S. J., & Tieleman, D. P. (2013). Perspective on the MARTINI model. *Chemical Society Reviews*, *42*, 6801–6822.
Martin, C., Berridge, G., Higgins, C. F., Mistry, P., Charlton, P., & Callaghan, R. (2000). Communication between multiple drug binding sites on p-glycoprotein. *Molecular Pharmacology*, *58*(3), 624–632.
Martínez, L., Andrade, R., Birgin, E. G., & Martínez, J. M. (2009). PACKMOL: A package for building intial configurations for molecular dynamics simulations. *J. Comput. Chem.*, *30*, 2157–2164.
Martí-Renom, M. A., Stuart, A. C., Fiser, A., Sánchez, R., Melo, F., & Šali, A. (2000). Comparative protein structure modeling of genes and genomes. *Annual Review of Biophysics and Biomolecular Structure*, *29*, 291–325.
Martyna, G. J., Tobias, D. J., & Klein, M. L. (1994). Constant pressure molecular dynamics algorithms. *Journal of Chemical Physics*, *101*(5), 4177–4189.
Mayne, C. G., Saam, J., Schulten, K., Tajkhorshid, E., & Gumbart, J. C. (2013). Rapid parameterization of small molecules using the Force Field Toolkit. *Journal of Computational Chemistry*, *34*, 2757–2770.
McGibbon, R. T., Beauchamp, K. A., Harrigan, M. P., Klein, C., Swails, J. M., Hernández, C. X., ... Pande, V. S. (2015). MDTraj: A modern open library for the analysis of molecular dynamics trajectories. *Biophysical Journal*, *109*(8), 1528–1532.
McGreevy, R., Singharoy, A., Li, Q., Zhang, J., Xu, D., Perozo, E., & Schulten, K. (2014). xMDFF: Molecular dynamics flexible fitting of low-resolution X-Ray structures. *Acta Crystallographica D*, *70*, 2344–2355.
McHaourab, H. S., Steed, P. R., & Kazmier, K. (2011). Toward the fourth dimension of membrane protein structure: Insight into dynamics from spin-labeling EPR spectroscopy. *Structure*, *19*(11), 1549–1561.
Medalia, O. (2002). Macromolecular architecture in eukaryotic cells visualized by cryo-electron tomography. *Science*, *298*(5596), 1209–1213.
Meng, X.-Y., Zhang, H.-X., Mezei, M., & Cui, M. (2011). Molecular docking: A powerful approach for structure-based drug discovery. *Current Computer-Aided Drug Design*, 7(2), 146–157.
Michaud-Agrawal, N., Denning, E. J., Woolf, T. B., & Beckstein, O. (2011). MDAnalysis: A toolkit for the analysis of molecular dynamics simulations. *Journal of Computational Chemistry*, *32*(10), 2319–2327.
Mihasan, M. (2012). What in silico molecular docking can do for the 'bench-working biologists'. *Journal of Biosciences*, *37*(6), 1089–1095.
Minoukadeh, K., Chipot, C., & Leliévre, T. (2010). Potential of mean force calculations: A multiple-walker adaptive biasing force approach. *Journal of Chemical Theory and Computation*, *6*(4), 1008–1017.
Mobley, D. L., & Guthrie, J. P. (2014). FreeSolv: A database of experimental and calculated hydration free energies, with input files. *Journal of Computer-Aided Molecular Design*, *28*(7), 711–720.
Mobley, D. L., Liu, S., Cerutti, D. S., Swope, W. C., & Rice, J. E. (2011). Alchemical prediction of hydration free energies for SAMPL. *Journal of Computer-Aided Molecular Design*, *26*, 551–562.

Monticelli, L., & Tieleman, D. P. (2013). Force fields for classical molecular dynamics. In L. Monticelli & E. Salonen (Eds.), *Biomolecular simulations: Methods and protocols* (pp. 197–213). Totowa, NJ: Humana Press.

Moradi, M., Enkavi, G., & Tajkhorshid, E. (2015). Atomic-level characterization of transport cycle thermodynamics in the glycerol-3-phosphate:phosphate transporter. *Nature Communications*, *6*, 8393.

Moradi, M., & Tajkhorshid, E. (2013). Mechanistic picture for conformational transition of a membrane transporter at atomic resolution. *Proceedings of the National Academy of Sciences, USA*, *110*(47), 18916–18921.

Moradi, M., & Tajkhorshid, E. (2014). Computational recipe for efficient description of large-scale conformational changes in biomolecular systems. *Journal of Chemical Theory and Computation*, *10*(7), 2866–2880.

Morris, G. M., Huey, R., Lindstrom, W., Sanner, M. F., Belew, R. K., Goodsell, D. S., & Olson, A. J. (2009). Autodock4 and autodocktools4: Automated docking with selective receptor flexibility. *Journal of Computational Chemistry*, *30*(16), 2785–2791.

Morris, G. M., & Lim-Wilby, M. (2008). Molecular docking. In A. Kukol (Ed.), *Molecular modeling of proteins* (pp. 365–382). Totowa, NJ: Humana Press.

Murali, S., Wallner, B., Trudell, J. R., Bertaccini, E., & Lindahl, E. (2011). Microsecond simulations indicate that ethanol binds between subunits and could stabilize an open-state model of a glycine receptor. *Biophysical Journal*, *100*, 1642–1650.

Murray, D. T., Das, N., & Cross, T. A. (2013). Solid state NMR strategy for characterizing native membrane protein structures. *Accounts of Chemical Research*, *46*(9), 2172–2181.

Neale, C., Rodinger, T., & Pomès, R. (2008). Equilibrium exchange enhances the convergence rate of umbrella sampling. *Chemical Physics Letters*, *460*(1–3), 375–381.

Ohkubo, Y. Z., Pogorelov, T. V., Arcario, M. J., Christensen, G. A., & Tajkhorshid, E. (2012). Accelerating membrane insertion of peripheral proteins with a novel membrane mimetic model. *Biophysical Journal*, *102*, 2130–2139.

Olsson, M. H., Søndergaard, C. R., Rostkowski, M., & Jensen, J. H. (2011). Propka3: Consistent treatment of internal and surface residues in empirical pKa predictions. *Journal of Chemical Theory and Computation*, *7*, 525–537.

Oostenbrink, C., Villa, A., Mark, A. E., & Van Gunsteren, W. F. (2004). A biomolecular force field based on the free enthalpy of hydration and solvation: The GROMOS force-field parameter sets 53A5 and 53A6. *Journal of Computational Chemistry*, *25*, 1656–1676.

Oxenoid, K., & Chou, J. J. (2013). The present and future of solution NMR in investigating the structure and dynamics of channels and transporters. *Current Opinion in Structural Biology*, *23*(4), 547–554.

Paloncýová, M., DeVane, R., Murch, B., Berka, K., & Otyepka, M. (2014). Amphiphilic drug-like molecules accumulate in a membrane below the head group region. *Journal of Physical Chemistry B*, *118*(4), 1030–1039.

Pan, A. C., Sezer, D., & Roux, B. (2008). Finding transition pathways using the string method with swarm of trajectories. *Journal of Physical Chemistry B*, *112*(11), 3432–3440.

Papaleo, E. (2015). Integrating atomistic molecular dynamics simulations, experiments, and network analysis to study protein dynamics: Strength in unity. *Frontiers in Molecular Biosciences*, *2*.

Paradies, G., Paradies, V., Benedictis, V. D., Ruggiero, F. M., & Petrosillo, G. (2014). Functional role of cardiolipin in mitochondrial bioenergetics. *Biochimica et Biophysica Acta – Bioenergetics*, *1837*, 408–417.

Pérez, A., Marchán, I., Svozil, D., Sponer, J., Cheatham, T. E., Laughton, C. A., & Orozco, M. (2007). Refinement of the AMBER Force Field for Nucleic Acids: Improving the Description of α/γ Conformers. *Biophysical Journal*, *92*, 3817–3829.

Perilla, J. R., Goh, B. C., Cassidy, C. K., Liu, B., Bernardi, R. C., Rudack, T., … Schulten, K. (2015). Molecular dynamics simulations of large macromolecular complexes. *Current Opinion in Structural Biology*, *31*, 64–74.

Pettersen, E. F., Goddard, T. D., Huang, C. C., Couch, G. S., Greenblatt, D. M., Meng, E. C., & Ferrin, T. E. (2004). UCSF Chimera—A visualization system for exploratory research and analysis. *Journal of Computational Chemistry*, *25*(13), 1605–1612.

Phillips, J. C., Braun, R., Wang, W., Gumbart, J., Tajkhorshid, E., Villa, E., … Schulten, K. (2005). Scalable molecular dynamics with NAMD. *Journal of Computational Chemistry*, *26*, 1781–1802.

Phillips, R., Ursell, T., Wiggins, P., & Sens, P. (2009). Emerging roles for lipids in shaping membrane-protein function. *Nature*, *459*, 379–384.

Plewczynski, D., Łaźniewski, M., Augustyniak, R., & Ginalski, K. (2011). Can we trust docking results? Evaluation of seven commonly used programs on pdbbind database. *Journal of Computational Chemistry*, *32*(4), 742–755.

Pohorille, A., Jarzynski, C., & Chipot, C. (2010). Good practices in free-energy calculations. *Journal of Physical Chemistry B*, *114*, 10235–10253.

Ponder, J. W., & Case, D. A. (2003). Force fields for protein simulations. In F. M. Richards, D. S. Eisendberg, & J. Kuriyan (Eds.), *Vol. 66 Protein Simulations. Advances in protein chemistry* (pp. 27–85). Cambridge, MA: Elsevier Academic Press.

Pronk, S., Páll, S., Schulz, R., Larsson, P., Bjelkmar, P., Apostolov, R., … Lindahl, E. (2013). Gromacs 4.5: A high-throughput and highly parallel open source molecular simulation toolkit. *Bioinformatics*, *29*(7), 845–854.

Qi, Y., Cheng, X., Lee, J., Vermaas, J. V., Pogorelov, T. V., Tajkhorshid, E., … Im, W. (2015). CHARMM-GUI HMMM builder for membrane simulations with the highly mobile membrane-mimetic model. *Biophysical Journal*, *109*, 2012–2022.

Raiteri, P., Laio, A., Gervasio, F., Micheletti, C., & Parrinello, M. (2006). Efficient reconstruction of complex free energy landscapes by multiple walkers metadynamics. *Journal of Physical Chemistry B*, *110*(8), 3533–3539.

Raman, E. P., Guvench, O., & Mackerell, A. D., Jr. (2010). CHARMM additive all-atom force field for glycosidic linkages in carbohydrates involving furanoses. *Journal of Physical Chemistry B*, *40*, 12981–12994.

Reif, M. M., Winger, M., & Oostenbrink, C. (2013). Testing of the GROMOS force-field parameter set 54A8: Structural properties of electrolyte solutions, lipid bilayers, and proteins. *Journal of Chemical Theory and Computation*, *9*(2), 1247–1264.

Reyes, N., Ginter, C., & Boudker, O. (2009). Transport mechanism of a bacterial homologue of glutamate transporters. *Nature*, *462*, 880–885.

Rinne, A., Mobarec, J. C., Mahaut-Smith, M., Kolb, P., & Bünemann, M. (2015). The mode of agonist binding to a g protein–coupled receptor switches the effect that voltage changes have on signaling. *Science Signaling*, *8*(401).

Rodríguez, D., Ranganathan, A., & Carlsson, J. (2015). Discovery of GPCR ligands by molecular docking screening: Novel opportunities provided by crystal structures. *Current Topics in Medicinal Chemistry*, *15*(24), 2484–2503.

Rodriguez, R., Chinea, G., Lopez, N., Pons, T., & Vriend, G. (1998). Homology modeling, model and software evaluation: Three related resources. *CABIOS*, *14*, 523–528.

Ruscio, J. Z., Kumar, D., Shukla, M., Prisant, M. G., Murali, T., & Onufreiv, A. V. (2008). Atomic level computational identification of ligand migration pathways between solvent and binding site in myoglobin. *Proceedings of the National Academy of Sciences, USA*, *105*, 9204–9209.

Saam, J., Ivanov, I., Walther, M., Holzhutter, H., & Kuhn, H. (2007). Molecular dioxygen enters the active site of 12/15 lipoxygenase via dynamic oxygen access channels. *Proceedings of the National Academy of Sciences, USA*, *104*, 13319–13324.

Saam, J., Rosini, E., Molla, G., Schulten, K., Pollegioni, L., & Ghisla, S. (2010). O_2-reactivity of flavoproteins: Dynamic access of dioxygen to the active site and role of a H^+ relay system in D-amino acid oxidase. *Journal of Biological Chemistry*, *285*, 24439–24446.

Sabri Dashti, D., & Roitberg, A. E. (2013). Optimization of umbrella sampling replica exchange molecular dynamics by replica positioning. *Journal of Chemical Theory and Computation*, *9*(11), 4692–4699.

Safa, A. R. (2004). Identification and characterization of the binding sites of p-glycoprotein for multidrug resistance-related drugs and modulators. *Current Medicinal Chemistry - Anti-Cancer Agents*, *4*(1).

Saier, M. H., Reddy, V. S., Tamang, D. G., & Vastermark, A. (2014). The transporter classification database. *Nucleic Acids Research*, *42*(D1), D251–D258.

Sangster, J. (1989). Octanol-water partition coefficients of simple organic compounds. *Journal of Physical and Chemical Reference Data*, *18*(3), 1111–1229.

Saunders, M. G., & Voth, G. A. (2012). Coarse-graining of multiprotein assemblies. *Current Opinion in Structural Biology*, *22*(2), 144–150.

Saunders, M. G., & Voth, G. A. (2013). Coarse-graining methods for computational biology. *Annual Review of Biophysics*, *42*(1), 73–93.

Schlitter, J., Engels, M., Krüger, P., Jacoby, E., & Wollmer, A. (1993). Targeted molecular dynamics simulation of conformational change—Application to the T ⇔ R transition in insulin. *Molecular Simulation*, *10*(2–6), 291–308.

Schneider, G., Guttmann, P., Heim, S., Rehbein, S., Mueller, F., Nagashima, K., … McNally, J. G. (2010). Three-dimensional cellular ultrastructure resolved by X-ray microscopy. *Nature Methods*, 7(12), 985–987.

Schreiner, E., Trabuco, L. G., Freddolino, P. L., & Schulten, K. (2011). Stereochemical errors and their implications for molecular dynamics simulations. *BMC Bioinformatics*, *12*, 190. 9. pp.

Schrödinger, L. L. C. (2015). *The PyMOL molecular graphics system, version 1.8*.

Schüettelkopf, A. W., & van Aalten, D. M. (2004). PRODRG: A tool for high-throughput crystallography of protein-ligand complexes. *Acta Crystallographica D*, *D60*, 1355–1363.

Seyler, S. L., Kumar, A., Thorpe, M. F., & Beckstein, O. (2015). Path similarity analysis: A method for quantifying macromolecular pathways. *PLoS Computational Biology*, *11*(10), e1004568.

Sharom, F. J. (2014). Complex interplay between the p-glycoprotein multidrug efflux pump and the membrane: Its role in modulating protein function. *Frontiers in Oncology*, *4*, 1–19.

Shaw, D. E., Grossman, J. P., Bank, J. A., Batson, B., Butts, J. A., Chao, J. C., & Young, C. (2014). Anton 2: Raising the bar for performance and programmability in a special-purpose molecular dynamics supercomputer. In *Proceedings of the international conference for high performance computing, networking, storage and analysis* (pp. 41–53): IEEE Press.

Shen, L., Bassolino, D., & Stouch, T. (1997). Transmembrane helix structure, dynamics, and interactions: Multi-nanosecond molecular dynamics simulations. *Biophysical Journal*, *73*, 3–20.

Shirts, M. R., & Chodera, J. D. (2008). Statistically optimal analysis of samples from multiple equilibrium states. *Journal of Chemical Physics*, *129*(12), 124105.

Simmons, K. J., Jackson, S. M., Brueckner, F., Patching, S. G., Beckstein, O., Ivanova, E., … Henderson, P. J. (2014). Molecular mechanism of ligand recognition by membrane transport protein, Mhp1. *EMBO Journal*, *33*(16), 1831–1844.

Siu, S. W. I., Pluhackova, K., & Böckmann, R. A. (2012). Optimization of the OPLS-AA force field for long hydrocarbons. *Journal of Chemical Theory and Computation*, *8*(4), 1459–1470.

Skjevik, Å. A., Made, B. D., Walker, R. C., & Teigen, K. (2012). LIPID11: A modular framework for lipid simulations using amber. *Journal of Physical Chemistry B*, *116*, 11124–11136.

Soares, T. A., Hünenberger, P. H., Kastenholz, M. A., Kräutler, V., Lenz, T., Lins, R. D., … Van Gunsteren, W. F. (2005). An improved nucleic acid parameter set for the GROMOS force field. *Journal of Computational Chemistry*, *26*, 725–737.

Søndergaard, C. R., Olsson, M. H., Rostkowski, M., & Jensen, J. H. (2011). Improved treatment of ligands and coupling effects in empirical calculation and rationalization of pKa values. *Journal of Chemical Theory and Computation*, 7(7), 2284–2295.

Stachowiak, J. C., Schmid, E. M., Ryan, C. J., Ann, H. S., Sasaki, D. Y., Sherman, M. B., … Hayden, C. C. (2012). Membrane bending by protein-protein crowding. *Nature Cell Biology*, *14*(9), 944–949.

Stansfeld, P. J., Goose, J. E., Caffrey, M., Carpenter, E. P., Parker, J. L., Newstead, S., & Sansom, M. S. (2015). MemProtMD: Automated insertion of membrane protein structures into explicit lipid membranes. *Structure*, *23*, 1350–1361.

Stansfeld, P. J., & Sansom, M. S. (2011). From coarse grained to atomistic: A serial multiscale approach to membrane protein simulations. *Journal of Chemical Theory and Computation*, 7(4), 1157–1166.

Strynadka, N., Eisenstein, M., Katchalski-Katzir, E., Shoichet, B., Kuntz, I., Abagyan, R., … James, M. (1996). Molecular docking programs successfully predict the binding of a *β*-lactamase inhibitory protein to TEM-1 *β*-lactamase. *Nature Structural Biology*, *3*, 233–239.

Sugita, Y., Kitao, A., & Okamoto, Y. (2000). Multidimensional replica-exchange method for free-energy calculations. *Journal of Chemical Physics*, *113*(15), 6042–6051.

Sun, D., Forsman, J., & Woodward, C. E. (2015). Evaluating force fields for the computational prediction of ionized arginine and lysine side-chains partitioning into lipid bilayers and octanol. *Journal of Chemical Theory and Computation*, *11*(4), 1775–1791.

Tavoulari, S., Margheritis, E., Nagarajan, A., DeWitt, D. C., Zhang, Y.-W., Rosado, E., … Rudnick, G. (2016). Two Na+ sites control conformational change in a neurotransmitter transporter homolog. *Journal of Biological Chemistry*, *291*(3), 1456–1471.

Tembre, B. L., & Cammon, J. A. M. (1984). Ligand-receptor interactions. *Computers & Chemistry*, *8*(4), 281–283.

Thangapandian, S., John, S., Lee, Y., Arulalapperumal, V., & Lee, K. W. (2012). Molecular modeling study on tunnel behavior in different histone deacetylase isoforms. *PLoS One*, 7(11), e49327.

Thangapandian, S., John, S., Sakkiah, S., & Lee, K. W. (2010). Docking-enabled pharmacophore model for histone deacetylase 8 inhibitors and its application in anticancer drug discovery. *Journal of Molecular Graphics and Modelling*, *29*(3), 382–395.

Thompson, A. N., Kim, I., Panosian, T. D., Iverson, T. M., Allen, T. W., & Nimigean, C. M. (2009). Mechanism of potassium-channel selectivity revealed by Na^+ and Li^+ binding sites within the KcsA pore. *Nature Structural & Molecular Biology*, *16*(12), 1317–1324.

Thomson, W. (1874). Kinetic theory of the dissipation of energy. *Nature*, *9*(232), 441–444.

Tian, S., Sun, H., Pan, P., Li, D., Zhen, X., Li, Y., & Hou, T. (2014). Assessing an ensemble docking-based virtual screening strategy for kinase targets by considering protein flexibility. *Journal of Chemical Information and Modeling*, *54*(10), 2664–2679. PMID: 25233367.

Tieleman, D. P., & Berendsen, H. J. C. (1998). A molecular dynamics study of the pores formed by *Escherichia coli* OmpF porin in a fully hydrated palmitoyloleoylphosphatidylcho line bilayer. *Biophysical Journal*, *74*, 2786–2801.

Torrie, G. M., & Valleau, J. P. (1977). Nonphysical sampling distributions in Monte Carlo free-energy estimation: Umbrella sampling. *Journal of Computational Physics*, *23*, 187–199.

Touw, W. G., Joosten, R. P., & Vriend, G. (2015). Detection of *trans-cis* flips and peptide-plane flips in protein structures. *Acta Crystallographica D, 71*(8), 1604–1614.

Trabuco, L. G., Villa, E., Schreiner, E., Harrison, C. B., & Schulten, K. (2009). Molecular Dynamics Flexible Fitting: A practical guide to combine cryo-electron microscopy and X-ray crystallography. *Methods, 49*, 174–180.

Tribello, G. A., Bonomi, M., Branduardi, D., Camilloni, C., & Bussi, G. (2014). Plumed 2: New feathers for an old bird. *Computer Physics Communications, 185*(2), 604–613.

van Meer, G., Voelker, D. R., & Feigenson, G. W. (2008). Membrane lipids: Where they are and how they behave. *Nature Reviews Molecular Cell Biology, 9*(2), 112–124.

Vanommeslaeghe, K., Hatcher, E., Acharya, C., Kundu, S., Zhong, S., Shim, J., MacKerell, A. D., Jr. ... (2010). CHARMM General Force Field: A force field for drug-like molecules compatible with the CHARMM all-atom additive biological force fields. *Journal of Computational Chemistry, 31*(4), 671–690.

Vanommeslaeghe, K., & MacKerell, A. D., Jr. (2012). Automation of the CHARMM General Force Field (CGenFF) I: Bond perception and atom typing. *Journal of Chemical Information and Modeling, 52*(12), 3144–3154.

Vanommeslaeghe, K., Raman, E. P., & MacKerell, A. D., Jr. (2012). Automation of the CHARMM General Force Field (CGenFF) II: Assignment of bonded parameters and partial atomic charges. *Journal of Chemical Information and Modeling, 52*(12), 3155–3168.

Vanommeslaeghe, K., Yang, M., & MacKerell, A. D., Jr. (2015). Robustness in the fitting of molecular mechanics parameters. *Journal of Computational Chemistry, 36*, 1083–1101.

Vanquelef, E., Simon, S., Marquant, G., Garcia, E., Klimerak, G., Delepine, J. C., ... Dupradeau, F.-Y. (2011). R.E.D. Server: A web service for deriving RESP and ESP charges and building force field libraries for new molecules and molecular fragments. *Nucleic Acids Research, 39*, W511–W517.

Vassiliev, S., Zaraiskaya, T., & Bruce, D. (2013). Molecular dynamics simulations reveal highly permeable oxygen exit channels shared with water uptake channels in photosystem ii. *Biochimica et Biophysica Acta – Bioenergetics, 1827*, 1148–1155.

Vergara-Jaque, A., Fenollar-Ferrer, C., Kaufmann, D., & Forrest, L. R. (2015). Repeat-swap homology modeling of secondary active transporters: Updated protocol and prediction of elevator-type mechanisms. *Frontiers in Pharmacology, 6*, 183.

Vermaas, J. V., Baylon, J. L., Arcario, M. J., Muller, M. P., Wu, Z., Pogorelov, T. V., & Tajkhorshid, E. (2015). Efficient exploration of membrane-associated phenomena at atomic resolution. *Journal of Membrane Biology, 248*(3), 563–582.

Vermaas, J. V., Taguchi, A. T., Dikanov, S. A., Wraight, C. A., & Tajkhorshid, E. (2015). Redox potential tuning through differential quinone binding in the photosynthetic reaction center of *Rhodobacter sphaeroides*. *Biochemistry, 54*(12), 2104–2116.

Vinothkumar, K. R. (2015). Membrane protein structures without crystals, by single particle electron cryomicroscopy. *Current Opinion in Structural Biology, 33*, 103–114.

Vitrac, H., MacLean, D. M., Jayaraman, V., Bogdanov, M., & Dowhan, W. (2015). Dynamic membrane protein topological switching upon changes in phospholipid environment. *Proceedings of the National Academy of Sciences, USA, 112*(45), 13874–13879.

Waldher, B., Kuta, J., Chen, S., Henson, N., & Clark, A. E. (2010). ForceFit: A code to fit classiccal force fields to quantum mechanical potential energy surfaces. *Journal of Computational Chemistry, 31*, 2307–2316.

Wang, J., Wang, W., Kollman, P. A., & Case, D. A. (2006). Automatic atom type and bond type perception in molecular mechanical calculations. *Journal of Molecular Graphics and Modelling, 25*, 247–260.

Wang, J., Wolf, R. M., Caldwell, J. W., Kollman, P. A., & Case, D. A. (2004). Development and testing of a general amber force field. *Journal of Computational Chemistry, 25*, 1157–1174.

Wang, L.-P., Chen, J., & Voorhis, T. V. (2013). Systematic parameterization of polarizable force fields from quantum chemistry data. *Journal of Chemical Theory and Computation*, *9*, 452–460.

Wang, P. H., Best, R. B., & Blumberger, J. (2011). Multiscale simulation reveals multiple pathways for H_2 and O_2 transport in a [NiFe]-hydrogenase. *Journal of the American Chemical Society*, *133*, 3548–3556.

Wang, R. Y.-R., Kudryashev, M., Li, X., Egelman, E. H., Basler, M., Cheng, Y., ... DiMaio, F. (2015). De novo protein structure determination from near-atomic-resolution cryo-EM maps. *Nature Methods*, *12*(4), 335–338.

Wang, Y., Cohen, J., Boron, W. F., Schulten, K., & Tajkhorshid, E. (2007). Exploring gas permeability of cellular membranes and membrane channels with molecular dynamics. *Journal of Structural Biology*, *157*, 534–544.

Wang, Y., Ohkubo, Y. Z., & Tajkhorshid, E. (2008). Gas conduction of lipid bilayers and membrane channels. In S. Feller (Ed.), *Vol. 60 Current topics in membranes: Computational modeling of membrane bilayers* (pp. 343–367). Cambridge, MA/Amsterdam: Elsevier. Chapter 12.

Wang, Y., & Tajkhorshid, E. (2010). Nitric oxide conduction by the brain aquaporin AQP4. *PROTEINS: Structure, Function, and Bioinformatics*, *78*, 661–670.

Ward, A., Reyes, C. L., Yu, J., Roth, C. B., & Chang, G. (2007). Flexibility in the ABC transporter MsbA: Alternating access with a twist. *Proceedings of the National Academy of Sciences, USA*, *104*(48), 19005–19010.

Wassenaar, T. A., Ingólfsson, H. I., Böckmann, R. A., Tieleman, D. P., & Marrink, S. J. (2015). Computational lipidomics with insane: A versatile tool for generating custom membranes for molecular simulations. *Journal of Chemical Theory and Computation*, *11*(5), 2144–2155.

Wassenaar, T. A., Pluhackova, K., Böckmann, R. A., Marrink, S. J., & Tieleman, D. P. (2014). Going backward: A flexible geometric approach to reverse transformation from coarse grained to atomistic models. *Journal of Chemical Theory and Computation*, *10*(2), 676–690.

Watanabe, A., Choe, S., Chaptal, V., Rosenberg, J. M., Wright, E. M., Grabe, M., & Abramson, J. (2010). The mechanism of sodium and substrate release from the binding pocket of vSGLT. *Nature*, *468*, 988–991.

Webb, B., & Sali, A. (2014). Comparative protein structure modeling using modeller. *Current Protocols in Bioinformatics*, *47*(5.6), 5.6.1–5.6.32.

Weber, J., Wilke-Mounts, S., & Senior, A. E. (2003). Identification of the F_1-binding surface on the δ-subunit of ATP synthase. *Journal of Biological Chemistry*, *278*, 13409–13416.

White, S. H. (2009). Biophysical dissection of membrane proteins. *Nature*, *459*(7245), 344–346.

Wisedchaisri, G., Park, M. S., Iadanza, M. G., Zheng, H., & Gonen, T. (2014). Proton-coupled sugar transport in the prototypical major facilitator superfamily protein XylE. *Nature Communications*, *5*, 4521.

Wolf, M. G., Hoefling, M., Aponte-Santamaría, C., Grubmüller, H., & Groenhof, G. (2010). g_membed: Efficient insertion of a membrane protein into an equilibrated lipid bilayer with minimal perturbation. *Journal of Computational Chemistry*, *31*, 2169–2174.

Wu, E., Cheng, X., Jo, S., Rui, H., Song, K., Dávila-Contreras, E., ... Im, W. (2014). CHARMM-GUI Membrane Builder toward realistic biological membrane simulations. *Journal of Computational Chemistry*, *35*(27), 1997–2004.

Wu, H., Mey, A. S. J. S., Rosta, E., & Noé, F. (2014). Statistically optimal analysis of state--discretized trajectory data from multiple thermodynamic states. *Journal of Chemical Physics*, *141*, 214106.

Yernool, D., Boudker, O., Jin, Y., & Gouaux, E. (2004). Structure of a glutamate transporter homologue from *Pyrococcus horikoshii*. *Nature*, *431*, 811–818.

Yesselman, J. D., Price, D. J., Knight, J. L., & Brooks, III, C. L. (2012). MATCH: An atom-typing toolset for molecular mechanics force fields. *Journal of Computational Chemistry*, *33*, 189–202.

Yin, Y., Jensen, M. Ø., Tajkhorshid, E., & Schulten, K. (2006). Sugar binding and protein conformational changes in lactose permease. *Biophysical Journal*, *91*, 3972–3985.

Zhang, J., Tuguldur, B., & van der Spoel, D. (2015). Force field benchmark of organic liquids. 2. Gibbs energy of solvation. *Journal of Chemical Information and Modeling*, *55*, 1192–1201.

Zhao, Y., Terry, D. S., Shi, L., Quick, M., Weinstein, H., Blanchard, S. C., & Javitch, J. A. (2011). Substrate-modulated gating dynamics in a Na^+-coupled neurotransmitter transporter homologue. *Nature*, *474*, 109–113.

Zomot, E., & Bahar, I. (2010). The sodium/galactose symporter crystal structure is a dynamic, not so occluded state. *Molecular BioSystems*, *6*, 1040–1046.

Zomot, E., & Bahar, I. (2012). A conformational switch in a partially unwound helix selectively determines the pathway for substrate release from the carnitine/γ-butyrobetaine antiporter CaiT. *Journal of Biological Chemistry*, *287*, 31823–31832.

Zomot, E., Gur, M., & Bahar, I. (2015). Microseconds simulations reveal a new sodium-binding site and the mechanism of sodium-coupled substrate uptake by LeuT. *Journal of Biological Chemistry*, *290*(1), 544–555.

Zwanzig, R. W. (1954). High-temperature equation of state by a perturbation method. I. Nonpolar gases. *Journal of Chemical Physics*, *22*(8), 1420–1426.

CHAPTER SEVENTEEN

Detecting Allosteric Networks Using Molecular Dynamics Simulation

S. Bowerman, J. Wereszczynski[1]

Center for Molecular Study of Condensed Soft Matter, Illinois Institute of Technology, Chicago, IL, United States

[1]Corresponding author: e-mail address: jwereszc@iit.edu

Contents

Abstract

Allosteric networks allow enzymes to transmit information and regulate their catalytic activities over vast distances. In principle, molecular dynamics (MD) simulations can be used to reveal the mechanisms that underlie this phenomenon; in practice, it can be difficult to discern allosteric signals from MD trajectories. Here, we describe how MD simulations can be analyzed to reveal correlated motions and allosteric networks, and provide an example of their use on the coagulation enzyme thrombin. Methods are discussed for calculating residue-pair correlations from atomic fluctuations and mutual information, which can be combined with contact information to identify allosteric networks and to dynamically cluster a system into highly correlated communities. In the case of thrombin, these methods show that binding of the antagonist hirugen

Methods in Enzymology, Volume 578
ISSN 0076-6879
http://dx.doi.org/10.1016/bs.mie.2016.05.027

significantly alters the enzyme's correlation landscape through a series of pathways between Exosite I and the catalytic core. Results suggest that hirugen binding curtails dynamic diversity and enforces stricter venues of influence, thus reducing the accessibility of thrombin to other molecules.

1. INTRODUCTION

Many enzymes have evolved complex control mechanisms that involve the binding of an effector molecular at one location regulating substrate recognition or catalysis at a distant functional site. This phenomenon of allostery is central to the function of several critical protein families, including kinases, proteases, G-protein coupled receptors, and transcription factors (Beckett, 2009; Conn, Christopoulos, & Lindsley, 2009; Kornev & Taylor, 2015; Merdanovic, Monig, Ehrmann, & Kaiser, 2013). Indeed, the modification of allosteric mechanisms by naturally occurring mutations or targeted drug binding has been shown to alter cellular networks and is a prominent mechanism for both the cause and treatment of disease (Nussinov & Tsai, 2013; Nussinov, Tsai, & Ma, 2013). Therefore, understanding the physical basis for allostery has been a central goal of enzymology research over the past 50 years (Huang et al., 2014), resulting in several general models that describe allosteric effects. For brevity, an overview of these models is presented here, however the reader is directed to some of the many excellent reviews on the topic for further details (Cui & Karplus, 2008; Hilser, Wrabl, & Motlagh, 2012; Motlagh, Wrabl, Li, & Hilser, 2014; Ribeiro & Ortiz, in press).

The classical view of allostery involves structural transitions of the target protein through either induced fits or conformational selection models. When an enzyme is allosterically enhanced via the induced fit mechanism, agonist binding forces the enzyme to undergo a conformational change into a new state that is more beneficial to substrate binding and/or catalysis (Koshland, Némethy, & Filmer, 1966; Sullivan & Holyoak, 2008). In the conformational selection approach, the favorable state is already accessible to the enzyme, but agonist binding drastically increases the population of the improved substrate binding conformation (Changeux, 2013; Monod, Wyman, & Changeux, 1965). Both of these models can be considered specific cases of a more general ensemble-based allostery model that treats conformational transitions with a statistical, instead of a deterministic, approach (Hilser et al., 2012).

More recently, the idea of allosteric influences without large-scale conformational transitions has been proposed (Allain et al., 2014; Cooper & Dryden, 1984; McLeish, Cann, & Rodgers, 2015; McLeish, Rodgers, & Wilson, 2013). In these cases, networks created by the cumulative perturbation of residue-pair correlations propagate dynamic changes between substrate and allosteric sites (del Sol, Fujihashi, Amoros, & Nussinov, 2006; Tsai, del Sol, & Nussinov, 2008; Tzeng & Kalodimos, 2011; Van Wart, Durrant, Votapka, & Amaro, 2014). This understanding compliments conformational techniques by providing insight to systems with little structural change (Motlagh et al., 2014) or even those without well-defined structures (Hilser & Thompson, 2007). In many cases, one or several residues within a protein act as allosteric "hotspots" and play a prominent role in its dynamic network structure (Amitai et al., 2004; Bhattacharya & Vaidehi, 2014; Bowerman & Wereszczynski, 2016; Scarabelli & Grant, 2014). Mutations along these allosteric networks are often linked to clinically relevant mutations, as was recently shown in the case of the kinesin-5 motor domain (Scarabelli & Grant, 2014).

Although general models for allosteric mechanisms have been extensively developed, the molecular mechanisms that underlie these effects are highly specific to the system of interest. In theory, the detailed information necessary for describing these processes is contained in a well-converged molecular dynamics (MD) trajectory. In practice, it can be difficult to filter these signals from the high-dimensional and noisy dynamics that are inherent to MD. Here, we describe several methods for detecting and analyzing allosteric effects from MD trajectories. In particular, methods for calculating correlations and contacts are compared, and the processes for mapping dynamic networks and creating coarse-grain representations of these interactions are presented. These methods are illustrated on the serine protease thrombin, and it is shown that binding of the antagonist hirugen enhances the allosteric connection between Exosite I and the catalytic core.

2. THEORY

2.1 Calculating Residue–Residue Correlations

The calculation of correlated motions between residues can be performed at multiple mathematical levels. Historically, the most straightforward and widely used approach is the dynamic *cross-correlation* of atomic fluctuations (Hünenberger, Mark, & van Gunsteren, 1995), which is a Pearson correlation calculated from covariance matrix elements:

$$\mathbf{C}_{i,j} = \frac{\langle (\mathbf{r}_i - \langle \mathbf{r}_i \rangle) \cdot (\mathbf{r}_j - \langle \mathbf{r}_j \rangle) \rangle}{\sqrt{\left(\langle \mathbf{r}_i^2 \rangle - \langle \mathbf{r}_i \rangle^2 \right) \left(\left\langle \mathbf{r}_j^2 \right\rangle - \langle \mathbf{r}_j \rangle^2 \right)}} \quad (1)$$

where bracket-enclosed quantities represent time-averaged values, and $\mathbf{r}_i$ and $\mathbf{r}_j$ are the positional vectors of atoms i and j, respectively. Cross-correlation values span the range of -1 (perfectly anticorrelated) to $+1$ (perfectly correlated). Although ubiquitous, this metric has two significant limitations. First, the dot product of vectors in the numerator results in orthogonal motions always yielding a correlation of zero. Therefore, by this metric, even perfectly correlated motions may be unobserved if they are perpendicular to one another. Second, the averaging assumes that the correlations are linear in time. For example, two perfectly correlated oscillators that have a phase separation of $\pi/2$ are considered noncorrelated in the cross-correlation method $(\langle \sin(t) \sin(t + \pi/2) \rangle = 0)$. These limitations may cause the cross-correlation to undervalue slowly propagating, long-range connections that are important to allosteric signaling.

To overcome these shortcomings, methods have been developed to make use of the mutual information metric of information theory. While extensive explanations of the underlying theory can be found elsewhere (Kraskov, Stogbauer, & Grassberger, 2004; Lange & Grubmuller, 2006), the main concepts are summarized here. In general, the mutual information ($I_{i,j}$) between two atoms can be determined via the equation:

$$I_{i,j} = \iint p(x_i, x_j) \log \left(\frac{p(x_i, x_j)}{p(x_i) p(x_j)} \right) dx_i dx_j \quad (2)$$

where $p(x_i)$ and $p(x_j)$ are the marginal distributions of x_i and x_j and $p(x_i,x_j)$ is the observed joint distribution. While mathematically more complex than the standard cross-correlation, this method does not rely on the resulting geometries of the correlated motions. Rather, Eq. (2) answers the question: "How does knowledge of atom x_i improve knowledge of x_j (and vice versa)?" For linearly independent data, $p(x_i,x_j) = p(x_i)(x_j)$ and $I_{i,j} = 0$. As $p(x_i)$ and $p(x_j)$ become increasingly correlated, the value of $I_{i,j}$ diverges toward infinity. A Pearson-like correlation can be computed from mutual information using the following relationship:

$$\mathbf{C}_{i,j} = \sqrt{1 - \mathrm{e}^{-(2/d) I_{i,j}}} \quad (3)$$

where d is the dimensionality of the data ($d = 3$ for Cartesian trajectories).

Mutual information based correlations are typically conducted at two levels of theory: *linear*, which only accounts for correlations that are temporally in-phase, and *generalized correlation*, which includes out of phase contributions. The linear mutual information is solved analytically by the equation:

$$I_{i,j} = \frac{1}{2}\left[\log \mathbf{C}_i + \log \mathbf{C}_j - \log \mathbf{C}_{i,j}\right] \tag{4}$$

where $\mathbf{C}_i$ and $\mathbf{C}_j$ are marginal–covariance matrices and $\mathbf{C}_{i,j}$ is the pair-covariance matrix, respectively. While this linear approximation yields only the lower limit of the mutual information between two atoms, the computation time is comparable to that of the cross-correlation. The generalized correlation calculation is significantly more taxing and must be solved numerically. However, in many cases the added cost is justified as the generalized correlation can identify physically relevant allosteric connections that escape the other two methods (see Section 4.1).

2.2 Identifying Allosteric Pathways

In biomolecular systems, the motions of adjacent residues are often highly correlated with one another. This can produce a "domino effect" wherein perturbations to one residue create long-range allosteric influences by propagating through networks of highly correlated neighbors (McLeish et al., 2015; Van Wart et al., 2014). One class of methods for identifying and studying these allosteric propagations is with a graph theory approach, where each protein residue forms a "node" in the graph and "edges" connect nodes representing residues that are considered to be in contact with one another. These edges are weighted according to residue-pair correlations (Van Wart, Eargle, Luthey-Schulten, & Amaro, 2012):

$$d_{i,j} = -\log\left|C_{i,j}\right| \tag{5}$$

where $d_{i,j}$ is the "distance" between contacting nodes i and j and $C_{i,j}$ is the pairwise correlation between them. Eq. (5) produces a graph in which strongly correlated residues are separated by short distances, whereas residues with a weak correlation have longer distances between them. In this graph theory approach, the likely allosteric pathway between residues s and t is described by the shortest path between their respective nodes. The shortest path between nodes can be found quickly via search heuristics, such as Dijkstra's algorithm (Dijkstra, 1959), whereas longer pathways, termed the "suboptimal pathways," can be identified by other search methods, eg, Yen's algorithm (Yen, 1971).

The importance of an individual residue to *all* of the optimal dynamic networks in a system can be expressed through its *centrality* (Brandes, 2001):

$$c_i = \frac{1}{N} \sum_{i \neq s \neq t} \sigma_i(s, t) \tag{6}$$

where N is the total number of pathways, $\sigma_i(s,t)$ equals one if residue i is in the shortest pathway between residues s and t and zero otherwise, and the summation is carried out for all paths not starting or ending at residue i. Similarly, the collective importance of a residue across all of the suboptimal pathways connecting two nodes can be expressed by its *prominence*, calculated using Eq. (6) for a fixed s and t.

2.3 Constructing Dynamic Communities

The data resulting from interresidue correlation or allosteric network analyses can often be difficult to interpret. Therefore, in many cases, a coarse-grain representation of the network graph can provide unique insights into the dynamic regulation of a system (Sethi, Eargle, Black, & Luthey-Schulten, 2009). In this method, called "community structure analysis" a large number of residue nodes are clustered into "communities," which are then connected to one another through edges that are weighted to reflect the strength of their interactions. This reduction can be done using the iterative algorithm of Girvan and Newman (2002):

1. Represent the system as a graph (as described in Section 2.2).
2. Calculate the edge "betweenness" by counting the number of shortest paths traversing each edge.
3. Remove the edge with the largest betweenness from the graph.
4. Cluster nodes based on edge distances.
5. Calculate the "modularity" of the graph using Eq. (7).
6. Repeat Steps 2–5 until all communities are connected by only a single edge.

The modularity (Q) of the graph is calculated by:

$$Q = \sum \left(e_i - a_i^2\right) \tag{7}$$

where e_i is the percent of total edges in community i and a_i^2 is the expected percentage if the edges were randomly associated. Graphs with high modularity possess a higher organization above random noise (Newman, 2006).

For an initial graph with N edges and M nodes, the earlier algorithm will produce $(N-M)$ unique graphs in which there are no isolated communities. The model with the largest modularity, or the most structure above random noise, is typically chosen as the one that best fits the data. The result of this algorithm is a new graph in which nodes represent communities of highly intercorrelated residues, and the strength of the linkage between two communities is dictated by the number of shortest paths that traverse their connecting edges.

3. METHODS

3.1 System Construction and Simulation Details

The analyses of a series of MD simulations of the enzyme thrombin are presented as a case study. This well-studied protein possesses two distinct allosteric binding sites, Exosites I and II (Fuglestad et al., 2012; Stubbs & Bode, 1993). Here, two systems were constructed: isolated thrombin (PDB: 1PPB) (Bode, Turk, & Karshikov, 1992) and thrombin complexed with hirugen (PDB: 1HAH) (Vijayalakshmi, Padmanabhan, Mann, & Tulinsky, 1994). In both systems, the PPACK molecule was removed from the catalytic pocket, and each system was neutralized by either Na^+ or Cl^- ions and solvated in a box of TIP3P water that extended at least 10 Å from the solute (Jorgensen, Chandrasekhar, Madura, Impey, & Klein, 1983). Atoms were modeled by the Amber ff14SB force field, with altered monovalent ion parameters (Hornak et al., 2006; Joung & Cheatham, 2009; Maier et al., 2015). Hirugen was parameterized using GAFF (Wang, Wang, Kollman, & Case, 2006; Wang, Wolf, Caldwell, Kollman, & Case, 2004). Simulations were conducted with the GPU accelerated pmemd.cuda program in the AMBER suite (v14) (Salomon-Ferrer, Gotz, Poole, Le Grand, & Walker, 2013).

Each system was geometrically minimized for 10,000 steps (5000 with protein heavy atoms restrained and 5000 without restraints), then gradually heated three separate times from 10 to 300 K in the NVT ensemble while restraining protein heavy atoms. Following heating, the restraints in each simulation were slowly released over 150 ps, and the simulations were extended in the NPT ensemble for an additional 200 ns. Each trajectory was observed to equilibrate in 30 ns based upon RMSD calculations. This resulted in three separate 170 ns coordinate sets (510 ns total) for use in the analysis of each system.

3.2 Correlation Calculations

Residue–residue correlations were calculated at three levels of theory: the cross-correlation, the linear mutual information, and the (nonlinear) generalized correlation. Each calculation was done using GROMACS tools (*g_covar* for cross-correlations and *g_correlation* otherwise) (Lange & Grubmuller, 2006; Lindahl, Hess, & van der Spoel, 2001). Mutual information calculations were converted to their analogous correlation values using Eq. (3). Cα coordinates from every 100 ps of the simulation were used to improve calculation efficiency. Each set of coordinates was least squares fit to the average structure prior to correlation calculations. The generalized correlation matrix was converted to a "correlation distance" matrix using Eq. (5).

3.3 Graph Construction and Calculations

There is no single, rigorous definition for a contact in MD simulations. Therefore, we tested an exhaustive range of distance cutoffs using both Cα–Cα and heavy atom separations. Separation distances ranged from 8 to 12 Å for Cα–Cα and 3 to 6 Å for heavy atoms. Contacts were only defined between two residues that satisfied these distances in at least 75% of frames. The Cα–Cα metric routinely identified nonphysical contacts (see Section 4.2); therefore, graphs were created using only heavy atom contact matrices. Each graph was formed by an element-wise multiplication of the corresponding contact matrix and the edge matrix calculated from the generalized correlations. Any zero-value elements (self- and noncontacts) were removed, and the matrix was converted into a graph format for analysis in the NetworkX Python module (Hagberg, Schult, & Swart, 2008).

The community structure of each graph was constructed using the Girvan–Newman algorithm. The contact definition of 5.5 Å, which produced the smallest number of communities in both thrombin and thrombin–hirugen, was selected for further analysis. Shortest pathways between all residue-pairs were calculated using Dijkstra's algorithm in NetworkX, and allosteric hotspots in each system were identified using Eq. (6). Furthermore, the 150 suboptimal paths between Exosite I residue T74 and the catalytic core H57 residue were calculated using the Weighted Implementation of Suboptimal Pathways method (Van Wart et al., 2014).

4. RESULTS AND DISCUSSION

4.1 Comparison of Correlation Methods

Comparison of the cross-correlation, linear, and nonlinear mutual information approaches shows that the level of interresidue correlation is consistently weakest in the cross-correlation and strongest when using the nonlinear generalized method (Fig. 1). All three methods correctly identify strong correlations in the catalytic core (residues 79 and 135) and across the disulfide bond connecting the light and heavy chains (residues 9 and 155). In addition, all three methods show the expected strong correlation between neighboring β-sheet residues (ie, L40–F34 (residues 62 and 55) and F34–L64 (residues 55 and 95)).

However, cross-correlation and linear mutual information fail to capture a number of correlations that are both physically intuitive and biologically relevant. In contrast, these couplings are readily identified by the generalized correlation method to have strong correlations ($c > 0.5$). For example, only the generalized correlation displays a significant signal between the catalytic core (residue 135) and Exosite II (residue 214), which is the binding site for the allosteric inhibitor heparin (Yang, Sun, Gailani & Rezaie, 2009).

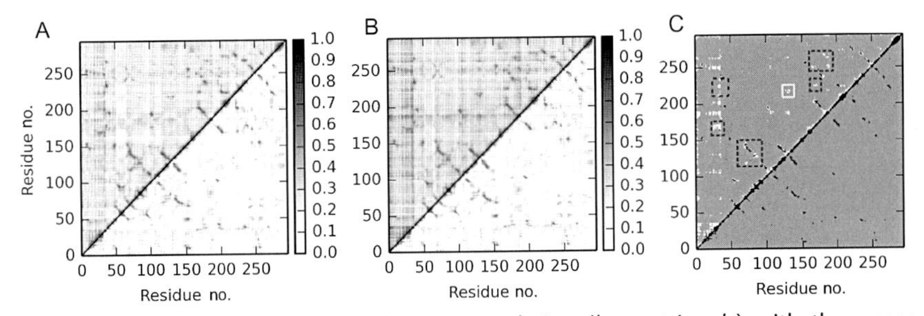

Fig. 1 Comparison of the standard cross-correlation (*lower triangle*) with the *upper triangle* populated by the (A) linear mutual information metric and (B) nonlinear generalized correlation. (C) A filtered comparison between the cross-correlation (*lower triangle*) and the generalized correlation (*upper triangle*), where only correlations >0.5 are shown. *Black regions* denote strong correlations that were identified by both methods, while *white regions* show sections of strong correlation that are identified by the general correlation but not the cross-correlation. In particular, the allosteric connection between Exosite II and the catalytic core is highlighted (*solid yellow* (*white* in the print version) *box*), along with several expected interactions between contacting residues and three-body interactions (*dotted red* (*gray* in the print version) *boxes*).

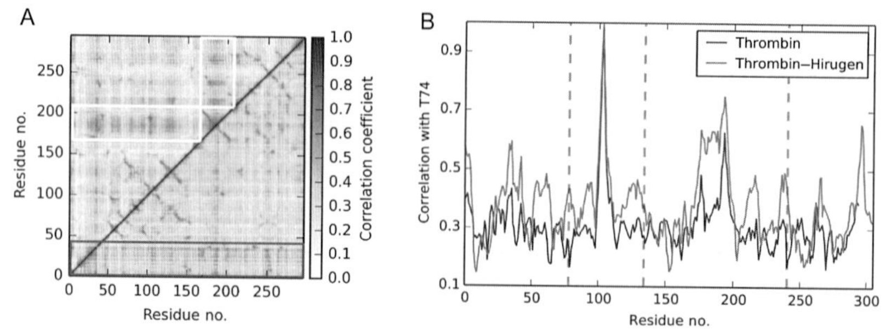

Fig. 2 (A) Comparison of generalized correlations in isolated thrombin (*lower triangle*) and thrombin–hirugen (*upper triangle*). The prominent role of the light chain in isolated thrombin is outlined by the *red* (*dark gray* in the print version) *box*, and the global increase of correlation with the 148_{CT} Loop in thrombin–hirugen is highlighted by *white boxes*. (B) The pairwise correlations between thrombin residues and Exosite I residue T74. The binding of hirugen raises correlations with T74 system wide, but the catalytic residues (*vertical dotted lines*) experience a significant increase in correlation strength.

Furthermore, generalized correlation calculations highlight several three-body correlations, such as the F34-mediated interaction between L40 and L64 and the K107-mediated interaction between R50 (residue 72) and K87 (residue 119).

Comparison of isolated thrombin with the thrombin–hirugen complex reveals several notable differences (Fig. 2). First, the average strength of correlations in thrombin is reduced upon the addition of hirugen. Second, the light chain, which has been shown to have an allosteric role in normal thrombin function (Carter, Vanden Hoek, Pryzdial, & Macgillivray, 2010; Gasper, Fuglestad, Komives, Markwick, & McCammon, 2012), has more prominent correlations in isolated thrombin. A similar decrease in light chain correlations has been observed upon binding of a truncated thrombomodulin molecule, which also binds to Exosite I (Gasper et al., 2012). This common trend highlights the importance of the site in allosteric signaling. Furthermore, the binding of hirugen produces a system-wide increase in correlation with the 148_{CT} loop and with Exosite I, most notably residue T74. Interestingly, residues in and neighboring the catalytic triad experience significant increases in correlations with T74, showing that the binding of hirugen to Exosite I acts as an allosteric signal to regulate catalytic activity.

4.2 Contact Definition Analysis

Graph-based allosteric methods require a consistent definition of two residues that are in contact with one another. Typically, residues are considered to be in contact if either their Cα–Cα distance or their minimum heavy atom separation is closer than a specified cutoff value for more than a certain percentage of the simulation.

Although commonly used, we note that the choice of Cα–Cα distance may be prone to produce false contacts. For example, the Cα-based metric may identify interactions of nonneighboring β-strands as direct contacts, even if a third β-strand is located between them (Fig. 3A). These errant contacts can produce "short-circuits" in a network graph, and thus significantly alter the resulting allosteric pathway or community structure. While these false positives can be avoided by reducing the contact distance, decreasing the distance will conversely increase the number of false negatives, since residue contacts formed by the interaction of long side chains will be missed if one measures only the Cα–Cα distance. For an example in isolated thrombin, a Cα distance of 9.0 Å is stringent enough to avoid the false contact shown in Fig. 3A, but it also fails to identify the electrostatic contact between K202 and E14C (Fig. 3B). Therefore, we used the heavy atom contact definition, which is able to correctly identify contacts similar to those shown in Fig. 3 and avoid both false positives and negatives.

The choice of a cutoff distance is also not well defined, thus a series of values from 3.0 to 6.0 Å were tested. A distance of 5.5 Å was chosen as it minimizes the number of communities in both thrombin systems (Table 1). Finally, we tested different values for the contact frequency criterion, but this parameter had minimal impact on the final results and a value of 75% was chosen.

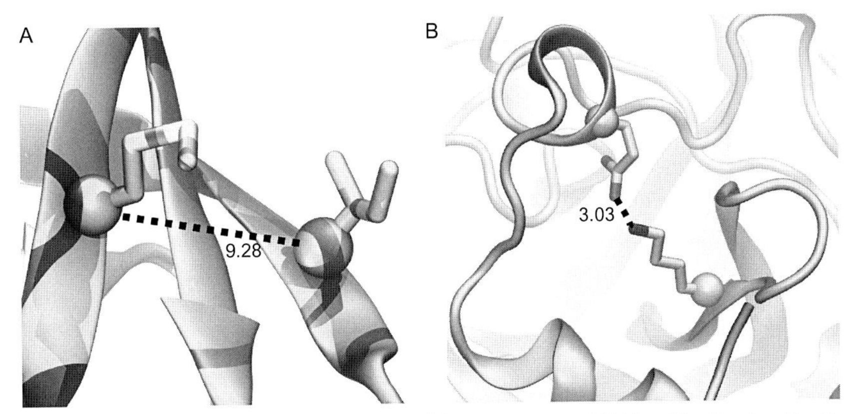

Fig. 3 Examples of a false contact (A) and a false noncontact (B) identified by the Cα–Cα distance definition. If a heavy atom definition is used instead, then the false positive in (A) is removed and the contact in (B) is properly identified. (See the color plate.)

Table 1 The Number of Communities Identified in Isolated Thrombin and Thrombin–Hirugen Using the Specified Heavy Atom Contact Distance

Cut-Off Distance (Å)	Thrombin	Thrombin–Hirugen
3.0	19	16
3.5	18	17
4.0	14	10
4.5	15	10
5.0	14	11
5.5	12	10
6.0	13	11

4.3 Allosteric Pathways in Thrombin

The optimal and suboptimal pathways between T74 and catalytic H57 show that hirugen binding strengthens the correlation between Exosite I and the catalytic core. Although the shortest path between T74 and the catalytic site traverses the same nodes in both systems, the length is shorter in thrombin–hirugen, suggesting stronger correlations between these sites in the bound state. Furthermore, there is a reduction in pathway diversity in the suboptimal pathways connecting these regions (Fig. 4). In isolated thrombin, pathways traverse directly to the catalytic core and through the surface exposed 60s and 70s loops, resulting in a total of 42 residues accessed. In the thrombin–hirugen complex, the hirugen molecule replaces the 70s loop and pathways never access the 60s loop, resulting in the 150 pathways accessing only 29 residues in a direct path to the catalytic core. This strong intercorrelation between the 29 nodes in thrombin–hirugen streamlines the transfer of dynamics from Exosite I to the catalytic core, thus strengthening the net correlation between the regions and overpowering other allosteric influences. This allows the dynamics of Exosite I to be the prominent allosteric signal to catalytic dynamics.

The locations of allosteric hotspots further suggest that hirugen binding may serve to focus the diverse dynamic networks of isolated thrombin into shorter pathways between Exosite I and the catalytic core. While the 70s loop is the most central region in isolated thrombin, hirugen replaces its role in local networks and reduces its centrality to the average peak height (Fig. 5). Furthermore, hirugen binding reduces the centrality scores of the light chain and Exosite II, signifying their decreased importance in

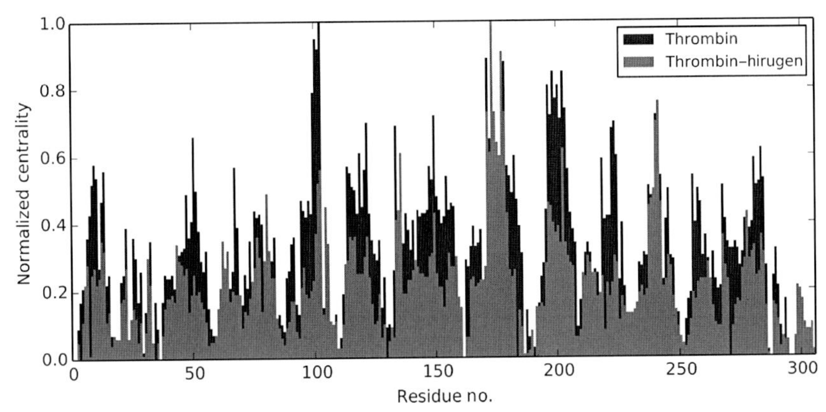

Fig. 4 Normalized centrality values in isolated thrombin (*blue* (*black* in the print version)) and thrombin–hirugen (*green* (*dark gray* in the print version)). While the trend in peak values is the same, the average centrality score in thrombin–hirugen is less than that of isolated thrombin, which is indicative of a reduction in diversity in global pathways.

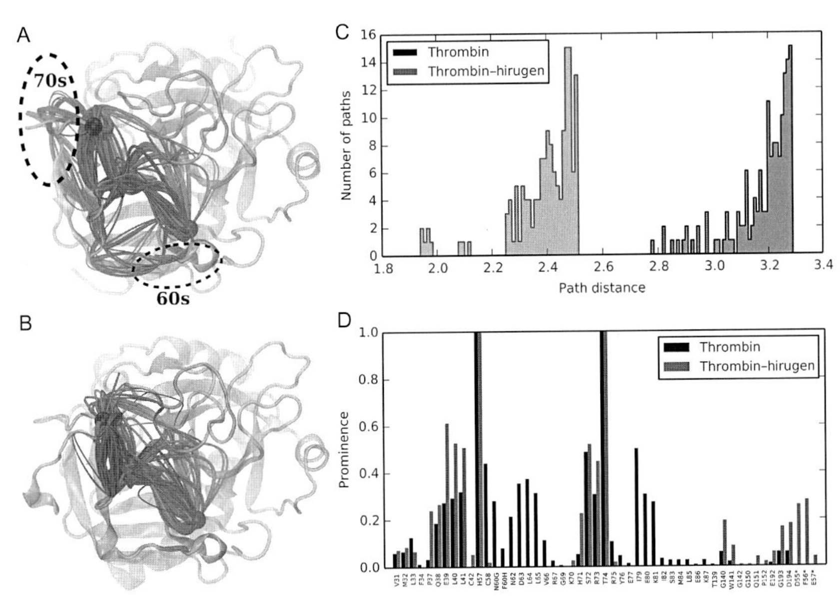

Fig. 5 The suboptimal pathways in (A) isolated thrombin and (B) thrombin–hirugen. T74 is identified by the *black sphere*, and H57 is shown as a *blue* (*dark gray* in the print version) *sphere*. (C) Path length histograms for the two systems. (D) The prominence of residues in the pathways of each system. The pathways of isolated thrombin "detour" through the 70s and 60s loops, thus increasing their path lengths and decreasing their strength of interaction.

allosteric signaling. Ultimately, the general trend of centrality values is similar within each system, but the average score is higher in isolated thrombin. This narrowed route of communication in thrombin–hirugen may serve to combat the effects from binding other allosteric molecules or substrates.

4.4 Community Analysis

Comparison of the community structures of isolated thrombin and thrombin–hirugen reveals several noteworthy changes (Fig. 6). While the catalytic triad is unified in isolated thrombin, H57 and D102 are in a separate, but strongly linked, community from S195 in thrombin–hirugen. Both systems show a connection between the catalytic triad communities and Exosite I. In the thrombin–hirugen complex, the connection is especially strong between Exosite I and the S195 community, in agreement with

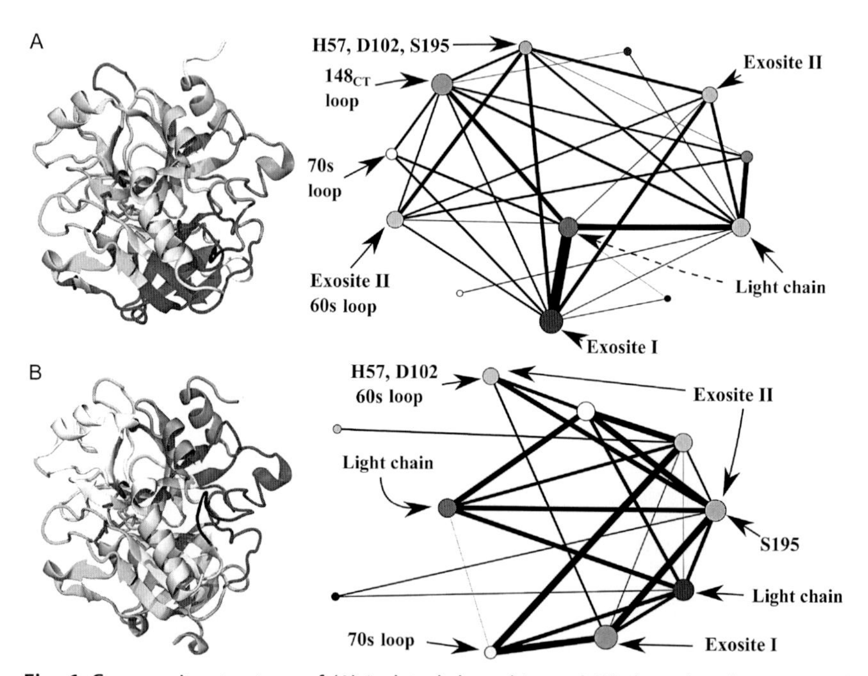

Fig. 6 Community structure of (A) isolated thrombin and (B) thrombin–hirugen with the 3D structures colored according to community membership. Catalytic residues are also shown as *purple sticks*. Relevant regions of the community graphs are labeled accordingly. While the catalytic residues in thrombin–hirugen are split between two communities, the interaction strength between their respective groups and Exosite I is increased. (See the color plate.)

our correlation observations, but the strength of direct interaction between Exosite I and the H57/60s loop community appears to be quite weak. This agrees with the pathway analysis (Section 4.3) which showed the allosteric signal between Exosite I and H57 in thrombin–hirugen propagating through the S195 community and minimally accessing regions around the 60s loop. These results support the claim that hirugen binding creates a strong allosteric signal between Exosite I and the catalytic core to inhibit thrombin function. Lastly, the thrombin–hirugen complex consistently forms fewer communities than isolated thrombin, regardless of the choice of contact distance (Table 1). The decreased number of communities suggests that the thrombin–hirugen system may be harder to dynamically perturb, as the structural regions move as highly correlated domains and reduce the ability of further binding factors altering the activity of the molecule.

5. CONCLUSION

In this chapter, multiple techniques for identifying allosteric effects in enzymes are presented and applied to the sample case of hirugen binding to thrombin. We have focused on methods grounded in interresidue correlations and contacts; however, we note that complementary techniques may provide additional insights (Allain et al., 2014; LeVine & Weinstein, 2014; McClendon, Hua, Barreiro, & Jacobson, 2012). The generalized correlation calculation provides a better understanding of the correlation landscape than the standard cross-correlation, justifying its increased computational cost. Furthermore, the importance of using heavy atom separation as a contact definition rather than the colloquially used Cα separation is highlighted. In thrombin, these methods show that binding of hirugen significantly changes the residue-pair correlations, increasing the importance of the 148_{CT} loop and strengthening the correlation between Exosite I and the catalytic core. Mapping communication pathways between this exosite and the catalytic core show that hirugen streamlines the propagation of dynamics, creating shorter pathways, and stronger correlations. Furthermore, binding of hirugen reduces the total number of communities in thrombin and causes Exosite I to act as a single dynamic community, increasing the strength of interaction between Exosite I and catalytic communities. The combination of methods presented here show that hirugen binding reduces the dynamic diversity in thrombin and restricts the possible venues of influence by interaction with substrates or other allosteric molecules.

ACKNOWLEDGMENTS

The authors thank Dr. Paul Gasper for invaluable conversations in preparing this manuscript. Work reported in this publication was supported by the National Institute of General Medical Sciences of the National Institutes of Health (Grant No. R15GM114578). The content is solely the responsibility of the authors and does not necessarily represent the official views of the National Institutes of Health. The simulations presented here were conducted, in part, on XSEDE supercomputing resources supported by the National Science Foundation (Grant No. ACI-1053575).

REFERENCES

Allain, A., Chauvot de Beauchene, I., Langenfeld, F., Guarracino, Y., Laine, E., & Tchertanov, L. (2014). Allosteric pathway identification through network analysis: From molecular dynamics simulations to interactive 2D and 3D graphs. *Faraday Discussions*, *169*, 303–321. http://dx.doi.org/10.1039/c4fd00024b.

Amitai, G., Shemesh, A., Sitbon, E., Shklar, M., Netanely, D., Venger, I., & Pietrokovski, S. (2004). Network analysis of protein structures identifies functional residues. *Journal of Molecular Biology*, *344*(4), 1135–1146. http://dx.doi.org/10.1016/j.jmb.2004.10.055.

Beckett, D. (2009). Regulating transcription regulators via allostery and flexibility. *Proceedings of the National Academy of Sciences of the United States of America*, *106*(52), 22035–22036. http://dx.doi.org/10.1073/pnas.0912300107.

Bhattacharya, S., & Vaidehi, N. (2014). Differences in allosteric communication pipelines in the inactive and active states of a GPCR. *Biophysical Journal*, *107*(2), 422–434. http://dx.doi.org/10.1016/j.bpj.2014.06.015.

Bode, W., Turk, D., & Karshikov, A. (1992). The refined 1.9-Å X-ray crystal structure of D-Phe-Pro-Arg chloromethylketone-inhibited α-thrombin: Structure analysis, overall structure, electrostatic properties, detailed active-site geometry, and structure–function relationships. *Protein Science*, *1*(4), 426–471.

Bowerman, S., & Wereszczynski, J. (2016). Effects of macroH2A and H2A.Z on nucleosome dynamics as elucidated by molecular dynamics simulations. *Biophysical Journal*, *110*, 327–337. http://dx.doi.org/10.1016/j.bpj.2015.12.015.

Brandes, U. (2001). A faster algorithm for betweenness centrality. *Journal of Mathematical Sociology*, *25*(2), 163–177.

Carter, I. S., Vanden Hoek, A. L., Pryzdial, E. L., & Macgillivray, R. T. (2010). Thrombin a-chain: Activation remnant or allosteric effector? *Thrombosis*, *2010*, 416167. http://dx.doi.org/10.1155/2010/416167.

Changeux, J.-P. (2013). 50 years of allosteric interactions: The twists and turns of the models. *Nature Reviews. Molecular Cell Biology*, *14*, 819–829.

Conn, J. P., Christopoulos, A., & Lindsley, C. W. (2009). Allosteric modulators of GPCRs: A novel approach for the treatment of CNS disorders. *Nature Reviews. Drug Discovery*, *8*(1), 41–54.

Cooper, A., & Dryden, D. T. (1984). Allostery without conformational change: A plausible model. *European Biophysics Journal*, *11*(2), 103–109.

Cui, Q., & Karplus, M. (2008). Allostery and cooperativity revisited. *Protein Science*, *17*(8), 1295–1307. http://dx.doi.org/10.1110/ps.03259908.

del Sol, A., Fujihashi, H., Amoros, D., & Nussinov, R. (2006). Residues crucial for maintaining short paths in network communication mediate signaling in proteins. *Molecular Systems Biology*, *2*(2006), 19. http://dx.doi.org/10.1038/msb4100063.

Dijkstra, E. W. (1959). A note on two problems in connexion with graphs. *Numerische Mathematik*, *1*, 269–271.

Fuglestad, B., Gasper, P. M., Tonelli, M., McCammon, J. A., Markwick, P. R., & Komives, E. A. (2012). The dynamic structure of thrombin in solution. *Biophysical Journal, 103*(1), 79–88. http://dx.doi.org/10.1016/j.bpj.2012.05.047.

Gasper, P. M., Fuglestad, B., Komives, E. A., Markwick, P. R., & McCammon, J. A. (2012). Allosteric networks in thrombin distinguish procoagulant vs. anticoagulant activities. *Proceedings of the National Academy of Sciences of the United States of America, 109*(52), 21216–21222.

Girvan, M., & Newman, M. E. (2002). Community structure in social and biological networks. *Proceedings of the National Academy of Sciences of the United States of America, 99*(12), 7821–7826. http://dx.doi.org/10.1073/pnas.122653799.

Hagberg, A. A., Schult, D. A., & Swart, J. P. (2008). Exploring network structure, dynamics, and function using networkX. In *Paper presented at the proceedings of the 7th python in science conference*.

Hilser, V. J., & Thompson, E. B. (2007). Intrinsic disorder as a mechanism to optimize allosteric coupling in proteins. *Proceedings of the National Academy of Sciences of the United States of America, 104*(20), 8311–8315. http://dx.doi.org/10.1073/pnas.0700329104.

Hilser, V. J., Wrabl, J. O., & Motlagh, H. N. (2012). Structural and energetic basis of allostery. *Annual Review of Biophysics, 41*, 585–609. http://dx.doi.org/10.1146/annurev-biophys-050511-102319.

Hornak, V., Abel, R., Okur, A., Strockbine, B., Roitberg, A., & Simmerling, C. (2006). Comparison of multiple amber force fields and development of improved protein backbone parameters. *Proteins, 65*(3), 712–725. http://dx.doi.org/10.1002/prot.21123.

Huang, Z., Mou, L., Shen, Q., Lu, S., Li, C., Liu, X., … Zhang, J. (2014). ASD v2.0: Updated content and novel features focusing on allosteric regulation. *Nucleic Acids Research, 42*(Database issue), D510–D516. http://dx.doi.org/10.1093/nar/gkt1247.

Hünenberger, P. H., Mark, A. E., & van Gunsteren, W. F. (1995). Fluctuation and cross-correlation analysis of protein motions observed in nanosecond molecular dynamics simulations. *Journal of Molecular Biology, 252*, 492–503.

Jorgensen, W. L., Chandrasekhar, J., Madura, J. D., Impey, R. W., & Klein, M. L. (1983). Comparison of simple potential functions for simulating liquid water. *The Journal of Chemical Physics, 79*(2), 926. http://dx.doi.org/10.1063/1.445869.

Joung, I. S., & Cheatham, T. E. I. (2009). Molecular dynamics simulations of the dynamic and energetic properties of alkali and halide ions using water-model-specific ion parameters. *Journal of Physical Chemistry B, 113*, 13279–13290.

Kornev, A. P., & Taylor, S. S. (2015). Dynamics-driven allostery in protein kinases. *Trends in Biochemical Sciences, 40*(11), 638–647.

Koshland, D. E., Jr., Némethy, G., & Filmer, D. (1966). Comparison of experimental binding data and theoretical models in proteins containing subunits. *Biochemistry, 5*(1), 365–385.

Kraskov, A., Stogbauer, H., & Grassberger, P. (2004). Estimating mutual information. *Physical Review. E, Statistical, Nonlinear, and Soft Matter Physics, 69*(6 Pt 2), 066138. http://dx.doi.org/10.1103/PhysRevE.69.066138.

Lange, O. F., & Grubmuller, H. (2006). Generalized correlation for biomolecular dynamics. *Proteins, 62*(4), 1053–1061. http://dx.doi.org/10.1002/prot.20784.

LeVine, M. V., & Weinstein, H. (2014). NbIT—A new information theory-based analysis of allosteric mechanisms reveals residues that underlie function in the leucine transporter LeuT. *PLoS Computational Biology, 10*(5), e1003603. http://dx.doi.org/10.1371/journal.pcbi.1003603.

Lindahl, E., Hess, B., & van der Spoel, D. (2001). GROMACS 3.0: A package for molecular simulation and trajectory analysis. *Journal of Molecular Modeling, 7*, 306–317. http://dx.doi.org/10.1007/s008940100045.

Maier, J. A., Martinez, C., Kasavajhala, K., Wickstrom, L., Hauser, K. E., & Simmerling, C. (2015). ff14SB: Improving the accuracy of protein side chain and backbone parameters from ff99SB. *Journal of Chemical Theory and Computation, 11*(8), 3696–3713. http://dx.doi.org/10.1021/acs.jctc.5b00255.

McClendon, C. L., Hua, L., Barreiro, A., & Jacobson, M. P. (2012). Comparing conformational ensembles using the Kullback–Leibler divergence expansion. *Journal of Chemical Theory and Computation, 8*(6), 2115–2126. http://dx.doi.org/10.1021/ct300008d.

McLeish, T. C., Cann, M. J., & Rodgers, T. L. (2015). Dynamic transmission of protein allostery without structural change: Spatial pathways or global modes? *Biophysical Journal, 109*(6), 1240–1250. http://dx.doi.org/10.1016/j.bpj.2015.08.009.

McLeish, T. C., Rodgers, T. L., & Wilson, M. R. (2013). Allostery without conformation change: Modelling protein dynamics at multiple scales. *Physical Biology, 10*(5), 056004. http://dx.doi.org/10.1088/1478-3975/10/5/056004.

Merdanovic, M., Monig, T., Ehrmann, M., & Kaiser, M. (2013). Diversity of allosteric regulation in proteases. *ACS Chemical Biology, 8*(1), 19–26. http://dx.doi.org/10.1021/cb3005935.

Monod, J., Wyman, J., & Changeux, J.-P. (1965). On the nature of allosteric transitions: A plausible model. *Journal of Molecular Biology, 12*, 88–118.

Motlagh, H. N., Wrabl, J. O., Li, J., & Hilser, V. J. (2014). The ensemble nature of allostery. *Nature, 508*(7496), 331–339. http://dx.doi.org/10.1038/nature13001.

Newman, M. E. (2006). Modularity and community structure in networks. *Proceedings of the National Academy of Sciences of the United States of America, 103*(23), 8577–8582. http://dx.doi.org/10.1073/pnas.0601602103.

Nussinov, R., & Tsai, C. J. (2013). Allostery in disease and in drug discovery. *Cell, 153*(2), 293–305. http://dx.doi.org/10.1016/j.cell.2013.03.034.

Nussinov, R., Tsai, C. J., & Ma, B. (2013). The underappreciated role of allostery in the cellular network. *Annual Review of Biophysics, 42*, 169–189. http://dx.doi.org/10.1146/annurev-biophys-083012-130257.

Ribeiro, A. A., & Ortiz, V. (in press). A chemical perspective on allostery, *Chemical Reviews*. http://dx.doi.org/10.1021/acs.chemrev.5b00543.

Salomon-Ferrer, R., Gotz, A. W., Poole, D., Le Grand, S., & Walker, R. C. (2013). Routine microsecond molecular dynamics simulations with AMBER on GPUs. 2. Explicit solvent particle mesh Ewald. *Journal of Chemical Theory and Computation, 9*(9), 3878–3888. http://dx.doi.org/10.1021/ct400314y.

Scarabelli, G., & Grant, B. J. (2014). Kinesin-5 allosteric inhibitors uncouple the dynamics of nucleotide, microtubule, and neck-linker binding sites. *Biophysical Journal, 107*(9), 2204–2213. http://dx.doi.org/10.1016/j.bpj.2014.09.019.

Sethi, A., Eargle, J., Black, A. A., & Luthey-Schulten, Z. (2009). Dynamical networks in tRNA: Protein complexes. *Proceedings of the National Academy of Sciences of the United States of America, 106*(16), 6620–6625. http://dx.doi.org/10.1073/pnas.0810961106.

Stubbs, M. T., & Bode, W. (1993). A player of many parts: The spotlight falls on thrombin's structure. *Thrombosis Research, 69*(1), 1–58.

Sullivan, S. M., & Holyoak, T. (2008). Enzymes with lid-gated active sites must operate by an induced fit mechanism instead of conformational selection. *Proceedings of the National Academy of Sciences of the United States of America, 105*(37), 13829–13834. http://dx.doi.org/10.1073/pnas.0805364105.

Tsai, C. J., del Sol, A., & Nussinov, R. (2008). Allostery: Absence of a change in shape does not imply that allostery is not at play. *Journal of Molecular Biology, 378*(1), 1–11. http://dx.doi.org/10.1016/j.jmb.2008.02.034.

Tzeng, S.-R., & Kalodimos, C. G. (2011). Protein dynamics and allostery: An NMR view. *Current Opinion in Structural Biology, 21*(1), 62–67.

Van Wart, A. T., Durrant, J., Votapka, L., & Amaro, R. E. (2014). Weighted implementation of suboptimal paths (WISP): An optimized algorithm and tool for dynamical network analysis. *Journal of Chemical Theory and Computation, 10*(2), 511–517. http://dx.doi.org/10.1021/ct4008603.

Van Wart, A. T., Eargle, J., Luthey-Schulten, Z., & Amaro, R. E. (2012). Exploring residue component contributions to dynamical network models of allostery. *Journal of Chemical Theory and Computation, 8*(8), 2949–2961. http://dx.doi.org/10.1021/ct300377a.

Vijayalakshmi, J., Padmanabhan, K. P., Mann, K. G., & Tulinsky, A. (1994). The isomorphous structures of prethrombin2, hirugen-, and PPACK-thrombin: Changes accompanying activation and exosite binding to thrombin. *Protein Science, 3*, 2254–2271.

Wang, J., Wang, W., Kollman, P. A., & Case, D. A. (2006). Automatic atom type and bond type perception in molecular mechanical calculations. *Journal of Molecular Graphics & Modelling, 25*(2), 247–260. http://dx.doi.org/10.1016/j.jmgm.2005.12.005.

Wang, J., Wolf, R. M., Caldwell, J. W., Kollman, P. A., & Case, D. A. (2004). Development and testing of a general amber force field. *Journal of Computational Chemistry, 25*(9), 1157–1174.

Yang, L., Sun, M. F., Gailani, D., & Rezaie, A. R. (2009). Characterization of a heparin-binding site on the catalytic domain of factor XIa: Mechanism of heparin acceleration of factor XIa inhibition by the serpins antithrombin and C1-inhibitor. *Biochemistry, 48*, 1517–1524.

Yen, J. Y. (1971). Finding the K shortest loopless paths in a network. *Management Science, 17*(11), 712–716.

AUTHOR INDEX

Note: Page numbers followed by "*f*" indicate figures, and "*t*" indicate tables.

A

B

D

E

F

G

H

I

J

K

L

M

N

O

P

Q

R

S

T

U

V

W

X

Y

Z

SUBJECT INDEX

Note: Page numbers followed by "*f*" indicate figures, "*t*" indicate tables, and "*s*" indicate schemes.

D

F

G

I

J

K

L

M

Q

R

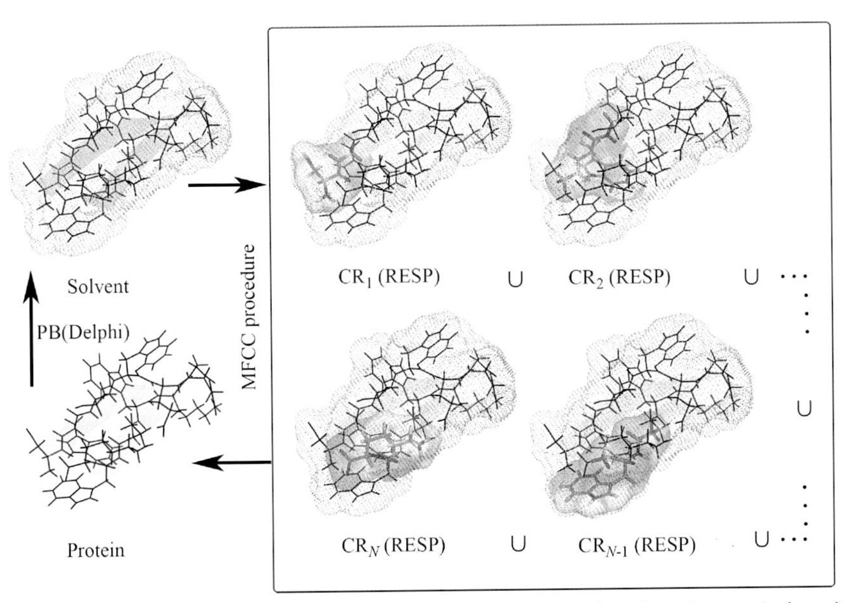

X. Wang *et al.*, Fig. 3 Illustration of the PPC approach. Induced surface charges (solvent) were obtained by solving Poison–Boltzmann equation using Delphi. Other part of the protein was treated as background charges in QM calculation of center residue (CR; *colored surfaces*). Each fragment calculation was performed in parallel.

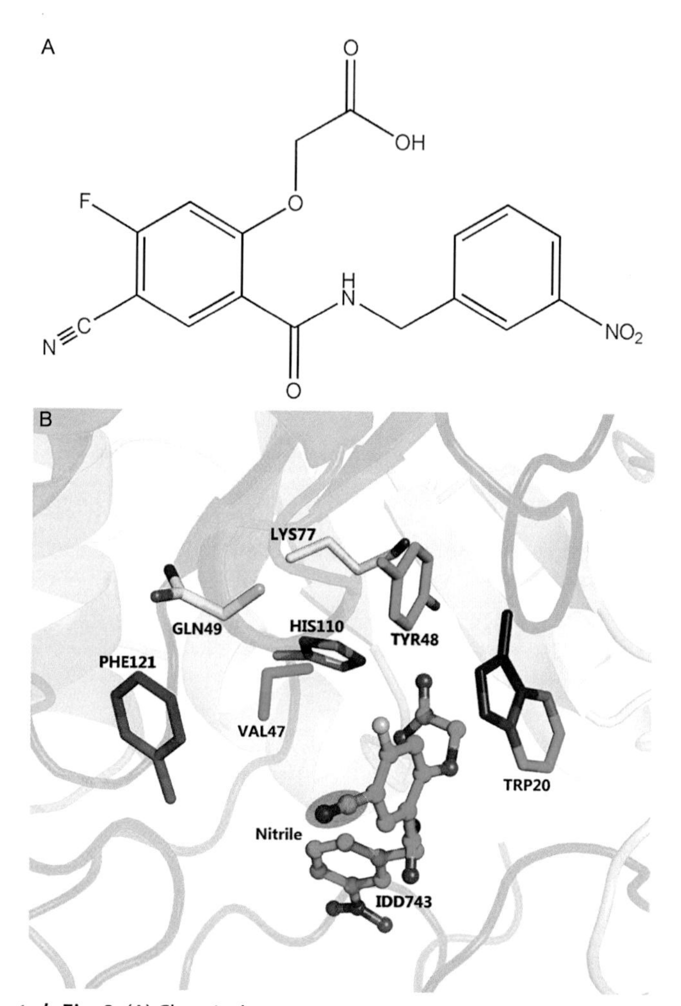

X. Wang *et al.*, Fig. 8 (A) Chemical structures IDD743. (B) Structural model of the hALR2/NADP$^+$/IDD743 complex. IDD743 and hALR2 are shown in ball-and-stick and cartoon models, respectively. The *pink ellipse* highlights the position of the nitrile probe. The side chains of eight mutated residues are shown as sticks. All the mutated residues are within 11 Å from the midpoint of the nitrile of IDD743, where the electric field was calculated. *Adapted with permission from Wang, X. W., He, X., & Zhang, J. Z. H. (2013). Predicting mutation-induced Stark shifts in the active site of a protein with a polarized force field.* Journal of Physical Chemistry A, 117*(29), 6015–6023. doi:10.1021/Jp312063h. Copyright 2013 American Chemical Society.*

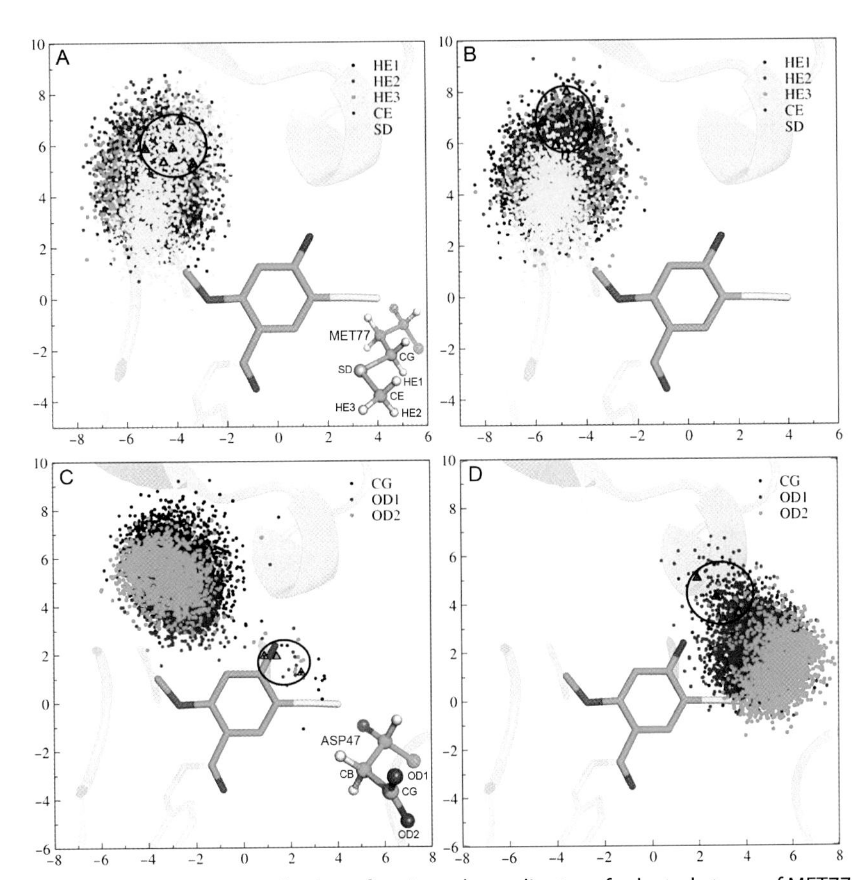

X. Wang *et al.*, Fig. 11 Distribution of projected coordinates of selected atoms of MET77 (*top*) and ASP47 (*bottom*) onto the plane of the phenyl ring of IDD743 during MD simulations based on the Amber ff99SB (A, C) and PPC (B, D) models. The *triangles* in the *black circles* represent the initial projected positions of the corresponding atoms. *Reprinted with permission from Wang, X. W., He, X., & Zhang, J. Z. H. (2013). Predicting mutation-induced Stark shifts in the active site of a protein with a polarized force field.* Journal of Physical Chemistry A, 117*(29), 6015–6023. doi:10.1021/Jp312063h. Copyright 2013 American Chemical Society.*

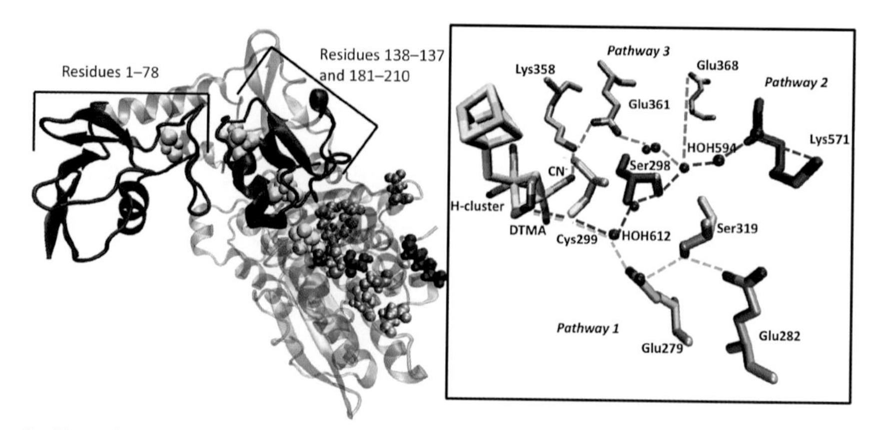

B. Ginovska *et al.*, Fig. 1 Proton pathways in *Clostridium pastuerianum* [FeFe]-hydrogenase proposed based on experimental and theoretical studies. *Pathway 1* (*cyan*) has been experimentally demonstrated to almost fully block catalytic activity when any one of the four residues is mutated to a nonhydrogen bonding residue. *Pathway 2* (*blue*) and *Pathway 3* (*orange*), which are largely water-based pathways, are additional proposed pathways based on sequence analysis. *Reprinted from Ginovska-Pangovska, B., Ho, M.-H., Linehan, J. C., Cheng, Y., Dupuis, M., Raugei, S., & Shaw, W. J. (2014). Molecular dynamics study of the proposed proton transport pathways in [FeFe]-hydrogenase.* Biochimica et Biophysica Acta, 1837, *131–138 with permission from Elsevier.*

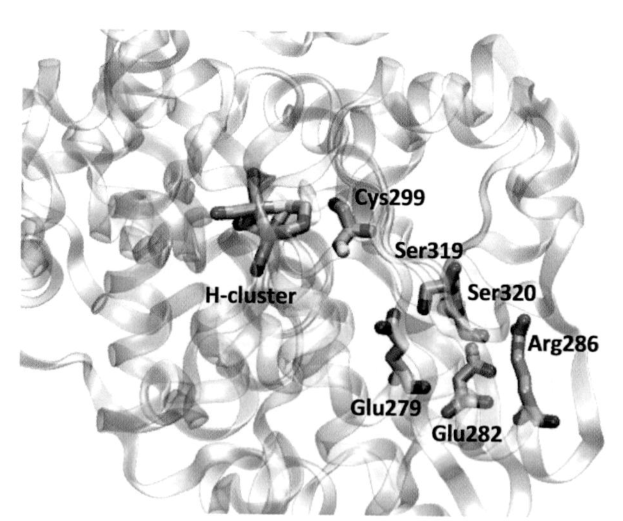

B. Ginovska *et al.*, Fig. 8 Proton pathway and active site in [FeFe]-hydrogenase, showing the proposed entrance to the proton pathway as well as the relative orientations of Ser320, Arg286, and Glu282 to each other and to the surface of the protein.

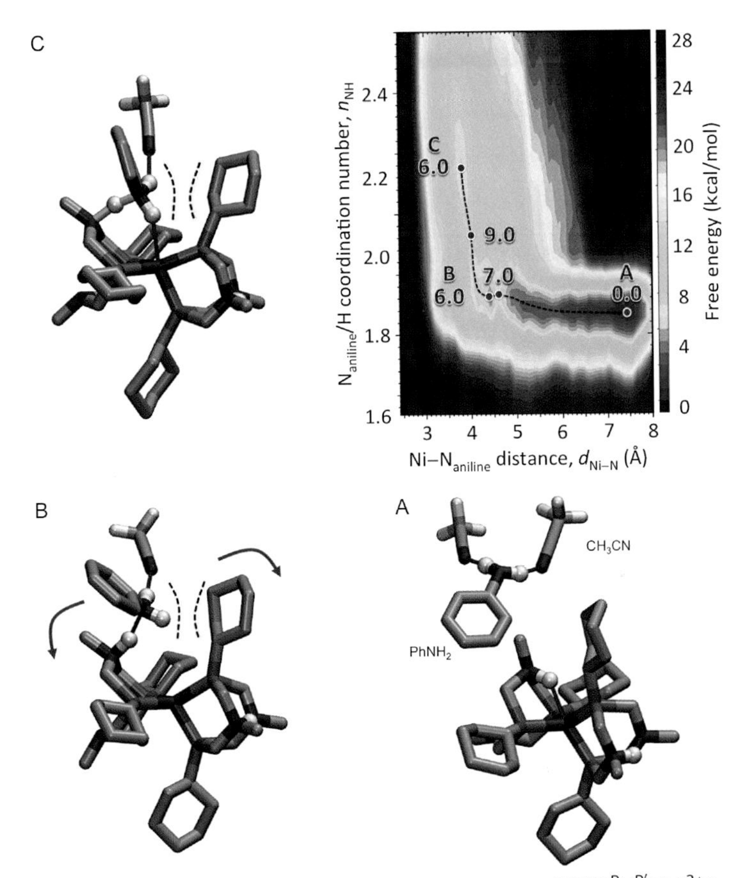

B. Ginovska *et al.*, Fig. 12 Initial steps of the deprotonation of $[Ni(P^R_2N^{R'}_2H)_2]^{2+}$ by aniline in acetonitrile solutions: base association and proton transfer. *Top, right panel*: Free energy for the binding of aniline as a function of the distance (d_{Ni-N}) between the Ni center and the N atom of aniline, and as a function of the fractional number (n_{NH}) of hydrogen atoms around the N atom of aniline (coordination number) obtained from hybrid QM/MM metadynamics simulations. Catalyst and base were described within the DFT framework, and the solvent and counterions with classical force field. Local minima and transition states are marked with *blue* and *red circles*, respectively. Free energies at minima and transition states are also reported using the same colors. The *dotted lines* indicate the lowest-free energy pathway. Representative configurations of the minima on the free energy surface are shown in (A)–(C). Hydrogen bond interactions are shown as *red lines*; in (C) the *green sticks* indicate that the H atom is shared between the two N atoms. The *red arrows* in (C) indicate the overall movement of the pendant amine and the proximal cyclohexyl group upon binding of the base. In all of the figures, acetonitrile molecules hydrogen bonded to aniline are also shown. For clarity, nonprotic H atoms on the Ni complex and aniline are not shown. *Reprinted with permission from O'Hagan, M., Ho, M. H., Yang, J. Y., Appel, A. M., Rawkowski DuBois, M., Raugei, S., … Bullock, R. M. (2012). Proton delivery and removal in [Ni((P2N2R)-N-R')](2) hydrogen production and oxidation catalysts.* Journal of the American Chemical Society, *134(47), 19409–19424. doi:10.1021/Ja307413x. Copyright 2012 American Chemical Society.*

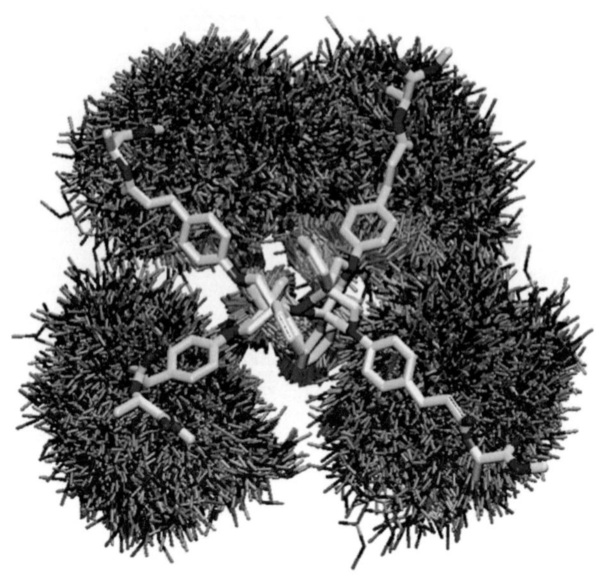

B. Ginovska *et al.*, Fig. 16 A static $[Ni(P_2^{Ph}N_2^{Dipeptide})_2]^{2+}$ complex is shown overlaid on top of the full molecular dynamics trajectory. While the dipeptide did result in enhanced catalytic activity, the significant mobility of the dipeptides demonstrates that there is too much mobility in this outer coordination *sphere* to provide the interactions so important in the functions of enzymes. *Adapted from reference Reback, M. L., Ginovska-Pangovska, B., Ho, M.-H., Jain, A., Squier, T. C., Raugie, S., … Shaw, W. J. (2013). The role of a dipeptide outer-coordination sphere on H2-production catalysts: influence on catalytic rates and electron transfer.* Chemistry—A European Journal, 19, *1928–1941 with permission from John Wiley and Sons.*

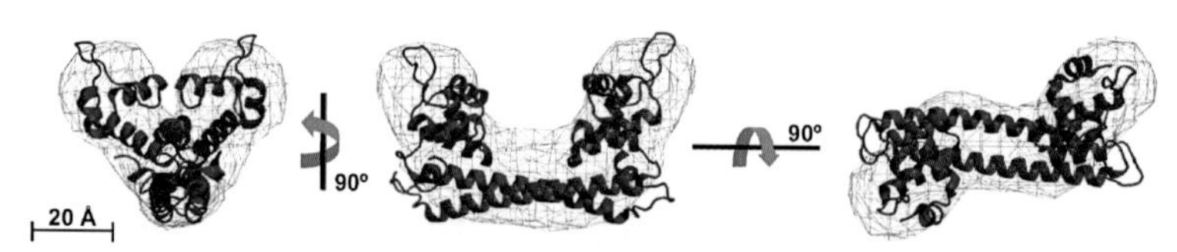

J.M. Parks and J.C. Smith, Fig. 3 Best-fitting conformation of Hg(II)-MerR obtained from MD simulations superimposed onto the three-dimensional molecular envelope (*blue mesh*) as determined by SAXS. *Reprinted from Guo, H.-B., Johs, A., Parks, J. M., Olliff, L., Miller, S. M., Summers, A. O., … Smith, J. C. (2010). Structure and comformational dynamics of the metalloregulator MerR upon binding of Hg(II).* Journal of Molecular Biology, 398, *555–568. doi: 10.1016/j.jmb.2010.03.020. Copyright 2010, with permission from Elsevier.*

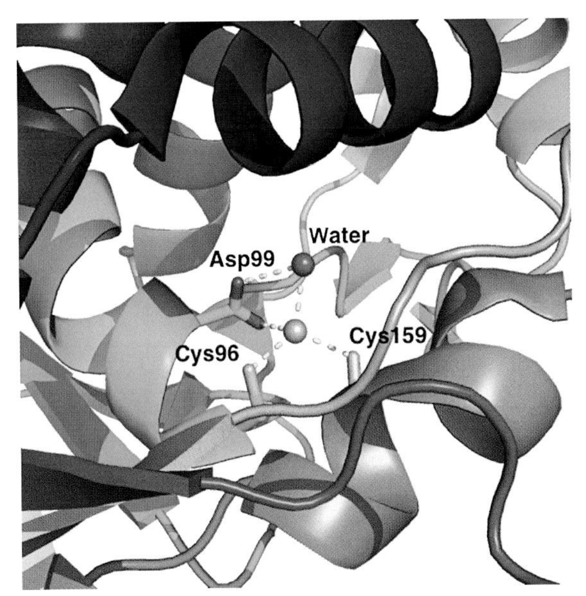

J.M. Parks and J.C. Smith, Fig. 4 Active site of Hg^{2+}-bound MerB (PDB entry 3F2F (Lafrance-Vanasse et al., 2009)). *Adapted with permission from Parks, J. M., Guo, H., Momany, C., Liang, L. Y., Miller, S. M., Summers, A. O., & Smith, J. C. (2009). Mechanism of Hg–C protonolysis in the organomercurial lyase MerB.* Journal of the American Chemical Society, 131*(37), 13278–13285. doi:10.1021/Ja9016123. Copyright 2009 American Chemical Society.*

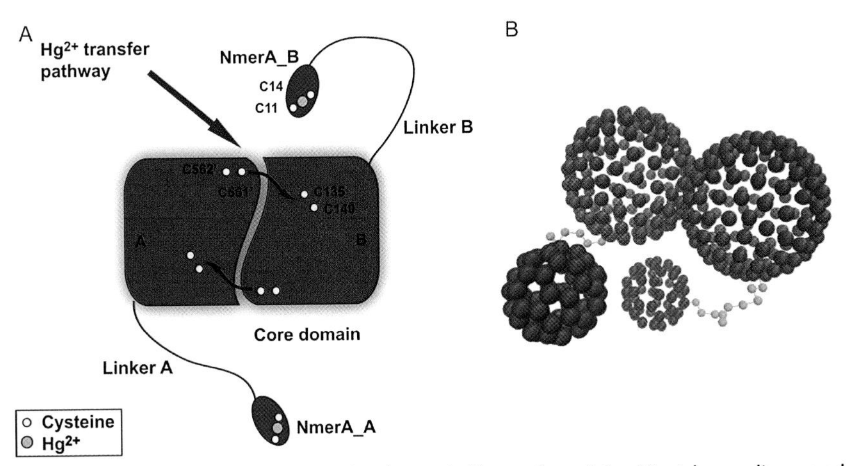

J.M. Parks and J.C. Smith, Fig. 7 (A) Schematic illustration of the MerA homodimer and scheme for intramolecular Hg^{2+} transfer. Hg^{2+} is first bound by the NmerA cysteines (C11 and C14), and is then delivered to the C-terminal cysteines of the other monomer (C561′ and C562′). Lastly, Hg^{2+} is transferred to a pair of cysteines in the catalytic site of the core (C135 and C140) where it is then reduced to Hg^0. (B) Simplified coarse-grained model for MerA with a rigid two-sphere core (*red*) and small spheres for the NmerA domains (*blue*), which are connected to the core by flexible linkers (*green*). *Reprinted from Hong, L., Sharp, M. A., Poblete, S., Biehl, R., Zamponi, M., Szekely, N., … Smith, J. C. (2014). Structure and dynamics of a compact state of a multi-domain protein, the mercuric ion reductase.* Biophysical Journal, 107, *2014, 393–400. doi: 10.1016/j.bpj.2014.06.013. Copyright 2014, from Elsevier.*

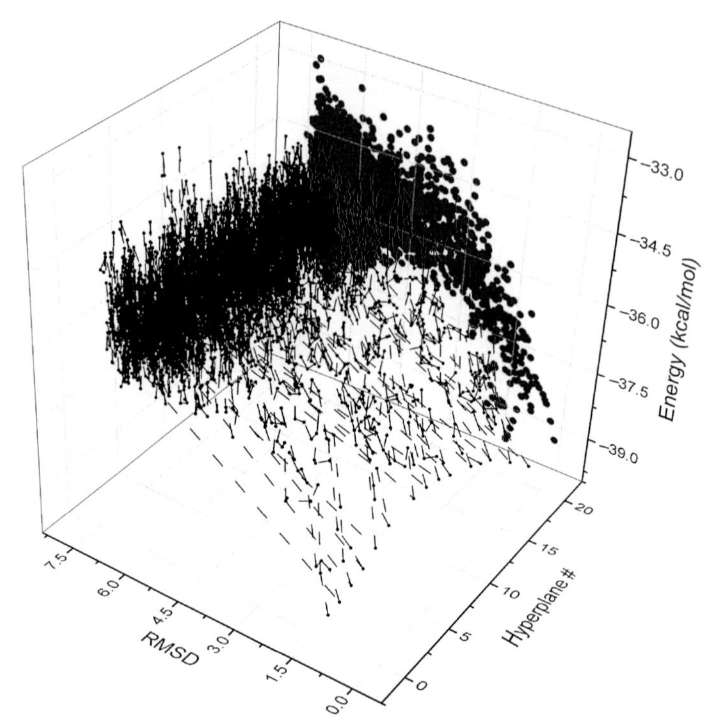

D. Weber *et al.*, Fig. 7 The obtained minima of the constraint optimization on 20 intervening hyperplanes are shown by their energy in kcal/mol and with respect to the RMSD values, which are referenced to the NEB pathway starting structures.

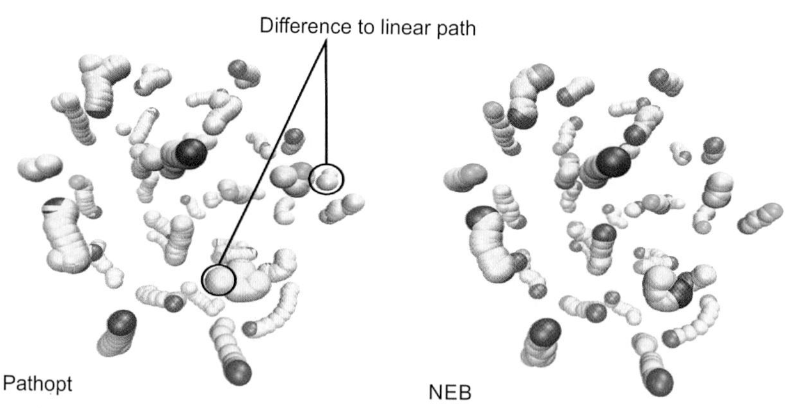

D. Weber *et al.*, Fig. 11 On the left side, the structures building the PO path 1 are shown. In *green* and *red* the start and final structures of the simulation are presented. On the right side, the NEB pathway structures are depicted. Two differences between the PO and the NEB pathway are highlighted.

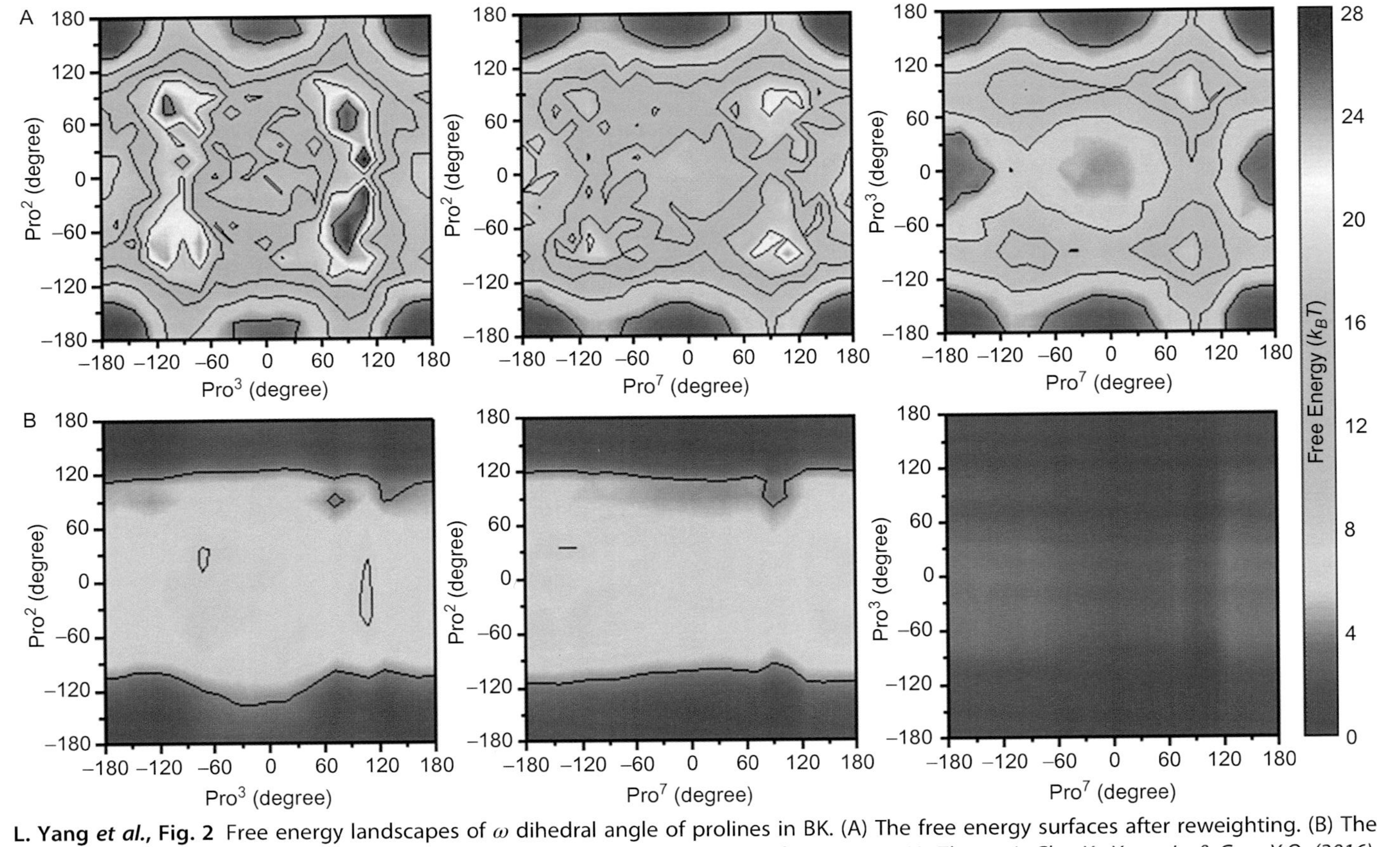

L. Yang *et al.*, Fig. 2 Free energy landscapes of ω dihedral angle of prolines in BK. (A) The free energy surfaces after reweighting. (B) The effective free energy surfaces without reweighting. *Reprinted with permission from Yang, I. Y., Zhang, J., Che, X., Yang, L., & Gao, Y.Q. (2016). Efficient sampling over rough energy landscapes with high barriers: A combination of metadynamics with integrated tempering sampling.* The Journal of Chemical Physics, *144, 094105. Copyright (2016) AIP Publishing LLC.*

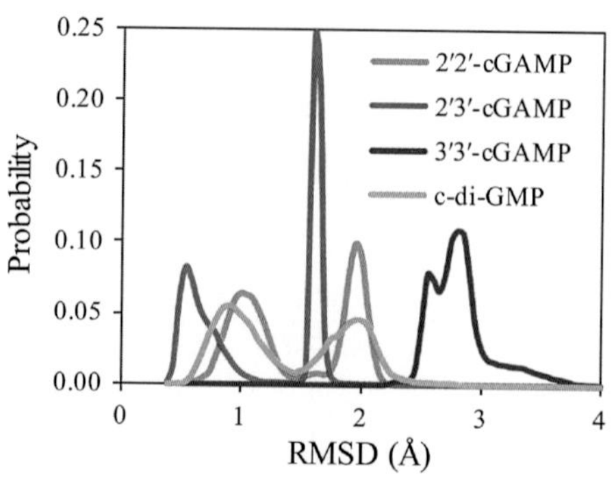

L. Yang *et al.*, Fig. 7 Structural deviation of free CDNs in solution with reference to the protein-bound crystal structure. All heavy atoms of CDNs are included in the RMSD calculations. *Reprinted with permission from Che, X., Zhang, J., Zhu, Y., Yang, L., Quan, H., & Gao, Y.Q. (2016). Structural flexibility and conformation features of cyclic dinucleotides in aqueous solutions.* The Journal of Physical Chemistry B, 120*(10), 2670–2680. Copyright (2016) American Chemical Society.*

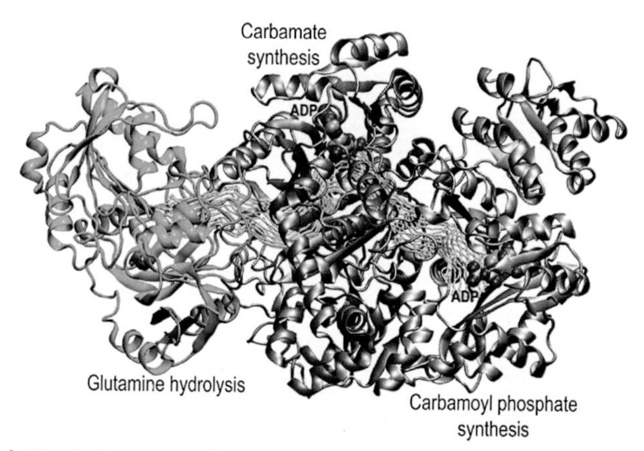

L. Yang *et al.*, Fig. 9 Structure of carbamoyl phosphate synthetase. The small subunit that contains the active site for the hydrolysis of glutamine is shown in *green*. The N-terminal domain of the large subunit that contains the active site for the synthesis of carboxy phosphate and carbamate is shown in *red*. The C-terminal domain of the large subunit that contains the active site for the synthesis of carbamoyl phosphate is shown in *blue*. The two molecular tunnels for the translocation of ammonia and carbamate are shown in *yellow dotted lines*. The image was constructed from PDB file: 1c30. *Reprinted with permission from Fan, Y. B., Lund, L., Shao, Q., Gao, Y. Q., & Raushel, F. M. (2009). A combined theoretical and experimental study of the ammonia tunnel in carbamoyl phosphate synthetase.* Journal of the American Chemical Society, 131*(29), 10211–10219. doi: 10.1021/ja902557r. Copyright (2009) American Chemical Society.*

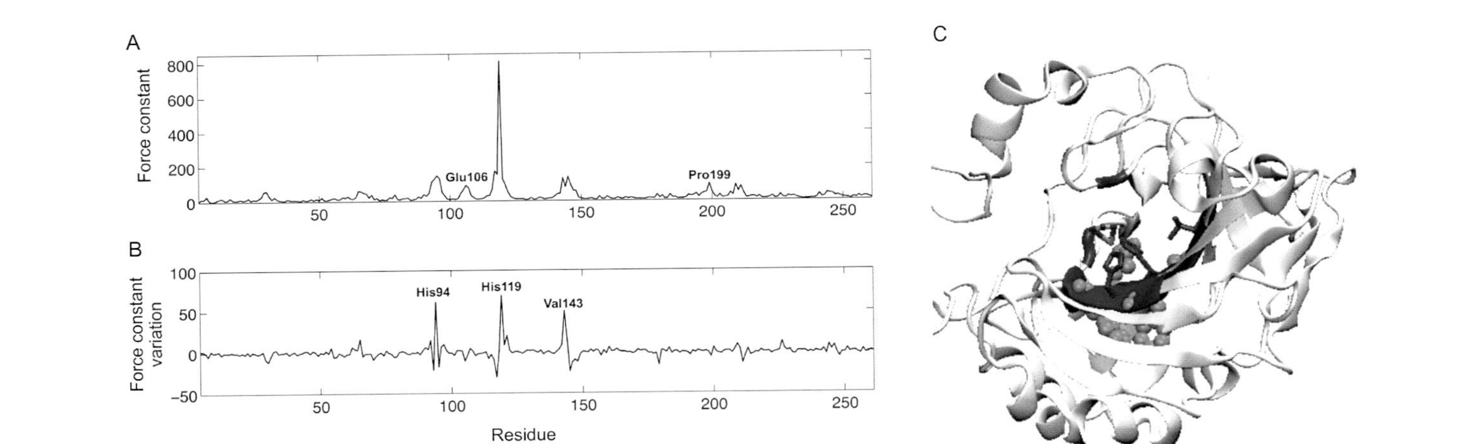

S. Sacquin-Mora, Fig. 1 Carbonic anhydrase. (A) Rigidity profile. (B) Force constant variation upon ligand binding. Force constants in Figs. 1–7 are in kcal mol^{-1} $Å^{-2}$. (C) Cartoon representation with the most rigid residues highlighted in *blue*, residues undergoing a rigidity increase upon ligand binding are shown as *red sticks* and residues undergoing a rigidity decrease as *green* van der Waals spheres. This *color code* is also valid for Figs. 2–7 which were all prepared using Visual Molecular Dynamics (Humphrey, Dalke, & Schulten, 1996).

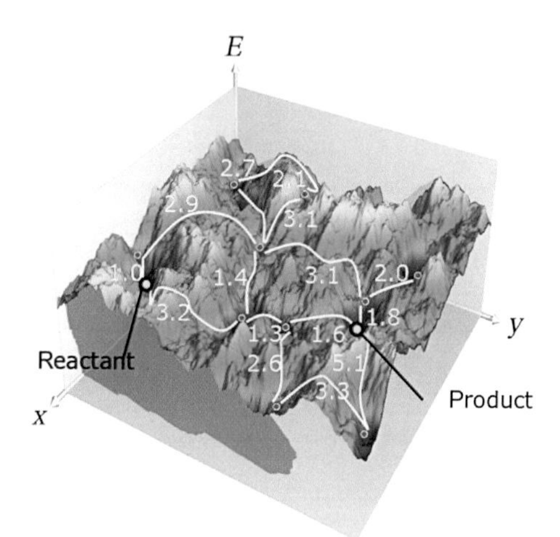

P. Imhof, Fig. 1 Schematic energy landscape with "valleys" (*blue*) and "mountains" (*yellow/orange*). *Yellow points* mark end states of a reaction, *green dots* are intermediate states. *White connections* with transition barriers indicate a variety of possible pathways. *Adapted with permission from Noé, F., Krachtus, D., Smith, J. C., & Fischer, S. (2006). Transition networks for the comprehensive characterization of complex conformational change in proteins.* Journal of Chemical Theory and Computation, 2, *840–857. Copyright (2006) American Chemical Society.*

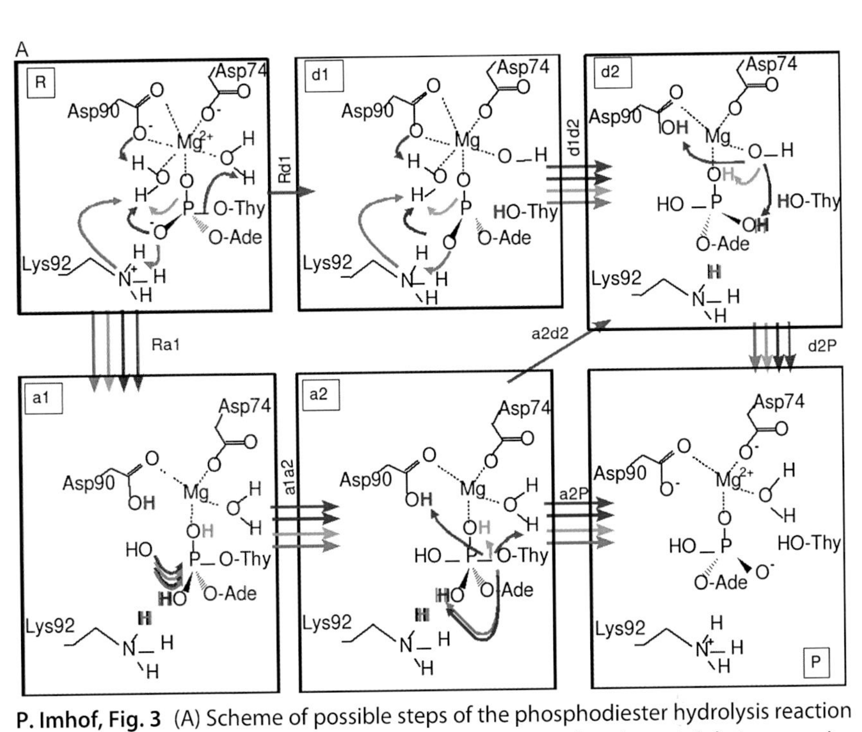

P. Imhof, Fig. 3 (A) Scheme of possible steps of the phosphodiester hydrolysis reaction in the enzyme EcoRV. (B) (see p. 254) Network of computed pathways. Relative energies (in kcal/mol) are given in *blue numbers inside the circles* for minima and in *red numbers next to the connection lines* for transition states. The states are labeled according to the scheme in (A). The labels Asp, Lys, O1, and O2 refer to the different hydrogen atom positions, *colored red, green, yellow,* and *blue*, respectively, in subfigure (A). Note that the transition from preproduct preP to product P is not shown in the scheme. *Reprinted with permission from Imhof, P., Fischer, S., & Smith, J. C. (2009). Catalytic mechanism of DNA backbone cleavage by the restriction enzyme Eco RV: A quantum mechanical/molecular mechanical simulations analysis.* Biochemistry, 48, *9061–9075. Copyright (2009) American Chemical Society.*

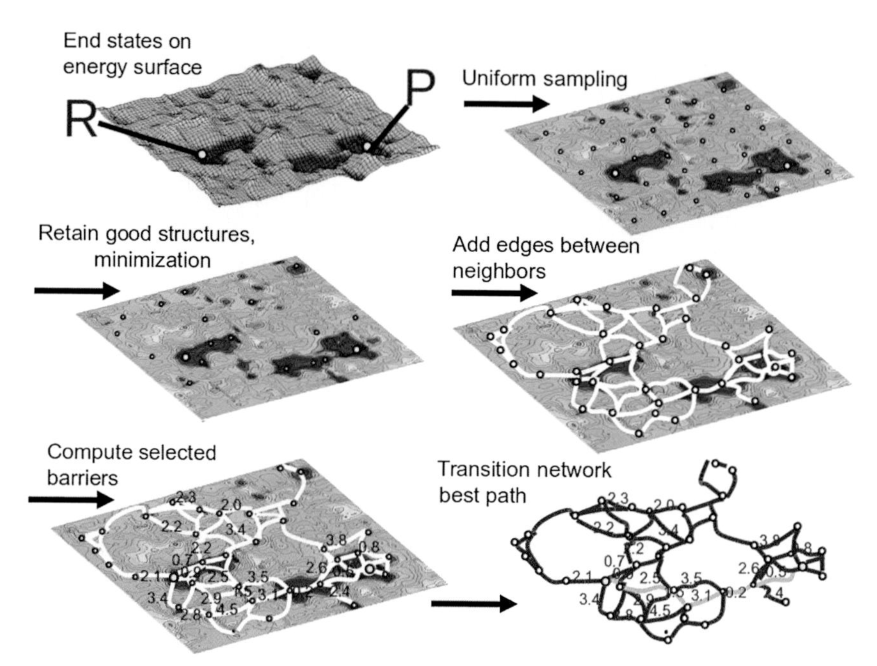

P. Imhof, Fig. 8 Steps to generate a transition network between two end states (reactant, R; product, P) of a reaction, schematically shown in a two-dimensional energy landscape. The underlying potential energy landscape is sampled around discrete positions by uniform sampling which are subsequently minimized. Nodes identified as neighbors are connected by edges, and transitions are computed along discrete pathways between neighbors. The edge weights are given by the relative energies of the transition state. On the transition network, the best path is identified by graph theoretical algorithms. *Adapted with permission from Noé, F., Krachtus, D., Smith, J. C., & Fischer, S. (2006). Transition networks for the comprehensive characterization of complex conformational change in proteins.* Journal of Chemical Theory and Computation, 2, *840–857. Copyright (2006) American Chemical Society.*

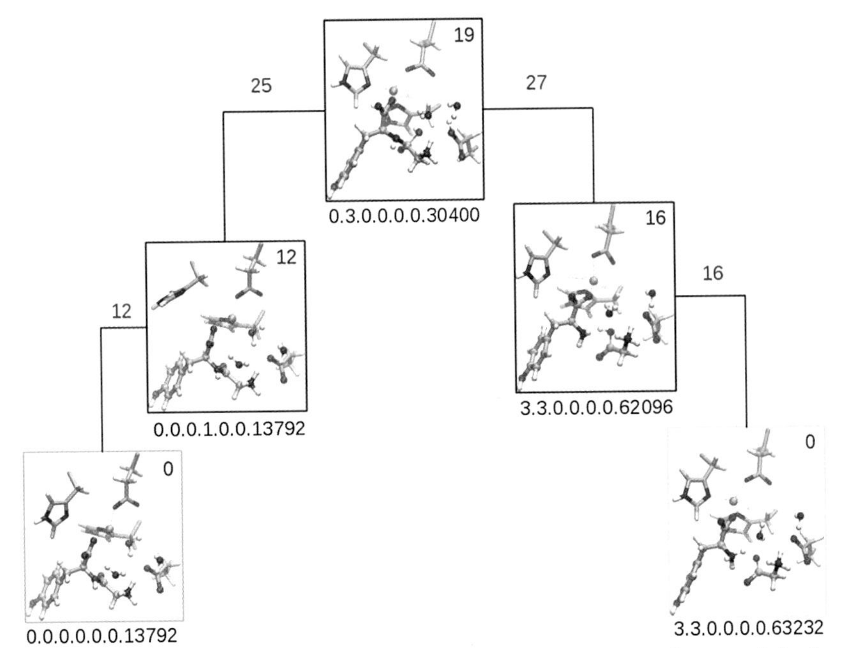

P. Imhof, Fig. 10 Best pathway on the transition network computed for peptide hydrolysis in carboxy peptidase. *Black digits* encode the state of the nodes (minima) as displacement steps relative to the reactant state: oldBond.newBond.χH196 .χH69 .χE270 .χE72 .protonation. *Blue numbers inside the boxes* are relative minimum energies and *red numbers next to the connection lines* are relative transition state energies for the connection between to nodes, as indicated by the *lines*. All energies are in kcal/mol.

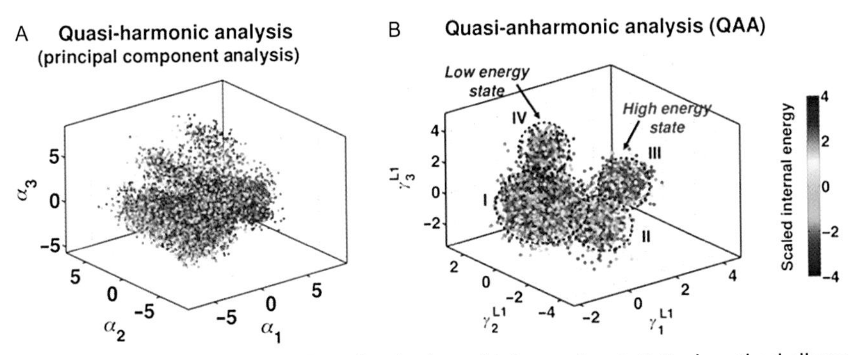

P.K. Agarwal *et al.*, Fig. 6 The benefit of using a higher-order statistical method allows identification of conformational substates with homogeneous properties. Conformational substates identification for protein ubiquitin was performed using two different methods. (A) The conformational substates identified by second-order methods such as quasi-harmonic analysis and principal component analysis do not achieve clear of separation and the population of the conformation show mixed properties. (B) Using a higher-order method such as quasi-anharmonic analysis (QAA) allows identification of substates that are clearly separated and conformations have homogeneous properties. In both panels, *each dot* corresponds to a single conformation and the *coloring* is by scaled internal conformational energy.

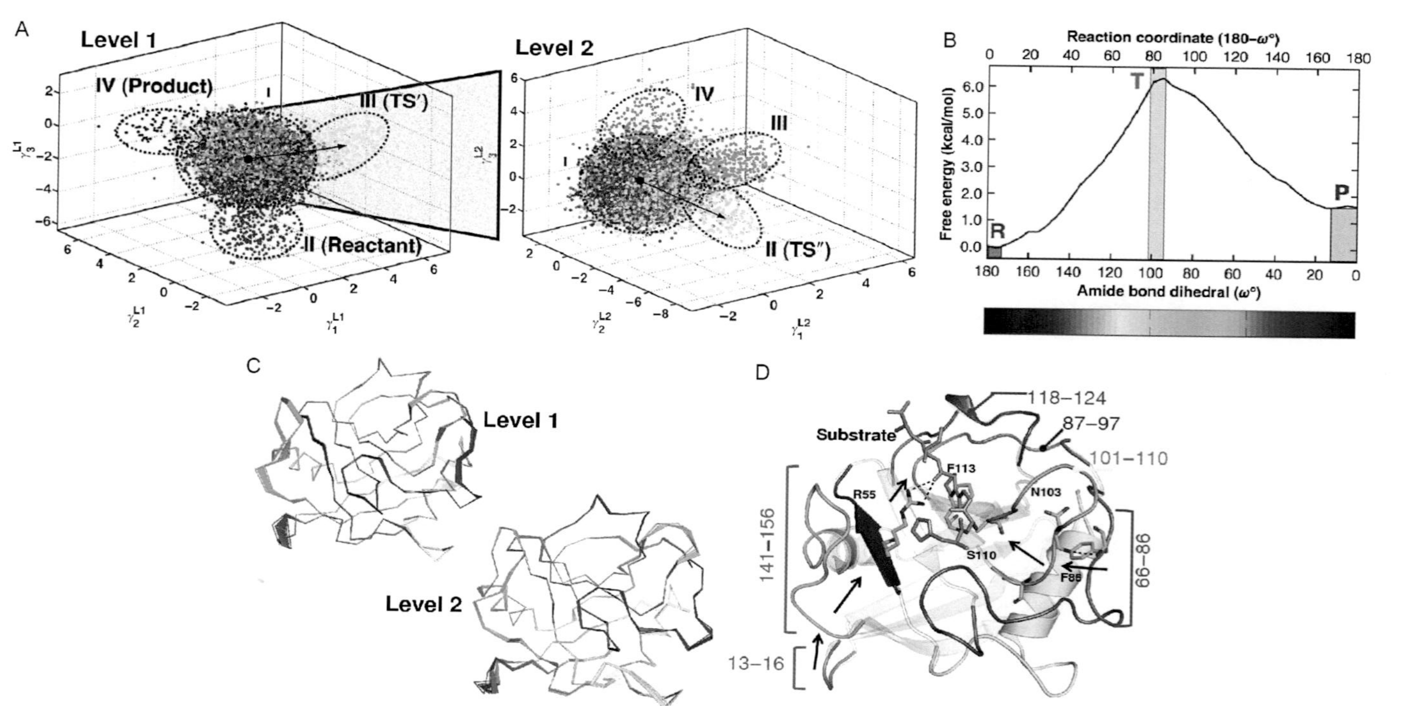

P.K. Agarwal *et al.*, Fig. 8 Computational method QAA allows identification of multiscale hierarchy associated with catalysis by enzyme CypA. (A) Multilevel (two levels shown) hierarchy of conformational substates, *each dot* is a conformations. Each *colored dot* represents a single sampled conformation; *ellipses* indicate substates; TS′, TS″, and T indicate transition state area. (B) The free energy profile and conformations in (A) are colored according to reaction coordinate, (C) conformational change between substates corresponding to *black arrow* in (A), and (D) impact of identified motions on CypA's mechanism. *Adapted from Ramanathan, A., Savol, A. J., Langmead, C. J., Agarwal, P. K., & Chennubhotla, C. S. (2011). Discovering conformational sub-states relevant to protein function.* PLoS ONE, 6*(1), e15827.*

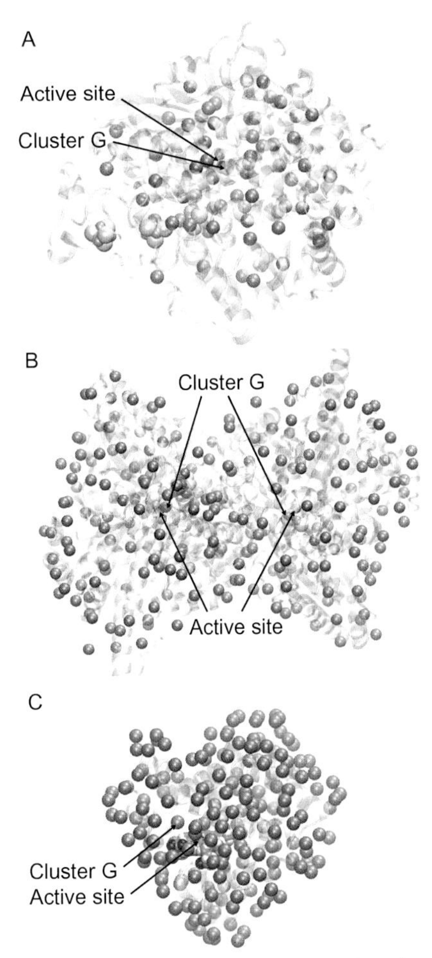

P.-H. Wang *et al.*, Fig. 2 Location of microstates (clusters) and active sites in (A) [NiFe]-hydrogenase, (B) CODH, and (C) myoglobin. Protein secondary structures are shown in *white ribbon* representation, *while clusters* are indicated as *dark yellow spheres. Color codes*: Fe, *pink*; Ni, *blue*; S, *yellow*; O, *red*; C, *cyan. Blue spheres* in (C) denote N atoms. *Panel (A): Adapted with permission from Wang, P., Best, R. B., & Blumberger, J. (2011b). Multiscale simulation reveals multiple pathways for H_2 and O_2 transport in a [NiFe]-hydrogenase.* Journal of the American Chemical Society, 133, *3548–3556. Copyright 2011 American Chemical Society. Panel (B): Adapted with permission from Wang, P., Bruschi, M., De Gioia, L., & Blumberger, J. (2013). Uncovering a dynamically formed substrate access tunnel in carbon monoxide dehydrogenase/acetyl-CoA synthase.* Journal of the American Chemical Society, 135*(25), 9493–9502. Copyright 2013 American Chemical Society. Panel (C): Adapted with permission from De Sancho, D., Kubas, A., Wang, P., Blumberger, J., & Best, R. B. (2015). Identification of mutational hot spots for substrate diffusion: Application to myoglobin.* Journal of Chemical Theory and Computation, 11*(4), 1919–1927. Copyright 2015 American Chemical Society.*

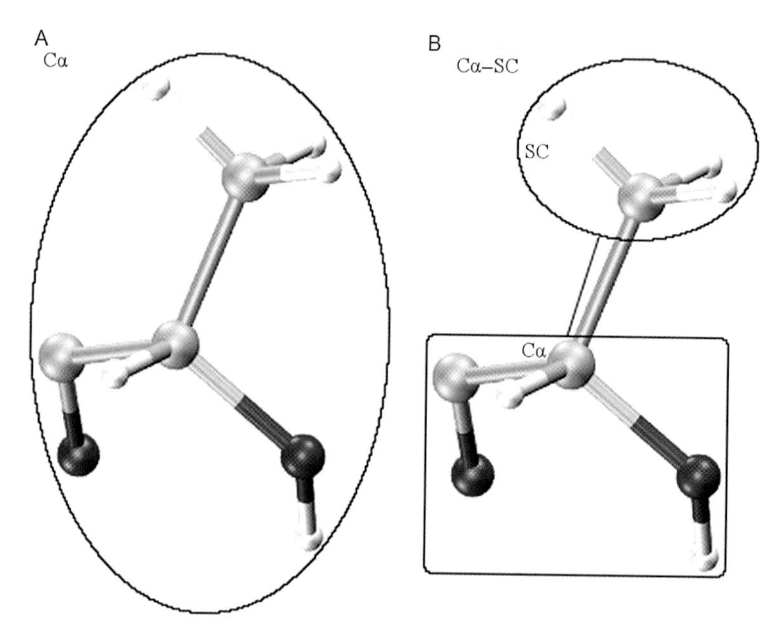

T.H. Click *et al.*, Fig. 1 Coarse-grain model of an alanine. (A) The Cα model utilizes the Cα coordinates of each residue and the total mass of the residue. (B) The Cα–SC model separates the model into the backbone (Cα coordinates and total mass of square region) and the sidechain (sidechain center of mass and total mass of circular region).

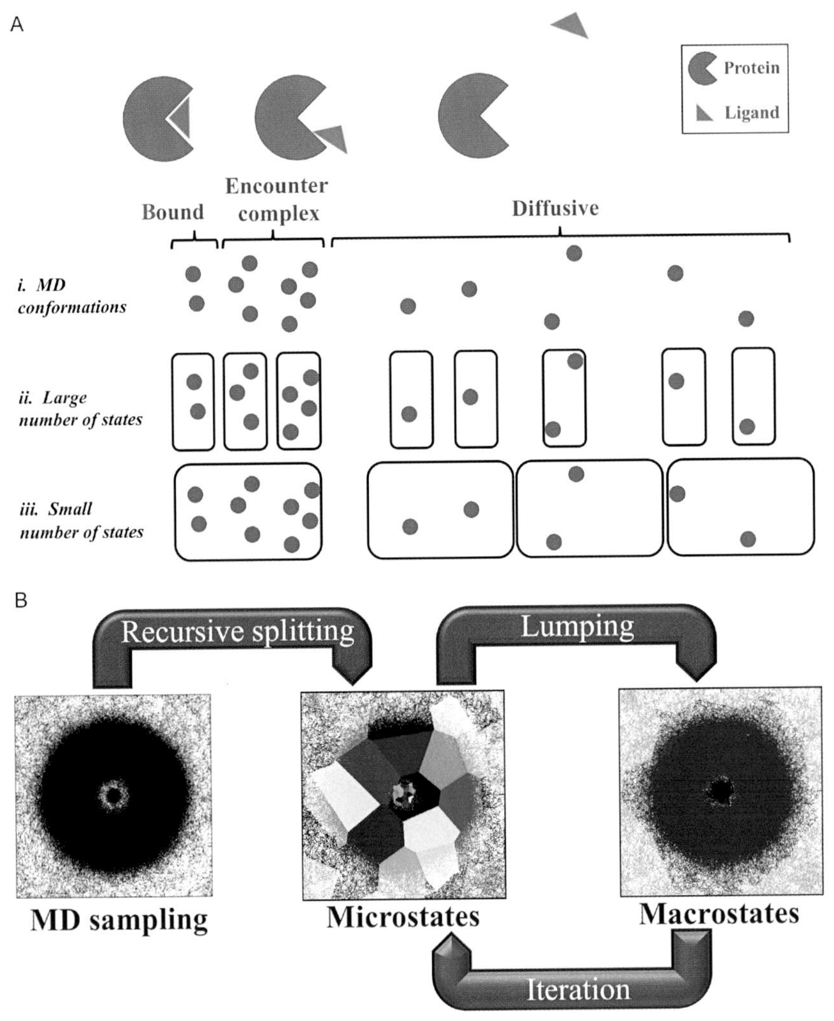

L. Zhang *et al.*, Fig. 3 See legend on opposite page.

L. Zhang *et al.*, Fig. 3 Illustration of the automatic state partitioning for multibody (APM) algorithm to address the limitation of k-centers algorithm in two-body system. (A) Limitation of k-center algorithm used to investigate the protein–ligand binding mechanisms. There are three types of interaction between protein and ligand: bounded form when the ligand is correctly oriented with the protein binding site, encounter complex form when the ligand is not correctly oriented with the protein binding site, and diffusive form when the ligand is far from the protein. In the bounded and encounter complex forms, the dynamics of the protein and ligand is much slower than that in the diffusive form. This difference in dynamics will usually result in a large deviation in sampling densities as shown in (i) (each MD conformation is represented by a *blue sphere*). Under such condition, the k-centers algorithm is not appropriate: as shown in (ii), partitioning the conformational space with a large number of clusters using the k-centers algorithm will generate many states with poor statistics in the diffusive region; as shown in (iii), low clustering resolution due to the small number of states will lead to the mixing of the bounded and the encounter complex forms. (B) Illustration of APM algorithm. We recursively split the MD conformations into microstates such that the escape probability is $>1-\frac{1}{e}$. This state decomposition method will generate microstates with different sizes. Microstates are then lumped into a few macrostates, and the recursive splitting is repeated for a predetermined number of times for each macrostate. *Panel B adapted with permission from Sheong, F. K., Silva, D.-A., Meng, L., Zhao, Y., & Huang, X. (2015). Automatic state partitioning for multibody systems (APM): An efficient algorithm for constructing Markov state models to elucidate conformational dynamics of multibody systems.* Journal of Chemical Theory and Computation, 11*(1), 17–27. doi:10.1021/Ct5007168. Copyright (2015) American Chemical Society.*

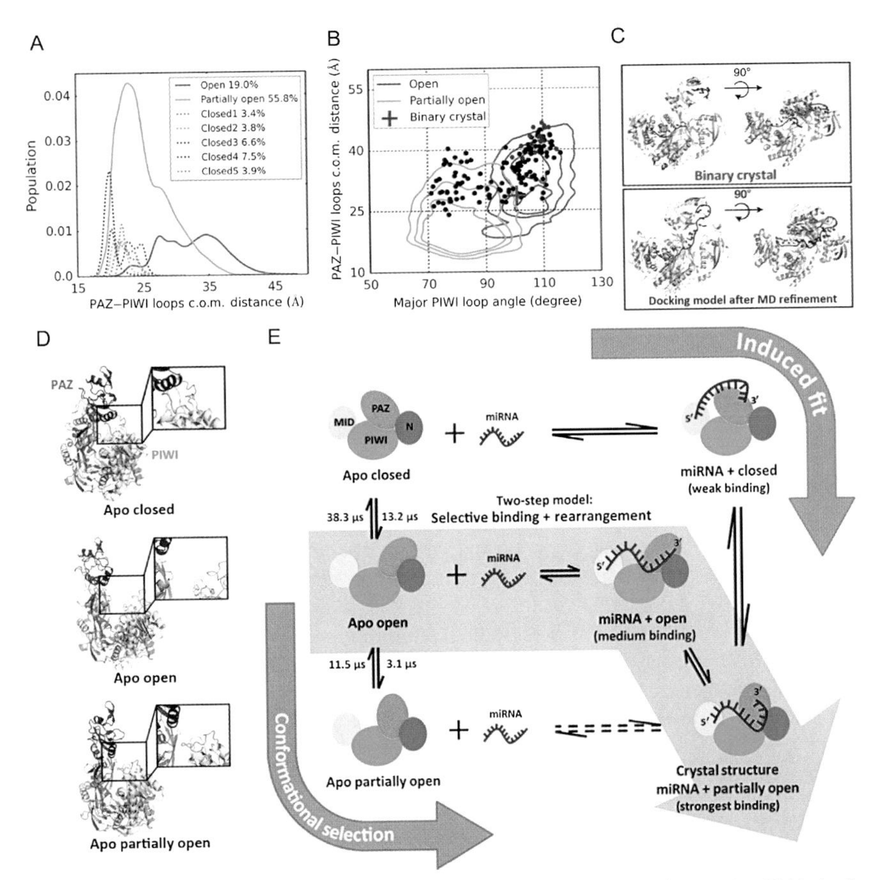

L. Zhang *et al.*, Fig. 4 MSMs based on MD simulations coupled with protein–RNA docking elucidate a two-step mechanism of hAgo2–miRNA recognition. (A) Distribution of distances between the center-of-mass (c.o.m.) of the PAZ domain and PIWI loops for each macrostate. The equilibrium population of each macrostate is presented. A large distance implies an open conformation and small distance implies a closed conformation. (B) Projections of hAgo2–miRNA successful (*red dots*) and unsuccessful (*black dots*) docking models built from the selected structures of open microstates. (C) Structural comparison between the binary crystal structure (*upper panel*) and a representative model after refinement by MD simulations (*lower panel*) with hAgo2 colored in *gray* and miRNA in *red.* (D) Representative structures of closed, open, and partially open states of apo hAgo2 from the MSM. Enlarged view of the interdomain region between PAZ (*pink*) and PIWI (*green*) of each structure is presented in the *inset panels.* (E) The proposed two-step model (highlighted by the *cyan arrow*) of miRNA loading into hAgo2: selective binding followed by structural rearrangement. The two-step model is compared with the induced fit mechanism (highlighted by the *upper right gray arrow*) and the conformational selection mechanism (highlighted by the *lower left gray arrow*). The average transition times between states are denoted next to the *arrows. Panel A adapted from fig. 1A of, panel B adapted from fig. 2 of, panel C adapted from fig. 3C of, panel D adapted from fig. 1C of, and panel E adapted from fig. 4 of Jiang, H., Sheong, F. K., Zhu, L., Gao, X., Bernauer, J., & Huang, X. (2015). Markov state models reveal a two-step mechanism of miRNA loading into the human argonaute protein: Selective binding followed by structural re-arrangement.* PLoS Computational Biology, 11*(7), e1004404. doi:10.1371/journal.pcbi.1004404.*

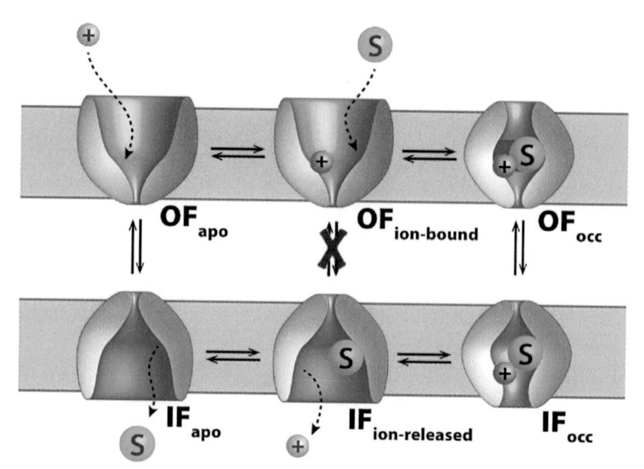

J.V. Vermaas *et al.*, Fig. 1 An exemplary scheme of the alternating access mechanism adopted by a cation-coupled symporter, where ions and substrate move in tandem. Here, the coupling mechanism only permits conformational transitions when the binding site is either completely vacant or bound with both chemical species, forbidding transitions of partially bound states (red X). Different types of transporters have different coupling mechanisms; however, all share this feature of certain forbidden transitions to regulate substrate transit and to prevent draining membrane potential.

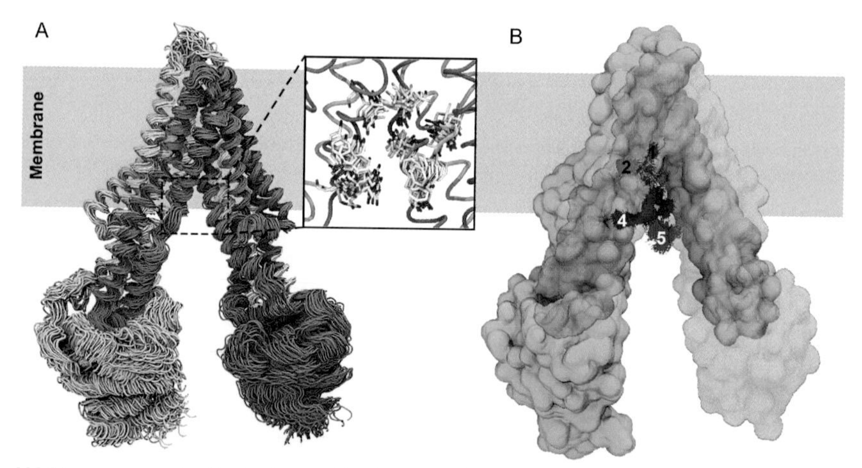

J.V. Vermaas *et al.*, Fig. 5 Ensemble docking of a drug substrate into the translocation lumen of P-gp. (A) MD simulation-generated ensemble of conformations used in molecular docking. Two pseudosymmetric halves of P-gp are colored in cyan and orange, respectively. Only 100 out of 1000 conformations used in ensemble docking are shown. A set of translocation path residues is shown to display the diverse side chain conformations represented by the MD-generated ensemble (inset). (B) Top 5 RMSD-based clusters of docked poses of the drug molecule are shown in different colors with P-gp in surface representation. One half of the protein is shown transparent to clearly show the clusters of the docked drug molecule.

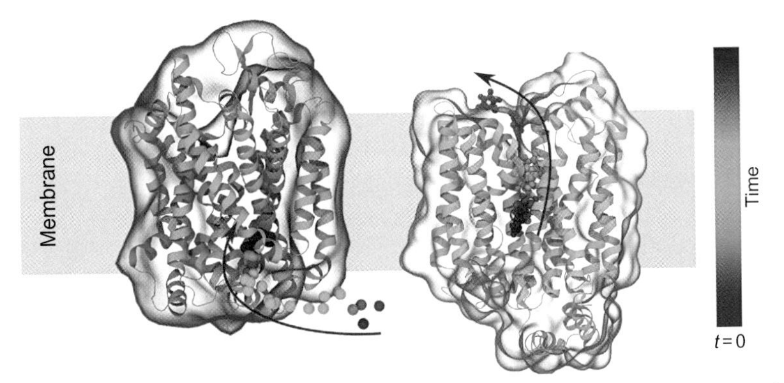

J.V. Vermaas *et al.*, Fig. 6 Binding of an ion (left) or unbinding of substrate (right) from membrane transporters captured in unbiased equilibrium simulations. The color of the ion and substrate changes with time to show movement in the direction of the arrows.

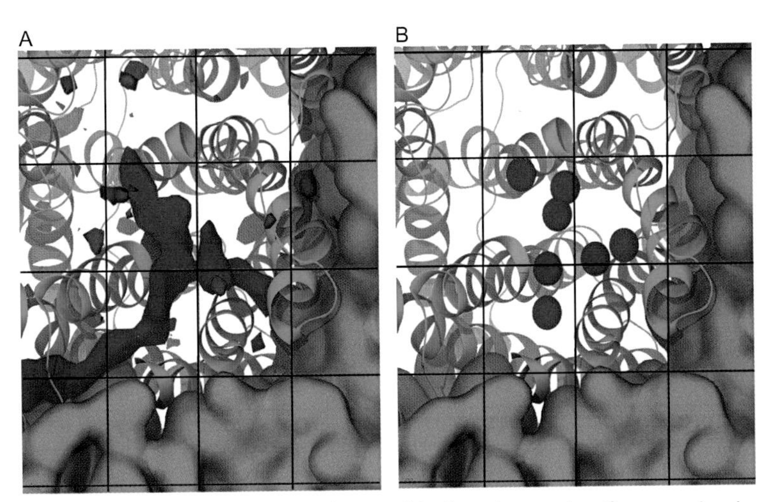

J.V. Vermaas *et al.*, Fig. 7 Probing substrate binding sites and pathways using implicit ligand sampling (ILS) technique. (A) Substrate binding sites predicted by ILS are shown in red isosurfaces. Blue helices represent the protein. The brown surface represents lipid bilayer. Black lines represent the grid on which the substrate is placed systematically during ILS analysis. (B) Experimental observed substrate binding sites taken from crystal structures. Substrates are shown in balls.

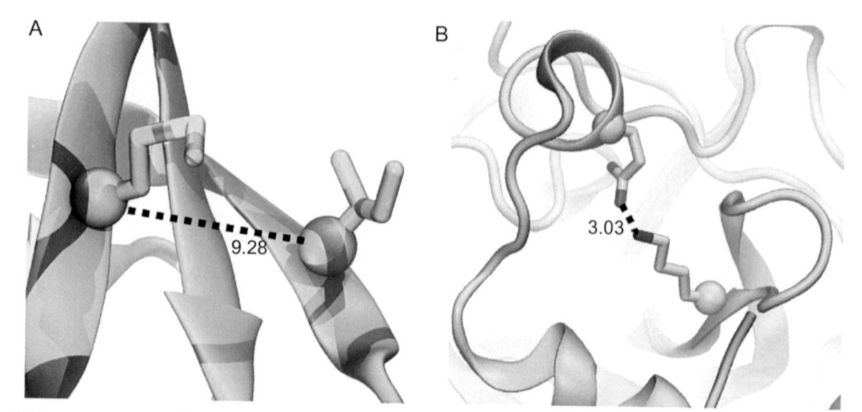

S. Bowerman and J. Wereszczynski, Fig. 3 Examples of a false contact (A) and a false noncontact (B) identified by the Cα–Cα distance definition. If a heavy atom definition is used instead, then the false positive in (A) is removed and the contact in (B) is properly identified.

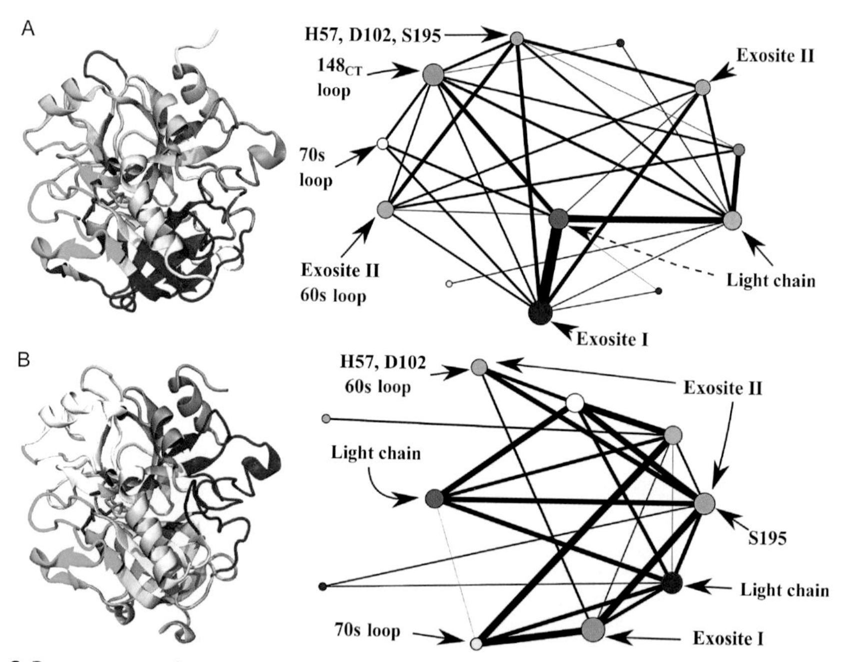

S. Bowerman and J. Wereszczynski, Fig. 6 Community structure of (A) isolated thrombin and (B) thrombin–hirugen with the 3D structures colored according to community membership. Catalytic residues are also shown as *purple sticks*. Relevant regions of the community graphs are labeled accordingly. While the catalytic residues in thrombin–hirugen are split between two communities, the interaction strength between their respective groups and Exosite I is increased.

Edwards Brothers Malloy
Ann Arbor MI. USA
August 18, 2016